Audel™

HVAC Fundamentals Volume 2

Heating System Components, Gas and Oil Burners, and Automatic Controls

All New 4th Edition

James E. Brumbaugh

Wiley Publishing, Inc.

Vice President and Executive Group Publisher: Richard Swadley
Vice President and Executive Publisher: Robert Ipsen
Vice President and Publisher: Joseph B. Wikert
Executive Editor: Carol A. Long
Acquisitions Editor: Katie Feltman
Editorial Manager: Kathryn A. Malm
Senior Production Manager: Fred Bernardi
Development Editor: Kenyon Brown
Production Editor: Vincent Kunkemueller
Text Design & Composition: TechBooks

Published simultaneously in Canada

For general information on our other products and services, please contact our Customer Care Department within the United States at (800) 762-2974, outside the United States at (317) 572-3993 or fax (317) 572-4002.

Wiley also publishes its books in a variety of electronic formats. Some content that appears in print may not be available in electronic books.

Library of Congress Cataloging-in-Publication Data:

ISBN: 0-7645-4207-9

Printed in the United States of America

10 9 8 7 6 5 4 3 2 1

For Laura, my friend, my daughter.

Contents

Introduction xv

About the Author xvi

Chapter 1 Oil Burners 1

Gun-Type Oil Burners 3
  Construction Details 3
  Operating Principles 10
Flame-Retention Head Burners 16
High-Static Oil Burners 16
Rotary Oil Burners 16
Vaporizing (Pot-Type) Oil Burners 18
Combination Oil and Gas Burners 18
Fuel Pump 19
  Single-Stage Fuel Pump 21
  Two-Stage Pump 25
  Fuel Pump Capacity 26
  Fuel Pump Service and Maintenance 26
  Priming Fuel Pumps 29
  Adjusting Fuel Pump Pressure 29
  Troubleshooting Fuel Pumps 31
Fuel Supply Tank and Line 32
Oil Burner Nozzles 32
Electrodes 36
  Troubleshooting Electrodes 37
  Servicing Electrodes 37
Oil Burner Air System 37
Primary Safety Control Service 40
Installing an Oil Burner 40
Starting an Oil Burner 42
Air Delivery and Blower Adjustment 43
Combustion Testing and Adjustments 43
Troubleshooting Oil Burners 48

Chapter 2	**Gas Burners**	57
	Operating Principles	57
	Electrical Circuits	61
	Automatic Controls	61
	Types of Gas Burners	63
	Integral-Type Gas Burners	65
	Gas Conversion Burners	66
	Gas Conversion Burner Combustion Chambers	67
	Gas Piping for Conversion Burners	68
	Venting and Ventilation	71
	Safety Precautions	72
	Troubleshooting Gas Burners	73
Chapter 3	**Coal Firing Methods**	77
	Coal-Firing Draft Requirements	77
	Firing Anthracite Coal	78
	Firing Bituminous Coal	80
	Firing Semibituminous Coal	81
	Stoker Firing	81
	Stoker Construction	84
	Stoker Automatic Controls	86
	Stoker Operating Instructions	90
	Coal Selection	91
	Starting the Fire	91
	Natural Stack Draft	91
	Manual Air Adjustment	92
	Automatic Air Control	92
	Changing Coal Feeds	92
	Motor Overload Protection	92
	Transmission Overload Protection	93
	Removal of Obstruction	93
	Lubrication	93
	Summer Service	93
	How to Remove Clinkers	94
	How to Adjust Coal Feed	94

How to Adjust Air Supply 94
Troubleshooting Coal Stokers 94

Chapter 4 Thermostats and Humidistats 99
Automatic Control Systems 99
Temperature Control Circuits 100
Thermostats 100
Thermostat Components 105
Thermostat Terminal Identification 109
Thermostat Anticipators 109
Types of Thermostats 119
Room Thermostats 119
Programmable Thermostats 125
Insertion Thermostats 125
Immersion Thermostats 126
Cylinder Thermostats 127
Boiler Thermostats 129
Remote-Bulb Thermostats 129
Proportional Thermostats 132
Outdoor Thermostats 132
Troubleshooting Thermostats 134
Humidistats 134
Location of Room Humidistats 140
Troubleshooting Humidistats 142

Chapter 5 Gas and Oil Controls 145
Gas Controls 145
Gas Control Circuits 146
Gas Burner Primary Control 146
Servicing a Gas Burner Primary Control 151
Gas Valves 153
Solenoid Gas Valves 153
Solenoid Coils 158
Direct-Acting Heat Motor Valves 163
Diaphragm Valves 164
Pressure Regulators 166
Pressure Switches 170

Automatic Pilot Safety Valve 174
Thermopilot Valves 178
Thermocouples 181
Troubleshooting Thermocouples 183
Thermopiles (Pilot Generators) 184
Pilot-Operated Diaphragm Valves 185
Combination Gas Valves 187
Standing Pilot Combination Gas Valves 187
Continuous Pilot Dual Automatic Gas Valve 191
Universal Electronic Ignition Combination Gas Valve 194
Pilot Burners 194
Installing a Pilot Burner 198
Replacing the Pilot Burner Orifice 200
Lighting the Pilot 201
Pilot Flame Adjustment 202
Main Burner Ignition 202
Pilot-Pressure Switch 203
Electronic Ignition Modules 203
Intermittent Pilot Ignition Module 204
Direct-Spark Ignition Module 207
Hot-Surface Ignition Module 208
Igniters 211
Flame Sensors 214
Mercury Flame Sensors 216
Oil Controls 217
Oil Valves 217
Oil Burner Primary Control 219
Cadmium Cell Primary Controls 220
Stack Detector Primary Control 223
Combination Primary Control and Aquastat 227
Troubleshooting the Oil Burner Primary Control 231

Chapter 6 **Other Automatic Controls** 233
Fan Controls 233
Fan Control 233
Air Switch 236
Fan Relays 237
Fan Center 239
Fan Manager 241
Fan Timer Switch 241
Fan Safety Cutoff Switch 242
Limit Controls 244
Limit Control 244
Secondary High-Limit Switch 248
Combination Fan and Limit Control 251
Switching Relays 256
Impedance Relays 259
Heating Relays/Time-Delay Relays 261
Potential Relay 263
Pressure Switches 265
Sail Switches 266
Other Switches and Relays 268
Sequence Controllers 269
Contactors 275
Troubleshooting Contactors 277
Cleaning Contactors 280
Replacing Contactors 280
Motor Starter 281
Overload Relay Heater 281
Inherent Protector 282
Pilot Duty Motor Protector 283
Capacitors 284
Troubleshooting Capacitors 287
Replacing Capacitors 287
High-Pressure Cutout Switch 288
Low-Pressure Cutout Switch 289
Transformers 290
Sizing Transformers 291

	Installing Transformers	291
	Control Panels	293
Chapter 7	**Ducts and Duct Systems**	**295**
	Codes and Standards	295
	Types of Duct Systems	295
	Perimeter Duct Systems	296
	Extended Plenum Systems	297
	Crawl-Space Plenum Systems	297
	Duct Materials	298
	Duct System Components	299
	Supply Air Registers, Grilles, and Diffusers	301
	Return Air and Exhaust Air Inlets	302
	Duct Run Fittings	303
	Air Supply and Venting	305
	Duct Dampers	305
	Damper Motors and Actuators	313
	Installing Damper Motors	316
	Troubleshooting Damper Motors	320
	Blowers (or Fans) for Duct Systems	321
	Designing a Duct System	322
	Duct System Calculations	323
	Duct Heat Loss and Gain	324
	Air Leakage	325
	Duct Insulation	325
	Equal Friction Method	326
	Balancing an Air Distribution System	331
	Duct Maintenance	331
	Roof Plenum Units	332
	Mobile Home Duct Systems	333
	Proprietary Air Distribution Systems	336
	Duct Furnaces	338
	Electric Duct Heaters	347

Chapter 8 Pipes, Pipe Fittings, and Piping Details 355
Types of Pipe Materials 355
Wrought-Iron Pipe 356
Wrought-Steel Pipe 363
Galvanized Pipe 363
Copper and Brass Pipes and Tubing 363
Plastic Tubing 367
Synthetic Rubber Hose 369
Composite Tubing 369
Pipe Fittings 369
Classification of Pipe Fittings 370
Extension or Joining Fittings 370
Reducing or Enlarging Fittings 378
Directional Fittings 380
Branching Fittings 380
Shutoff or Closing Fittings 382
Union or Makeup Fittings 382
Flanges 382
Pipe Expansion 382
Valves 384
Pipe Threads 384
Pipe Sizing 384
Sizing Steam Pipes 385
Sizing Hot-Water (Hydronic) Pipes/Tubing 393
Pipe Fitting Measurements 396
Calculating Offsets 397
First Method 400
Second Method 401
Third Method 401
Fourth Method 403
Pipe Supports 403
Joint Compound 403
Pipe Fitting Wrenches 406
Pipe Vise 409

	Installation Methods	410
	Pipe Cutting	410
	Pipe Threading	412
	Pipe Reaming	414
	Pipe Cleaning	414
	Pipe Tapping	414
	Pipe Bending	415
	Assembling and Make–Up	415
	Nonferrous Pipes, Tubing, and Fittings	420
	Soldering Pipe	420
	Brazing Pipes	424
	Braze Welding Pipe	425
	Welding Pipe	425
	Gas Piping	429
	Insulating Pipes	429
	Piping Details	430
	Connecting Risers to Mains	431
	Connections to Radiators or Convectors	431
	Lift Fittings	431
	Drips	432
	Dirt Pockets	434
	Siphons	434
	Hartford Connections	434
	Making Up Coils	434
	Relieving Pipe Stress	436
	Swivels and Offsets	439
	Eliminating Water Pockets	440
	Pressure Tests	444
Chapter 9	**Valves and Valve Installation**	**445**
	Valve Components and Terminology	445
	Valve Materials	451
	Globe and Angle Valves	454
	Gate Valves	456
	Check Valves	458

	Stop Valves	463
	Butterfly Valves	465
	Two-Way Valves	467
	Three-Way Valves	469
	Y Valves	469
	Valve Selection	469
	Troubleshooting Valves	472
	Valve Stuffing-Box Leakage	474
	Valve Seat Leakage	474
	Damaged Valve Stems	475
	Automatic Valves and Valve Operators	475
	Valve Pipe Connections	487
	Valve Installing Pointers	489
	Soldering, Brazing, and Welding Valves to Pipes	492
	Soldering or Silver-Brazing Procedure	494
	Butt-Welding Procedure	495
	Socket-Welding Procedure	496
Chapter 10	**Steam and Hydronic Line Controls**	**497**
	Steam and Hydronic System Pumps	497
	Condensate Pumps	497
	Circulators (Water-Circulating Pumps)	505
	Circulator Selection	511
	Steam Traps	518
	Sizing Steam Traps	519
	Steam Trap Maintenance	520
	Automatic Heat-Up	520
	Installing Steam Traps	522
	Float Traps	523
	Thermostatic Traps	524
	Balanced-Pressure Thermostatic Steam Traps	525
	Maintenance	526
	Float and Thermostatic Traps	526
	Thermodynamic Steam Traps	529

	Bucket Traps	530
	Flash Traps	534
	Impulse Traps	534
	Tilting Traps	536
	Lifting Traps	537
	Boiler Return Traps	537
	Expansion Tanks	540
	Closed Steel Expansion Tanks	541
	Diaphragm Expansion Tanks	543
	Sizing Expansion Tanks	543
	Troubleshooting Expansion Tanks	544
	Air Eliminators	545
	Pipeline Valves and Controls	547
	Temperature Regulators	548
	Electric Control Valves (Regulators)	548
	Water-Tempering Valves	550
	Hot-Water Heating Control	554
	Flow Control Valve	558
	Electric Zone Valve	559
	Balancing Valves, Valve Adapters, and Filters	561
	Manifolds	564
	Pipeline Strainers	565
Appendix A	**Professional & Trade Associations**	**567**
Appendix B	**Manufacturers**	**579**
Appendix C	**Data Tables**	**591**
Appendix D	**Conversion Tables**	**629**
Index		**639**

Introduction

The purpose of this series is to provide the layman with an introduction to the fundamentals of installing, servicing, troubleshooting, and repairing the various types of equipment used in residential and light-commercial heating, ventilating, and air conditioning (HVAC) systems. Consequently, it was written not only for the HVAC technician and others with the required experience and skills to do this type of work but also for the homeowner interested in maintaining an efficient and trouble-free HVAC system. A special effort was made to remain consistent with the terminology, definitions, and practices of the various professional and trade associations involved in the heating, ventilating, and air conditioning fields.

Volume 1 begins with a description of the principles of thermal dynamics and ventilation, and proceeds from there to a general description of the various heating systems used in residences and light-commercial structures. Volume 2 contains descriptions of the working principles of various types of equipment and other components used in these systems. Following a similar format, Volume 3 includes detailed instructions for installing, servicing, and repairing these different types of equipment and components.

The author wishes to acknowledge the cooperation of the many organizations and manufacturers for their assistance in supplying valuable data in the preparation of this series. Every effort was made to give appropriate credit and courtesy lines for materials and illustrations used in each volume.

Special thanks is due to Greg Gyorda and Paul Blanchard (Watts Industries, Inc.), Christi Drum (Lennox Industries, Inc.), Dave Cheswald and Keith Nelson (Yukon/Eagle), Bob Rathke (ITT Bell & Gossett), John Spuller (ITT Hoffman Specialty), Matt Kleszezynski (Hydrotherm), and Stephanie DePugh (Thermo Pride).

Last, but certainly not least, I would like to thank Katie Feltman, Kathryn Malm, Carol Long, Ken Brown, and Vincent Kunkemueller, my editors at John Wiley & Sons, whose constant support and encouragement made this project possible.

James E. Brumbaugh

About the Author

James E. Brumbaugh is a technical writer with many years of experience working in the HVAC and building construction industries. He is the author of the *Welders Guide, The Complete Roofing Guide,* and *The Complete Siding Guide.*

Chapter 1

Oil Burners

An oil burner is a mechanical device used to prepare the oil for burning in heating appliances such as boilers, furnaces, and water heaters. The term *oil burner* is somewhat of a misnomer because this device does not actually burn the oil. It combines the fuel oil with the proper amount of air for combustion and delivers it to the point of ignition, usually in the form of a spray.

The fuel oil is prepared for combustion either by vaporization or by atomization. These two methods of fuel oil preparation are used in the three basic types of oil burners employed in commercial, industrial, and residential heating. The following are the three basic types of oil burners:

1. Gun-type (atomizing) oil burners.
2. Vaporizing (pot-type) oil burners.
3. Rotary oil burners.

Gun-type atomizing oil burners are available as either low-pressure or high-pressure types (see Figures 1-1, 1-2, and 1-3). Both are used in residential heating applications with the latter being by far the more popular of the two. The remainder of this chapter is devoted to a description of the gun-type high-pressure atomizing oil burners used in residential and light commercial oil heating systems.

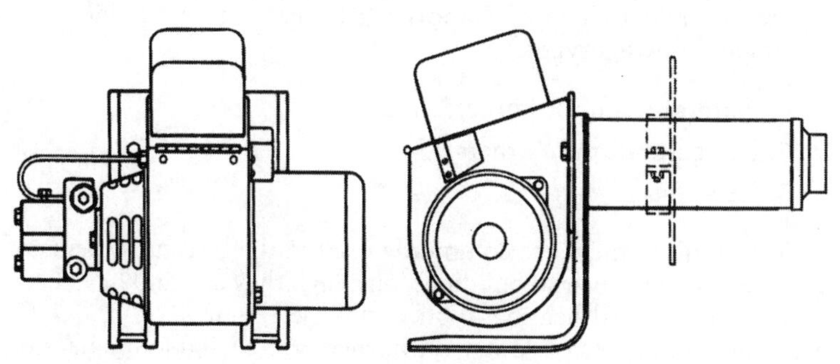

Figure 1-1 Basic shape of a gun-type oil burner.
(Courtesy Stewart-Warner Corp.)

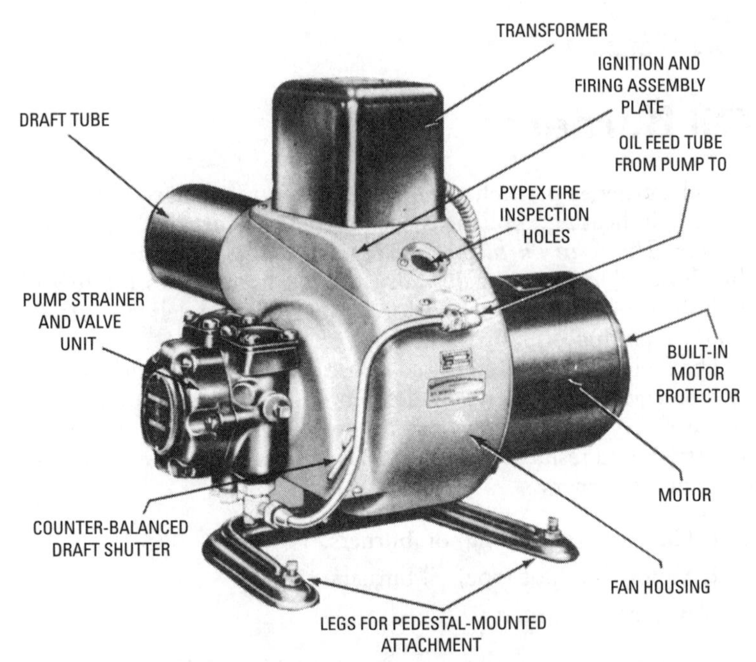

Figure 1-2 Principal components of an S.T. Johnson gun-type oil burner. *(Courtesy S.T. Johnson Company)*

The advantage of the vaporizing (pot-type) oil burner is its low operating cost. It is the least expensive to use, but it has limited heating applications. It is currently used only in small structures located in milder climates. Vaporizing burners can be divided into the three following types:

1. Natural-draft pot burners.
2. Forced-draft pot burners.
3. Sleeve burners.

Rotary oil burners are commonly used in the heating systems of commercial or industrial buildings, although they can and have been used for residential heating applications (see Figures 1-4 and 1-5). The following types of rotary oil burners are available for heating purposes:

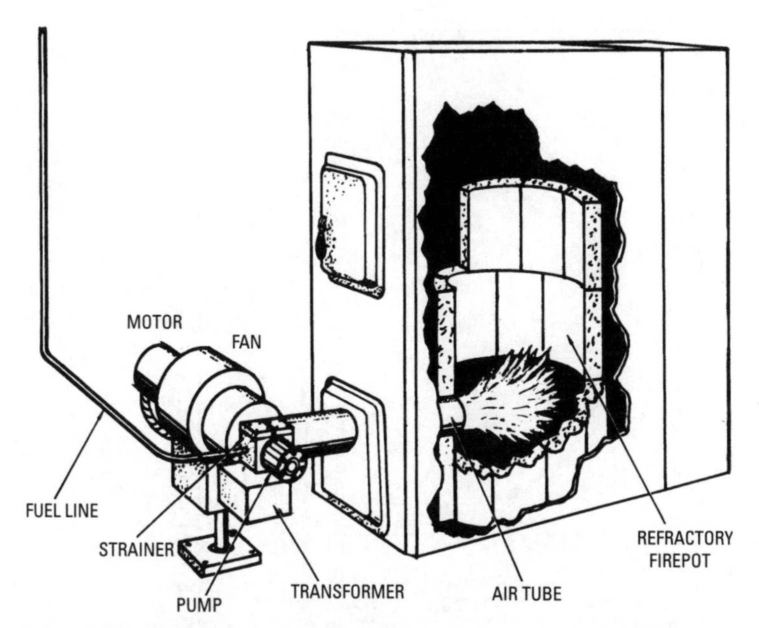

Figure 1-3 Gun-type oil burner firing into furnace combustion chamber. *(Courtesy U.S. Department of Agriculture)*

- Vertical rotary burners
- Horizontal rotary burners
- Wall-flame rotary burners

Gun-Type Oil Burners

Gun-type, high-pressure atomizing oil burners are sometimes called *sprayers* or *atomizing burners* because they spray the fuel oil instead of vaporizing it. They are also referred to as *gun* or *pressure* oil burners because the oil is forced under pressure through a special gun-like atomizing nozzle. The liquid fuel is broken up into minute liquid particles or globules to form the spray.

Construction Details

The principal components and parts of a gun-type, high-pressure atomizing oil burner used in residential and light commercial oil heating systems are illustrated in Figures 1-6 and 1-7. The

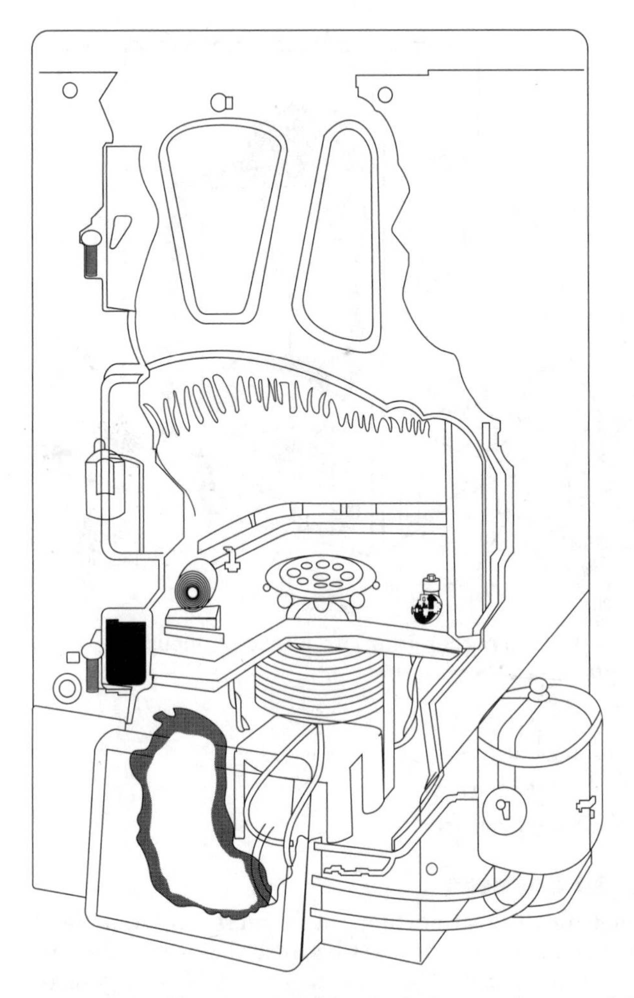

Figure 1-4 Cutaway view of a vertical rotary burner of the vaporizing or wall-flame type. *(Courtesy Integrated Publishing)*

construction details of gun-type oil burners will vary somewhat in different makes and models, but the overall design of these burners is now nearly standardized. The components and parts of a typical gun-type oil burner can be divided into the following categories:

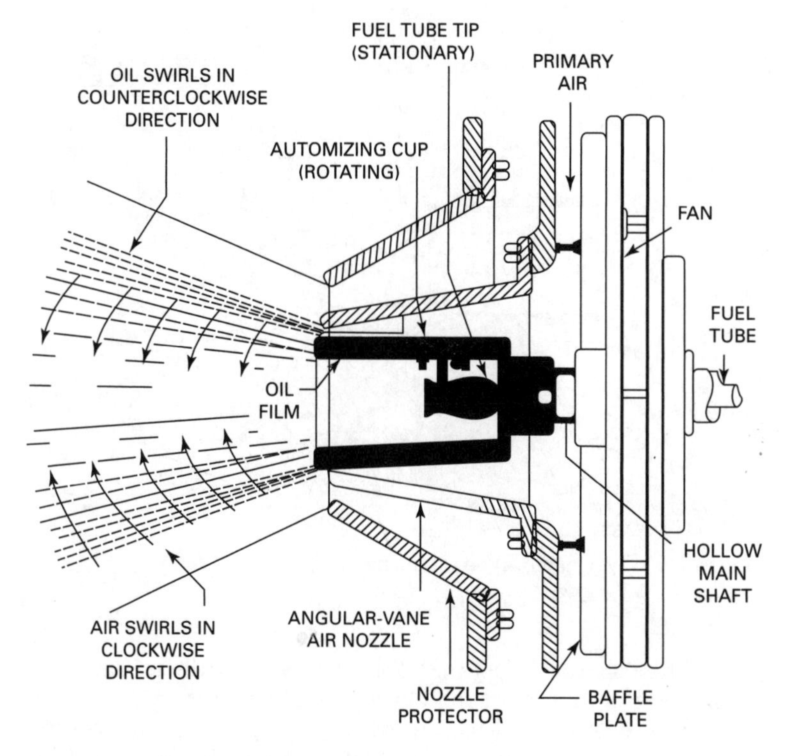

Figure 1-5 Horizontal rotary burner. *(Courtesy Integrated Publishing)*

1. Burner control.
2. Primary safety control.
3. Gun assembly.
4. Ignition transformer.
5. Burner motor and coupling.
6. Fuel pump.
7. Combustion air blower.

Burner Control

The *burner control* is the operational control center of the burner. As shown in Figures 1-6 and 1-7, it is located on the right side of the burner assembly directly above the combustion air blower housing. It operates in conjunction with the primary control and a bimetallic

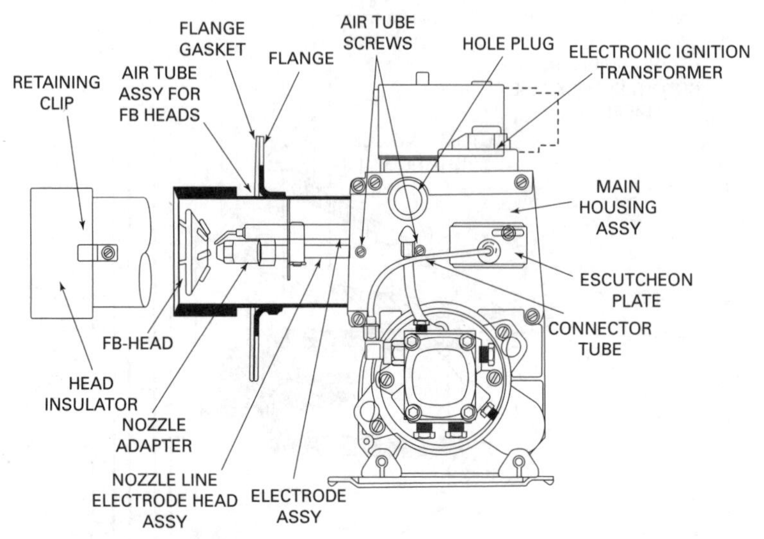

Figure 1-6 Typical gun-type oil burner (side view).
(Courtesy Lennox Industries Inc.)

temperature sensor. When the room thermostat calls for heat and the ignition cycle begins, the burner control will start the burner *only* when the cad cell detects (proves) a flame. The burner control shuts off the burner if the cad cell fails to prove the flame or if the bimetallic sensor detects a temperature too high for safe operation.

Primary Safety Control

The *primary safety control* is an automatic safety device designed to stop the flow of fuel oil at the burner should ignition or flame failure occur. Modern oil-fired furnaces and boilers use a cad cell as the primary control to prove the flame; older ones were equipped with a stack detector primary control. The former is mounted inside the burner behind the access door (see Figure 1-8), and the latter is located in the stack.

Gun Assembly

The oil burner gun assembly consists of a burner nozzle, the electrodes, and a tube connecting the electrodes to the fuel pump (see Figure 1-9). The burner nozzle changes the fuel oil into a form that can be burned in the combustion chamber. It accomplishes this by forcing the oil under pressure through a small hole at the end of the nozzle. The atomized fuel oil is ignited by spark from the electrodes.

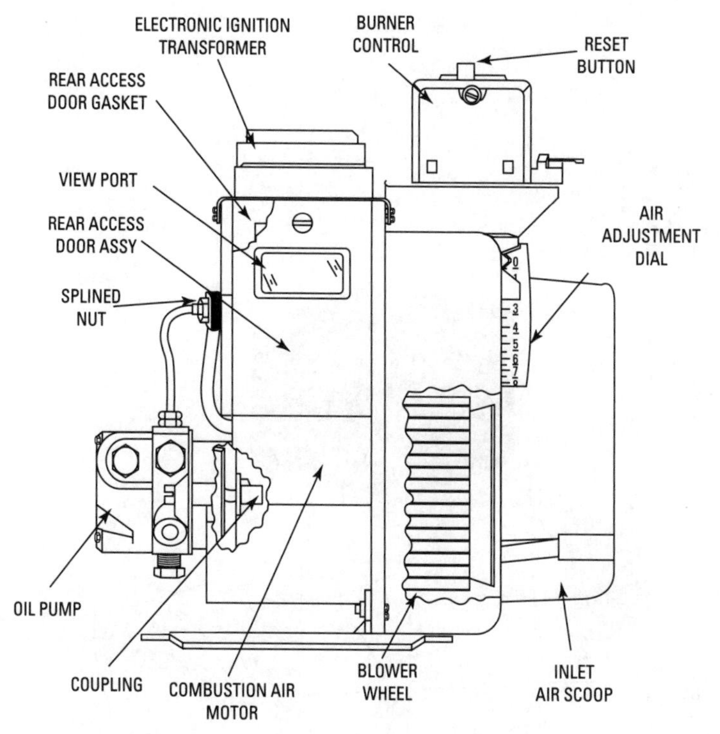

Figure 1-7 Typical gun-type oil burner (front view).
(Courtesy Lennox Industries Inc.)

Ignition Transformer

A step-up ignition transformer located on top of the burner assembly produces the voltage used by the electrodes to ignite the fuel oil. This type of transformer is designed to increase the voltage of a high-voltage (110 VAC) circuit to the ultrahigh 14,000 volts required to ignite the fuel oil.

Burner Motor and Coupling

As shown in Figure 1-5, the burner motor is located on the right side of the oil burner assembly. The drive shaft of the burner motor is connected to both the fuel pump and the combustion air blower by a coupling that functions as the drive shaft for both of these units. A burner motor is also sometimes called an *oil pump motor* or a *pump motor* because it is connected to and drives the fuel (oil) pump.

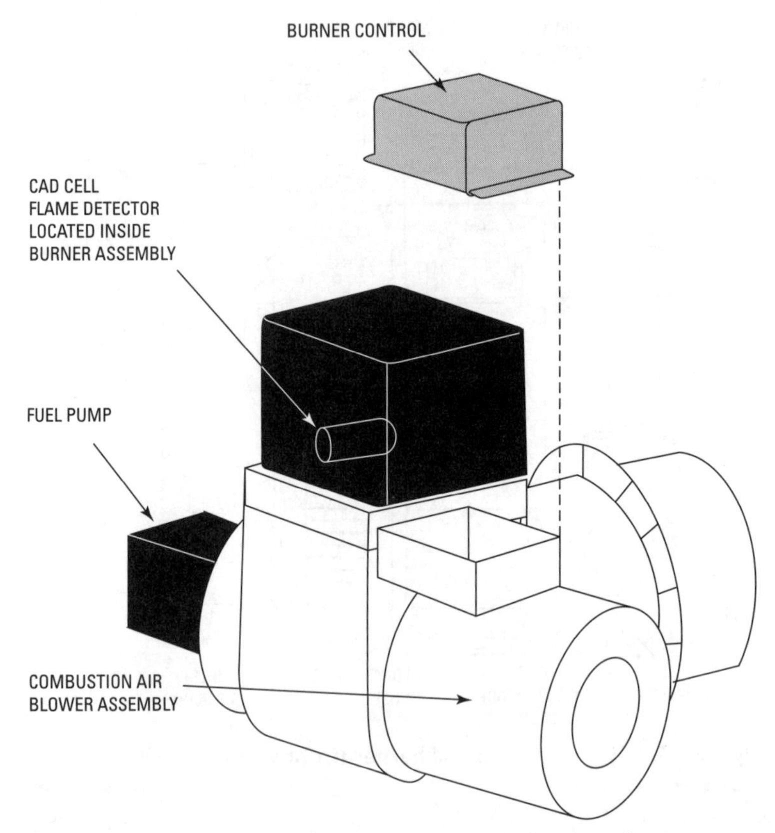

Figure 1-8 Locations of burner control and cadmium cell primary safety control.

Fuel Pump

The fuel pump (also called an *oil pump* or a *fuel unit*) is used to draw fuel oil from the storage tank and deliver it under high pressure (100 to 140 psi) to the nozzle assembly (see Figure 1-11). It is driven by the burner motor and coupling and is located on the left side of the oil burner.

Combustion Air Blower

The combustion air blower is also driven by the burner motor and coupling. It is located between the burner motor and the fuel pump. Its function is to introduce the required amount of air for the

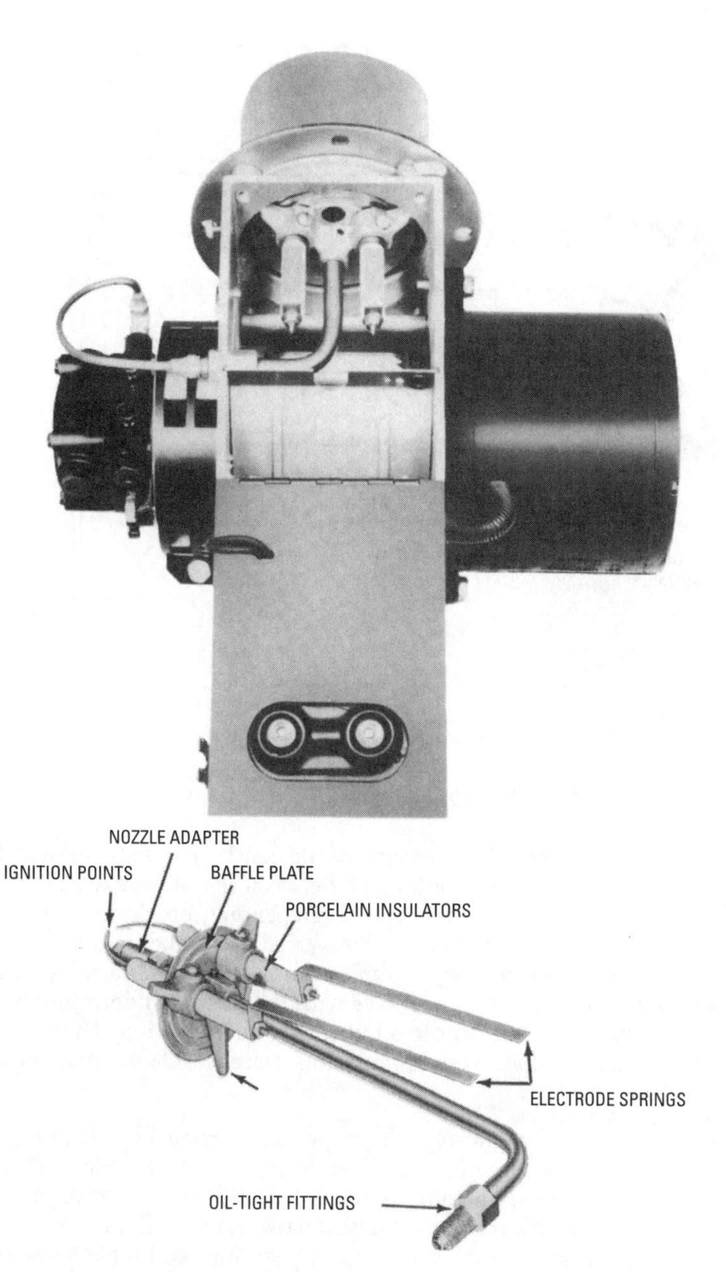

Figure 1-9 Oil burner with transformer removed revealing the gun assembly. *(Courtesy Wayne Home Equipment Co., Inc.)*

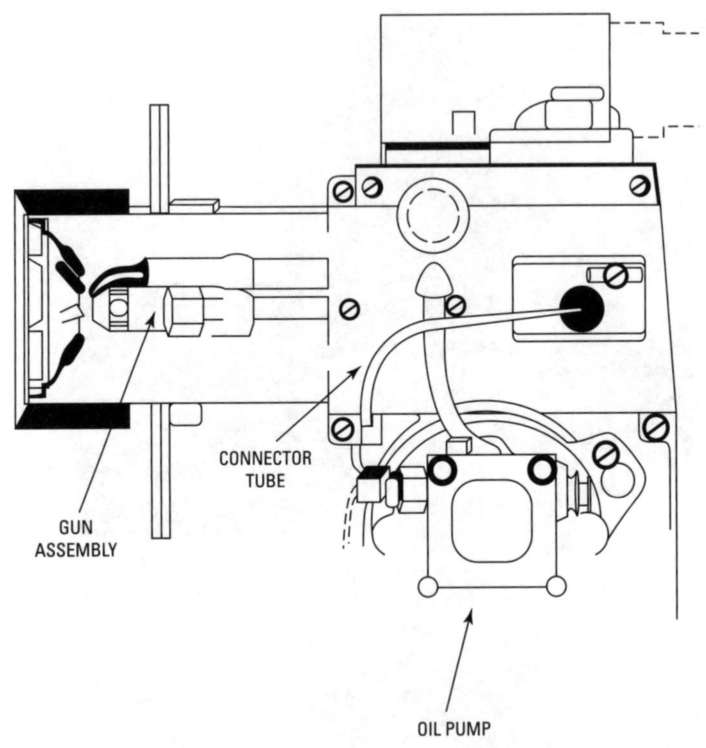

Figure 1-10 Gun assembly details. *(Courtesy Lennox Industries Inc.)*

combustion process. The amount of air can be manually adjusted by an air adjustment gauge located between the blower wheel and the inlet air scoop (see Figure 1-7). Depending on the oil burner manufacturer, a combustion air blower is also sometimes called a *blower wheel*, a *burner motor fan*, or an *induction blower.* Do not confuse the combustion air blower with the furnace indoor blower. The former delivers air to the oil burner for combustion. The latter delivers the heated air to the rooms and spaces inside the structure.

Operating Principles

The operation of a gun-type, high-pressure atomizing oil burner can be traced in Figure 1-12. The fuel oil is drawn through a strainer from the supply tank by the fuel pump and is forced under pressure past the pressure relief cutoff valve via the oil line where it eventually passes through the fine mesh strainer and into the nozzle. The amount of pressure required to pump the fuel oil through

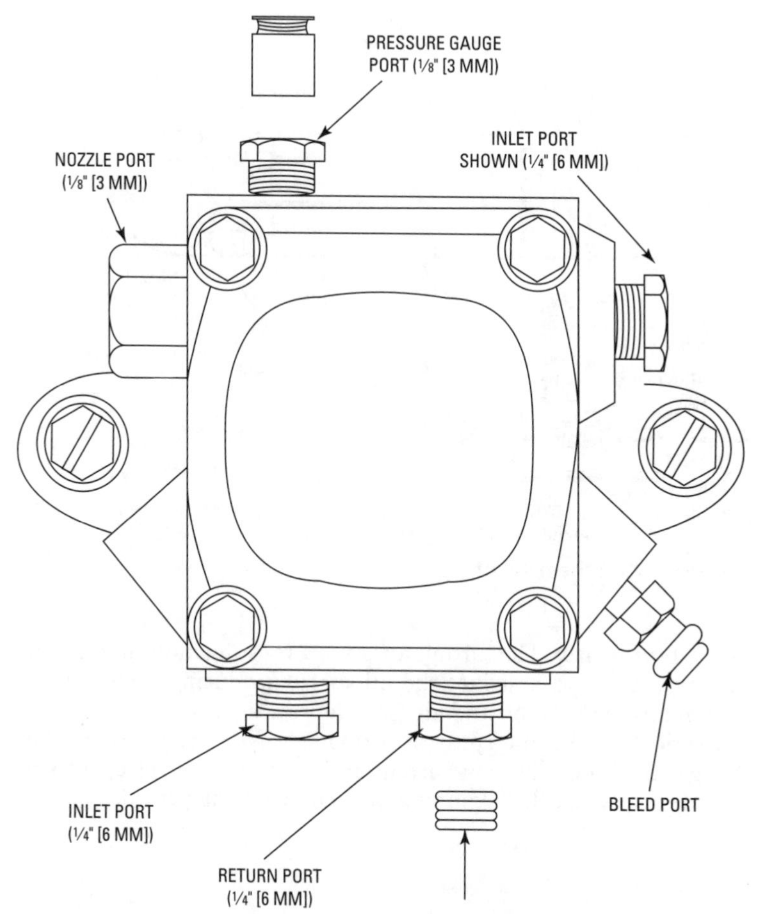

Figure 1-11 Typical fuel pump. *(Courtesy Lennox Industries Inc.)*

the line depends on the size and capacity of the oil burner and the purpose for which it is used. For example, residential oil burners require 80 to 125 psi, whereas commercial and industrial oil burners operate on 100 to 300 psi.

As the fuel oil passes through the nozzle, it is broken up and sprayed in a very fine mist. The air supply is drawn in through the inlet air scoop opening (see Figure 1-5) and forced through the draft tube portion of the casing by the combustion air blower. This air mixes with the oil spray after passing through a set of vanes,

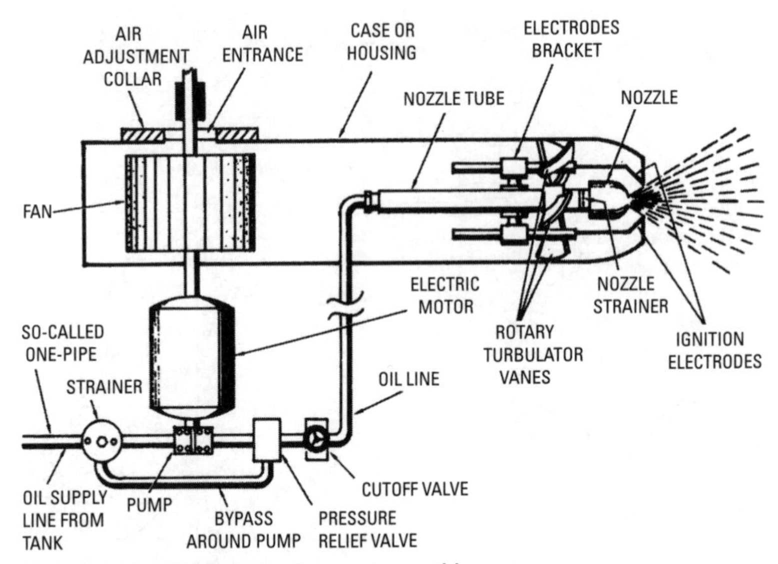

Figure 1-12 Schematic of a gun-type oil burner.

called a *turbulator.* The turbulator gives a twisting motion to the air stream just before it strikes the oil spray, producing a more thorough mixture of the oil and air (see Figure 1-13).

Ignition of the oil spray is provided by a transformer that changes the house lighting current and feeds it to the electrodes to provide a spark at the beginning of each operating period.

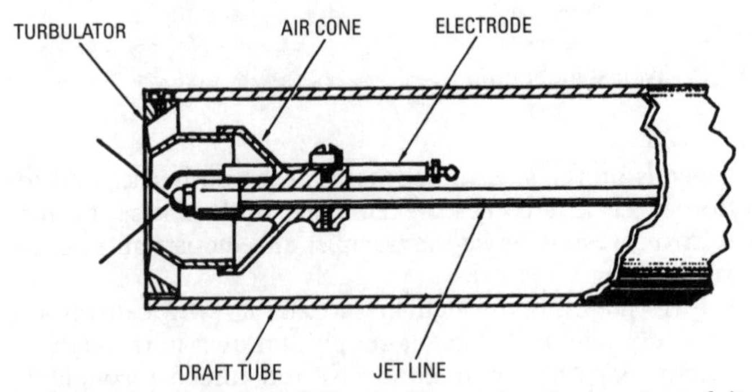

Figure 1-13 Details of draft tube illustrating the location of the turbulator, air cone, and electrode.

The starting cycle of the oil burner is initiated by the closing of the motor circuit. When the motor circuit is closed (automatically by room temperature control), the motor starts turning the fan and the pump. At the same time, the ignition transformer produces a spark at the electrodes ready to light the oil and air mixture.

The action of the pump draws the fuel oil from the tank through the strainer on the fuel line. Its flow is controlled by an oil cutoff valve, which prevents oil passing to the nozzle unless the pressure is high enough to spray the oil (approximately 60 lbs of pressure). Because the pump in the oil burner pumps oil much faster than it can be discharged through the nozzle at that pressure (i.e., 60 lbs of pressure), the oil pressure continues to rise very fast between the pump and the nozzle. When the pressure begins to rise above the normal operating pressure (100 lbs), a pressure relief valve opens and allows the excess oil to flow through the bypass line to the inlet, as in the so-called one-pipe system, or to flow through a second or return line to the supply tank. The pressure relief valve in either system maintains the oil at the correct operating pressure.

When the oil burner is turned off (i.e., when the burner motor stops), the oil pressure quickly drops below the operating pressure, and a pressure relief valve closes. The flame continues until the pressure drops below the setting of the cutoff valve.

The cutoff and pressure relief (regulating) valves may be either two separate units or combined into one unit. Figure 1-14 shows the essentials of the two-unit arrangement. These are, as shown, simply elementary schematics designed to illustrate basic operating principles. The cutoff needle valve is shown with a spring inside the bellows, and the pressure relief (mushroom) valve is shown with exposed spring. In the cutoff valve arrangement, the spring acts against oil pressure on the head of the bellows (tending to collapse it); in the pressure relief valve, the spring acts against the oil pressure, which acts on the lower face of the mushroom valve (tending to open it).

When the pump starts and the pressure in the line rises to about 60 lbs (depending on the spring setting), this pressure acting on the head of the bellows overcomes the resistance of the spring, causing the cutoff valve to open. Since the pump supplies more oil than the nozzle can discharge, the pressure quickly rises to 100 lbs, overcoming the resistance of the relief valve spring and causing the valve to open. This allows excess oil to bypass or return to the tank.

The relief valve will open high enough to maintain the working pressure constant at 100 lbs. When the oil burner is turned off, the oil pressure quickly drops, and the pressure relief valve closes.

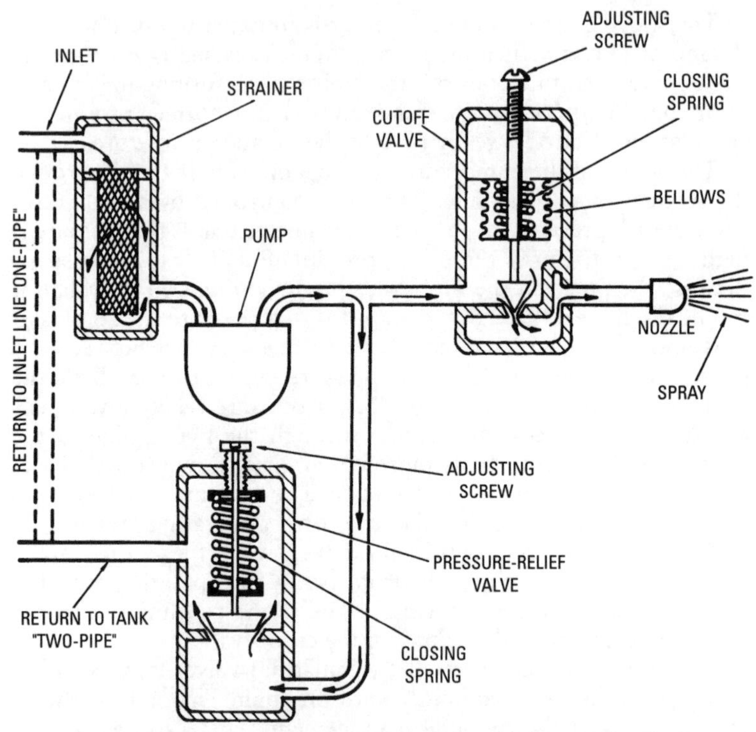

Figure 1-14 Schematic sectional view of separate unit cutoff valve and pressure relief valve showing strainer, pump, and piping.

However, oil will continue to discharge from the nozzle until the pressure drops below the cutoff valve setting when the cutoff valve closes and stops the nozzle discharge.

A passage to the return line is provided by a small slot cut in the seat of the mushroom valve. This causes any remaining pressure trapped in the line by the closing of the cutoff valve to be equalized.

Frequently the cutoff valve and pressure relief valve are combined in a compact cylindrical casing (see Figure 1-15). Here the two valves are attached to a common stem with a flange, which comes in contact with a stop when moved upward by the pressure of the valve actuating the spring.

The position of the stop limits the valve movements to proper maximum lift. A piston, free to move in the cylindrical casing, has an opening in its head that forms the valve seat for the pressure relief

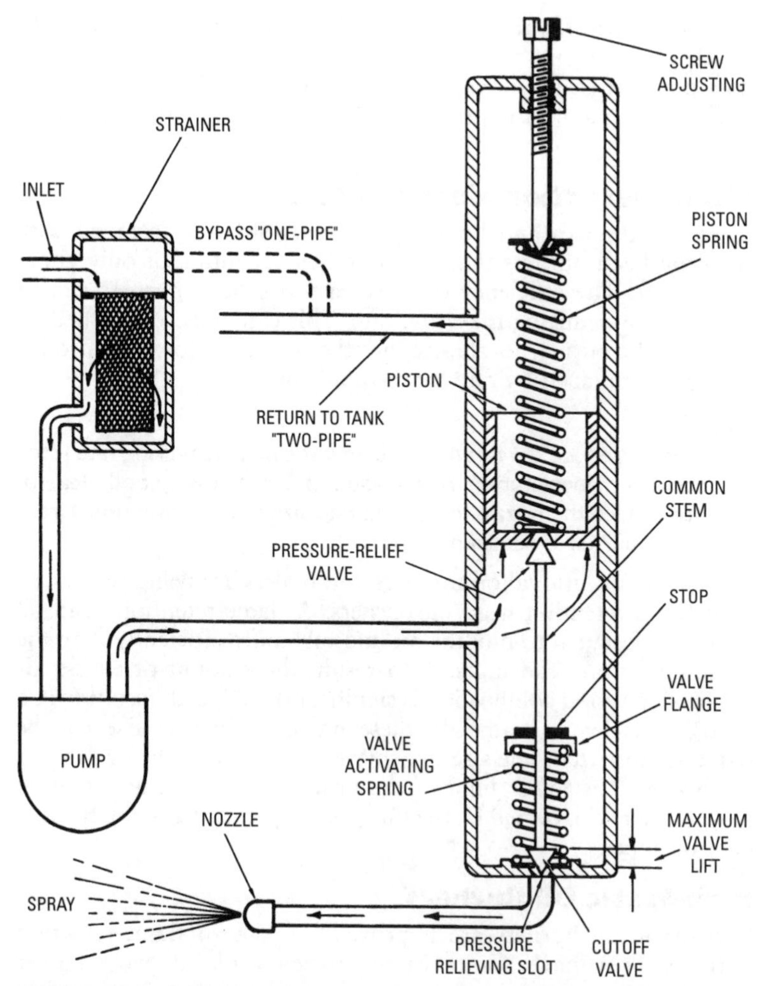

Figure 1-15 Schematic sectional view of combined cutoff valve and pressure relief valve showing the strainer, fuel pump, and piping locations.

valve. The strong piston spring tends to move the piston downward and close the pressure relief valve and then the cutoff valve.

When the pump starts and the pressure in the cylinder below the piston rises to about 60 lbs (depending on the piston spring setting), the piston and the two valves (i.e., the cutoff and pressure relief valve) rise until the valve flange contracts with the stop. At this instant, the cutoff valve is fully opened, allowing oil to flow to the nozzle, the pressure relief valve still being closed. Since the nozzle does not have sufficient capacity to discharge all the oil that is supplied by the pump, the pressure below the piston will continue to rise.

Flame-Retention Head Burners

Most oil furnaces and boilers prior to 1980 were installed with cast-iron head burners that had an efficiency rating of only about 60 percent. The efficiency of these cast-iron head burners can be increased by reducing the firing rate. This can be accomplished by reducing the burner nozzle size, but the size reduction is controlled by the minimum firing rate for the appliance.

Note

> Never reduce the nozzle size below the minimum firing rate listed on the manufacturer's rating plate. As a rule, it is a good idea not to reduce the nozzle more than one size if the conventional iron-head burner is retained.

Many conventional oil furnaces and boilers are being retrofitted with flame-retention head oil burners. A flame-retention head oil burner is designed to mix the air and fuel more efficiently than the traditional iron-head units. As a result, the amount of excess air required for good combustion is significantly reduced, resulting in a hotter and cleaner flame. In these units, the nozzle size can be reduced more than one size to achieve the maximum firing rate for the burner. The lower limit of the firing rate of a flame-retention head burner is governed by the flue gas temperature leaving the furnace or boiler.

High-Static Oil Burners

High-static oil burners are improved versions of flame-retention burners. They have an increased efficiency of 20 percent over flame-retention burners, and the high-static pressure developed in these burners allows them to run at even lower excess air levels.

Rotary Oil Burners

Rotary burners operate with low-pressure gravity and are available in a number of designs depending on the different conditions of use. In each case, the operating principle involves throwing the oil by centrifugal force.

Rotary oil burners can be classified either as rotary nozzle or rotary cup burners. The essential components of the rotary nozzle burner are shown in Figure 1-16. Air pressure acting on the pro-

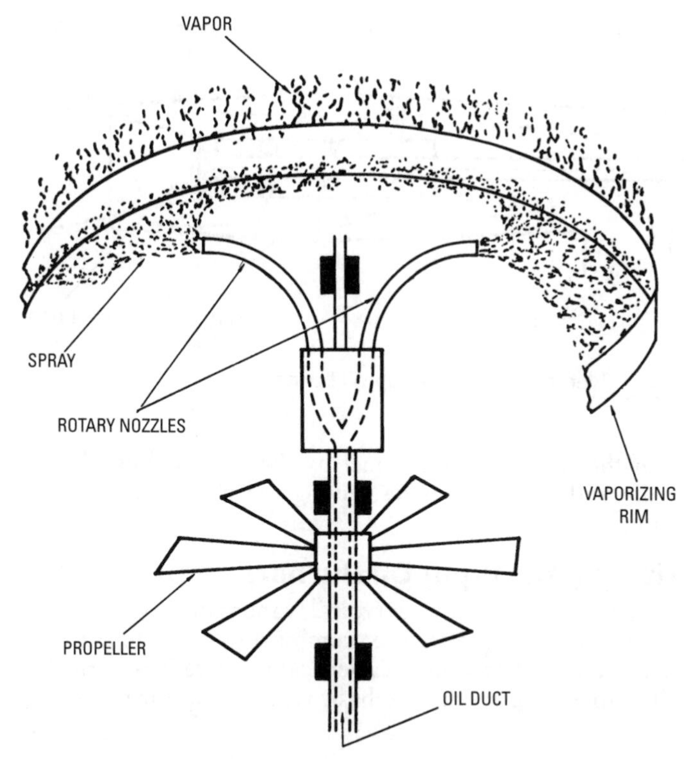

Figure 1-16 Elementary rotary nozzle oil burner.

peller causes the nozzle assembly to rotate at a very high speed. Oil is supplied through the hollow shaft to the nozzles, and the rotary motion causes the oil to be thrown off in a fine spray by centrifugal force. The flame from this spray heats up the metal vaporizing rim hot enough to vaporize the oil spray as it comes in contact with it. Being thoroughly mixed with air, a blue flame is produced. On some designs, the spray vaporized by the vaporizing rim is superheated by passing through grilles.

The rotary cup oil burner (see Figure 1-17) contains a cone-shaped cup that rotates on ball bearings carried by a central tube. The fuel is supplied to the cup through this tube. In operation, drops of oil, issuing from the oil feed tip, come into contact with the cup as shown; by centrifugal force the drops are both flattened into a film and projected toward and off the rim of the cup, as shown in Figure 1-18. Because the rim is surrounded by a concentric

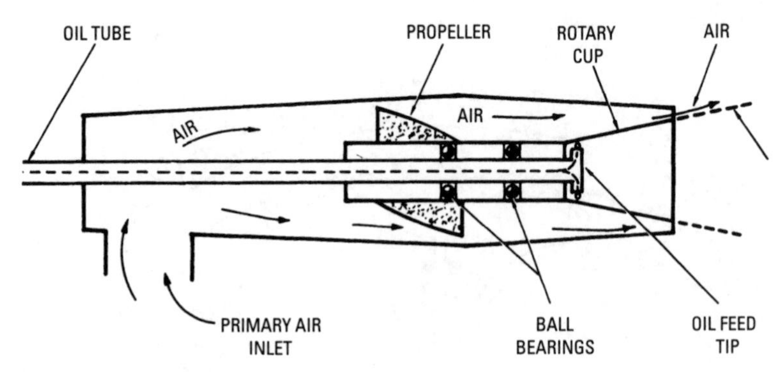

Figure 1-17 Elementary rotary cup oil burner.

opening of the casing, the oil is met by the surrounding blast of primary air with which it mixes, giving the proper mixture for combustion.

Vaporizing (Pot-Type) Oil Burners

Figures 1-19 and 1-20 show a typical vaporizing (pot-type) oil burner. The fuel oil is vaporized for combustion by heating it from below. The vaporized fuel oil rises vertically where it is burned at the top. The following are the two basic types of vaporizing, or pot, oil burners:

1. The natural-draft pot burner.
2. The forced-draft pot burner.

In the former, the air necessary for combustion is provided by the chimney. The forced-draft pot burner relies on both the chimney and a mechanical device (e.g., a fan) for the air supply.

Sleeve burners (also referred to as *perforated sleeve burners*) represent a third type of vaporizing, or pot, burner. Although these burners are used mostly in conjunction with small oil-fired equipment (e.g., kitchen ranges and space heaters), they can also be employed to heat a small house, if outside temperatures do not become too low.

Combination Oil and Gas Burners

Some oil burners are available with combination oil and gas firing accessories that make it possible to use either of these fuels in the

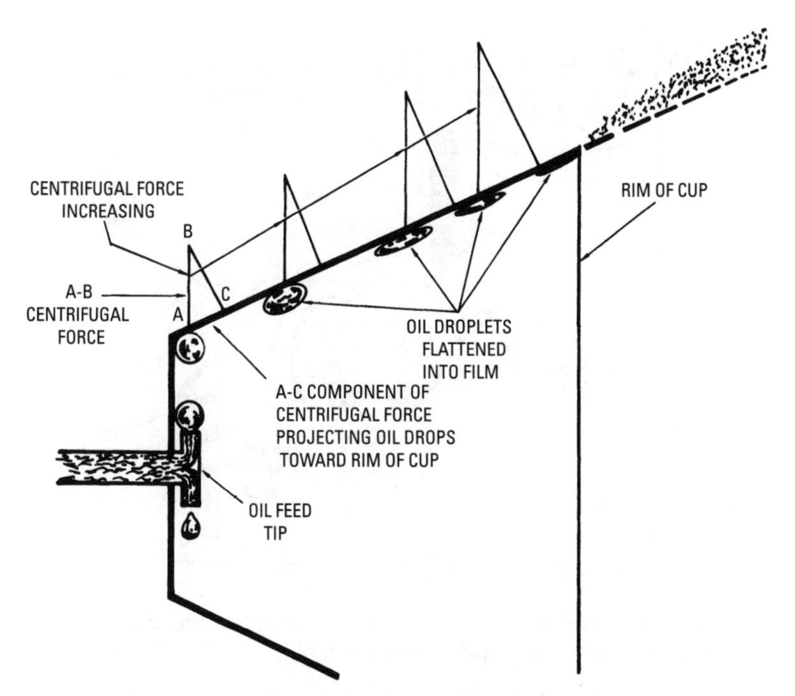

Figure 1-18 Detail of cup showing centrifugal forces acting on the droplets of oil, which flatten them into a film and project them toward and off of rim.

same burner. This is particularly advantageous in areas where low-cost gas is sometimes available.

The combination gas and oil burner illustrated in Figure 1-21 contains independent ignition and control systems for gas or oil. One convenience built into these combination burners is that the oil burner components and parts are standard and require only conventional service procedures. The safety features include a standard cadmium sulfide detection cell and primary relay control.

Fuel Pump

A wide variety of different makes and models of fuel pumps are available for use with oil burners. Both single-stage and two-stage

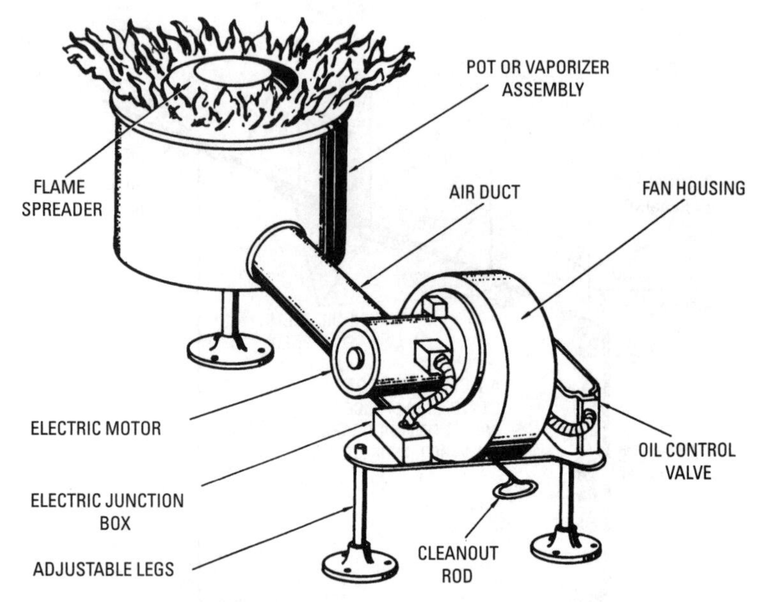

Figure 1-19 Vaporizing (pot-type) oil burner illustrating the connection to the pot or vaporizing assembly.

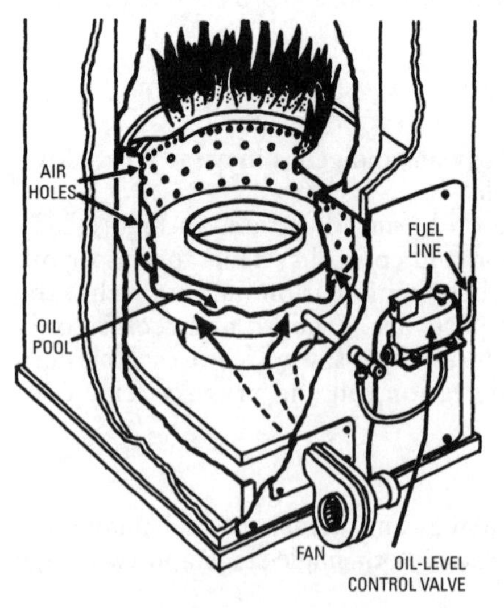

Figure 1-20 Vaporizing (pot-type) oil burner with cutaway of pot or vaporizing assembly. *(Courtesy U.S. Department of Agriculture)*

Figure 1-21 Combination oil and gas burner.
(Courtesy Wayne Home Equipment Co., Inc.)

fuel pumps are available in a number of different sizes and designs (see Figure 1-22).

Single-Stage Fuel Pump

A single-stage pump contains only one set of pumping gears (see Figure 1-23). These pumps are commonly used in single-pipe gravity-feed installations or two-pipe installations under low-lift conditions with up to 10 inches of vacuum. The following are the principal components of a single-stage fuel pump:

1. Pumping gears.
2. Cutoff valve.
3. Strainer.
4. Shaft seal.
5. Noise-dampening device.
6. Shaft bearing.
7. Body.
8. Bleed valve.

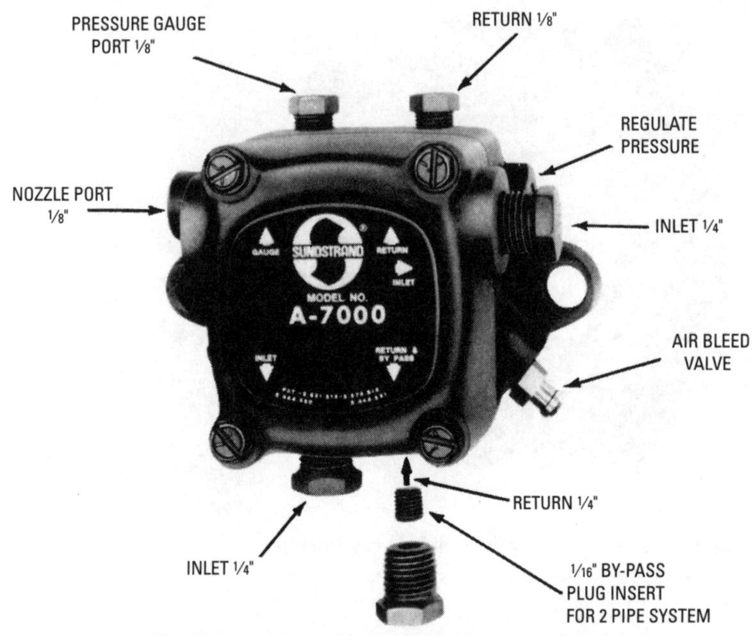

Figure 1-22 Fuel pump assembly. *(Courtesy Sundstrand Hydraulics)*

Fuels Used in Oil Burners

No. 1 and No. 2 fuel oil are both commonly used for residential heating purposes. The No. 2 is slightly more expensive, but the fuel oil gives more heat per gallon used. The lighter No. 1 fuel oil is used in vaporizing, or pot-type, oil burners. The No. 2 fuel oil is used in both atomizing and rotary oil burners.

The manufacturer of the oil burner will generally stipulate the grade of fuel oil to be used. If this information is unavailable, the label of Underwriters Laboratories, Inc., and the Underwriters Laboratories of Canada will stipulate the correct grade of fuel oil to be used.

The heavier the grade of fuel oil used in an oil burner, the greater the care that must be taken to ensure that the oil is delivered for combustion at the proper atomizing temperature. If the oil is not maintained at this temperature prior to delivery for combustion, the oil burner will fail to operate efficiently. An efficient oil burner is one that burns the fuel oil completely using the smallest amount of air necessary for combustion.

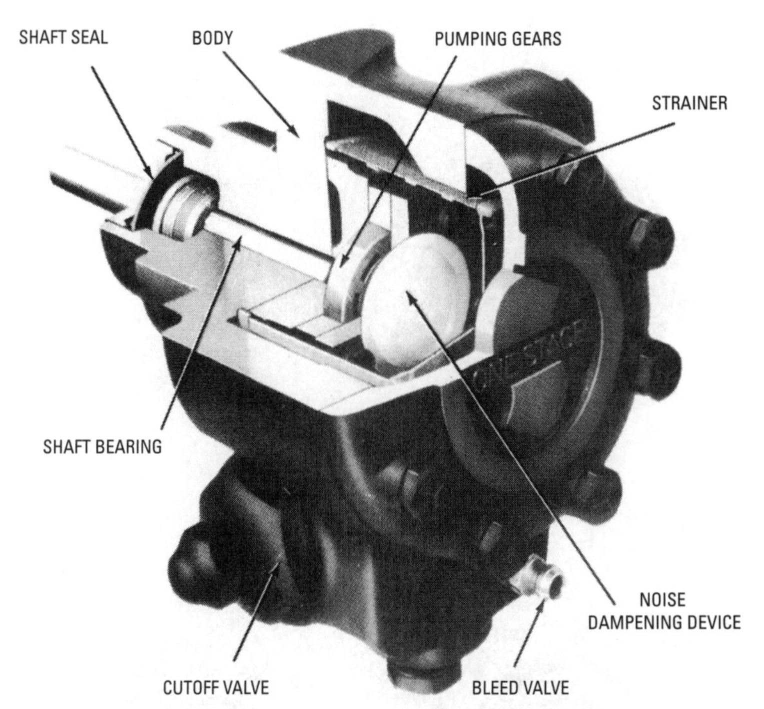

Figure 1-23 Cutaway view of a single-stage fuel pump.
(Courtesy Sundstrand Hydraulics)

The fuel oil first enters the unit by passing through the strainer, where foreign particles such as dirt and line filter fibers are removed. The fuel oil then moves through the hydraulically balanced pumping gears and is pumped under pressure to the valve (see circuit diagram in Figure 1-24). The pressure forces the piston away from the nozzle cutoff seat, and the fuel oil then flows out the nozzle port. Oil in excess of nozzle capacity is bypassed through the valve back to the strainer chamber in a single-pipe system or is returned to the tank in a two-pipe system. Pressure is reduced on the head of the piston when the pump motor is shut off. At this point, the piston snaps back, causing the nozzle port opening to close. A bleeder valve opening in the piston provides for automatic air purging on a two-pipe system, providing for fast cutoff.

(A) Single-stage unit.

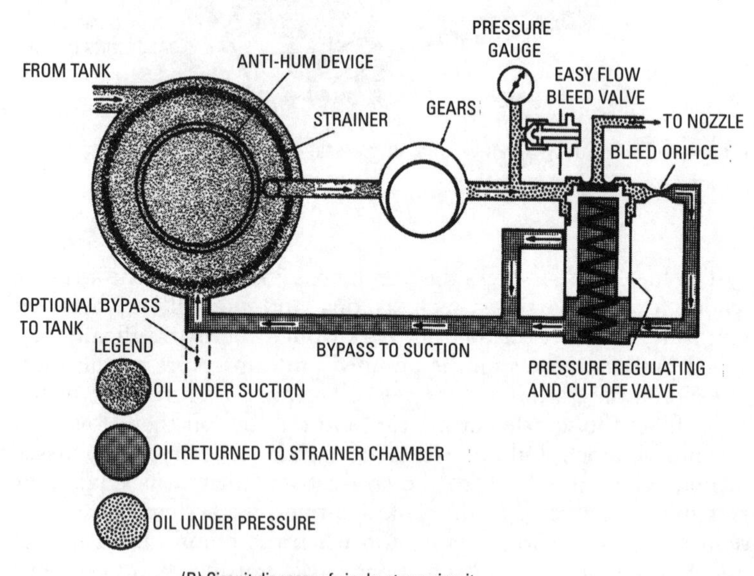

(B) Circuit diagram of single-stage circuit.

Figure 1-24 Circuit diagram of a single-stage fuel pump.

(Courtesy Sundstrand Hydraulics)

Two-Stage Pump

A two-stage pump has two sets of pumping gears. These pumps are used in installations with underground tanks where the combination of lift, horizontal run, fittings, and filters does not exceed the manufacturer's rating in inches of vacuum.

A major advantage of a two-stage fuel unit is that all air is eliminated from the oil being delivered to the nozzle. The inlet of the first stage is located *above* the inlet for the second stage. As a result, any air drawn into the fuel unit after priming is picked up by the first stage and discharged to the tank before it reaches the second stage. Consequently, the second stage draws completely air-free oil.

As the air is being discharged into the tank by the first stage, pressure begins to build up in the second stage, causing the regulating valve to bypass excess oil back into the strainer. These operating principles are illustrated by the circuit diagram in Figure 1-25.

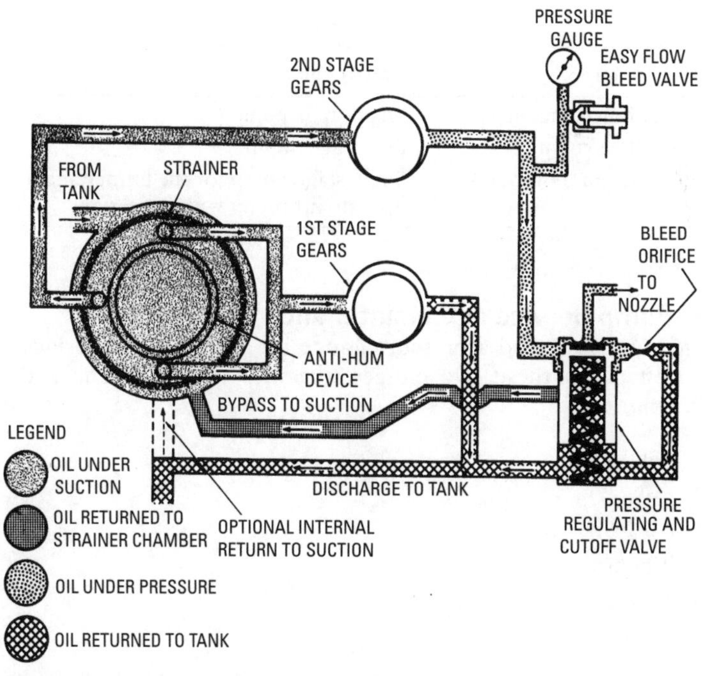

Figure 1-25 Circuit diagram of a system using a two-stage fuel pump.
(Courtesy Sundstrand Hydraulics)

Fuel Pump Capacity

The capacity of an oil burner fuel pump should be sufficient to handle the total vacuum in the system. The vacuum is expressed in inches and can be determined by the following procedure:

- 1 inch of vacuum for each foot of lift
- 1 inch of vacuum for each 90° elbow in either the suction or return lines
- 1 inch of vacuum for each 10 feet of horizontal run (3/8-inch OD line)
- 1 inch of vacuum for each 20 feet of horizontal run (1/2-inch OD line)

After you have calculated the total vacuum, you can use these data to select the most suitable pump for the burner. Table 1-1 lists various vacuums and suggests appropriate pump capacities.

Table 1-1 Types of Pumps Recommended for Different Vacuums

Total Vacuum	*Type of Pump*
Up to a 3-inch vacuum	Single-stage pump
4–13-inch vacuum	Two-stage pump
14-inch vacuum or more	Single-stage pump for the burner and a separate lift pump with a reservoir

(Courtesy National Fuel Oil Institute)

Fuel Pump Service and Maintenance

A vacuum gauge and a pressure gauge are both used to service a fuel unit. With these two gauges, the individual can check the following:

- Vacuum
- Lift
- Air leaks
- Pressure
- Cutoff
- Delivery

Figures 1-26 and 1-27 illustrate the attachment of the vacuum and pressure gauges to a fuel unit. The pressure gauge (shown as the upper gauge in Figure 1-26 and the gauge attached to the nozzle line opening in Figure 1-27) will indicate whether a positive cutoff is operable or whether an adequate and uniform buildup of pressure is present. When the pressure gauge is attached to the nozzle line opening, it should indicate a reading of 75 to 90 psi (see Figure 1-27). Any drop of the pressure gauge reading to zero indicates leaky cutoff and probable difficulty with the shutoff valve in the nozzle line.

The existence of air leaks in the supply line can be determined by vacuum gauge readings once the gauge is attached to the optional inlet connection (see Figure 1-26). An evaluation of the gauge reading is itself determined by the location of the oil storage tank. If the tank is located above the burner, and the oil is supplied by gravity flow, the vacuum gauge must show a reading of zero, unless there is a problem in the system. These problems can take the following forms:

- A partially closed cutoff valve of the oil supply tank
- A kinked or partially blocked oil supply line
- A blocked line filter

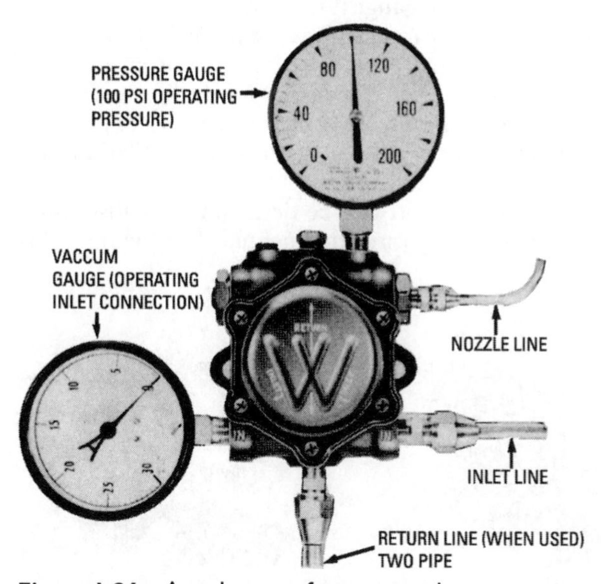

Figure 1-26 Attachment of vacuum and pressure gauge.
(Courtesy Wayne Home Equipment Co., Inc.)

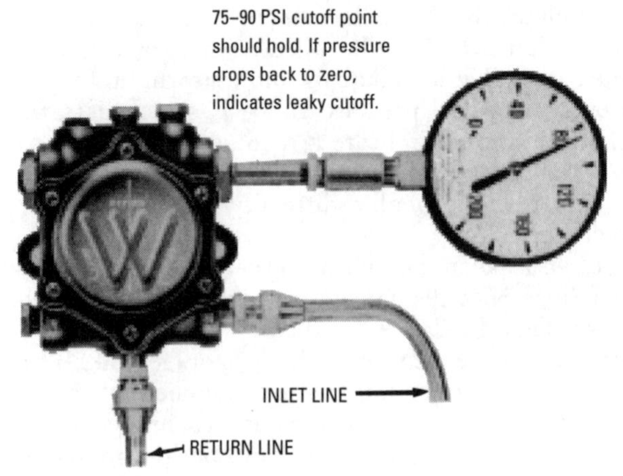

Figure 1-27 Pressure gauge attached to the nozzle line.
(Courtesy Wayne Home Equipment Co., Inc.)

A system with an oil supply tank located *below* the level of the oil burner that supplies the fuel oil through a line filter *must* produce a reading on the vacuum gauge if the system is operating properly. A zero reading will indicate the presence of an air leak.

Table 1-3 lists a few of the problems that may be encountered with fuel units and some suggested remedies for dealing with them. Most manufacturers of fuel units or gauges generally supply troubleshooting recommendations along with the installation and maintenance instructions.

The cause of improper cutoff can be determined by inserting a pressure gauge in the nozzle port of the fuel unit (see Figure 1-28).

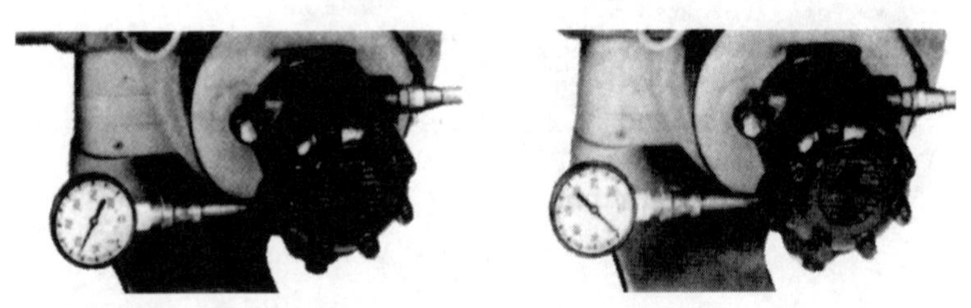

Figure 1-28 Checking fuel pressure with a gauge.
(Courtesy Sundstrand Corp.)

Allow the fuel unit to operate for a few minutes and then shut off the burner. If the pressure drops to 0 psi on the gauge, the fuel unit should be replaced. A gauge reading above 0 psi indicates that the fuel unit is operating properly. The probable cause of improper cut-off in those cases where a gauge reading is obtained is usually air in the system.

Priming Fuel Pumps

On occasion, the oil burner may fail to pump oil. When this occurs, check the oil supply line to the furnace for leaks. If there are no leaks, it may be necessary to prime the fuel pump. Pumps are self-priming for single-stage, two-pipe systems and for two-stage pumps. A single-stage pump (one-pipe system) should be primed as follows:

1. Turn off the electrical power supply to the unit.
2. Read and follow the priming instructions provided by the manufacturer.
3. Prime the pump until the oil is free of bubbles.

When a *new* pump fails to prime, it may be due to dry pump conditions, which can be corrected by removing the vent plug and filling the pressure cavity slowly so that the fuel oil wets the gears (see Figure 1-29). Other possible causes of the pump failing to prime include the following:

- Suction inlet vacuum is greater than 15 inches of vacuum.
- Suction line is incorrectly sized.
- Oil suction line strainer or filter capacity does not match the pump suction gear capacity.
- Bypass plug is not in position on two-pipe installations.
- Plug(s) and/or suction line connections are not airtight.

Adjusting Fuel Pump Pressure

The oil-pressure regulator on the fuel pump is generally factory-set to give nozzle oil pressures of 100 psig. The firing rate is indicated on the nameplate and can be obtained with standard nozzles by adjusting the pump pressures as follows:

1. Turn the adjusting screw clockwise to increase pressure.
2. Turn the adjusting screw counterclockwise to decrease pressure.
3. Never exceed the pressures indicated in Table 1-2.

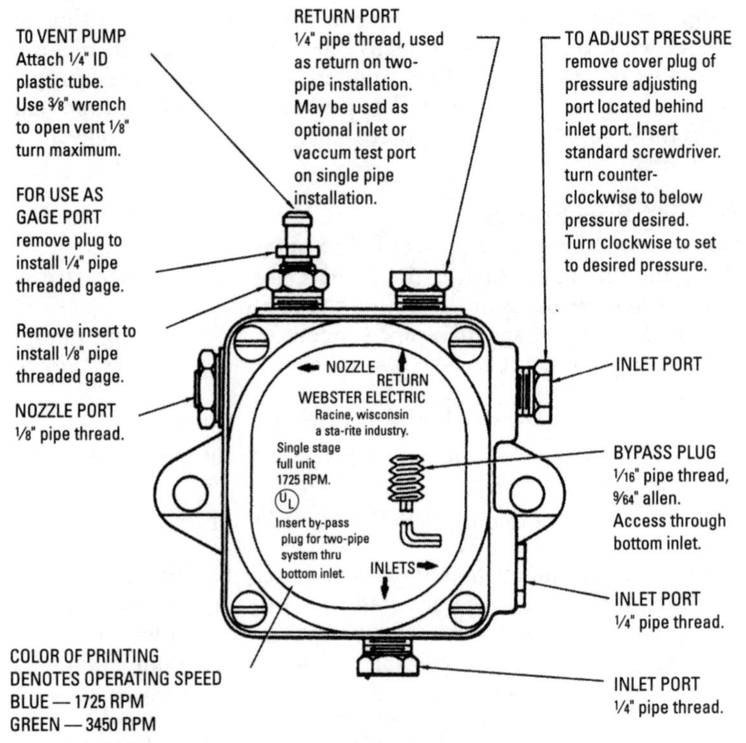

Figure 1-29 Webster model M series fuel pump.

(Courtesy Webster Electrical Co., Inc.)

Table 1-2 Maximum Recommended Pressures

Bonnet Capacity (1000 Btu/h)	*Firing Rate (gph)*	*Standard Nozzle Size*	*Pump Pressure (psig)*
85	0.76	0.75	103
100	0.90	0.90	100
125	1.12	1.10	104
150	1.35	1.35	100
200	1.80	1.75	112
250	2.25	2.25	100
335	3.00	1.50	100

(Courtesy Carrier Corp.)

Troubleshooting Fuel Pumps

The troubleshooting list in Table 1-3 contains the most common operating problems associated with fuel pumps and fuel units. Each problem is given in the form of a symptom, the possible cause, and a suggested remedy. The purpose of this list is to provide the operator with a quick reference to the cause and correction of a specific problem.

Table 1-3 Troubleshooting Fuel Pumps

Symptom and Possible Cause	*Possible Remedy*
No oil flow to nozzle.	
(a) Clogged strainer or filter.	(a) Remove and clean strainer; repack filter element.
(b) Air binding in two-pipe system.	(b) Check and insert bypass plug.
(c) Frozen pump shaft.	(c) Remove pump and return it to the manufacturer for repair or replacement.
Oil leak.	
(a) Loose plugs or fittings.	(a) Dope with good-quality thread sealer.
(b) Leak at pressure-adjusting end cap nut.	(b) Fiber washer may have been left out after adjustment of valve spring; replace washer.
(c) Blown seal.	(c) Replace fuel unit.
(d) Seal leaking.	(d) Replace fuel unit.
Noisy operation.	
(a) Air inlet line.	(a) Tighten all connections and fittings in the intake line and unused intake port plugs (see Figure 1-10).
(b) Bad coupling alignment.	(b) Loosen mounting screws and shift fuel pump to a position where noise is eliminated. Retighten mounting screws.
(c) Pump noise.	(c) Work in gears by continued running or replace.

(continued)

Table 1-3 (continued)

Symptom and Possible Cause	*Possible Remedy*
Pulsating pressure.	
(a) Air leak in intake line.	(a) Tighten all fittings and valve packing in intake line.
(b) Air leaking around strainer cover.	(b) Tighten strainer cover screws.
(c) Partially clogged strainer.	(c) Remove and clean strainer.
(d) Partially clogged filter.	(d) Replace filter element.
Low oil pressure.	
(a) Nozzle capacity is greater than fuel pump capacity.	(a) Replace fuel pump with one of correct capacity.
(b) Defective gauge.	(b) Check against another and replace if necessary.
Improper nozzle cutoff.	
(a) Filter leaks.	(a) Check face of filter cover and gasket for damage.
(b) Partially clogged nozzle strainer.	(b) Clean strainer or change nozzle.
(c) Air leak in intake line.	(c) Tighten intake fittings and packing nut on shutoff valve; tighten unused intake port plug.
(d) Strainer cover loose.	(d) Tighten screws.
(e) Air pockets between cutoff valve and nozzle.	(e) Start and stop burner until smoke and afterfire disappear.

Fuel Supply Tank and Line

The installation, maintenance, and troubleshooting of fuel supply tanks is described in Chapter 12 ("Oil Furnaces") in Volume 1.

Oil Burner Nozzles

An oil burner nozzle is a device designed to deliver a fixed amount of fuel to the combustion chamber in a uniform spray pattern and spray angle best suited to the requirements of a specific burner. The oil burner nozzle atomizes the fuel oil (i.e., breaks it down into extremely small droplets) so that the vaporization necessary for combustion can be accomplished more quickly.

The components in a typical nozzle (see Figure 1-30) include the following:

1. Orifice.
2. Swirl chamber.
3. Orifice disc.
4. Body.
5. Tangential slots.
6. Distributor.
7. Retainer.
8. Filter.

Fuel oil is supplied under pressure (100 psi) to the nozzle, where it is converted to velocity energy in the swirl chamber by directing it through a set of tangential slots. The centrifugal force caused within the swirl chamber drives the fuel oil against the chamber

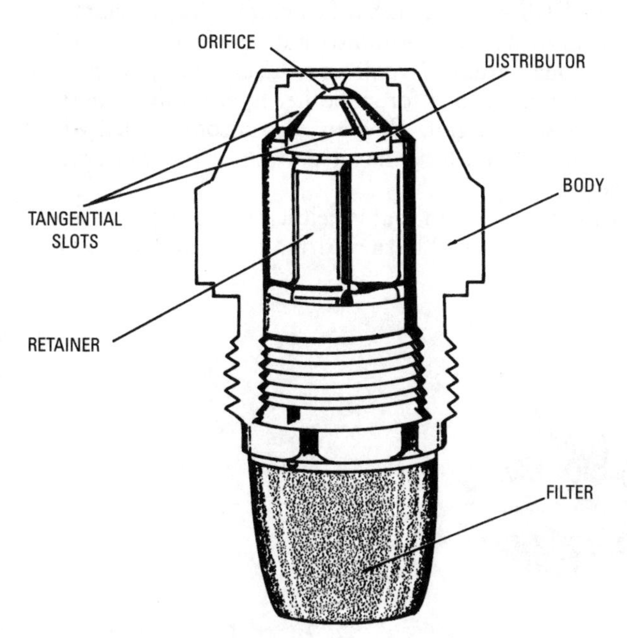

Figure 1-30 Details of an oil burner nozzle.
(Courtesy Wayne Home Equipment Co., Inc.)

walls, producing a core of air in the center. The latter effect moves the oil out through the orifice at the tip of the nozzle in a cone-shaped pattern.

The following are the two basic spray cone patterns:

1. The hollow cone.
2. The solid cone.

Each has certain advantages depending on its use.

The *hollow cone pattern* (see Figure 1-31) is recommended for use in smaller burners (those firing 1.00 gph and under). As shown in Figure 1-31, they are characterized by a concentration of fuel oil droplets all around the outer edge of the spray. There is little or no distribution of droplets in the center of the cone. The principal advantage of the hollow cone patterns is a more stable spray pattern and angle under adverse conditions than solid cone patterns operating under the same conditions and at the same flow rate.

The *solid cone pattern*, illustrated in Figure 1-32, is characterized by a uniform or near-uniform distribution of fuel oil droplets throughout the cone pattern. Nozzles producing this cone pattern are particularly recommended for smoother ignition in oil burners firing above 2.00 or 3.00 gph. They are also recommended where long fires are required or where the air pattern or the oil burner is heavy in the center.

A combination cone pattern that is neither a true cone nor a true hollow cone can be used in oil burners firing between 0.40 gph and 8.00 gph.

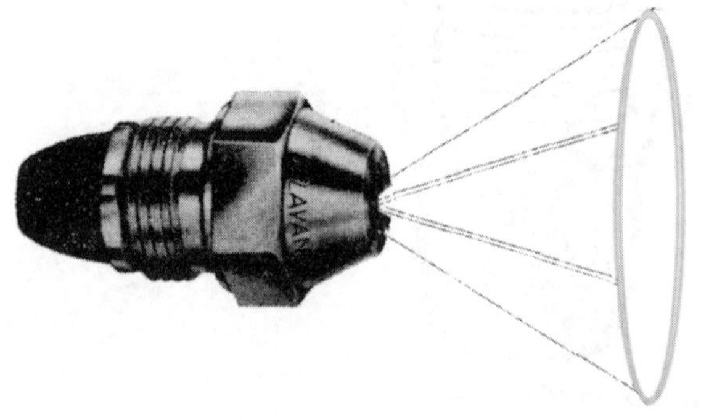

Figure 1-31 Hollow spray cone pattern.
(Courtesy Wayne Home Equipment Co., Inc.)

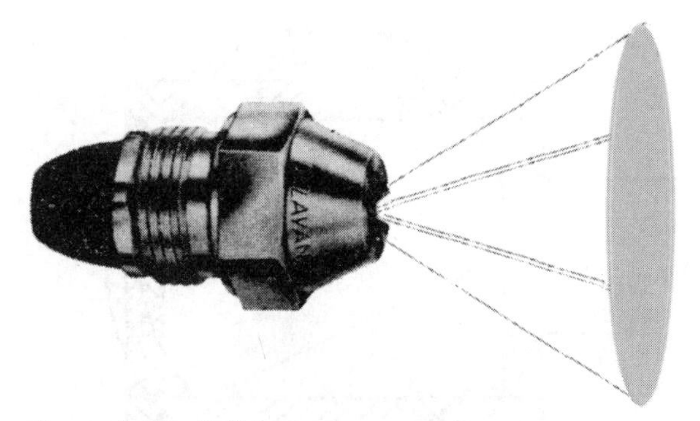

Figure 1-32 Solid spray cone pattern.
(Courtesy Wayne Home Equipment Co., Inc.)

Oil burner nozzles are also selected on the basis of the spray angle they produce (see Figure 1-33). The *spray angle* refers to the angle of the spray cone, and this angle will generally range from 30° to 90°. The angle selected will depend on the requirement of the burner air pattern and combustion chamber. For example, 70° to 90° spray angles are recommended for round or square combustion chambers (see Figure 1-34), and 30° to 60° spray angles are recommended for long, narrow chambers (see Figure 1-35). Recommended combustion chamber dimensions and spray angles for nozzles are given in Table 1-4.

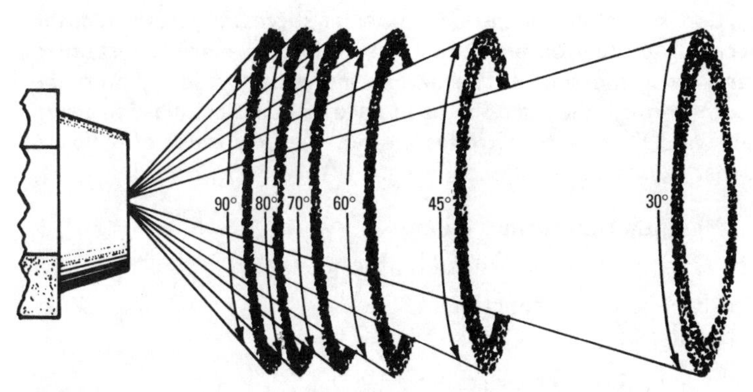

Figure 1-33 Varieties of spray angles. *(Courtesy Wayne Home Equipment Co., Inc.)*

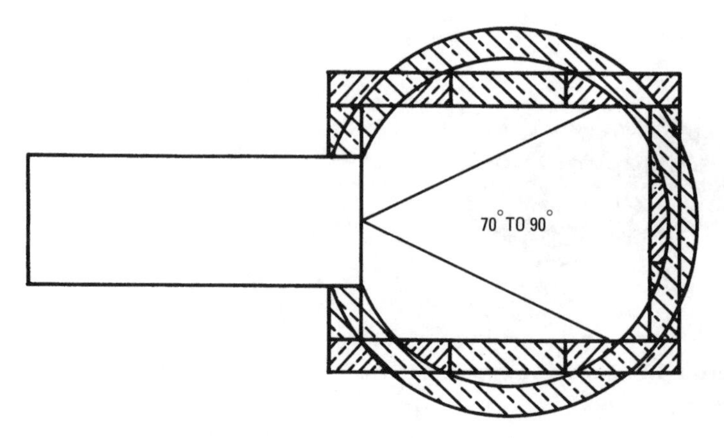

Figure 1-34 Spray angles (70° to 90°) suitable for round or square chambers. *(Courtesy Wayne Home Equipment Co., Inc.)*

Electrodes

The electrodes must be frequently checked and adjusted to ensure proper and efficient ignition of the fuel oil. Broken or malfunctioning electrodes can result in smoke leaking out into the rooms of the structure. This problem, called a *puffback*, is not an uncommon one in oil-fired appliances (see sidebar).

Puffbacks

A puffback, or the leaking of sooty smoke from the combustion chamber of an oil furnace or boiler, is caused by the accumulation of fuel oil in the combustion chamber of the furnace or boiler after an ignition failure. When the oil burner is successfully restarted, the accumulated fuel oil burns too rapidly for the exhaust system to carry away the smoke. The excess smoke is forced out into the rooms through the seams in the furnace or boiler combustion chamber walls. Damaged electrodes are not the only cause of puffback. Other causes include the following:

- Ignition transformer failure
- Contaminated or eroded oil burner nozzle
- Fuel pump malfunction
- Clogged oil filter
- Clogged burner air intake
- Damaged combustion chamber linings

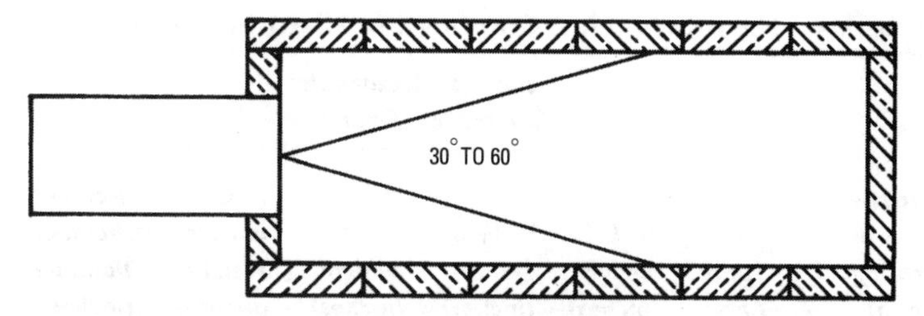

Figure 1-35 Spray angles (30° to 60°) suitable for long, narrow chambers. *(Courtesy Wayne Home Equipment Co., Inc.)*

Troubleshooting Electrodes

The gun assembly must be removed from the oil burner in order to inspect the electrodes. The following conditions require electrode replacement:

- Cracks in the electrode coating
- Dull electrode points
- Broken electrodes

Servicing Electrodes

Remove the electrodes by loosening the screw connecting them to the gun assembly. If the electrodes are round and appear worn, use a file to reestablish a good point. Reinstall them in the electrode holder and set the electrode gap at ½ inch. Set the distance between the center of the nozzle and the tips at ¼ inch.

Broken electrodes or electrodes with cracked ceramic coatings must be replaced with new ones.

Oil Burner Air System

The air system for the average oil burner is generally composed of the air shutter draft tube, the turbulator, and the fan. The draft tube and turbulator have already been shown (see Figures 1-2 and 1-12).

The fan construction consists of a (squirrel cage) series of vanes or blades mounted on the rim of a wheel. These vanes are slanted forward in such a manner as to provide the maximum discharge of air. Figure 1-36 shows the construction of a fan and flexible coupling.

The operating principles of the air system are fairly simple. The fan draws air into the fan housing and forces this air through the

Table 1-4 Recommended Combustion Chamber Dimensions

Nozzle Size of Rating (gph)	Spray Angle	*Square or Rectangular Combustion Chamber* L Length (inches)	W Width (inches)	H Height (inches)	C Nozzle Height (inches)	Round Chamber Diameter (inches)
0.50–0.65	80°	8	8	11	4	9
0.75–0.85	60°	10	8	12	4	*
	80°	9	9	13	5	10
1.00–1.10	45°	14	7	12	4	*
	60°	11	9	13	5	*
	80°	10	10	14	6	11
1.25–1.35	45°	15	8	11	5	*
	60°	12	10	14	6	*
	80°	11	11	15	7	12
1.50–1.65	45°	16	10	12	6	*
	60°	13	11	14	7	*
	80°	12	12	15	7	13
1.75–2.00	45°	18	11	14	6	*
	60°	15	12	15	7	*
	80°	14	13	16	8	15
2.25–2.50	45°	18	12	14	7	*
	60°	17	13	15	8	*
	80°	15	14	16	8	16
3.00	45°	20	13	15	7	*
	60°	19	14	17	8	*
	80°	18	16	18	9	17

**Recommend oblong chamber for narrow sprays.*
(Courtesy of Wayne Home Equipment Co., Inc.)

draft tube and turbulator and into the combustion chamber. The amount of incoming air can be regulated by adjusting the air shutter. As the air is forced through these vanes, it is given a swirling motion just before it strikes the oil spray. This motion provides a more thorough mixture of the oil and air, resulting in better combustion.

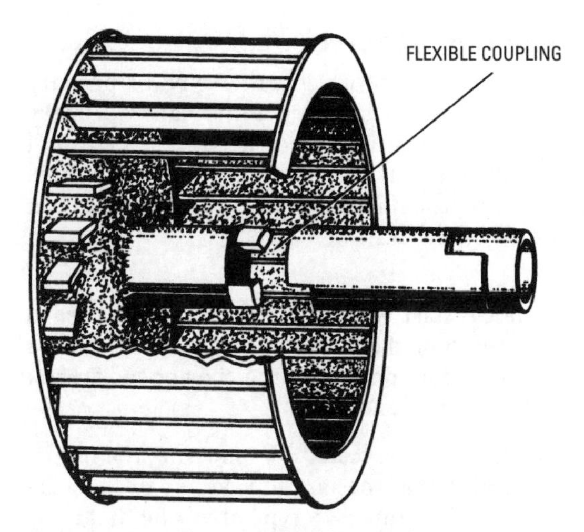

Figure 1-36 Fan and flexible coupling.

The shape of the turbulator varies in different models, but the purpose is the same: to thoroughly mix the air and oil spray. Figure 1-37 shows a double turbulator consisting of an air impeller and nose piece.

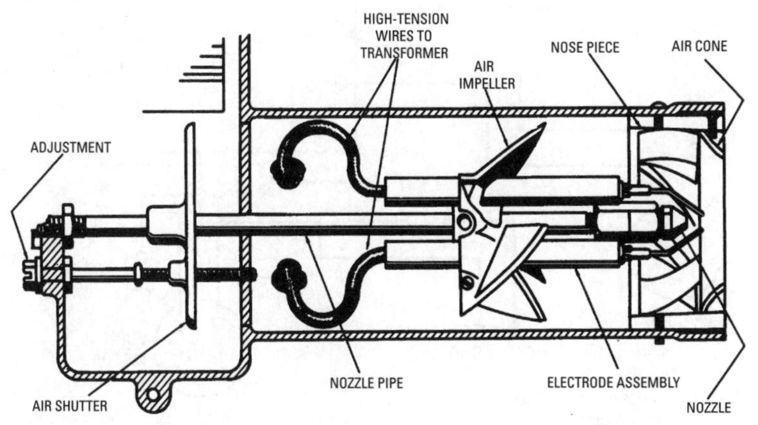

Figure 1-37 Details of a draft tube showing double turbulator consisting of air impeller and nose piece.

Primary Safety Control Service

The cadmium detection cell is the most effective type of primary safety control used on oil burners. Malfunctions cause primary safety control to build up electrical resistance across the cell until the burner is automatically shut off. As soon as the burner shuts off, a reset button pops up on the burner. The button must be reset (pushed down) to restart the burner.

Caution

> If the burner does not restart when the reset button is pushed down, do NOT keep resetting the button. Doing so will flood the firebox with oil. If ignition does not take place, the flooded firebox could result in a fire or an explosion.

The primary safety control can be tested by removing the motor lead from the burner and allowing the ignition circuit to be energized. Figures 1-38 and 1-39 illustrate two typical wiring diagrams for primary safety controls.

Installing an Oil Burner

Under most circumstances, oil burners and oil-fired units should be installed in rooms that provide adequate clearance from the combustible material. The only exception to this rule is when specific instructions are given otherwise. In this case, the manufacturer pro-

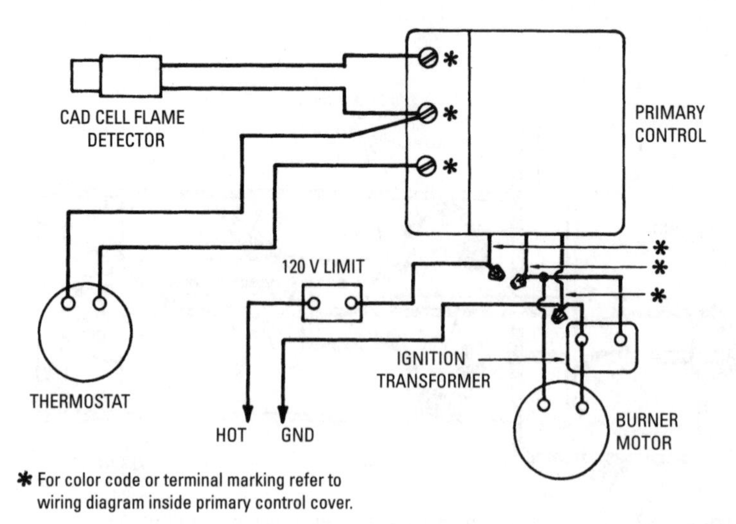

Figure 1-38 Constant-ignition wiring diagram. *(Courtesy Stewart-Warner Corp.)*

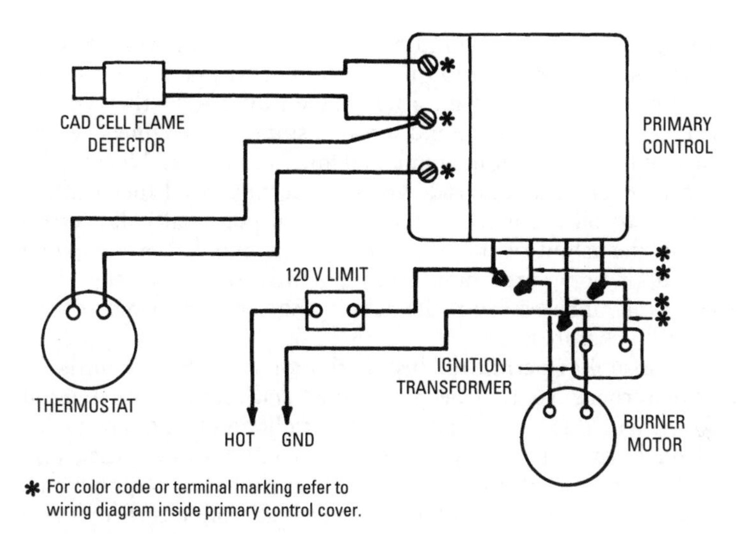

Figure 1-39 Intermittent-ignition wiring diagram.
(Courtesy Stewart-Warner Corp.)

vides or specifies a suitable combustion chamber (stainless steel, firebrick, etc.).

Note

All local codes and ordinances take precedence over the oil burner manufacturer's installation and operation manuals. Where local codes do not exist, install the oil burner in accordance with the most recent instructions and regulations of the National Fire Protection Association and the provisions of the *National Electrical Code* (ANSI/NFPA 70-199 or latest edition).

Warning

Only certified HVAC technicians or those with equivalent experience should attempt to install an oil burner.

Some sort of manual shutoff control should be provided for the oil burner in order to stop the flow of oil to the burner when the air supply is interrupted. This must be placed at a safe distance from the unit and in a convenient location. These manual shutoff valves generally consist of either a switch in the burner supply circuit (for electrically driven units) or a shutoff valve on the oil supply line.

Primary safety controls (automatic shutoff devices) must be provided for all oil burners and oil-fired units that operate automatically without the need of an attendant on duty—in other words, those types of equipment found where a stationary engineer would not be employed (i.e., noncommercial and nonindustrial locations).

One problem encountered when converting solid-fuel heating equipment to oil use is the accumulation of potentially dangerous vapors in the ashpit of the unit. This can be avoided by removing the ash door or by providing bottom ventilation to the unit. This precaution is unnecessary if the ashpit also serves as a part of the combustion chamber.

Never install or permit the installation of an oil burner until the boiler or furnace has first been inspected and found to be in good condition. The flue gas passages must be tight and free of any leaks.

All oil burners listed by Underwriters Laboratories, Inc., and Underwriters Laboratories of Canada meet the safety requirements detailed in the various booklets of the National Fire Protection Association.

Starting an Oil Burner

Oil burner manufacturers provide detailed starting and operating instructions for their burners in their user manuals. These instructions should be carefully followed when attempting to start an oil burner. If there is no available user manual, contact a local representative of the manufacturer or contact the manufacturer directly for a copy.

The procedure for starting an oil burner may be *summarized* as follows:

1. Open all warm-air registers.
2. Check to be sure all return air grilles are unobstructed.
3. Open the valve on the oil supply line.
4. Reset the burner primary relay.
5. Set the thermostat above the room temperature.
6. Turn on the electric supply to the unit by setting the main electrical switch to the *on* position.
7. Change the room thermostat setting to the desired temperature.

The oil burner should start after the electric power has been switched on (step 6). There is no pilot to light as is the case with gas-fired appliances. The spark for ignition is provided automatically on demand from the room thermostat.

Note

Allow the burner to operate at least 10 minutes before making any final adjustments. Whenever possible, use instruments to adjust the fire.

Air Delivery and Blower Adjustment

It is sometimes necessary to adjust the blower speed to produce a temperature rise through the furnace that falls within the limits stamped on the furnace nameplate. Blower adjustment procedures are described in Chapter 11 ("Gas Furnaces") of Volume 1.

Combustion Testing and Adjustments

The following instruments are recommended for combustion testing and adjustments:

- Draft gauge
- Smoke tester
- Carbon dioxide tester
- 200/1000°F stack thermometer
- 0/150-psig pressure gauge
- 0/30-inch mercury vacuum gauge

Smoky combustion indicates poor burner performance. The amount of smoke in the flue gas can be measured with a smoke tester (see Figure 1-40). The tube of the smoke tester is inserted through a 3⁄8-inch hole drilled in the flue pipe, and the test is run as shown in Figure 1-41. Any smoke in the air drawn into the smoke tester will register on a filter paper inserted in the device. The results are interpreted according to the smoke scale in Table 1-5.

One of the most common causes of smoky combustion is soot formation on the heating surfaces. This is easily corrected by cleaning. Other possible causes of smoky combustion include the following:

- Insufficient draft
- Poor fuel supply
- Fuel pump malfunctioning
- Defective firebox
- Incorrectly adjusted draft regulator
- Defective oil-burner nozzle
- Wrong size oil-burner nozzle

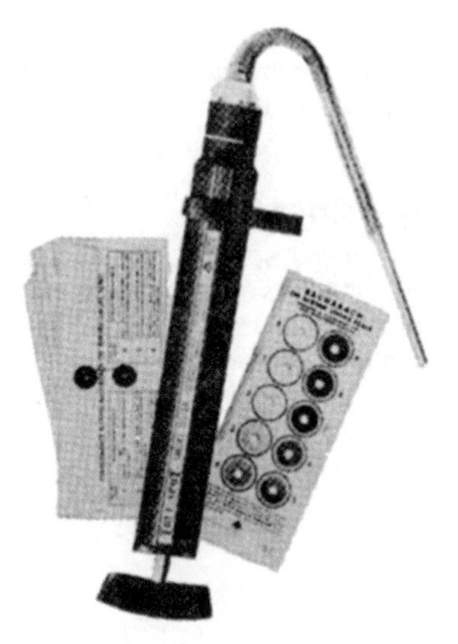

Figure 1-40 Bacharach true-spot smoke tester.

(Courtesy Bacharach Instrument Co.)

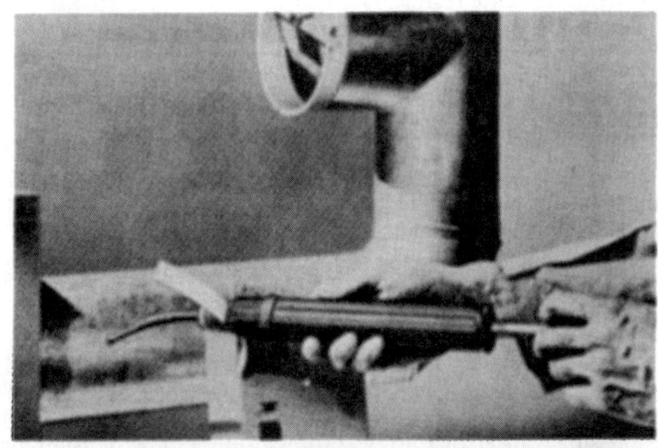

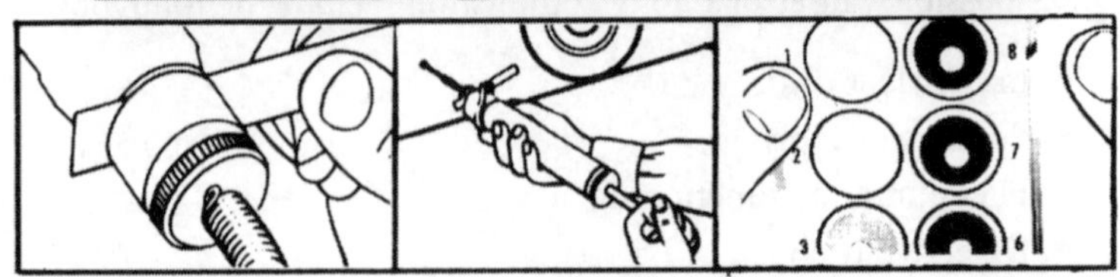

Insert filter test paper into TRUE-SPOT. | Withdraw gas sample from flue pipe by 10 pump strokes. | Grade soot spot test paper by comparison with shadings on scale.

Figure 1-41 A smoke tester in use.

(Courtesy Bacharach Instrument Co.)

Table 1-5 Smoke Scale

Bacharach Smoke Scale No.	*Rating*	*Sooting Produced*
1	Excellent	Extremely light, if at all.
2	Good	Slight sooting, which will not increase stack.
3	Fair	May be some sooting but will rarely require cleaning more than once a year.
4	Poor	Borderline condition; some units will require cleaning more than once a year.
5	Very poor	Soot produced rapidly and heavily.

(Courtesy Bacharach Instrument Co.)

- Improper fan delivery
- Excessive air leaks in boiler or furnace
- Unsuitable fuel-air ratio

Net stack temperatures in excess of 700°F for conversion units and 500°F for packaged units are considered abnormally high. The *net* stack temperature is the difference between the temperature of the flue gases inside the pipe and the room air temperature outside. For example, if the flue gas temperature is 600°F and the room temperature is 75°F, then the net stack temperature is 525°F (600°F – 75°F = 525°F).

A 200/1000°F stack thermometer is used to measure the flue gas temperature. The thermometer stem is inserted through a hole drilled in the flue pipe (see Figure 1-42). A high stack temperature may be caused by any of the following:

- Undersized furnace
- Defective combustion chamber
- Incorrectly sized combustion chamber
- Lack of sufficient baffling
- Dirty heating surfaces
- Excessive draft
- Boiler or furnace overfired

Figure 1-42 Running a stack-gas temperature test with a stack thermometer. *(Courtesy Bacharach Instrument Co.)*

- Unit unsuited to automatic firing
- Draft regulator improperly adjusted

When the carbon dioxide (CO_2) content of the flue gas is too low (less than 8 percent), heat is lost up the chimney and the unit operates inefficiently. This condition is usually caused by one of the following:

- Underfiring the combustion chamber
- Burner nozzle is too small
- Air leakage into the furnace or boiler

When the carbon dioxide content is too high, the furnace operation is generally characterized by excess smoke and/or pulsations and other noises. A high carbon dioxide content is usually caused by insufficient draft or an overfired burner.

The carbon dioxide reading is also taken through a hole drilled in the flue pipe with a CO_2 indicator (see Figure 1-43). The CO_2 indicator is used as shown in Figure 1-44. The results are indicated by a test liquid on a scale calibrated in $\%CO_2$.

A correct draft is essential for efficient burner operation. Insufficient draft can make it almost impossible to adjust the oil burner for its highest efficiency. Excessive draft can reduce the percentage of carbon dioxide in the flue gases and increase the stack temperature.

Figure 1-43 Bacharach Fyrite CO_2 indicator.
(Courtesy Bacharach Instrument Co.)

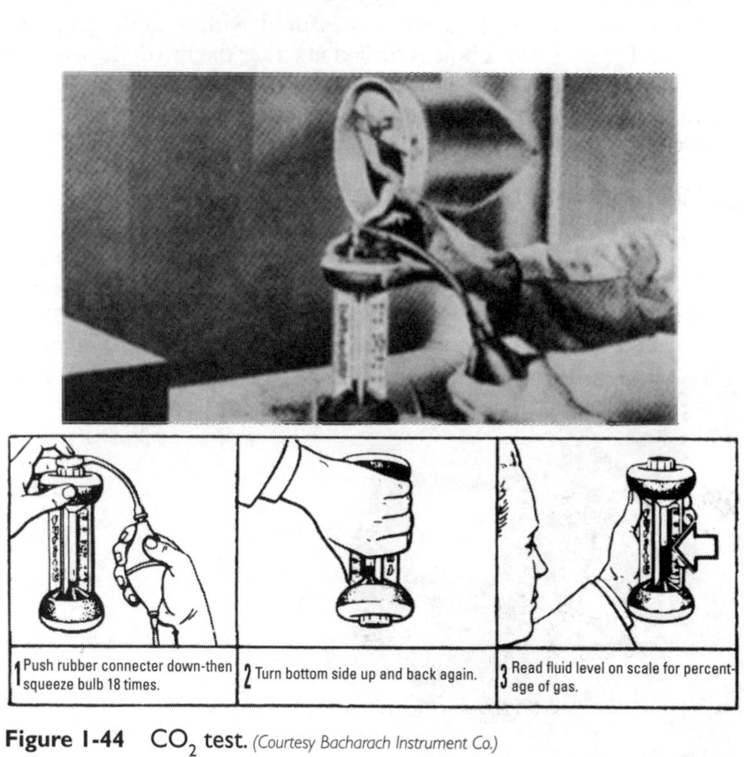

Figure 1-44 CO_2 test. *(Courtesy Bacharach Instrument Co.)*

For the most efficient operating characteristics, the overfire draft generally should be not less than 0.02 inch wg. Smoke and odor often occur when the overfire draft falls below 0.02 inch wg.

It may be necessary to adjust the barometric draft regulator to obtain the correct overfire draft. If it is not possible to adjust the overfire draft for a –0.01 to 0.02 inch wg, install a mechanical draft inducer between the chimney and the barometric draft regulator.

The primary air band should be adjusted to a 0+ smoke or until a hard clean flame is visible. A clean flame is preferred to one with high carbon dioxide. Adjust the overfire draft for a –0.01 to a –0.02 inch wg. An excessive overfire draft condition will cause high stack temperature and inefficient operation. A too low or positive draft over the fire will usually cause the flue gases and fumes to seep into the space upon startup or shutdown.

The flue pipe draft in most residential oil burners is between 0.04 and 0.06 inches of water. This is sufficient to maintain a draft of 0.02 inches in the firebox.

The furnace or boiler draft is measured with a draft gauge as shown in Figure 1-45. A hole is drilled in either the fire door (overfire draft measurement) or flue pipe (flue pipe draft measurement), and the unit is run for approximately 5 minutes. The draft tube is then inserted into the test hole, and the gauge is read (see Figure 1-46).

Troubleshooting Oil Burners

Individuals involved in the installing and repairing of oil burners should be aware of a number of different indicators of malfunctions in the equipment, their probable causes, and some suggested remedies.

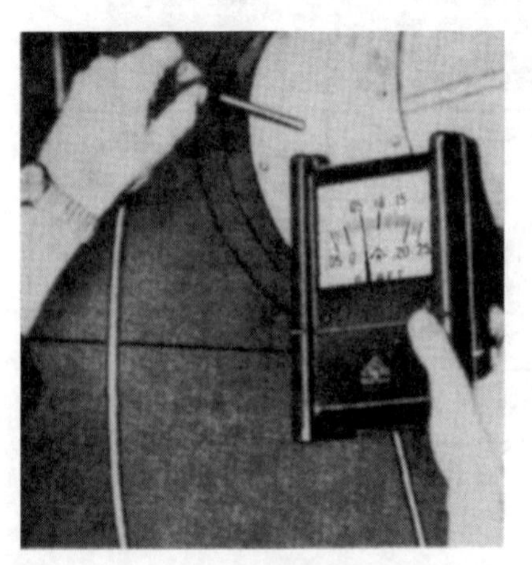

Figure 1-45 Bacharach model MZF draft gauge.

(Courtesy Bacharach Instrument Co.)

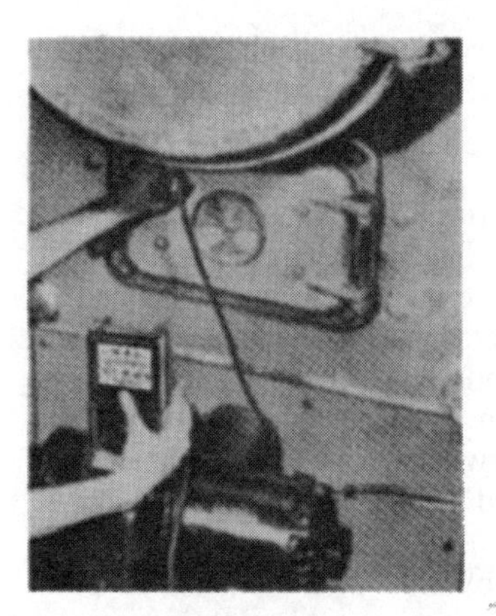

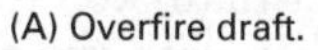

(A) Overfire draft.

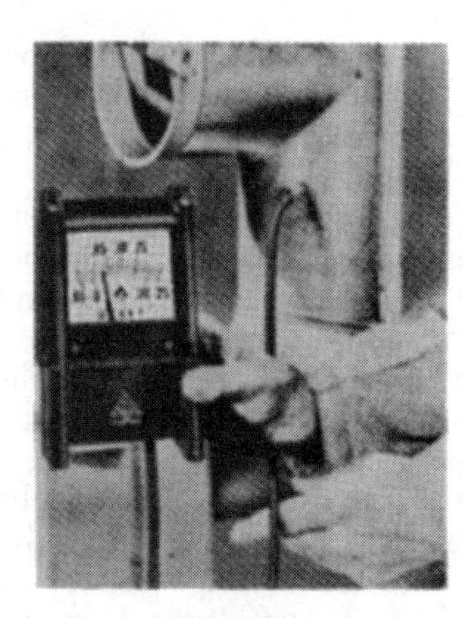

(B) Flue pipe draft.

Figure 1-46 Overfire and flue pipe draft tests.
(Courtesy Bacharach Instrument Co.)

The average individual is most aware of malfunctions that warn the senses through excessive noise, smoke, or odor. These are *external* warning signals that require immediate investigation. Their nature is such that tracing the probable cause of the malfunction is made easier.

Excessive noise (pulsation, thumping, rumbling, etc.) in the heating unit is generally caused by a problem with the oil burner nozzle. It can usually be corrected by any one of the following methods:

- Replace the nozzle with one having a wider spray angle.
- Replace the nozzle with one having the next size smaller opening.
- Install a delayed-opening solenoid on the nozzle line (this reduces pulsation).

Sometimes a noisy fire is caused by cold oil originating from outside storage tanks. This noise may be greatly reduced or eliminated by pumping the fuel oil under 120 to 125 psi through the next size smaller nozzle.

Excessive smoke has a number of possible causes, including the following:

- The air-handling parts of the oil burner may be too dirty to operate efficiently.

- The combustion chamber or burner tube may be damaged by burn-through or loose materials.
- The oil burner nozzle may be the wrong size.

The dirty air-handling parts (e.g., the fan blades, air intake, and air vanes in the combustion head) can be made to operate more efficiently by a thorough cleaning. If the excessive smoke is caused by the oil burner nozzle, this can be corrected by replacing the nozzle with one that is a size smaller or one having the next narrower spray angle. A damaged combustion chamber is a more difficult problem to correct than the other two. In any event, all leakage through the walls must be eliminated before the oil burner can be expected to operate efficiently.

Excessive odors can be caused by flue obstructions or poor chimney draft. If the draft over the fire is lower than 0.02 to 0.04, it is usually an indication that the problem lies with the flue or chimney draft. The cause is usually an obstruction in the flue or poor chimney draft. Other causes of excessive odor include the following:

- Delayed ignition
- Too much air through the burner

Delayed ignition is commonly traced to a problem with the electrodes. This condition can result from a variety of causes, including the following:

- Improper electrode setting
- Insulator cracks
- A coating of soot or oil on the electrode
- Incorrect pump pressure setting
- Incorrect spray pattern in the nozzle
- Clogged nozzle
- Air shutter open too far

Table 1-6 lists a number of recommended electrode settings that should eliminate delayed ignition if the electrode setting is the cause of the problem. The type of nozzle spray pattern can also result in delayed ignition. This is particularly true when using a hollow spray pattern in oil burners firing 2.00 gph and above. It can be corrected by replacing the nozzle with one having a solid spray pattern.

Table 1-6 Recommended Electrode Settings

Nozzle	*GPH*	*A*	*B*	*C*
45°	0.75 to 4.00	1/8″ to 3/16″	1/2″ to 9/16″	1/4″
60°	0.75 to 4.00	1/8″ to 3/16″	9/16″ to 5/8″	1/4″
70°	0.75 to 4.00	1/8″ to 3/16″	9/16″ to 5/8″	1/8″
80°	0.75 to 4.00	1/8″ to 3/16″	9/16″ to 5/8″	1/8″
90°	0.75 to 4.00	1/8″ to 3/16″	9/16″ to 5/8″	0

Table 1-7 lists a variety of problems encountered with oil burners, many of which are of an *internal* nature and require a great degree of experience and training to correct.

Table 1-7 Oil Burner Troubleshooting

Symptom and Possible Cause	*Possible Remedy*
No heat—circulator (pump) off and burner running.	
(a) Defective circulator.	(a) Replace circulator.
(b) Defective thermostat.	(b) Replace thermostat.
(c) Defective relay.	(c) Replace relay.
(d) Defective aquastat.	(d) Replace aquastat.
(e) Incorrect aquastat setting.	(e) Reset aquastat.
(f) Loose or disconnected wiring.	(f) Tighten or reconnect wiring.
(g) Defective zone valve.	(g) Replace zone valve.
No heat—both circulator (pump) and burner running.	
(a) Defective or loose circulator coupling.	(a) Repair or replace.
(b) Broken circulator impeller.	(b) Repair or replace circulator.
(c) Air trapped in lines.	(c) Locate point of entry and repair; purge air from lines.
(d) Loose or disconnected wiring.	(d) Tighten or reconnect wiring.
(e) Defective zone valve.	(e) Replace zone valve.
(f) Frozen flow valve.	(f) Repair or replace flow valve.
No oil flow at nozzle.	
(a) Oil level below intake line in oil storage tank.	(a) Fill tank with oil.
(b) Clogged strainer.	(b) Remove and clean strainer.
(c) Clogged filter.	(c) Replace filter element.

(continued)

Table 1-7 *(continued)*

Symptom and Possible Cause	*Possible Remedy*
(d) Clogged nozzle.	(d) Replace nozzle.
(e) Air leak in intake line.	(e) Tighten all fittings in intake line; tighten unused intake port plug; check filter cover and gasket.
(f) Restricted intake line (high vacuum reading).	(f) Replace any kinked tubing and check valves in intake line.
(g) Air-bound two-pipe system.	(g) Check for and insert bypass plug. Make sure return line is below oil level in tank.
(h) Air-bound single-pipe system.	(h) Loosen gauge port plug or easy-flow valve and bleed oil for 15 seconds after foam is gone in bleed hose. Check intake line fittings for tightness and tighten if necessary. Check all pump plugs for tightness and tighten if necessary.
(i) Slipping or broken coupling.	(i) Tighten or replace coupling.
(j) Rotation of motor and fuel unit is not the same as indicated by arrow on pad at top of unit.	(j) Install fuel unit with correct rotation.
(k) Frozen pump shaft.	(k) Check for water and dirt in tank and correct as necessary; return defective pump to manufacturer or service center for repair or to be replaced.
Noisy operation.	
(a) Bad coupling alignment at fuel unit.	(a) Loosen fuel unit mounting screws slightly and shift fuel unit in different positions until noise is eliminated. Retighten mounting screws.
(b) Air in inlet line.	(b) Check all connections for damage. Replace as necessary. Use only good flare fittings.
(c) Tank hum on two-pipe system and inside tank.	(c) Install return-line hum eliminator in return line.

(continued)

Table 1-7 *(continued)*

Symptom and Possible Cause	*Possible Remedy*
Pulsating pressure.	
(a) Partially clogged strainer.	(a) Remove and clean strainer.
(b) Partially clogged filter.	(b) Replace filter element.
(c) Air leak in intake line.	(c) Tighten all fittings; replace damaged fittings and/or damaged intake line.
(d) Air leaking around strainer cover.	(d) Check for loose cover screws and tighten securely. Check for damaged cover gasket and replace if necessary.
Low oil pressure.	
(a) Defective gauge.	(a) Replace defective gauge.
(b) Burner nozzle capacity is greater than fuel unit capacity.	(b) Replace fuel pump with one of correct capacity.
Improper nozzle cutoff.	
(a) Trapped air causing fuel pump operating problem.	(a) Insert pressure gauge in nozzle port of fuel pump. Run burner. If burner shuts down after a minute of operation and pressure drops from normal operating pressure and stabilizes, the fuel pump is running and the problem is with trapped air. Correct as necessary.
(b) Defective fuel pump.	(b) Insert pressure gauge in nozzle port of fuel pump. Run burner. If burner shuts down after a minute of operation and pressure drop is 0 psi, fuel pump is defective and should be replaced.
(c) Filter leaks.	(c) Check face of cover and gasket for damage and repair or replace as necessary.
(d) Loose strainer cover.	(d) Tighten strainer cover screws.

(continued)

Table 1-7 (continued)

Symptom and Possible Cause	*Possible Remedy*
(e) Air pocket between cutoff valve and nozzle.	(e) Run burner by stopping and starting unit until smoke and afterfire disappear.
(f) Partially clogged nozzle strainer.	(f) Clean strainer or change nozzle.
(g) Leak in nozzle adapter.	(g) Change nozzle and adapter.
Oil leak—oil leaking inside burner.	
(a) Seal leaking.	(a) Replace seal or pump.
(b) Blown seal in a single-pipe system.	(b) Check to see if bypass plug has been left in fuel pump. Replace pump.
(c) Blown seal in a two-pipe system.	(c) Check for kinked tubing or other obstructions in return line. Replace pump.
(d) Cracked nozzle adapter.	(d) Replace nozzle adapter.
(e) Defective pump piston.	(e) Replace pump.
(f) Loose fitting.	(f) Tighten or replace fitting.
(g) Loose fuel unit cover.	(g) Tighten cover screws.
(h) Loose plugs or fittings.	(h) Dope with good-quality thread sealer; retighten plugs or fittings.
(i) Leak at pressure adjustment screw or nozzle plug caused by damaged washer or O-ring.	(i) Replace washer or O-ring as necessary.
(j) Damaged gasket.	(j) Replace gasket.
Oil leak—oil leaking on outside of burner.	
(a) Loose fittings.	(a) Tighten fittings.
(b) Defective fittings.	(b) Replace fittings.
(c) Damaged gasket.	(c) Replace gasket.
Burner running—no oil pumping into combustion chamber and no fire in chamber.	
(a) No oil in storage tank.	(a) Fill storage tank.
(b) Clogged fuel pump.	(b) Repair or replace fuel pump.
(c) Defective fuel pump.	(c) Replace fuel pump.

(continued)

Table 1-7 *(continued)*

Symptom and Possible Cause	*Possible Remedy*
(d) Clogged nozzle.	(d) Clean or replace nozzle.
(e) Damaged or defective nozzle.	(e) Replace nozzle.
(f) Clogged filter.	(f) Clean or replace.
(g) Obstructed oil line.	(g) Remove obstruction or replace fuel line.
(h) Closed oil valve.	(h) Repair or replace oil valve.
(i) Loose or defective oil pump coupling.	(i) Tighten coupling or replace oil pump.
(j) Defective oil valve.	(j) Replace valve.
(k) Lost prime.	(k) Reestablish prime or replace pump.
Burner running—oil pumping into combustion chamber but no fire in chamber.	
(a) Chamber obstructed.	(a) Locate and remove obstruction.
(b) Too much air.	(b) Adjust to proper level.
(c) Water contaminating the oil.	(c) Locate point of contamination and repair; drain and replace oil.
(d) Defective or weak ignition transformer.	(d) Replace ignition transformer.
(e) Dirty electrodes.	(e) Clean electrodes.
(f) Cracked or broken electrodes.	(f) Replace electrodes.
(g) Loose wires.	(g) Tighten connection or replace wires.
(h) End cone obstruction.	(h) Repair.
Smoky fire.	
(a) Improper pump pressure.	(a) Set proper pump pressure or replace defective pump.
(b) Incorrect nozzle.	(b) Replace nozzle.
(c) Distorted and burnt end cone.	(c) Replace.
(d) Water contaminating the oil.	(d) Locate point of contamination and repair; drain and replace oil.
(e) Dirty boiler.	(e) Clean boiler.
(f) Dirty fan.	(f) Clean fan
(g) Defective combustion chamber.	(g) Replace.

(continued)

Table 1-7 (continued)

Symptom and Possible Cause	*Possible Remedy*
Burner fails to restart after resetting safety relay.	
(a) Power off.	(a) Restore electricity to burner.
(b) Defective burner motor.	(b) Replace burner motor or burner.
(c) Defective fuel pump.	(c) Replace fuel pump.
(d) Defective safety relay.	(d) Replace safety relay.
(e) Defective on-off switch.	(e) Replace switch.
(f) Tripped circuit breaker.	(f) Reset circuit breaker; call electrician if problem continues.
(g) Blown fuse.	(g) Replace fuse; call electrician if problem continues.
(h) Loose or disconnected wiring.	(h) Reconnect wires and tighten wiring connections.
Flame pattern not centered.	
(a) Improperly positioned burner.	(a) Reposition burner.
(b) Contaminated fuel.	(b) Locate source of contamination and repair; drain and replace fuel.
(c) Obstruction in combustion chamber.	(c) Locate and remove obstruction.
Water leaking from boiler pressure relief valve with pressure under 30 psi.	
(a) Defective pressure relief valve.	(a) Replace pressure relief valve.
Water leaking from boiler pressure relief valve with pressure at 30 psi or greater.	
(a) Defective feed valve.	(a) Replace valve.
(b) Holes or cracks in coils.	(b) Replace coils.
(c) Expansion tank full.	(c) Correct as necessary.
(d) Water temperature above 210°F.	(d) Reduce water temperature; replace valve.

Chapter 2

Gas Burners

A *gas burner* (see Figure 2-1) is a device for supplying gas, or a mixture of gas and air, to the combustion area. Those used in residential and light commercial heating are most commonly high-pressure, gun-type burners. Only those gas burners approved by the American Gas Association (AGA) should be used in a heating system.

Gas is used as a heating fuel in both urban and rural areas. Manufactured, natural, and bottled gas are the three types used as heating fuels. Each of these gases has different combustion characteristics and will have different heat values when burned. Because of this, a gas burner must be adjusted for each gas fuel, particularly when changing from one type to another. The principal types of bottled gas used as heating fuels are propane and butane. Bottled gas is frequently called LPG (liquefied petroleum gas) and is widely used as a heating fuel in rural areas. A more detailed description of heating fuels is found in Chapter 5 of Volume 1 ("Heating Fuels").

Operating Principles

The gas burners used in residential heating systems are most commonly the atmospheric injection type that operates on the same principle as the Bunsen burner.

The essential features of the Bunsen burner are shown in Figure 2-2. The burner consists of a small tube or burner, which is placed inside a larger tube. The latter has holes positioned slightly below the top of the small tube. The gas escaping from the small tube draws the air in through the holes and produces what is called an *induced current* of air in the large tube. This air enters through the holes and is mixed with the gas in the tube. The mixture is burned at the top of the larger tube. The flame from such a burner gives hardly any light, but the heat is intense. The intensity of the heat can be illustrated by holding a metal wire over the flame for a few seconds. It will glow with heat in a very short time.

The air supply in an atmospheric injection burner is classified as either *primary* or *secondary* air and is commonly introduced and mixed with the gas in the throat of the mixing tube. The cutaway of an upshot atmospheric gas burner in Figure 2-3 illustrates this principle of operation. The gas passes through the small orifice in the mixer head, which is shaped to produce a straight-flowing jet moving

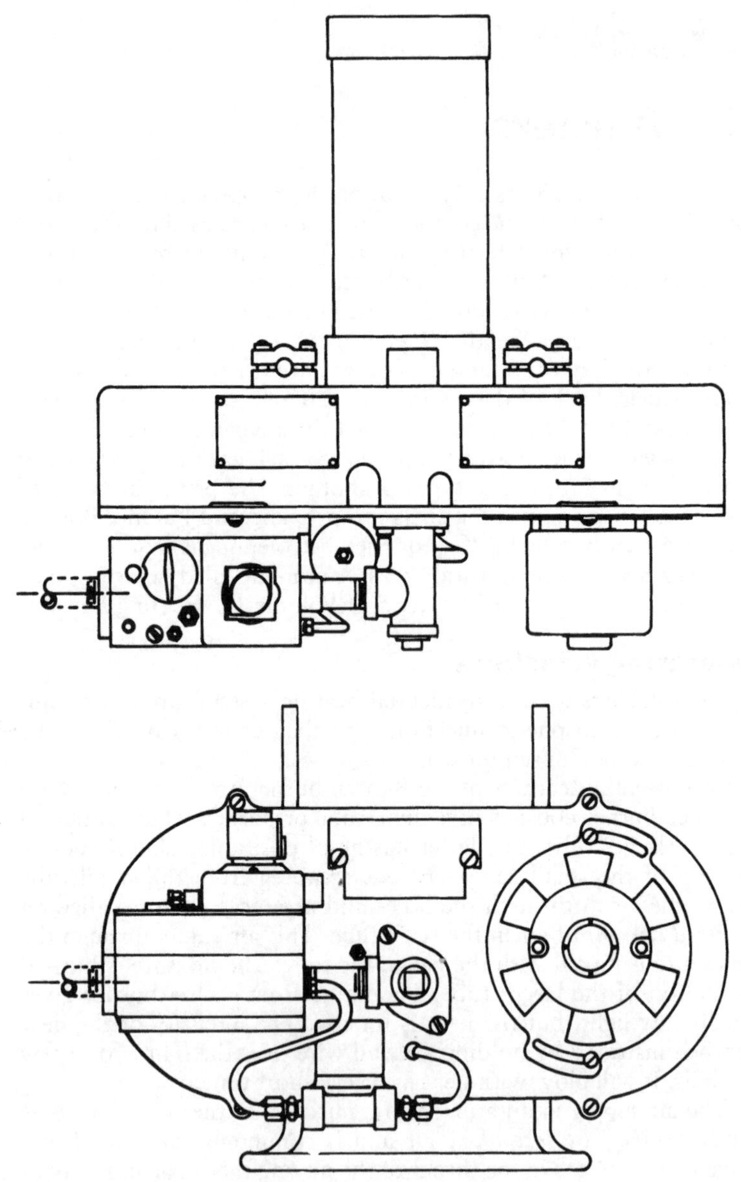

Figure 2-1 Power gas burner. *(Courtesy Nu-Way Burner)*

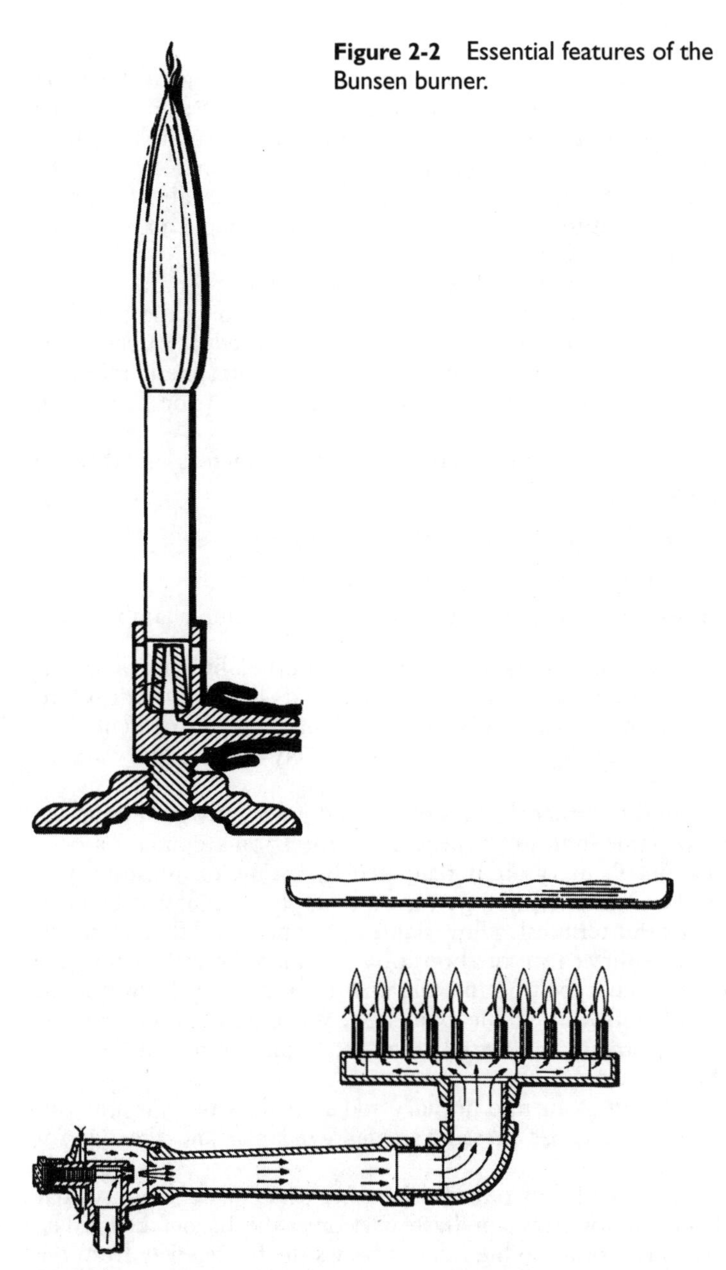

Figure 2-2 Essential features of the Bunsen burner.

Figure 2-3 Cutaway of the venturi, or mixing tube, of an upshot gas burner.

at high velocity. As the gas stream enters the throat of the venturi, or mixing tube, it tends to spread and induce air in through the opening at the adjustable air shutter. The energy in the gas stream forces the mixture through the mixing tube into the burner manifold casting, from which it issues through ports where additional air must be added to the flame to complete combustion. The air coming in through the venturi is the *primary* air and that supplied around the flame is the *secondary* air.

The primary air is admitted at a ratio of about 5 parts primary air to 1 part gas for manufactured gas, and a 10 to 1 ratio for natural gas. These ratios are generally used as theoretical values of air for purposes of complete combustion. Most atmospheric injection burners operate efficiently on 40 to 60 percent of the theoretical value.

The excess air required depends on several factors, notably the following:

- Uniformity of air distribution and mixing
- Direction of gas travel from the gas burner
- The height and temperature of the combustion chamber

The secondary air is drawn into the burner by natural draft. Excess secondary air constitutes a loss and should be reduced to a proper minimum (usually not less than 25 to 35 percent). All yellow-flame gas burners depend exclusively on secondary air for combustion.

The Bunsen burner flame is bluish and practically nonluminous. A yellow flame indicates dependence entirely on secondary air for combustion. Primary air is regulated by means of an adjustable shutter. For manufactured gas, the air supply is regulated by closing the air shutter until yellow flame tips appear and then by opening the air shutter to a final position at which the yellow tips just disappear. This type of flame obtains ready ignition from port to port and favors quiet flame extinction. When burning natural gas, the air adjustment is generally made to secure as blue a flame as possible.

The division of air into primary and secondary types is a matter of burner design, the pressure of gas available, and the type of flame desired.

The gas should flow out of the burner ports fast enough so that the flame cannot travel or flash back into the burner head. The velocity must not be so high that it blows the flame away from the port. In an all-yellow flame, flame flashback cannot occur, and a much higher velocity is needed to blow off the flame.

A draft hood is used to ensure the maintenance of constant low-draft conditions in the combustion chamber with a resultant stability of air supply. A draft hood will also control backdrafts that tend to extinguish the gas burner flame and the amount of excess air. These draft hoods must conform to American Standard Requirements.

Electrical Circuits

Each gas-fired appliance is wired according to the specific make and model. A wiring diagram is included in the appliance manufacturer's installation and operation manual. All electrical connections must be made in accordance with the manufacturer's installation instructions. Read these instructions and follow them carefully.

Caution

Only a certified HVAC technician or someone with similar qualifications and/or experience should attempt to wire a gas-fired appliance.

Note

All local codes and regulations for wiring gas-fired appliances must take precedence over the instructions in the manufacturer's installation and operation manuals. In the absence of local codes, all electrical wiring and connections should conform with the appropriate instructions and provisions found in the latest edition of the *National Electrical Code*.

A description of the electrical circuits used to operate modern gas-fired heating equipment is included in Chapter 5 ("Gas and Oil Controls").

Automatic Controls

The automatic controls are used to ensure the safe and efficient operation of a gas-fired appliance. They are mentioned only briefly here because detailed descriptions of gas system controls are found in other chapters of this volume. These controls can be roughly divided into the following six broad categories:

- Room thermostats
- Ignition (lighting) devices
- Main gas valves
- Flame-sensing devices
- Pressure regulators
- Safety valves and switches

A gas heating system is controlled by a centrally located room thermostat. The thermostat sends a call for heat to the furnace or boiler when the temperature of the air in the room reaches the *set-point* (heat setting) on the thermostat. This occurs automatically when the temperature falls below a preselected heat setting, or manually when the temperature adjustment dial or lever is moved up to a warmer heat setting. When the heat setpoint is reached, it closes an electrical circuit between the thermostat and the furnace or boiler, which, in turn, activates the furnace or boiler control circuit. (See Chapter 4, "Thermostats and Humidistats," in Volume 3 for a description of these thermostats.)

The *main gas valve* controls the flow of natural or propane gas to the burners when the thermostat calls for heat. In most heating systems, the main gas valve is combined with a pressure regulator to form a *combination gas valve*. Main gas valves are covered in Chapter 5 ("Gas and Oil Controls") of this volume.

Another important control is the device used to detect whether the gas in the burners has successfully lighted after the call for heat by the room thermostat. Some systems use a *thermocouple* to verify that a pilot burner is lit. If it can detect the pilot flame, it will open the main gas valve, allowing gas to flow to the main burners where the pilot flame will ignite it. If it cannot detect a pilot flame, it will not open the main gas valve. The *flame sensor* of an electronic ignition system performs the same function as the thermocouple in standing pilot systems, but it looks for the flame in the main gas burners. See Chapter 5 ("Gas and Oil Controls") for a description of flame-sensing devices.

The fan and limit control is another device that ensures the safe operation of a gas appliance. It controls the blower in a forced warm-air furnace. If the blower fails to operate during the heating cycle, the fan and limit control will shut off the main gas burners to prevent the furnace from overheating. Read the appropriate sections of Chapter 6 ("Other Automatic Controls") of this volume for a more detailed description of fan and limit controls.

The *pressure regulator*, which is commonly combined with the main gas valve in the form of a single combination, controls the amount and pressure of the gas used by the gas-fired appliance. See Chapter 5 ("Gas and Oil Controls").

A wide range of safety valves and switches have been created by different gas furnace and boiler manufacturers to ensure the safe and efficient operation of their appliances. Pilot safety valves, pressure switches, and other devices are described in the appropriate sections of Chapter 5 ("Gas and Oil Controls") of this volume.

Types of Gas Burners

Types of residential gas burners include atmospheric injection, yellow (luminous) flame, and power burner units. Their classification is determined by the firing method used. Gas burners can also be divided into two broad classifications based on whether they are specifically designed as integral parts of gas-fired heating equipment, as in Figure 2-4, or are used to convert a furnace or boiler from one fuel to another. The latter are called *conversion burners* and, at least outwardly, resemble the gun-type burners used in oil-fired appliances. Gas conversion burners are commonly designed and manufactured with integral controls so that they can be installed as a unit in the existing furnace or boiler.

Note

The burner(s) producing the heat in a gas-fired appliance is sometimes called the *main gas burner*. Do not confuse the main gas

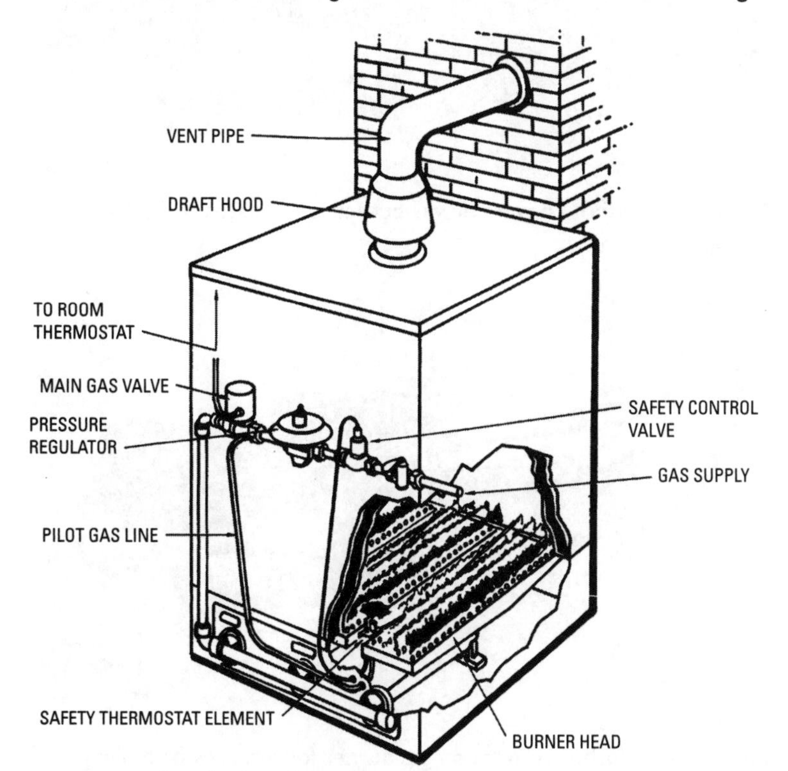

Figure 2-4 Standard gas burner components. *(Courtesy U.S. Department of Agriculture)*

burner with the pilot gas burner. The function of the latter (where it is used) is to light the gas flowing to the main gas burner.

Gas burners may also be classified as *inshot* and *upshot* types, depending on the design of the burner tube. The burner tube of an inshot gas burner is commonly a straight, adjustable venturi that extends horizontally from the unit (see Figure 2-5). An upshot gas burner is characterized by a burner tube that extends horizontally from the unit and then bends to assume a vertical position (see Figure 2-6).

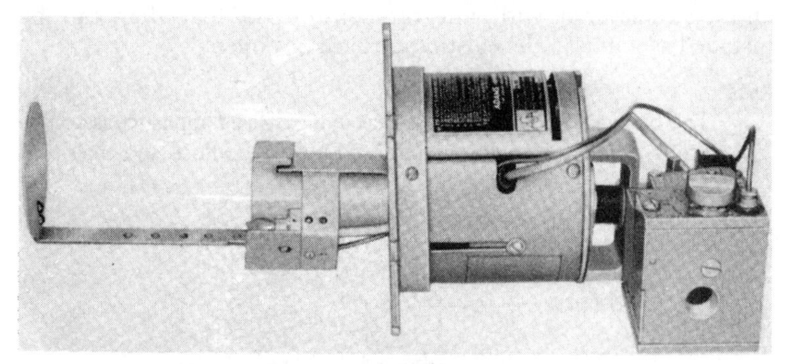

Figure 2-5 Inshot conversion gas burner for furnaces or boilers.
(Courtesy Adams Manufacturing Co.)

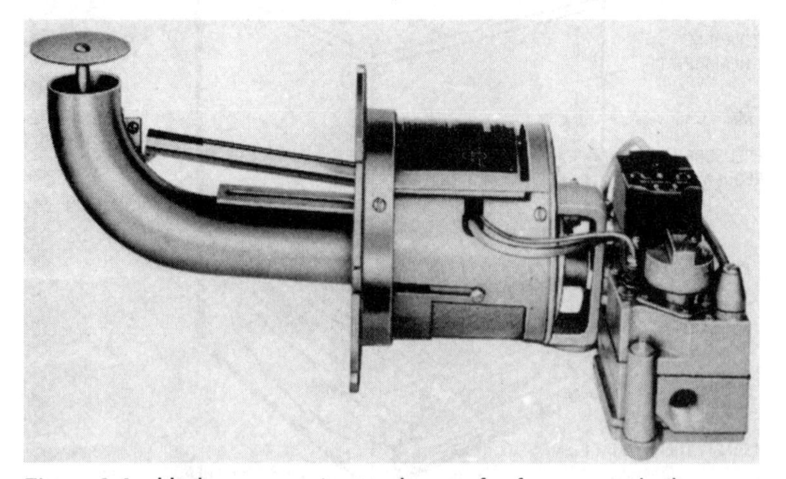

Figure 2-6 Upshot conversion gas burner for furnaces or boilers.
(Courtesy Adams Manufacturing Co.)

Integral-Type Gas Burners

An integral-type gas burner assembly consists of an array of parallel burner tubes connected by a manifold pipe running at a right angle to them. The burner tubes and manifold are part of a box/drawer assembly in modern furnaces and boilers. The entire assembly can be removed from the furnace or boiler for cleaning (see Figure 2-7). Each burner tube contains a series of orifices (openings) through which the gas flows. These orifices are sized to deliver the required amount of gas flow to achieve the maximum ratings at the rated pressure listed on the appliance nameplate.

Note

Instructions for servicing burner orifices are included in Chapter 11 ("Gas Furnaces") in Volume 1.

As shown in Figure 2-7, the burner manifold is connected at one end to the individual burner tubes and at its other end to the main gas valve. In other words, it functions as the bridge between the burner tubes and the main gas valve.

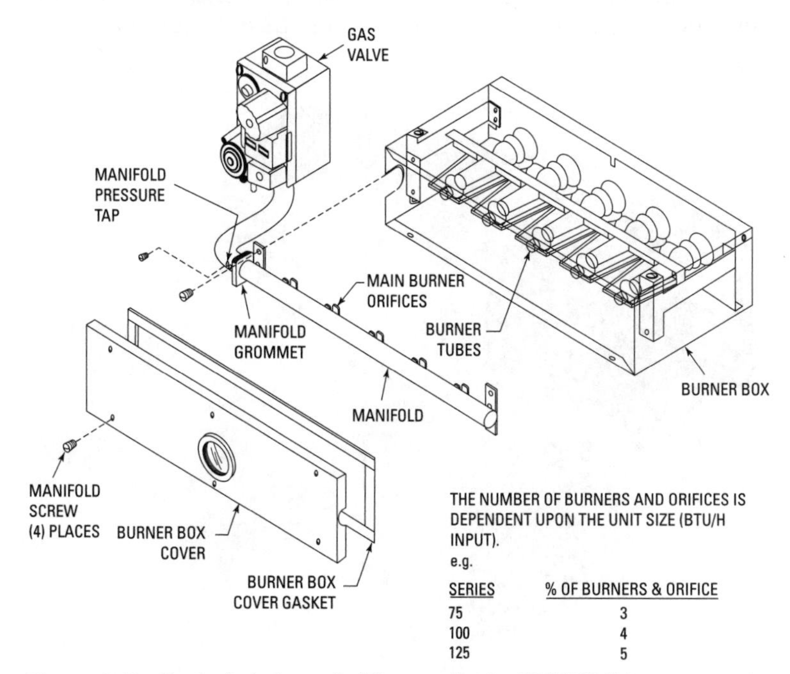

Figure 2-7 Exploded view of a Thermo Pride CDX/CHX gas control system. *(Courtesy Thermo Pride)*

In many furnaces, the main burner(s) can be manually adjusted. In others, no burner adjustment is required because burner aeration has been fixed at the factory. Natural gas burner flames should be well defined (but almost transparent) and should range from light to medium blue in color. Propane burner flames often have yellow- or orange-colored flame tips.

Manifold pressure adjustments, gas input adjustments, instructions for changing burner orifices, and other recommendations for gas burner maintenance and the improved operating efficiency of gas furnaces is covered in great detail in Chapter 11 ("Gas Furnaces") in Volume 1.

Gas Conversion Burners

A *gas conversion burner* is used to convert heating equipment designed for coal or oil to gas fuel use (see Figure 2-8). The boiler or furnace must be properly gastight and must have adequate heating surfaces.

A characteristic of gas conversion burners is that pressure will sometimes build up in a furnace due to puffs or backfire resulting

Figure 2-8 Residential spark-ignition gas conversion burner.
(Courtesy MIDCO International, Inc.)

from delayed ignition and other causes. Local heating codes and regulations usually stipulate that furnace doors be held tightly closed by spring tension only (in other words, not permanently closed) in order to provide a means for relieving pressure. Figure 2-9 shows an example of a door spring that can be used for this purpose.

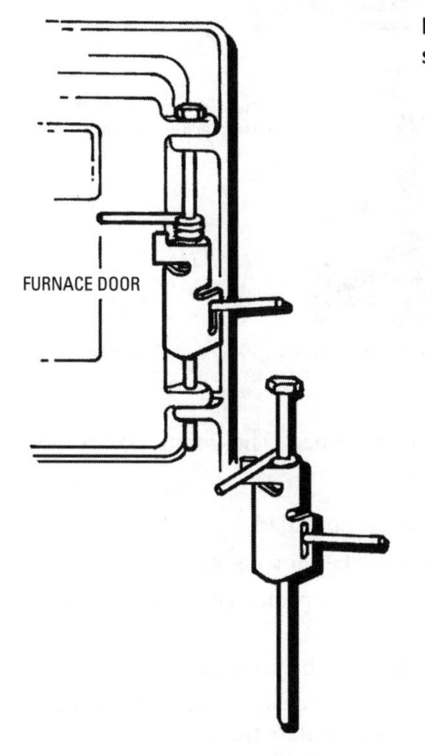

Figure 2-9 Door-closing springs for furnace doors.

(Courtesy Magic Servant Products Co.)

Gas Conversion Burner Combustion Chambers

The combustion chamber for a gas conversion burner is commonly located in the ashpit of a coal-fired boiler or furnace. Figure 2-10 illustrates the positioning of an upshot gas conversion burner. Note that the burner head port is located 1 inch (plus or minus ¼ in) *above* the grate level. This is a standard measurement when installing an upshot burner in the ashpit of a furnace.

The manufacturer's installation instructions provided with a conversion gas burner generally include specific instructions on the

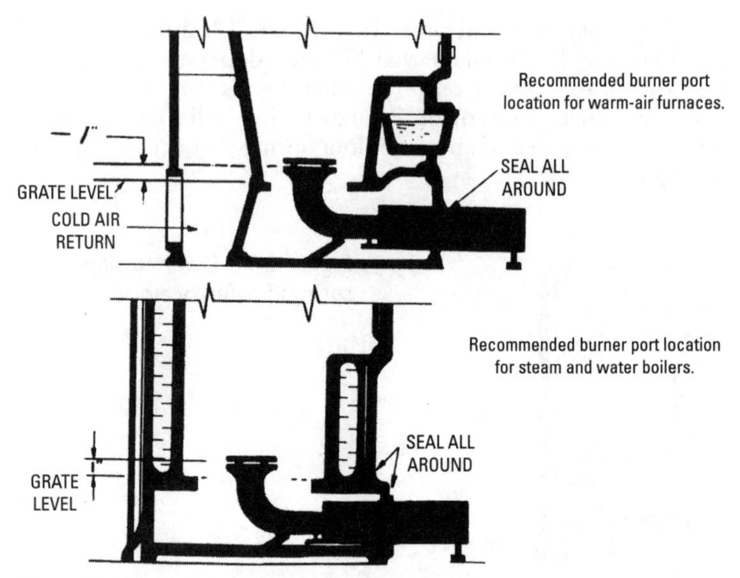

Figure 2-10 Positioning an upshot gas burner. *(Courtesy Magic Servant Products Co.)*

preparation of the combustion chamber. The main points you should remember are as follows:

- All openings in the boiler must be sealed.
- The combustion chamber must be thoroughly cleaned.
- Heat exchanger surfaces must be protected against concentrated heat.
- All nonheat transfer surfaces must be protected.
- The combustion chamber must be designed to contain combustion, to radiate heat, and to insulate the ashpit.

Additional information concerning gas conversion burners can be found in Chapter 16 ("Boiler and Furnace Conversion") in Volume 1.

Gas Piping for Conversion Burners

Gas piping is generally wrought iron or steel with malleable iron pipe fittings. Joint compound (pipe dope) is applied sparingly to the male threads only. Make certain before applying the joint compound that it is approved for all types of gas. Never use aluminum tubing or cast-iron fittings on the main gas line. Soldered or sweated connections are also not recommended.

Tables 2-1 and 2-2 can be used to determine the size of pipe to use from the meter to the gas burner. The correct number of threads for any particular length of pipe is given in Table 2-3.

Table 2-1 Pipe Capacity Table

	Nominal Diameter of Pipe in Inches				
	3/4	*1*	*1 1/4*	*1 1/2*	*2*
Length of Pipe in Feet	***Capacity—Cubic Feet per Hour with a 0.6 Specific Gravity (Sp Gr) Gas and Pressure Drop of 0.3″ Water Column***				
15	172	345	750		
30	120	241	535	850	
45	99	199	435	700	
60	86	173	380	610	
75	77	155	345	545	
90	70	141	310	490	
105	65	131	285	450	920
120		120	270	420	860
150		109	242	380	780
180		100	225	350	720

(Courtesy Magic Servant Products Co.)

Table 2-2 Multipliers for Various Specific Gravities

Specific Gravity	*Multiplier*	*Specific Gravity*	*Multiplier*
.35	1.31	1.00	.775
.40	1.23	1.10	.740
.45	1.16	1.20	.707
.50	1.10	1.30	.680
.55	1.04	1.40	.655
.60	1.00	1.50	.633
.65	.962	1.60	.612
.70	.926	1.70	.594
.75	.895	1.80	.577
.80	.867	1.90	.565
.85	.841	2.00	.547
.90	.817	2.10	.535

(Courtesy Magic Servant Products Co.)

Table 2-3 Specifications for Threading Pipe

Nominal Size of Pipe (inches)	*Approx. Length of Threaded Portion (inches)*	*Approx. Number of Threads to Be Cut*
$\frac{3}{4}$	$\frac{3}{4}$	10
1	$\frac{7}{8}$	10
$1\frac{1}{4}$	1	11
$1\frac{1}{2}$	1	11
2	1	11

**See AGA Requirements and Recommended Practice for House Piping and Appliance Installation.*
(Courtesy Magic Servant Products Co.)

Use pipe fittings at all turns in the gas line. Never bend or lap welded pipe because it will pinch or weaken it. Support the pipe with straps, bands, pipe hooks, or hangers (never allow one pipe to rest on another or to sag).

Pitch all horizontal pipe so that it grades toward the meter without the occurrence of sags. The piping should be protected against freezing and the accumulation of condensation.

Pipes and pipe fittings that are defective should be replaced, never repaired. Every effort must be made to eliminate any possibility of gas leakage. Gas leaks on pipes and pipe fittings should be located by spreading a soap solution over the surface. *Never* try to locate a gas leak with a flame. The results could be extremely hazardous.

Figure 2-11 illustrates the general configuration of the main shutoff valve, pilot shutoff valve, and riser installation for a gas conversion burner. The following suggestions are worth noting:

1. Install the main manual gas shutoff valve on the riser *at least 4 ft above the floor level.*
2. Install the pilot valve *on the inlet side* of the main manual gas shutoff valve.
3. Install a tee fitting at the bottom of the riser to catch any foreign matter in the pipe. The bottom of the tee fitting should be plugged or capped.
4. Install a ground joint union in the gas line between the burner air duct box and the tee fitting in the riser.
5. Install a pilot supply line ($\frac{1}{4}$-in OD tubing) between the pilot valve and a point located on the upper right side of the air duct box on the gas burner. This line will run parallel to the riser.

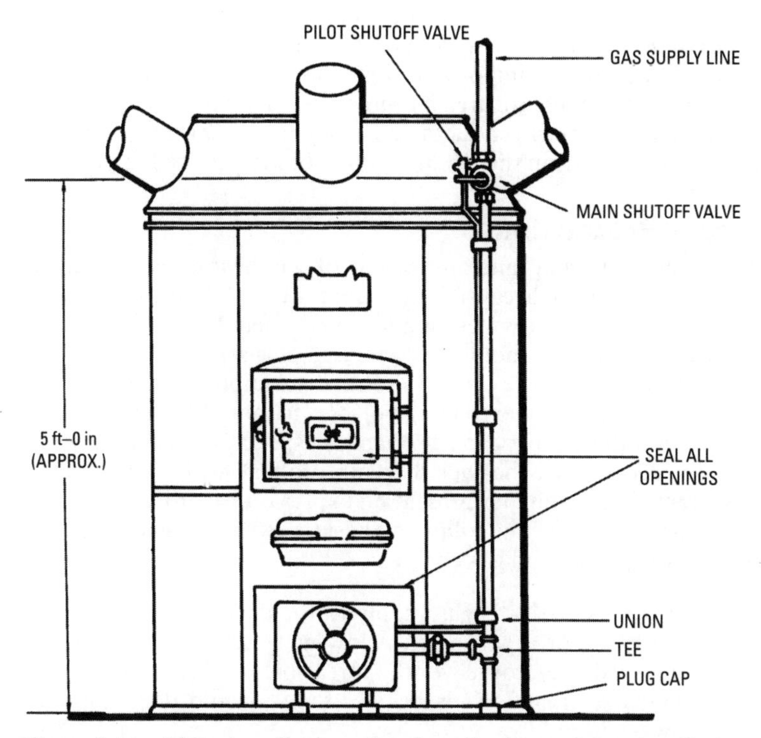

Figure 2-11 Main shutoff valve, pilot shutoff valve, and riser installation.

Venting and Ventilation

The venting system for gas heating equipment consists of the following:

1. The chimney or smoke elimination pipe.
2. The draft diverter or draft diverter hood.

Gas-burning equipment must be vented to the outside. All pipes leading from the equipment must be fitted so that the joints are tight and free of leaks.

A *draft diverter* or *draft hood* is a wind deflector placed in the chimney to prevent downdrafts of air (i.e., air moving down the chimney from the outside) from blowing out the pilot light. Many draft diverters are designed and positioned so that the downdraft is deflected into the room containing the heating unit.

Any room containing gas heating equipment must have adequate ventilation. Provision for incoming air (i.e., air necessary for combustion) is especially important in rooms or buildings of tight construction. The minimum area requirements (in relation to each 1000 Btu/h input) for both ventilating air openings and air inlet openings can be obtained from the manufacturers of the gas-fired appliance.

Safety Precautions

The annual cleaning and inspection of gas heating equipment is important not only because it contributes to its efficient operation but also because it provides an additional safety factor.

Electrical controls should be connected on a separate switch. This enables the circuit to be broken should the equipment malfunction.

Because fuel gas is extremely volatile, it should be handled and stored with the utmost care. Propane is especially dangerous when it leaks. Because it is heavier than air, propane will accumulate at low points in a room and present an explosion hazard.

Be sure to observe the following basic safety rules when working with heating gases and gas controls:

- Always shut off the gas supply to the device when installing, modifying, or repairing it. Allow at least 5 minutes for any unburned gas to leave the area before beginning work. Remember that LPG is heavier than air and does not vent upward naturally.
- Always conduct a *gas leak test* after completing the installation, modification, or repair. To test for a gas leak, coat the pipe joint, pilot gas tubing connections, and valve gasket lines with a soap-and-water solution. Then, with the main burner in operation, watch for bubbles at those points. The bubbles will indicate a gas leak, which can normally be eliminated simply by tightening joints or screws or by replacing the gasket.

Note

Never use a flame to check for a gas leak.

- Always disconnect the power supply to prevent electrical shock or equipment damage before connecting or disconnecting any wiring.
- Change the main burner and pilot orifice(s) to meet the appliance manufacturer's instructions when converting a gas system from one type of gas to another.

- Always read and carefully follow the installation and operating instructions supplied with the appliance or component. Failure to follow them could result in damage or cause a hazardous condition.
- Make certain that the appliance or component is designed for your application. Check the ratings given in the instructions and on the appliance or component.
- Check the operation of the appliance or component with the manufacturer's instructions after installation is completed.
- Do not bend the pilot tubing at the control after the compression nut has been tightened. This could cause a gas leak at the connection.
- Never jump (or short) the valve coil terminals on 24-volt controls. Doing so could short out the valve coil or burn out the heat anticipator in the thermostat.
- Never connect millivoltage controls to line voltage or to a transformer, because doing so will burn out the valve operator or the thermostat anticipator.
- Do not remove the seals covering control inlets or outlets until you are ready to connect the piping. The seals are there to prevent dirt and other materials from getting into the gas control and interfering with its operation.

Troubleshooting Gas Burners

As is the case with all mechanical and electrical equipment, it is recognized that occasional repair and adjustment may be necessary on any burner. Table 2-4 represents a very general list of troubles and causes that can occur with gas burners and gives some suggested remedies.

Table 2-4 Troubleshooting Gas Burners

Symptom and Possible Cause	*Possible Remedy*
Pilot does not light.	
(a) Air in gas line.	(a) Clear or replace line.
(b) High or low gas pressure.	(b) Check for possible gas supply problem with local gas company; replace defective burner.
(c) Blocked pilot orifice.	(c) Clean orifice or replace.
(d) Flame runner improperly located.	(d) Move to correct location.

(continued)

Table 2-4 *(continued)*

Pilot goes out frequently during standby or safety switch needs frequent resetting.	
(a) Restriction in pilot gas line.	(a) Clear or replace line.
(b) Low gas pressure.	(b) Check for possible gas supply problem first with local gas company; Replace defective burner.
(c) Blocked pilot orifice.	(c) Clear blockage or replace.
(d) Loose thermocouple connection on 100 percent shutoff.	(d) Secure connection or replace defective thermocouple.
(e) Defective thermocouple or pilot safety switch.	(e) Replace thermocouple or pilot safety switch.
(f) Poor draft connection.	(f) Correct.
(g) Draft tube set into or flush with inner wall of combustion chamber.	(g) Move tip of draft tube to proper location.
Pilot goes out when motor starts.	
(a) Restriction in pilot gas line.	(a) Remove restriction or replace line.
(b) High or low gas pressure.	(b) Check for possible gas supply problem with local gas company; replace defective burner.
(c) Excessive pressure drop when main gas valve opens.	(c) Check for possible gas supply problem with local gas company; test main gas valve and replace if defective.
Burner motor does not run.	
(a) Burned-out fuse or tripped circuit breaker.	(a) Replace fuse or reset circuit breaker. If problem continues, call an electrician.
(b) Thermostat or limit defective or improperly set.	(b) Reset thermostat or limit, or replace if defective.
(c) Relay or transformer defective.	(c) Replace relay or transformer.
(d) Motor burned out.	(d) Replace motor or burner.
(e) Tight motor bearings from lack of oil.	(e) Lubricate bearings; repair or replace damaged bearings.
(f) Improper wiring.	(f) Check wiring diagram for burner and rewire correctly.

(continued)

Table 2-4 (continued)

Burner motor running but no flame.	
(a) Pilot out.	(a) Relight pilot or replace defective pilot.
(b) Pilot safety switch needs to be reset.	(b) Reset switch.
(c) Thermocouple not generating sufficient voltage.	(c) Replace defective thermocouple.
(d) Very low or no gas pressure.	(d) Check for possible gas supply problem with local gas company; check for obstruction in gas line and correct.
(e) Motor running too slow.	(e) Replace defective motor.
Short and/or noisy burner flame.	
(a) Pressure regulator set too low.	(a) Change to proper setting.
(b) Air shutter open too wide.	(b) Correct opening size.
(c) Too much pressure drop in gas line.	(c) Check for possible gas supply problem with local gas company; check for obstruction in gas line and correct.
(d) Vent in regulator plugged.	(d) Clear vent.
(e) Defective regulator.	(e) Replace regulator.
Long yellow flame.	
(a) Air shutter not open enough.	(a) Adjust opening.
(b) Air openings or blower wheel clogged.	(b) Clear.
(c) Too much input.	(c) Reduce input.
Main gas valve does not close when blower stops.	
(a) Defective valve.	(a) Replace valve.
(b) Obstruction on valve seat.	(b) Remove obstruction.

Chapter 3

Coal Firing Methods

The fuel for a coal-fired furnace or boiler may be fed either automatically or by hand. Both methods have certain advantages and disadvantages. For example, automatic (stoker) firing is initially more expensive because it requires the purchase and installation of a suitable mechanical stoker to feed the coal to the furnace or boiler. As a result, stoker firing is a common practice in larger buildings (e.g., stores, hotels) where the initial high cost of the equipment can be more easily absorbed into the total cost of the structure. Despite the relative high cost of stokers, there are some designed for use in single-family residences. These will be considered in detail at a later point in this chapter.

Because there is no need to invest in special and expensive coal-handling equipment, hand-firing the coal has been the traditional method used for firing house heating furnaces and boilers. Although hand-firing coal *is* less expensive than stoker-firing for these smaller installations, the following objections to the hand-firing method should be noted:

- The frequent opening of the furnace or boiler doors allows a large excess of air to enter and chill the flame. The combustion efficiency of the flame therefore tends to fluctuate.
- The dumping of a lot of fuel at each firing results in a smoke period until normal combustion conditions are restored.
- Hand-firing coal is by its nature an intermittent firing method. The flame often reaches a low and inefficient level or is extinguished before new fuel is added.

Coal-Firing Draft Requirements

The amount of draft required for proper combustion is an important consideration, and it depends on a number of different considerations, including the following:

- Grate area
- Fuel size and type
- Fuel bed thickness
- Boiler pass resistance

The degree of resistance offered by the boiler passes to the flow of the gases is an important consideration in determining the required amount of draft. These gases *must* exist at a speed sufficient to prevent them from backing up into the combustion chamber and robbing the fire of necessary oxygen.

The total area of the grate, the type of coal burned (e.g., bituminous, semibituminous), the size of the coal, and the thickness of the coal bed all affect the amount of draft required for proper combustion.

Insufficient draft usually results in the accumulation of excess ashes in the ashpit. Moreover, it necessitates additional attention to the fire including more frequent cleaning. These and other aspects of improper firing contribute to fuel waste and higher operating costs.

Firing Anthracite Coal

Anthracite coal is preferred over other coal for domestic heating purposes because it produces a steadier and cleaner flame. Furthermore, it burns longer and with greater heat than the others. The principal objections to using anthracite coal are that it requires more heat than other coals to start combustion and is slightly more expensive.

Anthracite coal is available in a number of different standardized sizes, each suited to a different size of grate and firepot. The following are some of these coal sizes and their descriptive names:

- Buckwheat
- Egg
- Stove
- Chestnut
- Pea
- Broken

Buckwheat-size coal is available in five grades or sizes: Buckwheat No. 1; Buckwheat No. 2, or Rice size; Buckwheat No. 3, or Barley size; Buckwheat No. 4; and Buckwheat No. 5. Buckwheat Anthracite No. 2 (or Rice size) finds the widest use in automatic coal-firing equipment and is used in domestic, commercial, and industrial stokers.

There are certain recommended practices to be followed when firing buckwheat coal. For best results, the following techniques should be employed:

1. Always maintain a uniform low fire. This reduces clinker formation to a minimum and enables those clinkers that do form to be broken up more easily.
2. Use a smaller mesh grate when possible or a domestic stoker. Because of its relatively small size, buckwheat anthracite coal often falls through the openings on ordinary grates when they are shaken.
3. Immediately after coaling, push a poker down through the fresh bed of buckwheat anthracite coal and expose a portion of the hot fire. This tends to prevent delayed ignition and such undesirable accompanying side effects as furnace or boiler doors being blown open.
4. Keep the heating system warm at all times. Allowing it to cool down and then having to warm it up results in burning fuel at a higher rate.

Egg-size coal should be used in large firepots (24-inch grates or larger). This is a deep-firing coal, and the most suitable results are obtained with fuel beds that are at least 16 inches deep.

Stove-size coal was extensively used in heating buildings, although today it has been largely replaced by gas or oil. It is used on grates that are 16 inches or larger. The fuel bed should be at least 12 inches deep.

Chestnut-size coal is used for firepots as large as 20 inches in diameter. Fuel beds for this anthracite coal range in depth from 10 to 15 inches.

With careful firing, *pea-size coal* can be burned on standard grates. Care should be taken not to overshake the grates (shake only until the first bright coals begin to fall through the grates). After a pea-coal fire has been built, the thickness of the fuel bed should be increased by the addition of small charges until it is at least level with the sill of the fire door. A common method of firing pea coal consists of drawing the red coals toward the front of the firebox and piling fresh fuel toward the back of the firebox.

A strong draft is required when burning pea coal. Figure 3-1 illustrates a satisfactory method of burning this coal size in a boiler. The choke damper is kept open and regulated by means of the cold air check and air inlet dampers. In stoker firing, the air setting is generally kept lower for pea coal because an excess of air under this kind of coal will burn up the retort. Forced draft and small mesh grates are frequently used for burning buckwheat anthracite coal.

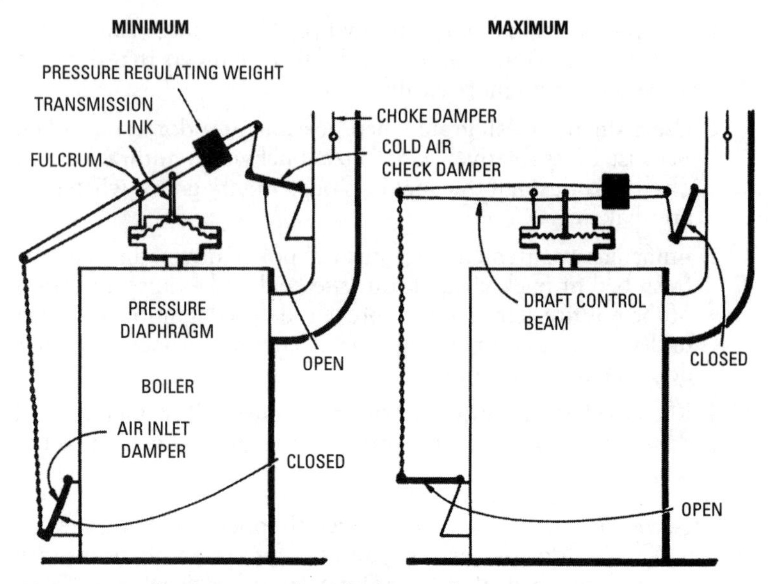

Figure 3-1 Dampers used on boilers and controls.

Firing Bituminous Coal

Bituminous coal is a broad category encompassing many different burning characteristics and properties. Generally speaking, bituminous coals ignite and burn easily with a relatively long flame. They are also characterized by excess smoke and soot when improperly fired.

The *side bank* method is commonly recommended for firing bituminous coal. It consists of moving the live coals to one side or the other of the grate and placing a fresh fuel charge on the opposite side. Variations of this firing method call for placing the live coals at the back of the grate or covering the fresh fuel charge with a layer of fine coal. The side bank method results in a slower and more uniform release of volatile gases.

Other recommendations that should be followed when firing bituminous coal include the following:

- Fire bituminous coal in small quantities at short intervals. This results in a better combustion because the fuel supply is maintained more nearly proportional to the air supply.
- *Never* fire bituminous coal over the entire fuel bed at one time. A portion of the glowing fuel should always be left exposed to ignite the gases leaving the fresh fuel charge.

- Use a stoking bar to break up a fresh charge of coking coal approximately 20 minutes to 1 hour after firing.
- Do *not* bring the stoking bar up to the surface of the fuel. Doing so will bring ash into the high-temperature zone at the top of the fire, where it will melt and form clinkers.

A stoking bar should always be kept as near the grate as possible and should be raised only enough to break up the fuel. The ash will usually be dislodged when stoking, making it unnecessary to shake the grates.

Alternate or *checker firing* is a bituminous coal firing method in which the fuel is fired alternately on separate sides of the grate. This method tends to decrease the amount of smoke and maintain a higher furnace or boiler temperature.

A similar effect is produced by the *coking method* of firing bituminous coal. The coal is first fired close to the firing door, and the coke is moved back into the furnace just before firing again.

Firing Semibituminous Coal

Semibituminous coal burns with far less smoke than the bituminous type. It ignites with more difficulty than bituminous coal but produces far less smoke.

The *central cone method* is recommended for firing semibituminous coal. In this method, the coal is heaped onto the center of the bed, forming a cone, the top of which should be level with the middle of the firing door. This allows the larger lumps to fall to the sides and the fine cones to remain in the center and be coked.

The poking should be limited to breaking down the coke without stirring, and gently rocking the grates. It is recommended that the slides in the firing door be kept closed, as the thinner fuel bed around the sides allows enough air to get through.

Stoker Firing

A *stoker* is a mechanical device designed and constructed to automatically feed fuel to a furnace. Stokers are used in commercial, industrial, and domestic heating systems. Their use results in more efficient combustion owing to constant instead of intermittent firing.

According to the *ASHRAE Guide* (1960), coal stokers can be divided on the basis of their coal-burning capacity into the following four classes:

- Class 1 stokers (10 to 100 lbs per hour)
- Class 2 stokers (100 to 300 lbs per hour)

- Class 3 stokers (300 to 1200 lbs per hour)
- Class 4 stokers (over 1200 lbs per hour)

Class 1 stokers are used most commonly in domestic heating installations. The other three classes of stokers are used in commercial and industrial heating systems.

Class 1 stokers are usually the underfeed type and are designed to burn anthracite, bituminous, semibituminous, and lignite coal, and coke. Ash can be removed automatically or manually, with the latter method being the most popular.

Stokers can also be classified on the basis of whether the coal is stored in a hopper or bin. The disadvantage of the hopper design (see Figure 3-2) is that it must be refilled at least once each day. The bin stoker design (see Figure 3-3) eliminates coal handling. The coal is delivered by the supplier and placed directly into the bin.

The underfeed stoker (see Figure 3-4) is generally used for house heating furnaces and boilers. This type of stoker is one in which the

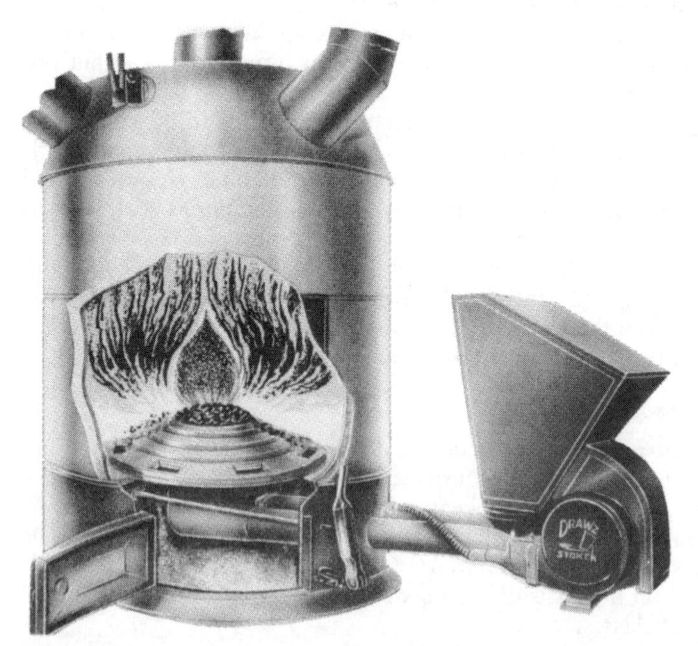

Figure 3-2 Hopper-fed conical grate. Coal is underfed into the furnace and overfed to the fire in a slow movement. *(Courtesy Drawz Stoker Mfg. Co.)*

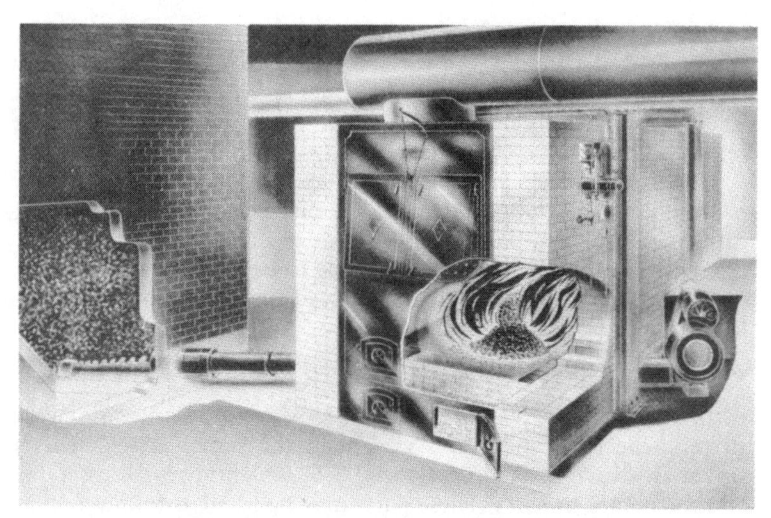

Figure 3-3 Bin-fed stoker equipped with conical grate.
(Courtesy Drawz Stoker Mfg. Co.)

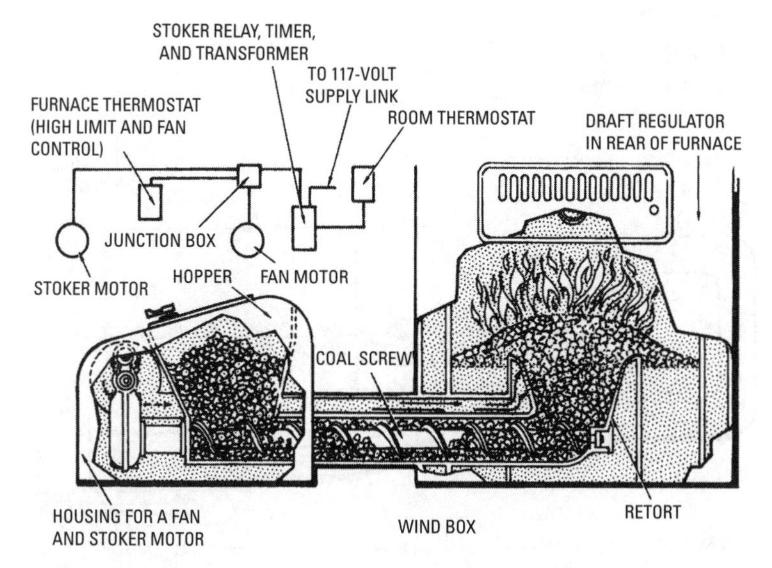

Figure 3-4 Underfeed stoker showing principal parts and typical wiring diagram.

fuel is fed upward from underneath the furnace or boiler. The action of a screw or worm carries the fuel back through a retort from which it passes upward as the fuel above is being consumed. The ash is generally deposited on dead plates on either side of the retort, from which it can be removed.

Underfeed stokers can be designed for use with either anthracite or bituminous coal, but the individual pieces of coal should be uniform in size and no larger than 1 inch in diameter. As mentioned elsewhere in this chapter, it is desirable to treat the coal with oil in order to eliminate dust. The worm feed mechanism can be regulated to feed coal at variable rates.

Stoker Construction

Although there are variations in the type and design of domestic stokers, the general features are much the same. An elementary stoker is shown in Figures 3-5 and 3-6, which gives the essentials and the names of parts. These parts may be listed as follows:

1. Retort.
2. Fan.
3. Motor.
4. Transmission.
5. Air duct.
6. Air control.

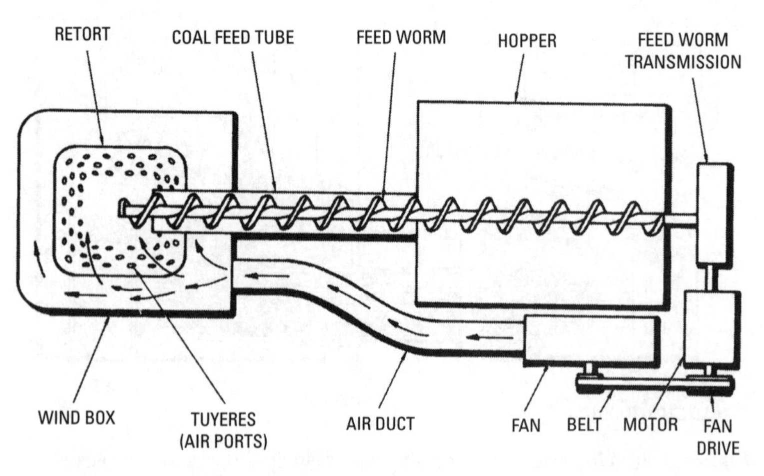

Figure 3-5 Underfeed stoker.

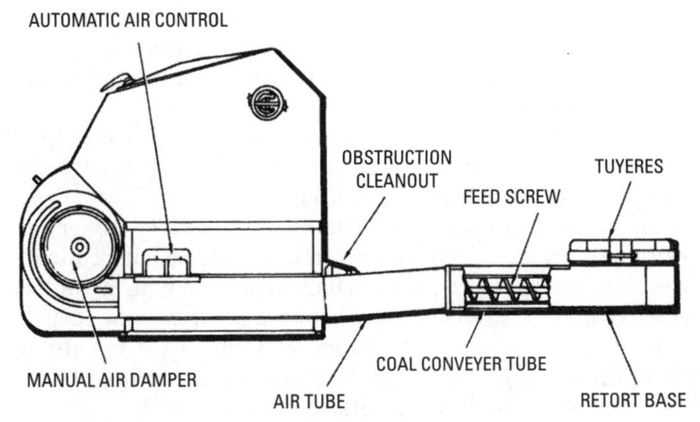

Figure 3-6 Domestic underfeed stoker.

7. Hopper.
8. Feed worm.
9. Bin.

The *retort* is a firepot cast in a round or rectangular trough-like shape in which the coal is burned. It is made of cast iron and is surrounded by the windbox. The retort is provided with a number of air ports, or tuyeres, through which air for combustion is supplied.

The purpose of the *fan* is to supply forced draft, which is directed to the windbox that surrounds the air ports in the retort. This fan is commonly of the squirrel-cage type.

The air enters the retort through the ports via the *air duct* from the fan and the windbox that surrounds the retort. The fan is equipped with either manual or automatic control in the form of a damper at either the discharge or intake end. The air supply is controlled by means of these fan controls.

Coal stokers are designed to operate on either high- or low-air-pressure systems. In the high-air-pressure system, the air is forced in small jets into the fire area. A major disadvantage of this type of system is that the coal sometimes tends to fuse, causing clinkers to form and wasting some of the combustible matter of the fuel. A low-air-pressure system tends to produce a more complete circulation of burning gases to all heat-inducting surfaces.

A stoker is usually powered with an *electric motor*, which operates both the coal feed worm and the fan. The stoker drive consists of a *transmission*, a shear pin (or clutch throw-out), pulleys, belts, and related components. The purpose of the shear pin is to protect the driving mechanism against damage in case large foreign objects get mixed up with the coal.

The transmission rotates the coal feed worm at the proper speed to feed the amount of coal required. The construction is such that the rate of feed can be changed as desired.

The two kinds of transmission usually employed in stokers are the *continuous drive*, which is operated by means of reduction gears, and the *intermittent drive*, which operates with a ratchet. Another drive used in stoker transmissions is the *hydraulic* (usually referred to as an *oil drive*), which operates by regulating the oil pressure on the driving mechanism to control the number of revolutions the feed screw makes per minute.

The *feed worm* (sometimes called the *feed screw*) carries the coal from the hopper to the retort (firepot). It is geared to the transmission, its rate of revolution depending on the desired feed rate. The feed worm extends from the coal supply in the hopper or bin, through the coal feed tube into the retort, where the coal it carries is discharged.

Ashpits can be constructed so that they are located directly below the furnace or boiler (see Figure 3-7). The ashes are automatically deposited into the pit as the coal is burned. If the pit is designed large enough, the ashes will need to be removed only once or twice a year. It is recommended that the ashpit be constructed so as to permit removal of ashes from outside the house. This will result in a much cleaner and more convenient operation in the long run. *Always* vent the ashpit to the chimney or outdoors.

Some stokers (e.g., the Drawz stoker illustrated in Figures 3-2 and 3-3) are designed to permit hand-firing in case of power failure. In these situations, a natural draft may be provided by opening the grate and ashpit door.

Stoker Automatic Controls

Figure 3-8 illustrates typical controls for a stoker in a forced warm-air heating system. There are approximately three *basic* automatic controls necessary for satisfactory operation of the stoker. These three controls are as follows:

1. Thermostat.
2. Limit control.
3. Hold-fire control.

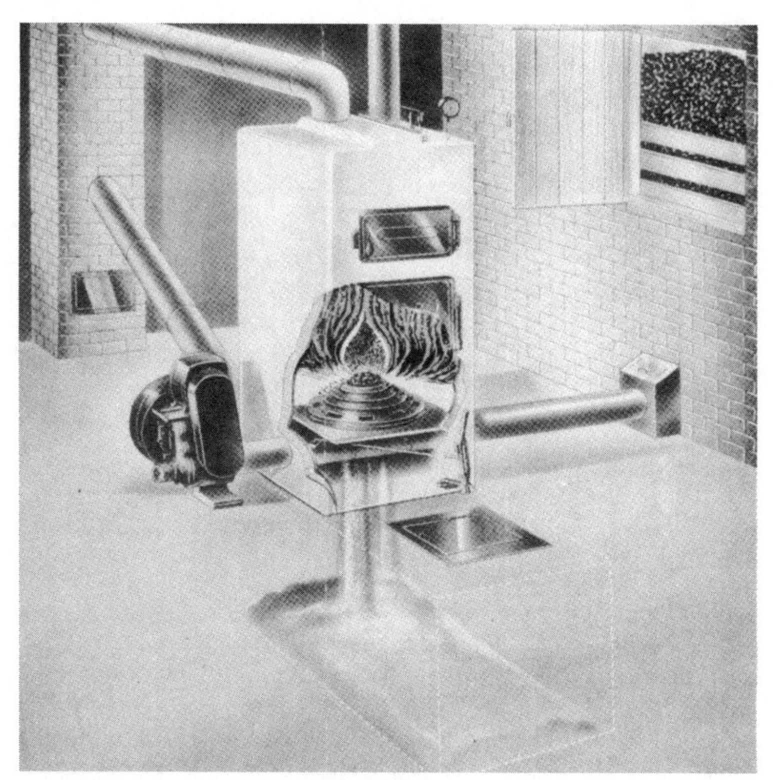

Figure 3-7 Bin-fed conical grate stoker.

The purpose of the *thermostat* is to start the fire when the room temperature falls below a predetermined point and to stop it when the temperature again rises to normal. The thermostat setting should be adjusted to give comfortable room temperature. Usually a setting between 72°F and 75°F is desirable.

The *limit control* stops the stoker should the furnace or boiler pressure become greater than the setting of the control. Furnace limit switches on warm-air gravity installations usually require settings above 300°F. Hot-water limit switches on hot-water systems usually require settings above 160°F. Steam pressure controls on steam pressure installations usually require settings of 2 to 5 lbs.

The purpose of the *hold-fire control* is to produce a stoker operation at intervals during mild weather in order to maintain fire when the thermostat is not demanding heat. Sometimes the

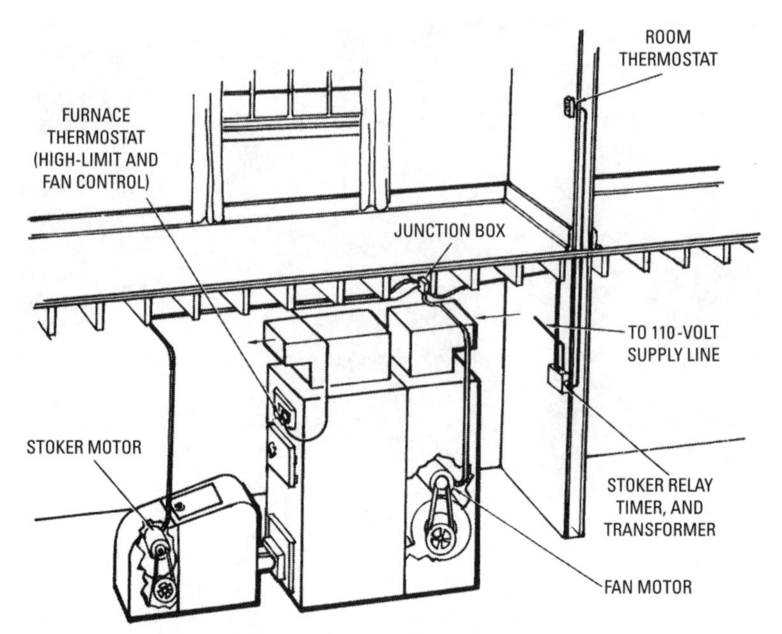

Figure 3-8 Typical controls for a stoker-fired coal burner in a forced warm-air heating system.

hold-fire control feeds either too much or too little coal to the retort. The former results in overheating, and the latter may cause the fire to go out. It is best to call a service representative of the stoker manufacturer to adjust the hold-fire control, because this is a complicated mechanism. The two types of hold-fire controls are interval timers and stack temperature control switches.

Timers may be adjusted to give various-length firing periods so that the stoker is operated for a few minutes at preset intervals. This is done to keep the fire alive during cool weather when little heat is required. The cycle of operation may be set for either 30-minute or 1-hour intervals.

Figures 3-9 and 3-10 show two examples of typical timers used on stokers. The combination switching relay and synchronous motor-driven timer shown in Figure 3-9 provides periodic burner operation so that the fire can be maintained during times when the thermostat is not demanding heat. It may be used with any two-wire, 24-volt thermostat or operating controller. This particular timer is adjustable from ½ to 7½ minutes every 30 or 60 minutes.

The timer shown in Figure 3-10 is designed for line voltage switching. When used with a line voltage controller, it maintains the stoker fire by providing short *on* periods between the controller *off* periods. Timing may be adjusted from 1 to 7½ minutes at 30- or 60-minute intervals.

The *stack switch* (or *stack thermostat*) starts the stoker when the stack temperature becomes lower than a predetermined point and operates it until the fire is again kindled to a degree that will guarantee that it will not go out.

Stack switches are not found on all stokers, but they should be required in areas where electric power failures are long enough to let the fire go out. The stack switch will keep the stoker from filling the cold firepot with coal as soon as the electricity goes on again.

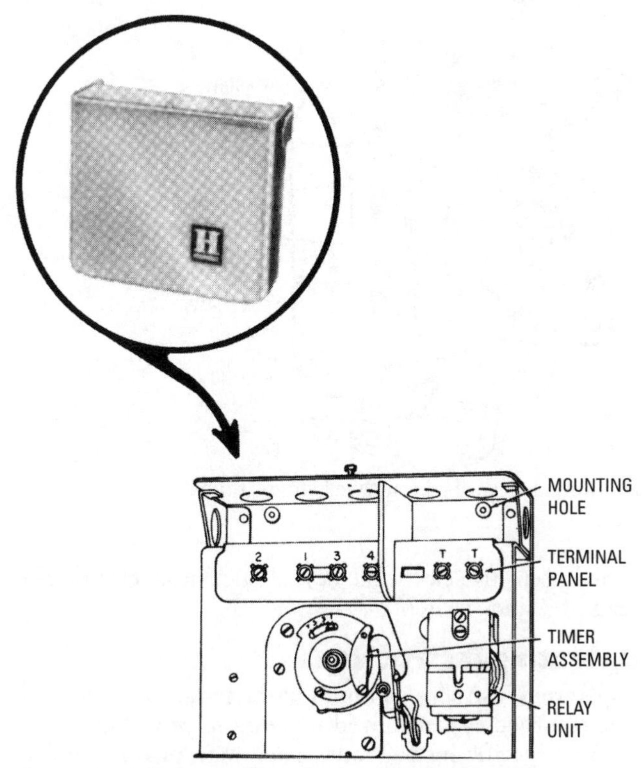

Figure 3-9 Honeywell combination switching relay and synchronous motor-driven timer. *(Courtesy Honeywell, Inc.)*

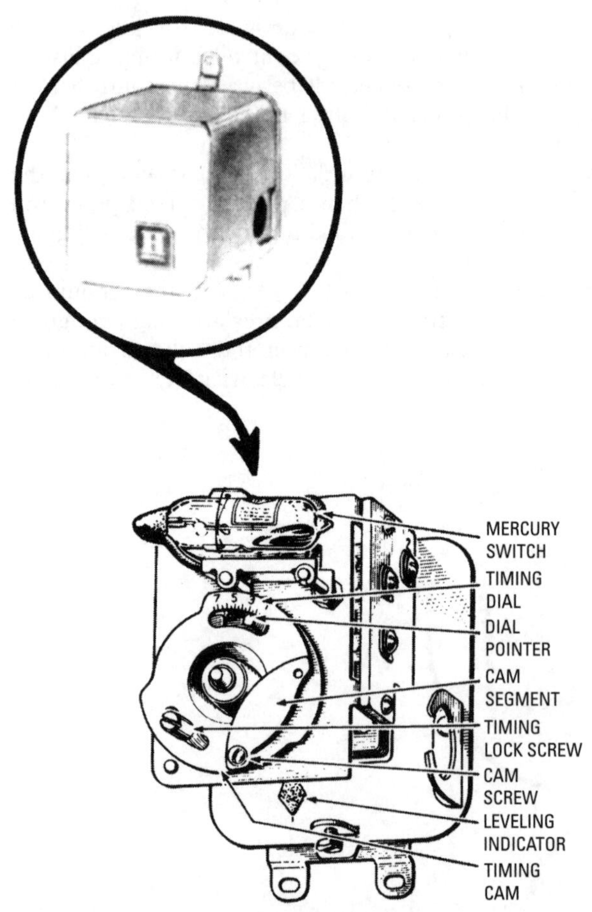

Figure 3-10 Internal view of a Honeywell stoker timer.
(Courtesy Honeywell, Inc.)

Sometimes a light-sensitive electronic device (such as an electric eye) is used instead of a stack switch.

Stoker Operating Instructions

The stoker operating instructions found in the paragraphs that follow should be regarded as generalized suggestions or recommendations rather than specific instructions. They may prove useful in those situations in which no operating manual from the manufacturer can be found. When possible, *always* consult the manufacturer's operating manual.

Coal Selection

A bituminous coal, low in ash (6 percent or less), with an ash fusion temperature of 2200°F to 2600°F and 1¼-inch to ⅜-inch maximum is ideal for stoker operation. Oil treatment of the coal is recommended in order to eliminate dust and add lifetime to the stoker. Generally speaking, in most sections of the country, a high-quality coal is generally most satisfactory and most economical. The annual coal tonnage for domestic stokers is usually low. Convenience and satisfaction are usually the primary factors considered when making the decision to install a stoker; therefore, good coal is recommended. Consult your local coal dealer or the stoker manufacturer for recommendations.

Starting the Fire

Generally, the procedure involved in starting a fire in a stoker-equipped heating installation includes the following steps:

1. Set the room thermostat above the room temperature.
2. Set the coal feed and air setting to the proper rate.
3. Throw the line switch to the *on* position so that the stoker starts.
4. Open the hopper lid, and watch the feed screw to make certain that it is turning. Sometimes in shipping or installing the stoker, the feed screw may slip off the shaft on the gear case. Be certain that the feed screw is engaged *before* putting any coal in the hopper.
5. Fill the hopper with coal.
6. Set the overfire air door on the furnace ¼ to ½ open, and lock in this position.
7. Let the stoker run until the retort (inside the furnace or boiler) is filled with coal.
8. Place a quantity of paper, kindling wood, and a small amount of coal on top of the retort and ignite it.

Natural Stack Draft

Natural draft has a decided effect on the operating economy of the stoker installation. Check the draft and baffle damper, which should be adjusted to give the lowest possible draft without causing smoking from the fire door.

The check damper in the smoke outlet to the furnace can also be used advantageously when extreme natural draft conditions exist. The ideal arrangement is obtained by limiting draft just to the point at which smoke or fumes are not emitted from the fire door when the stoker is in operation.

Manual Air Adjustment

As the fuel bed builds up to the desired condition, the air adjustment should be made in the following manner: Open or close the manual air damper to give a yellow and practically smokeless flame (not white-hot) and a fire fed with no intense blasts from air ports in the burner. Sufficient air must be delivered to maintain an even-burning fuel bed with a fairly consistent depth.

Automatic Air Control

Each stoker will usually have some means of automatically controlling the pressure and volume of air delivered by the fan so that the correct amount is supplied to the fire as burning conditions vary. Usually no adjustment is necessary, as the setting made at the factory will enable this control to function properly on most installations.

Changing Coal Feeds

On some stokers, the coal feed change is easily made by altering the position of the drive belt from the smaller to the larger or from the larger to the smaller pulleys of the motor and transmission.

Follow these instructions:

1. Cut off the stoker line switch.
2. Move the belt-change lever down to reduce the tension on the belt.
3. Move the belt to the pulley desired.
    a. Belt on the large pulley of the motor gives maximum feed.
    b. Belt on the center pulley of the motor gives intermediate feed.
    c. Belt on the small pulley of the motor gives minimum feed.
4. Move the belt-tightening lever up to the original position.
5. Throw in the line switch.

Motor Overload Protection

The stoker motor will have a built-in device for protection against excessive motor temperatures. Should the motor become overheated, the protection device on the motor will prevent damage by breaking the electrical circuit. Motor overloads are usually caused by lack of bearing lubrication, low voltage, or excessive belt tension. To reset, push the reset button on motor after the motor has cooled sufficiently.

Transmission Overload Protection

A stoker transmission will also include an overload protection device that automatically breaks the electrical circuit to the motor in the event that an obstruction should become lodged in the conveying mechanism of the unit. To reset (after removal of the obstruction), push in reset button on the side of the transmission.

Removal of Obstruction

Read the manufacturer's instructions for removing obstructions from the conveying mechanism of the transmission. If these are not available, then you will have to determine the best way to gain access to the obstruction and remove it. There is usually an obstruction cleanout panel located in the back of the hopper. Full access to the feed screw is obtained by removal of this panel. It may be necessary to reverse the rotation of the feed screw manually to relieve the obstruction. To do this, the transmission must be placed in neutral. This can be done by disengaging the transmission from the conveying mechanism.

Lubrication

The electric motor should be lubricated at the beginning of the heating season and twice during the season. Use a good grade of medium engine oil.

The transmission will require approximately one pint of a suitable grade of engine oil. This should be checked once each season. The oil should be removed and replaced at the end of two heating seasons provided there has been no flooding. Should the transmission become submerged in water, it is recommended that it be serviced by a representative of the stoker manufacturer.

Summer Service

It is recommended that the stoker be prepared for the next heating season just after the spring heating has been completed. The stoker should be prepared in the following manner:

1. Remove the coal from the hopper.
2. Paint or grease the inside of the hopper.
3. Open the hopper lid for air circulation.
4. Remove any siftings from the retort base, and remove any ash or clinker formation from the burner.

5. Clean and oil the electric motor and adjust the belts.
6. Oil the stoker screw (or worm).
7. Replace oil in the transmission, if necessary.
8. Run a heavily oiled coal or sawdust through the stoker, leaving the feed screw and coal tube full, over the summer. This prevents corrosion and rusting.

How to Remove Clinkers

You will find it easier to remove the clinker if you let the fire cool off for 5 to 10 minutes before removing it. Turn the stoker *off* and open the fire door to cool the fire. Fill the hopper while the clinker is cooling. The clinker normally forms *around* the retort. Use an iron bar or poker to raise the clinker. *Do not dig in the retort.* After you have raised the clinker, use the clinker tong to lift it from the furnace. It may be in one piece or several pieces, but remove all of it. Keep the fuel bed clean. Remove clinkers as often as necessary.

How to Adjust Coal Feed

Figure 3-11 illustrates the steps involved in a typical coal feed adjustment. Their order (in sequence) is as follows:

1. Select the proper amount of coal feed for the furnace (refer to the coal feed chart provided by the stoker manufacturer).
2. When the proper coal feed is selected, the opposite side of the pointer indicates the proper air setting (see Figure 3-11A).
3. When the coal meter is set on proper coal feed, lock the meter with a wrench at the locknut shown (see Figure 3-11B).

How to Adjust Air Supply

From information on the coal meter, set the air selector knob to the proper point and the automatic damper will furnish the proper amount of air for the amount of coal fed to the furnace. The automatic air damper opens slowly after the stoker starts feeding coal, thus preventing puffbacks out of the fire door, and closes when the stoker stops and automatically banks the fire. When this occurs, the motor stops running.

Troubleshooting Coal Stokers

All mechanical devices occasionally malfunction or operate below a commonly accepted level of efficiency. Coal stokers are no exception to this rule. Table 3-1 lists the conditions that indicate faulty operation:

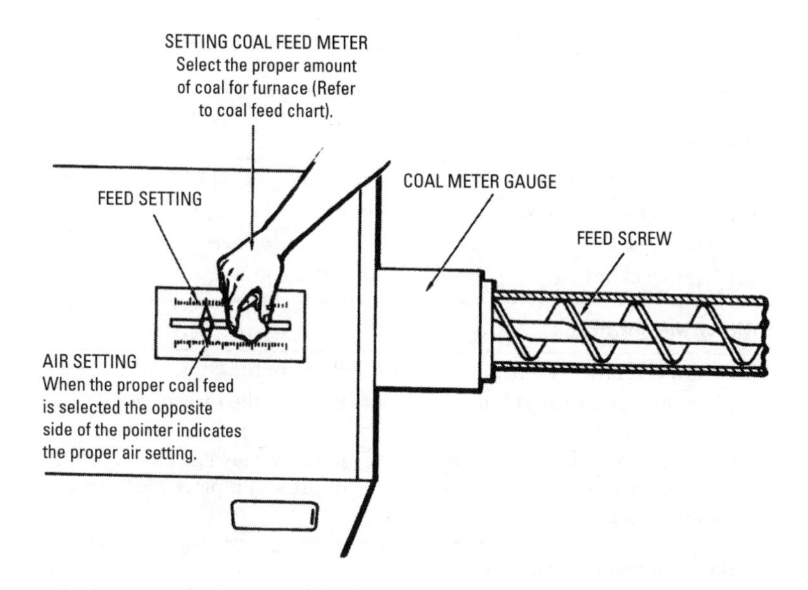

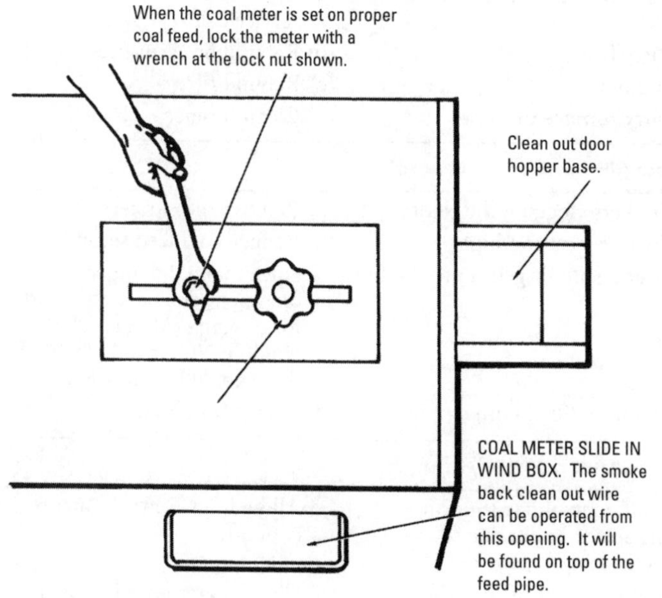

Figure 3-11 Adjustment of a coal feed.

Table 3-1 Troubleshooting Coal Stokers

Symptom and Possible Cause	*Possible Remedy*
Abnormal noises.	
(a) Loose pulleys or belt.	(a) Tighten or replace.
(b) Dry motor bearings.	(b) Oil the bearings.
(c) Worn gears.	(c) Oil or replace.
(d) Gears lack oil.	(d) Oil the gears.
Motor will not start.	
(a) Hard clinkers over or on retort.	(a) Remove the clinkers.
(b) Foreign matter caught in the feed screw.	(b) Remove the foreign matter.
(c) Packing of coal in the retort caused by the end of the feed screw being worn.	(c) Remove packed coal from the retort and replace feed screw.
Stoker operates continuously.	
(a) Controls out of adjustment.	(a) Contact manufacturer for a service call.
(b) Dirty fire.	(b) Rebuild or clean fire.
(c) Fire out.	(c) Rebuild fire.
(d) Dirty furnace or boiler.	(d) Clean furnace or boiler.
Furnace filled with unburned coal.	
(a) Clinkers clogging the retort.	(a) Remove the clinkers.
(b) Coal feed set too high.	(b) Reduce coal feed setting.
(c) Insufficient air getting to the fire.	(c) Open manual damper. If this does not help, check air ports for clogging. Also, check the windbox. If it is full of siftings, they should be removed.
(d) Windbox filled with siftings.	(d) Empty windbox.
Stoker will not run.	
(a) Limit control has shut off furnace or boiler due to overheating.	(a) Allow limit control time to cool off.
(b) Low-water cutoff has shut down the boiler.	(b) Check water level in boiler and correct.

(c) Gear case has been exposed to water.	(c) *Do not* try to operate the stoker. Drain and flush out the gear case immediately and refill with oil.
(d) Blown fuse.	(d) Replace fuse.
(e) Tripped circuit breaker.	(e) Reset circuit breaker.
Smoke backed into hopper.	
(a) Hopper empty or low in coal.	(a) Fill the hopper to the proper level.
(b) Clinker obstructing the retort.	(b) Remove clinker.
(c) Clogged smoke back connection.	(c) Remove obstruction.
(d) Fire burning down in the retort.	(d) Check air supply (fire may be getting too much) or rate of coal feed (may be too low).
Fire is out.	
(a) Empty hopper.	(a) Refill to proper level.
(b) Clinkers obstructing the retort.	(b) Remove clinkers.
(c) Switch may be off.	(c) Place in *on* position.
(d) Blown fuse.	(d) Replace fuse.
(e) Tripped circuit breaker.	(e) Reset circuit breaker.
(f) Failure in electric controls.	(f) Contact manufacturer for a service call.

Chapter 4

Thermostats and Humidistats

This chapter, and the two that immediately follow it, describe the principal components of the automatic control systems used in heating, ventilating, and air conditioning. The index should also be checked because additional information about automatic controls has been included in other chapters.

Automatic Control Systems

An automatic control system consists primarily of the following two basic components:

- Controller
- Controlled device

A *controller* is any device that can detect changes in temperature, humidity, or pressure and respond to these changes by activating a controlled device.

A *controlled device* may be a valve or a damper, or a motor that drives either of the two. It may also be a pump, fan, electric relay, or any other device used to regulate the flow of air, steam, water, gas, or oil.

Automatic control systems can be classified as either closed loop or open loop. A closed-loop system (see Figure 4-1) is the more common type and involves the following stages in the control sequence:

1. The controller measures a change in a variable condition (e.g., temperature) and actuates the controlled device.
2. The controlled device compensates for the change in the variable condition by regulating the flow rate of the medium (e.g., water, air, steam) carried in the system.
3. The result of the action of the controlled device is measured, and this information is fed back to the controller completing (i.e., closing) the loop.

An example of an open-loop system is one using an outdoor thermostat. It is characterized by having no means of feedback. In other words, room temperature has no effect on the operation of the controller.

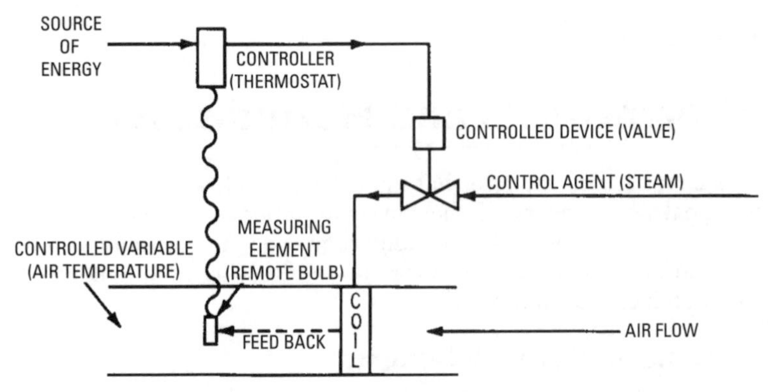

Figure 4-1 Closed-loop automatic control system.
(Courtesy ASHRAE 1960 Guide)

Temperature Control Circuits

The thermostat is the basic controller in the electrical control circuits used to operate a heating and/or cooling system. In systems using gas-fired heating equipment, there are three basic control circuits:

- Safety shutoff circuit
- Fan or circulator control circuit
- Temperature control circuit

There are three basic types of temperature control circuits used in heating and cooling systems. These three basic circuits are as follows:

- Low-voltage control circuit
- Line voltage control circuit
- Millivolt control circuit

Typical wiring diagrams for these three temperature control circuits are shown in Figures 4-2, 4-3, and 4-4.

Thermostats

A *thermostat* is an automatic device designed to maintain temperature control. It accomplishes this function by reacting to temperature changes with adjustments of a controlled device such as a damper or valve motor or the automatic firing equipment (gas

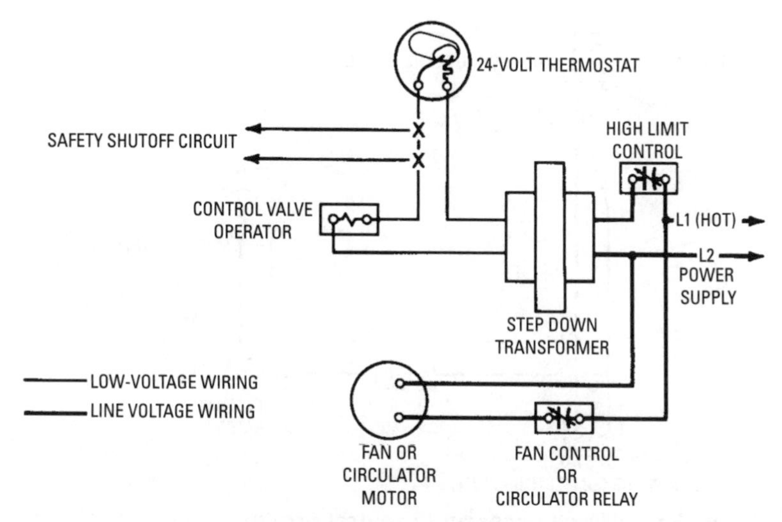

Figure 4-2 Low-voltage temperature control circuit.
(Courtesy Honeywell Tradeline Controls)

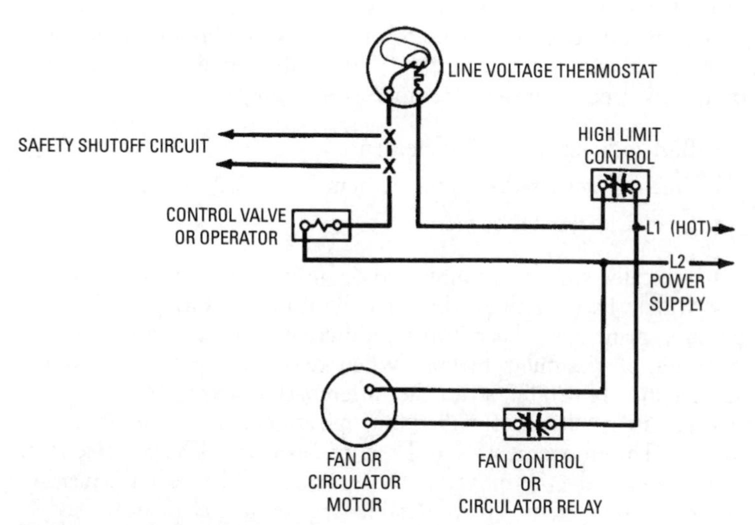

Figure 4-3 Line voltage temperature control circuit.
(Courtesy Honeywell Tradeline Controls)

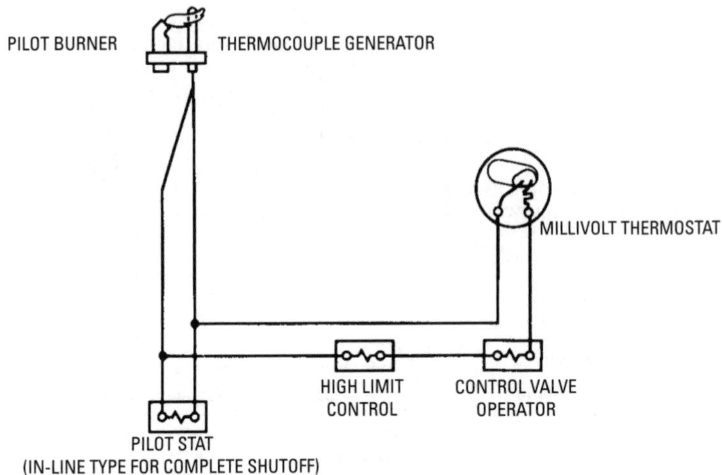

Figure 4-4 Millivolt temperature control circuit.
(Courtesy Honeywell Tradeline Controls)

burner, oil burner, or coal stoker) in space-heating furnaces and boilers. Because of its specific function, a thermostat is sometimes referred to as a *temperature controller.*

Thermostats can be classified on the basis of *how* they measure (or sense) temperature changes. The following devices are most commonly used to measure temperature changes:

- Bimetallic-strip sensing element
- Pressure-actuated sensing element
- Electrical resistance element

A bimetallic strip containing two dissimilar metals is probably the most widely used of these three temperature-measuring devices. Its operating principle is based on the different expansion and contraction rates of dissimilar metals. When two such metals are joined together in a bimetallic strip, the differences in expansion and contraction rates will cause a bending movement as the temperature changes. This movement is utilized to open or close an electrical circuit between the thermostat and the controlled device in the heating and/or cooling system. The bimetallic-strip sensing element is used in either snap-action switch thermostats or mercury-switch thermostats.

A thermostat that operates on the positive snap-action switching principle contains movable switch contacts. One of the contacts is connected to a movable switch armature; the other is fixed in position. An auxiliary armature attached to a bimetal coil responds to temperatures induced by the expansion or contraction of a moving magnet (see Figure 4-5). The magnet, attached to the auxiliary armature, controls the movement of the switch armature. When the switch armature moves toward the magnet, it causes the contacts to close (see Figure 4-6).

On some thermostats, the switch contacts are hermetically sealed in a glass enclosure to protect them from dust or moisture. The thermostats illustrated in Figures 4-5, 4-6, and 4-7 are of this design.

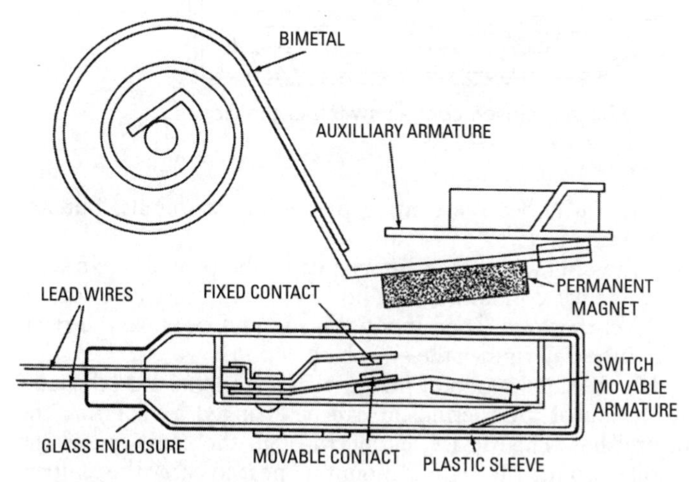

Figure 4-5 Glass-enclosed contact switch in the open position.
(Courtesy Robertshaw Controls Co.)

A mercury-switch thermostat contains fixed contacts sealed in a mercury-filled tube. The tube is attached to the end of a spiral bimetal element. When the temperature changes, the bimetal element tilts the tube and causes the mercury to shift its position, causing a definite opening and closing of the electrical circuit.

A two-wire thermostat using the mercury tube switch method is illustrated in Figure 4-8. A drop in temperature causes the mercury switch to complete (close) the circuit. The circuit is broken (opened) on a rise in temperature.

Figure 4-9 illustrates the application of the mercury tube switch method in a heating and cooling thermostat. A common terminal wire runs along the bottom of the mercury tube. Mercury can make

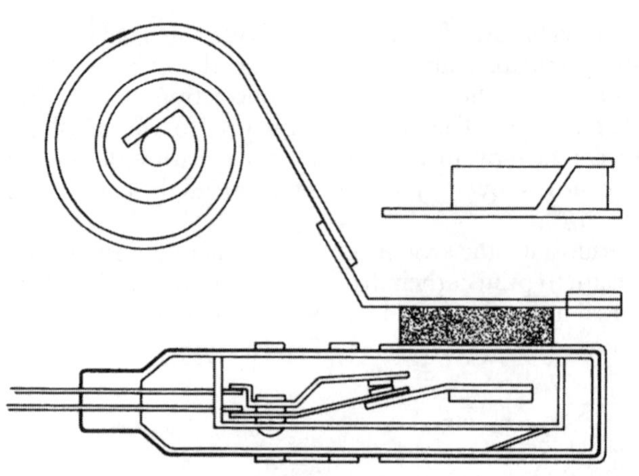

Figure 4-6 Glass-enclosed contact switch in the closed position. *(Courtesy Robertshaw Controls Co.)*

contact between either the heating or cooling terminals, but *not* both at the same time.

Sealing the contacts in a mercury-filled tube provides excellent protection against contamination; however, care must be taken to properly level the base when mounting the unit because the position of the tube determines the switching action.

On either the snap-action or mercury-switch thermostat, the temperature setting of the thermostat can be changed by rotating the temperature dial. This device, acting through the cam, causes the bimetal coil to rotate through its mounting post to carry the temperature setting (see Figure 4-10).

Another temperature-measuring device used in thermostats is the pressure-actuated sensing element. A liquid, gas, or vapor with a high coefficient of expansion is used to activate a bellows connected to a snap mechanism. A rise in temperature causes an expansion in the volume of the liquid, gas, or vapor. This expansion is transferred to the bellows, which activates the snap mechanism. See *Remote Bulb Thermostats* later in this chapter for additional details.

An electrical-resistance sensing element consists of a coil of wire with an electrical resistance that changes in direct proportion to temperature changes. This type of sensing element is commonly used in electronic controllers.

A thermocouple device consisting of two dissimilar electrical wires welded together at one end also serves as a temperature-sensing element in some thermostats. Temperature changes at the

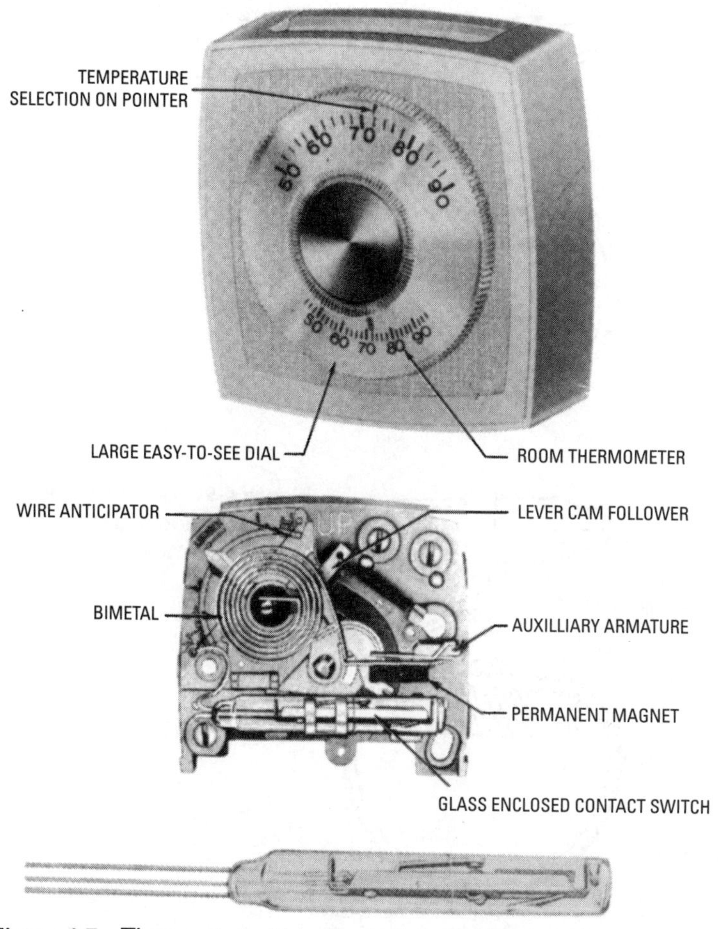

Figure 4-7 Thermostat with contacts in sealed glass enclosure.
(Courtesy Robertshaw Controls Co.)

welded juncture of the two wires cause electrical changes in the control circuit, which operates a regulatory device.

Thermostat Components

A thermostat consists of two basic parts: the base and the cover. A subbase can also be added to a thermostat to provide fan and various switch functions (see Figure 4-11).

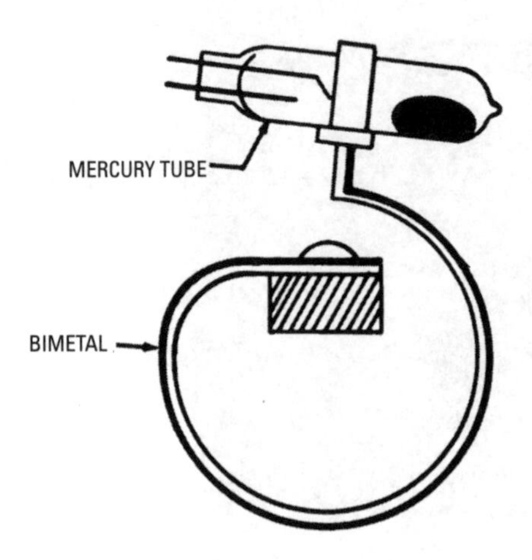

Figure 4-8 Mercury tube switch method in a two-wire thermostat. *(Courtesy Coleman Co., Inc.)*

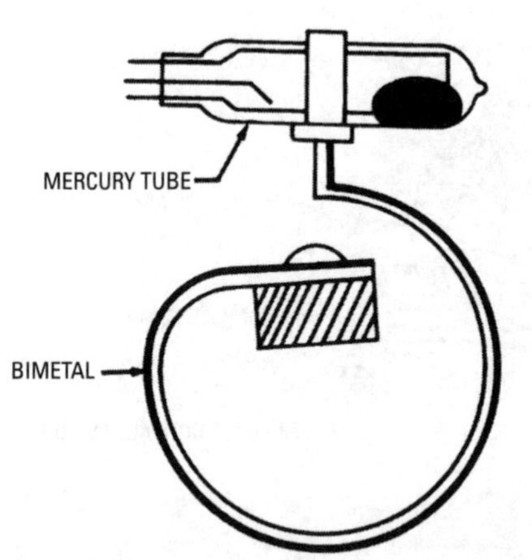

Figure 4-9 Mercury tube switch method in a three-wire thermostat. *(Courtesy Coleman Co., Inc.)*

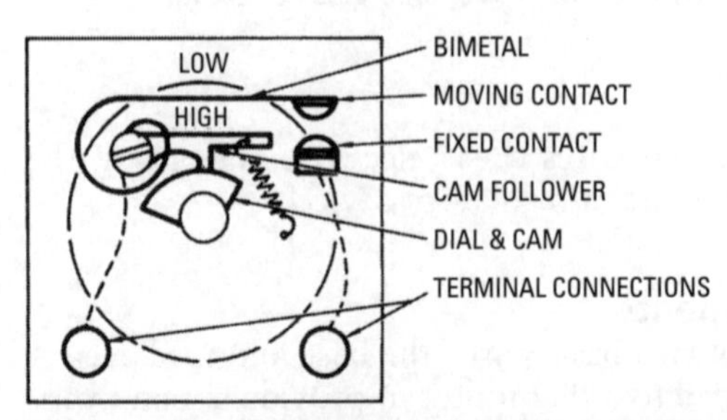

Figure 4-10 Temperature dial and cam mechanism.

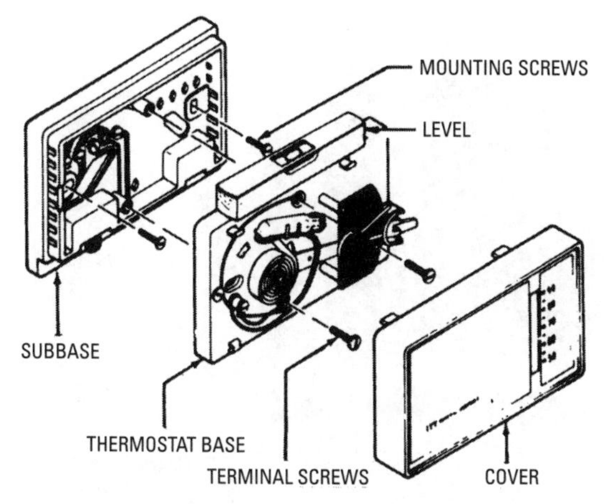

Figure 4-11 Thermostat cover, base, and subbase.
(Courtesy ITT General Controls)

The cover protects the internal wiring of the thermostat base from dust, lint, and other possible contaminants. It also contains the temperature-setting lever, the system-switching levers, and the temperature indicator (see Figure 4-12).

The cover is secured to the base either with a positive friction snap or screws. If the former is the case, the cover can be removed by grasping the base with one hand and pulling it off with the other (see Figure 4-13). Screws may require the use of an Allen wrench (see Figure 4-14), which is usually supplied by the thermostat manufacturer and shipped with the unit.

The base contains the internal wiring of the thermostat. Figure 4-15 shows the back and front views of typical base wiring for the low-voltage thermostat illustrated in Figure 4-12. This is *not* the only possible way the base can be wired. *How* the base is wired will depend on the particular application. A few of the possible variations are illustrated by the thermostat base wiring diagrams in Figure 4-16. The many variations in wiring will depend on which combination of the following features is required by the installation:

1. Type of switch and switching action (snap-action or mercury bulb; spst or spdt contact).
2. Number of field wires (two, three, four, or five).
3. Type of anticipator (fixed heating, adjustable heating, fixed cooling).

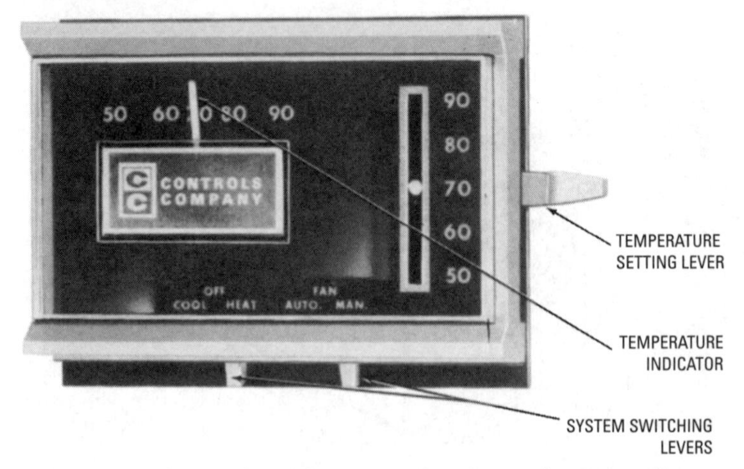

Figure 4-12 Low-voltage thermostat for electric heating and/or cooling system. *(Courtesy Singer Controls Co., of America)*

4. Use of fan and system switches.
5. Type of operating voltage (low voltage, line voltage, or millivolt circuit).

Adding a subbase to a thermostat provides fan and other switching functions. The wiring diagrams of the ITT General Controls

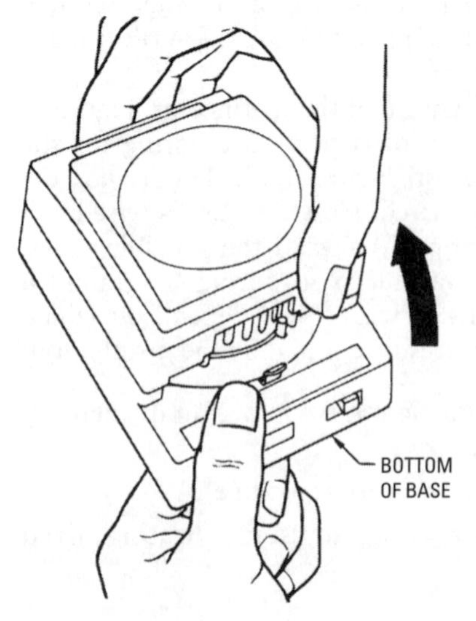

Figure 4-13 Removing a cover secured to the base with a positive friction snap. *(Courtesy Com-Stat Inc.)*

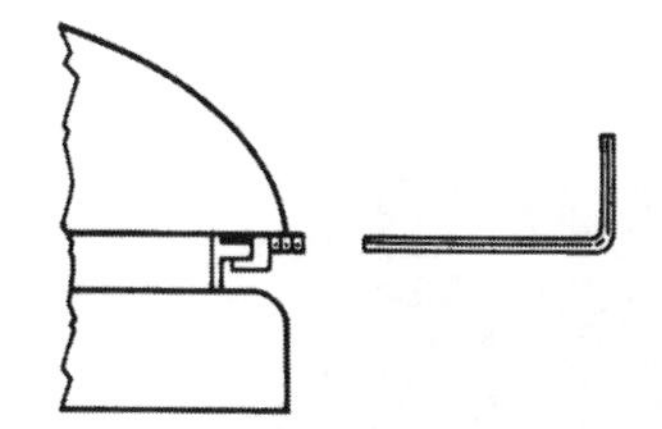

Figure 4-14 Loosening cover screws with an Allen wrench.
(Courtesy Honeywell Tradeline Controls)

thermostats and subbases shown in Figure 4-17 illustrate the variety of different switching combinations available.

Thermostat guards can be purchased to protect the thermostat from damage. This is particularly important in warehouses, stores, and other areas where there is a greater possibility of the thermostat being damaged. Some examples of how these guards are used are shown in Figure 4-18. Adapter plates (wall plates) are also available from thermostat manufacturers to cover electrical utility or junction boxes (see Figure 4-19).

Thermostat Terminal Identification

The National Electrical Manufacturers Association is at present attempting to standardize thermostat markings in order to aid the installer in wiring and servicing. Thermostat terminals have been given standard identification letters that specify the function of the terminal. These identification letters are also matched, in most cases, with the color-coding of the wire. A partial list of equivalent terminal markings is given in Table 4-1.

Thermostat installation literature will generally contain at least one internal view and/or wiring diagram in which the various terminals are identified by a specific letter.

Thermostat Anticipators

A *thermostat anticipator* is a device used to reduce the operating differential of the heating or cooling system. It is designed to enable the thermostat to shut off the furnace or boiler slightly in advance of the actual set temperature. As a result, the thermostat shuts off the heating equipment sooner than it would if it were affected by only the room temperature, thereby compensating for heat transfer lag. A thermostat may be equipped with a heat anticipator, a cold anticipator, or both.

A *heat anticipator* is a small resistor (resistive heater) connected in series with the switch inside the thermostat. Heat generated by the resistor when the switch is in the *on* position heats the thermostat bimetal actuator and causes the internal temperature of the thermostat to rise faster than the surrounding room temperature.

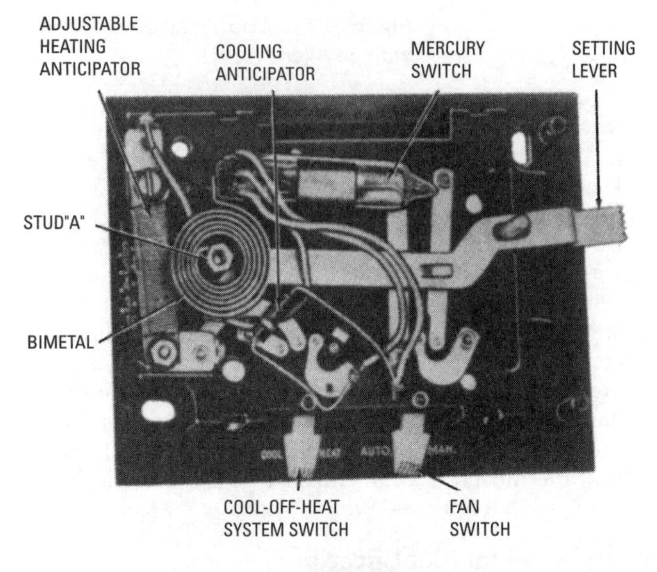

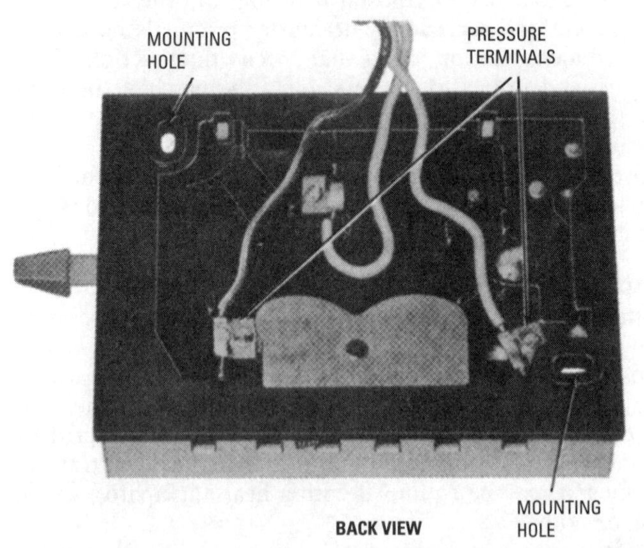

Figure 4-15 Front and back view of Singer Thermostat model 360.
(Courtesy Singer Controls Co., of America)

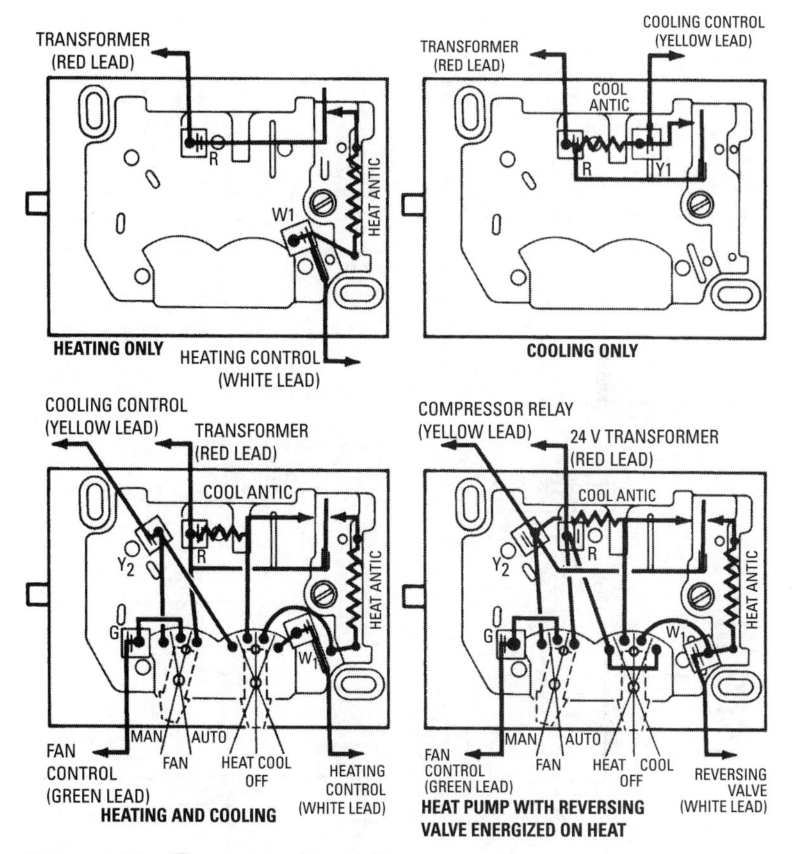

Figure 4-16 Thermostat base wiring. *(Courtesy Singer Controls Co., of America)*

Thermostats are available with fixed anticipators, plug anticipators, or variable anticipators.

A fixed heat anticipator must be sized in accordance with the amperage or current draw of the operating valve in order to secure the correct degree of heat anticipation. These anticipators are generally available in the range of 0.1 to 1.5 amperes. When using a thermostat equipped with a fixed heat anticipator, check the nameplate on the valve or relay to make certain the ampere rating does not exceed the maximum amp (current) draw. For example, if the thermostat is equipped with a 0.40- to 0.60-ampere fixed heat anticipator, the valve or relay should not exceed 0.60 ampere.

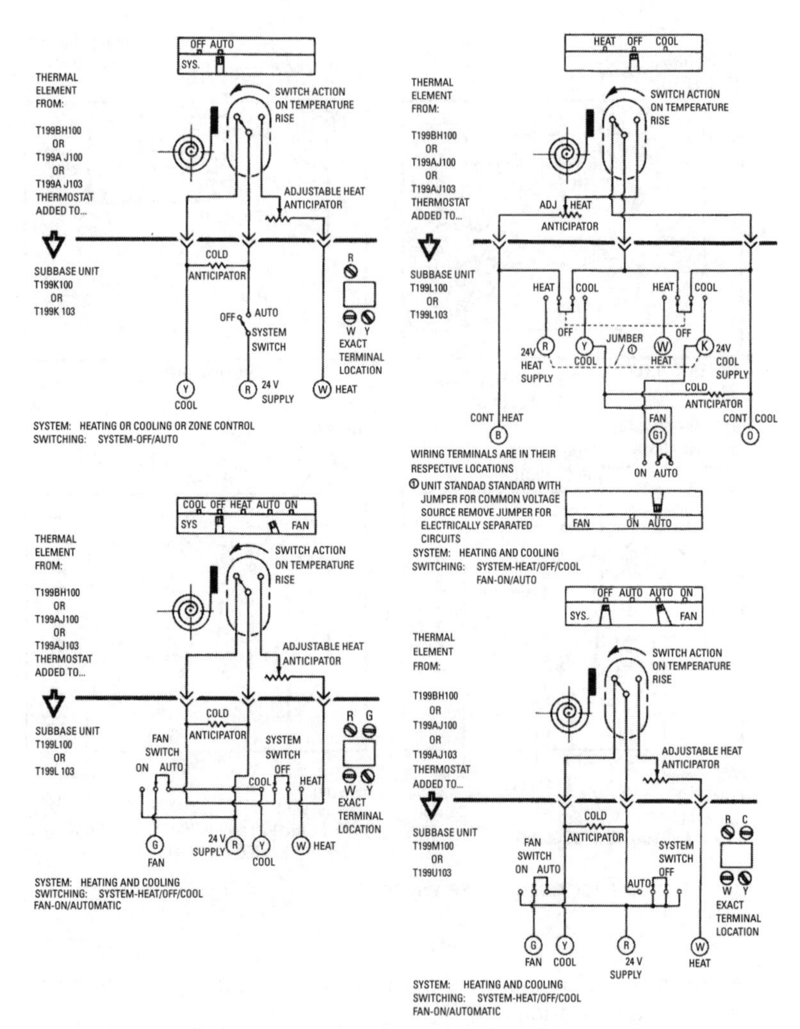

Figure 4-17 Thermostat switching combinations. *(Courtesy ITT General Controls)*

The thermostat shown in Figure 4-20 is an example of one equipped with a fixed heat anticipator. The anticipator is contained in the thermostat element along with the magnetic switch and a room temperature thermometer.

Figure 4-21 illustrates a thermostat equipped with a plug-type heat anticipator. This design offers some degree of latitude in selecting a

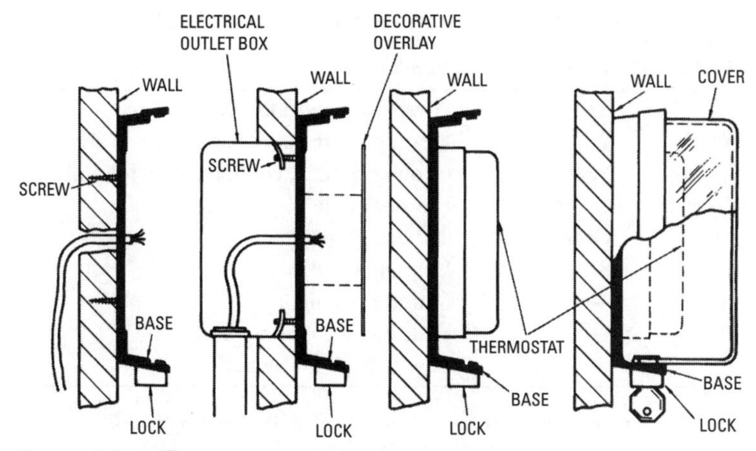

Figure 4-18 Thermostat guards. *(Courtesy ITT General Controls)*

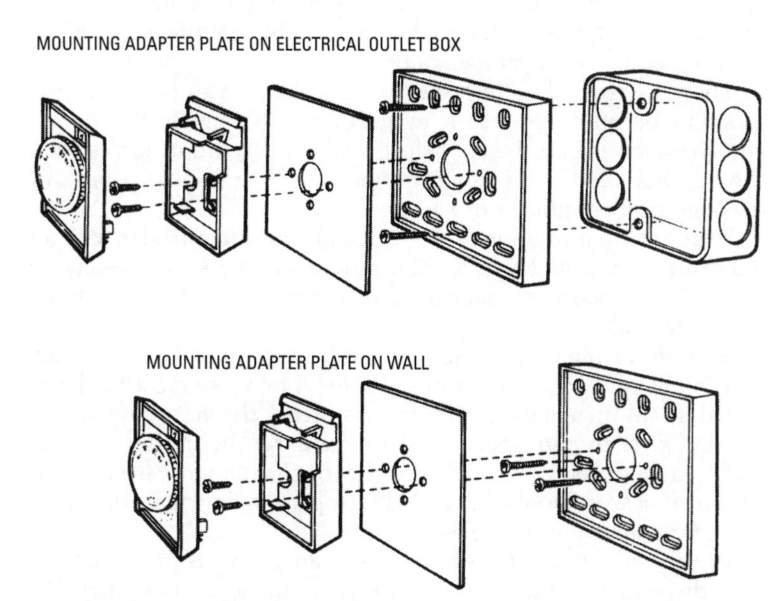

Figure 4-19 Adapter plate kit. *(Courtesy ITT General Controls)*

Table 4-1 Common Terminal Identification

Letter	Wire Color	Terminal Function
R	Red	Power supply; transformer
W	White	Heating control; heating relay or valve coil
Y	Yellow	Cooling control; cooling contactor coil
G	Green	Fan relay coil
O	Orange	Cooling damper
B	Brown	Heating damper
X	—	Malfunction light
P	—	Heat pump contactor coil
Z	—	Low-voltage fan switch

Note: The letters R_H (heating) and R_C (cooling) are used on thermostats with isolated circuits.

proper anticipator. The current (amps) drawn by the primary control or valves is first determined, and an anticipator having the proper value is selected from among those listed in Table 4-2. The heating cycle can be lengthened by selecting an anticipator one step above the proper value. Shorter cycles can be obtained by selecting an anticipator one step below the proper value.

Thermostats with adjustable (variable resistance) heat anticipators can be adjusted over a range of approximately 0.1 to 1.5 amperes (see Figure 4-22). Before making any adjustments, you should first read the equipment manufacturer's instructions for selecting proper anticipator values.

Heat anticipator adjustments are made on a thermostat with an indicator or adjustment lever that moves along a scale (see Figure 4-23). Adjustments are made in accordance with the values marked along this scale.

As with the plug-type heat anticipator (see Figure 4-21), the current (amps) drawn by the primary control or valve must be determined first. In a gas-fired heating system, the heat anticipator should be set to correspond to the secondary (thermostat) current of the valve or relay. In an oil-fired heating system, the heat anticipation indicator should be set 0.15 ampere higher than the rated secondary current of the relay.

The anticipator adjustment lever should be moved only ¼ to ½ scale division at a time. *Never* move or set the lever more than 1½ scale divisions under the valve or relay current ratings. Longer *on* periods can be obtained by setting the adjustment lever at a slightly

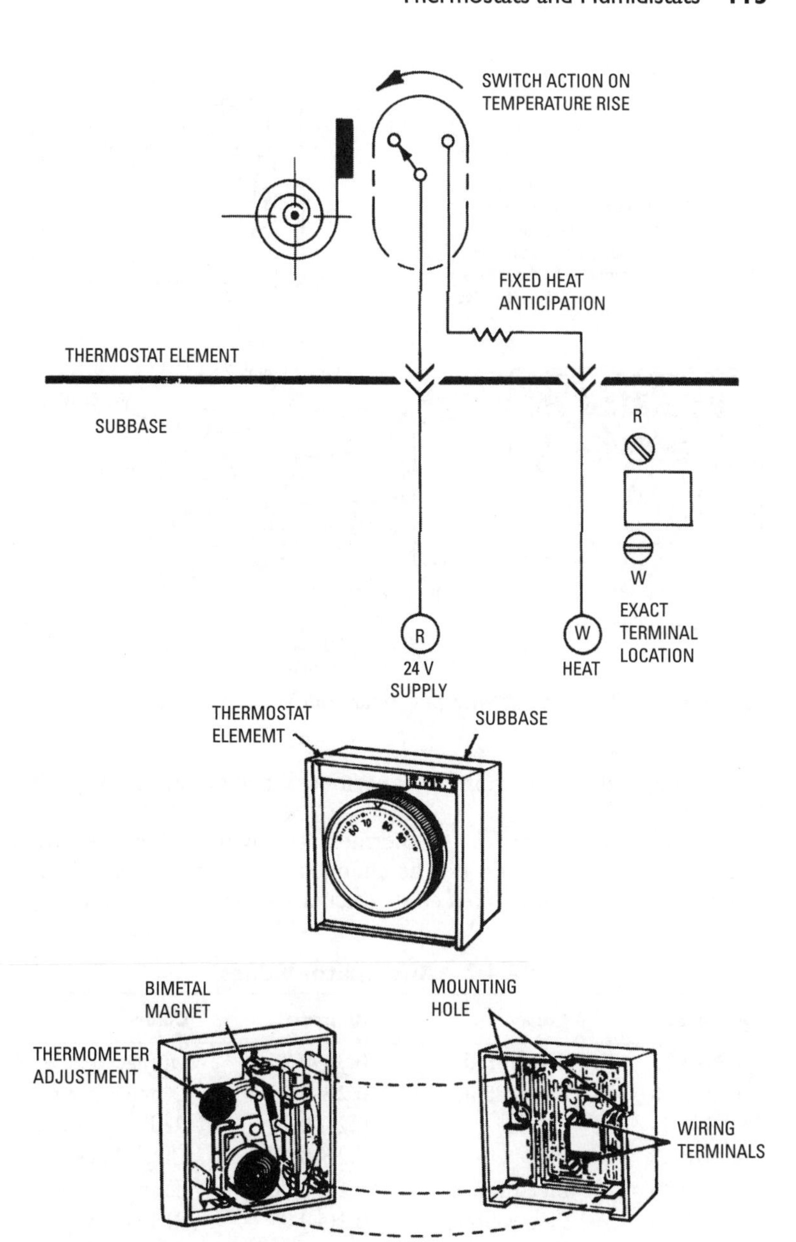

Figure 4-20 Thermostat with fixed heat anticipator. *(Courtesy ITT General Controls)*

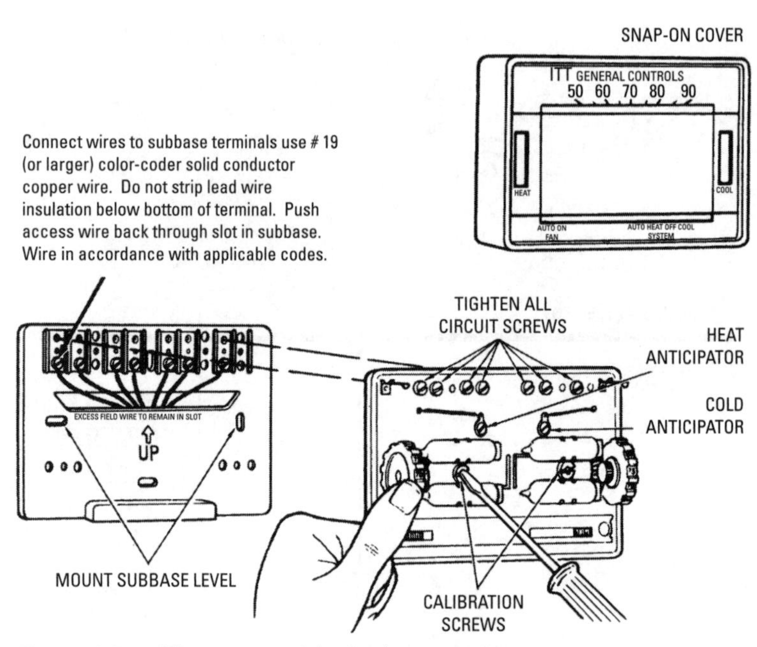

Figure 4-21 Thermostat with plug heat anticipator. *(Courtesy ITT General Controls)*

higher amp value. For shorter *on* periods, set the lever at a slightly lower amp value.

The cold anticipator for the thermostat shown in Figure 4-24 is not adjustable and should not be changed. The same is true of the other thermostats. Only *fixed* cold anticipation is used. Furthermore,

Table 4-2 Anticipator Values

Amperes	*Color*	*Amperes*	*Color*
0.83–0.72	Brown-red	0.33–0.29	Orange-yellow
0.72–0.68	Brown-blue	0.29–0.25	Orange-green
0.68–0.55	Orange	0.25–0.22	Red
0.55–0.48	Blue	0.22–0.19	Green-blue
0.48–0.41	Blue-orange	0.11–0.10	Orange-red
0.41–0.36	Blue-yellow	0.10–0.09	Red-yellow
0.36–0.33	Green	0.09–0.08	Green-yellow

(Courtesy ITT General Controls)

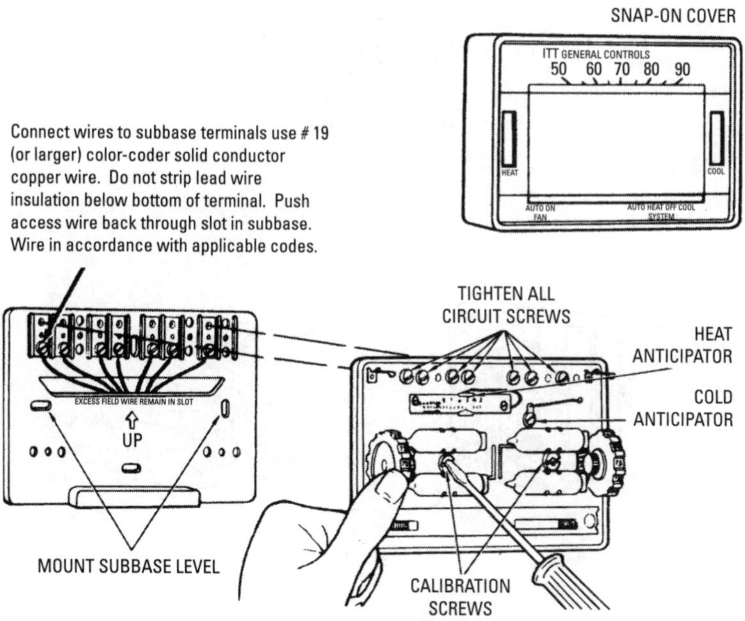

SNAP-ON COVER

ITT GENERAL CONTROLS
50 60 70 80 90
HEAT
COOL
AUTO ON
FAN
AUTO HEAT OFF COOL
SYSTEM

Connect wires to subbase terminals use # 19 (or larger) color-coder solid conductor copper wire. Do not strip lead wire insulation below bottom of terminal. Push access wire back through slot in subbase. Wire in accordance with applicable codes.

EXCESS FIELD WIRE REMAIN IN SLOT
UP
MOUNT SUBBASE LEVEL
TIGHTEN ALL CIRCUIT SCREWS
HEAT ANTICIPATOR
COLD ANTICIPATOR
CALIBRATION SCREWS

Figure 4-22 Thermostat with adjustable heat anticipator.

(Courtesy ITT General Controls)

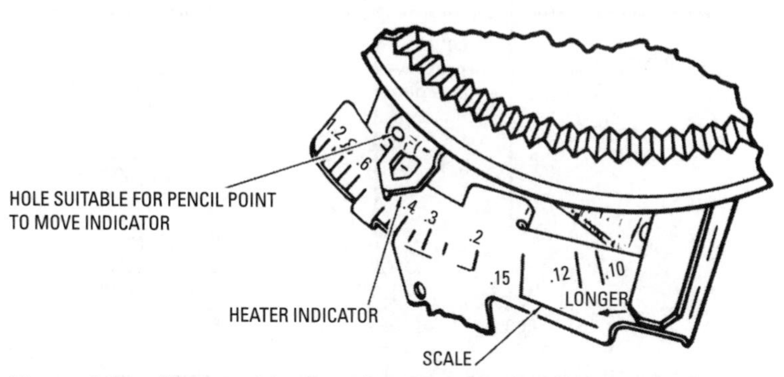

Figure 4-23 Heat anticipator adjustment lever and scale.

(Courtesy Honeywell Tradeline Controls)

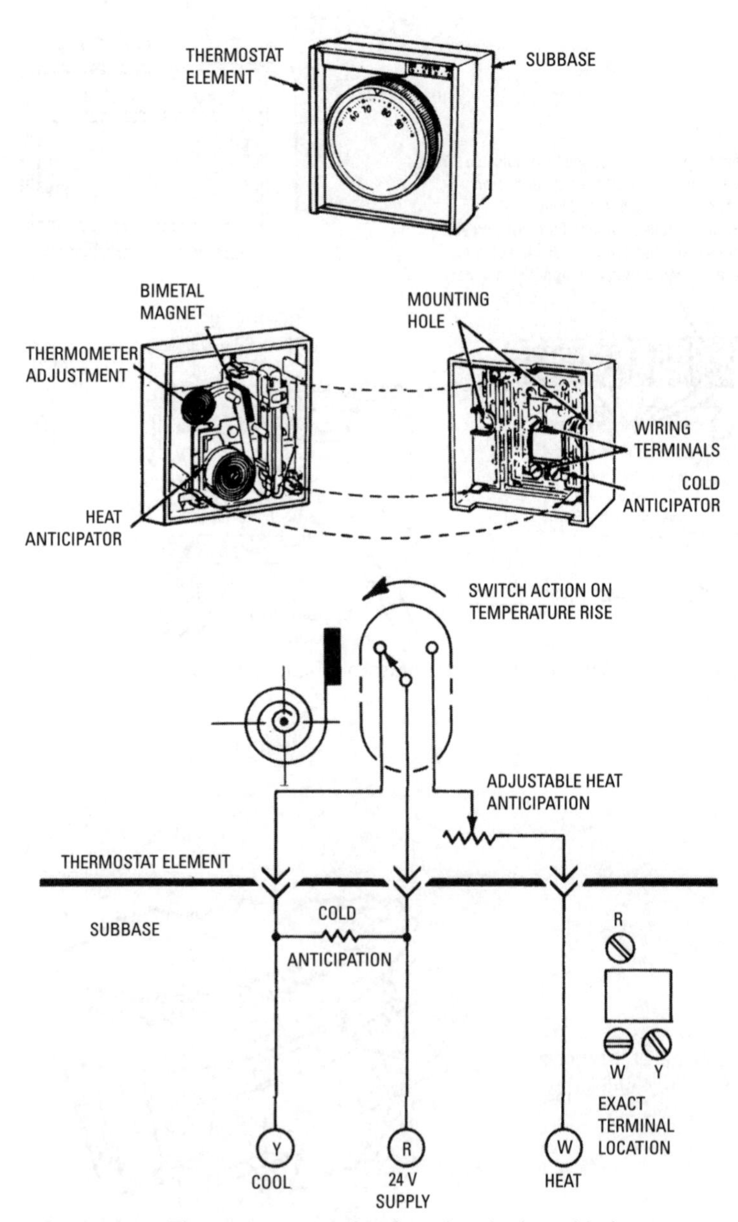

Figure 4-24 Thermostat with fixed cool and adjustable heat anticipator. *(Courtesy ITT General Controls)*

on cooling thermostats, the cold anticipation is in parallel with the cooling contacts so that anticipation heating occurs while the cooling unit is off.

Types of Thermostats

Many different thermostats are manufactured for use in heating and cooling systems. The design differences depend largely on the type of application.

The most common thermostat is the wall-mounted *room thermostat* used to control a heating and/or cooling system. The measuring element is contained in the thermostat unit itself. This distinguishes it from the *remote-bulb-type thermostat* used to measure temperatures in spaces separate from the location of the thermostat.

An *insertion-type thermostat* (or *duct thermostat*) is used to measure temperatures inside an air duct. The temperature-measuring element is contained in an insertion device that extends into the duct. The *immersion-type thermostat* is similar in design but is used to measure the temperature of fluids inside a pipe or tank. These thermostats are commonly used on water heaters.

A *heating-cooling thermostat* (also referred to as a *summer-winter thermostat*) is designed to be switched to either a heating or cooling application. The *day-night thermostat* (or *electric clock thermostat*) operates on a similar working principle except that it is designed to automatically switch from day to night operation and back again.

A *multistage thermostat* is designed to operate two or more circuits in sequence. These thermostats are used for line voltage or low-voltage temperature control of heating and cooling equipment. They are commonly used in heating and/or cooling systems where zone control is necessary.

A thermostat and humidistat can also be combined in the same control unit. These combined units sometimes also include the electronic air-cleaner control.

Room Thermostats

The *room thermostat* (see Figure 4-25) is regarded as the nerve center of the heating and cooling system because it controls the operation of the furnace, boiler, or air conditioner. Ideally, it should be mounted in an area of the living or working spaces where it is not subjected to temperature or moisture extremes (see the following section, *Location of Room Thermostats*).

Low-voltage room thermostats are recommended over the line voltage types for residential heating and/or cooling systems. The low-voltage thermostats respond more quickly to temperature

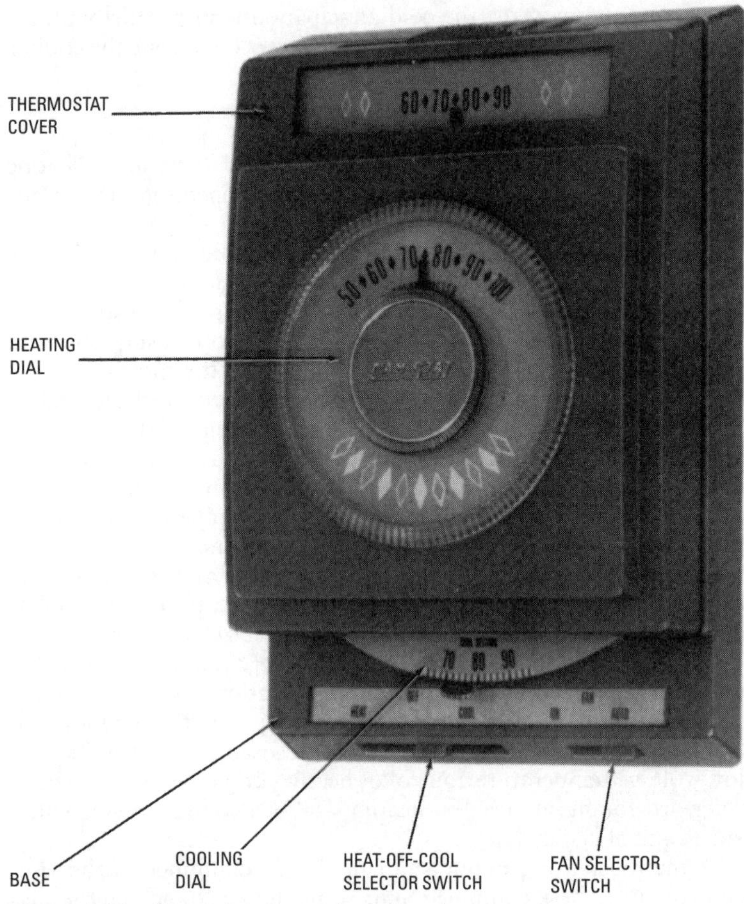

Figure 4-25 Room heating and cooling thermostat. *(Courtesy Com-Stat Inc.)*

changes and will maintain the temperature and humidity more closely than the line voltage types. A low-voltage thermostat requires the use of a transformer to reduce the line voltage for the control circuit, but the cost of the transformer is more than offset by the lower installation cost of this thermostat.

Location of Room Thermostats

The location of the room thermostat is very important to the efficient operation of the heating and/or cooling system. If the room thermostat is improperly located, it will often call for heat or cool air when neither is necessary. It is therefore important to locate the thermostat where it will measure the actual temperature conditions of the space.

The following recommendations are offered as a guide for the proper location of a room thermostat:

1. *Never locate the thermostat on the interior surface of an outside wall.* Outside walls are subject to temperature extremes caused by weather changes. Always locate the thermostat on a *suitable* inside partition.
2. *Never locate the thermostat in the path of warm or cold air drafts.* The thermostat should not be placed opposite warm air outlets or near a window or an outside door.
3. *Never locate the thermostat where it will be subjected to the direct rays of the sun or other forms of heat radiation.* Fireplaces, table lamps, and floor lamps are common sources of this type of heat.

When you select a location for the room thermostat on an inside wall, make certain that you are not placing it over a warm air duct, steam pipe, or hot-water pipe. The warmth from these ducts or pipes will interfere with the operation of the thermostat.

For the most satisfactory operation, locate the room thermostat about 5 feet above the floor on an inside wall where there is good natural air circulation and where the thermostat will be exposed to *average* room temperatures.

Installing a Room Thermostat

Before attempting to install the thermostat, read the manufacturer's installation instructions. These instructions should be followed as closely as possible to ensure efficient thermostat operation.

Always handle the thermostat carefully. Rough handling may decrease its accuracy. Inspect the thermostat carefully when you unpack it. Report any damage to the shipper if the damage was caused after it left the factory or distribution center. Their insurance should pay for any replacement. Any other damage or malfunction should be reported to the thermostat manufacturer or his field representative. The thermostat is the basic controller in any heating and/or cooling system. It must operate accurately and efficiently.

Make sure the thermostat selected for the heating and cooling system is the appropriate one for the job. Heat pumps require two-stage thermostats. Furnaces and boilers used with air-conditioning systems commonly require thermostats with extra terminals. Thermostats with changeover terminals are required for many zoned systems. Check the manufacturer's specification sheet before making a final selection.

The installation instructions differ from one thermostat manufacturer to another, usually on the basis of design and construction; however, the following points are common to all:

1. Mercury-switch thermostats are usually shipped with some form of protection around the mercury tube to prevent breakage. This *must* be removed before operating the thermostat.
2. Disconnect the power supply before connecting the wiring to the thermostat in order to prevent electrical shock and/or equipment damage. Low-voltage thermostats are used in heating and cooling systems. These low voltages are not fatal, but they can deliver a nasty shock and may damage the system controls.
3. All wiring must be done in accordance with local electrical codes and ordinances. Be sure to follow the thermostat manufacturer's wiring diagram when making the connections.
4. Use a plumb line or spirit level to accurately level the subbase or wall plate when mounting it on the wall (see Figure 4-26). Thermostat control deviations are often caused by inaccurate leveling.
5. Follow the manufacturer's instructions for making any internal wiring connections (e.g., connecting lead wires to terminal screws), connecting a heating thermostat to a cooling base, and so on.

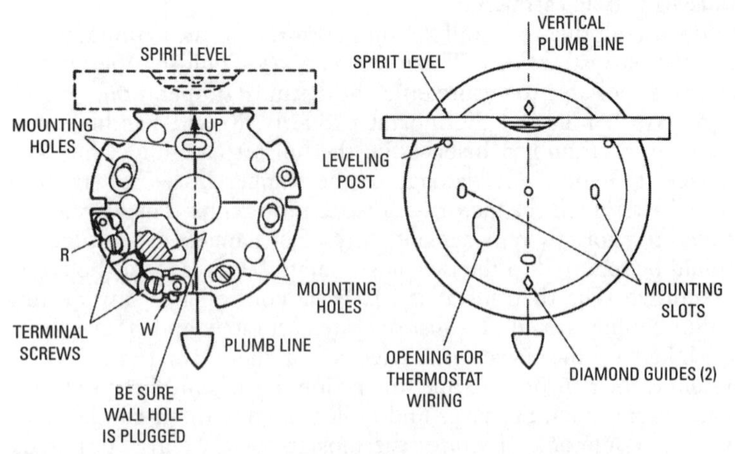

Figure 4-26 Using a plumb line and level. *(Courtesy Honeywell Tradeline Controls)*

6. Check out the installation to make sure the thermostat is operating correctly.

Thermostat Calibration

Thermostats are accurately calibrated at the factory under controlled conditions and should not require recalibration. Sometimes a thermostat will *appear* to require recalibration when the problem is actually a quite different one. For example, a thermostat that is not level or one subjected to a high degree of radiant heat from the sun, radiators, convectors, or other heat sources often fails to function properly. Before jumping to the conclusion that it needs to be recalibrated, check out the possibility that some external cause may be the source of the difficulty. If you are certain the problem is in the thermostat, then you should call a trained serviceman to recalibrate it. Do not attempt to recalibrate it yourself unless you have the necessary experience.

A new thermostat is generally shipped with complete installation literature. This literature also usually contains instructions for recalibrating the thermostat. By way of example, the instructions for recalibrating a Honeywell T87 Room Thermostat include the following steps:

1. Remove the thermostat cover ring (see Figure 4-27). The locking cover will require the use of an Allen wrench to loosen the screws securing the cover.
2. Set the thermostat below the room temperature, and allow it to remain in an *off* position for approximately 10 minutes.
3. Slowly raise the setting until the switch just makes contact. If the thermostat pointer and setting indicator do *not* read the same the instant the switch makes contact, then the thermostat requires recalibration.
4. Turn the setting dial a few degrees above room temperature.
5. Slip the calibration wrench onto the calibration (hex) nut under the bimetal coil (see Figure 4-28).
6. If the thermostat has a stationary pointer, hold the dial firmly and turn the calibration nut *counterclockwise* until the mercury breaks contact. If the thermostat has a movable pointer, turn the calibration nut *clockwise.*
7. Turn the thermostat dial to a low setting, and wait approximately 5 minutes.
8. Slowly turn the dial until the pointers read the same.

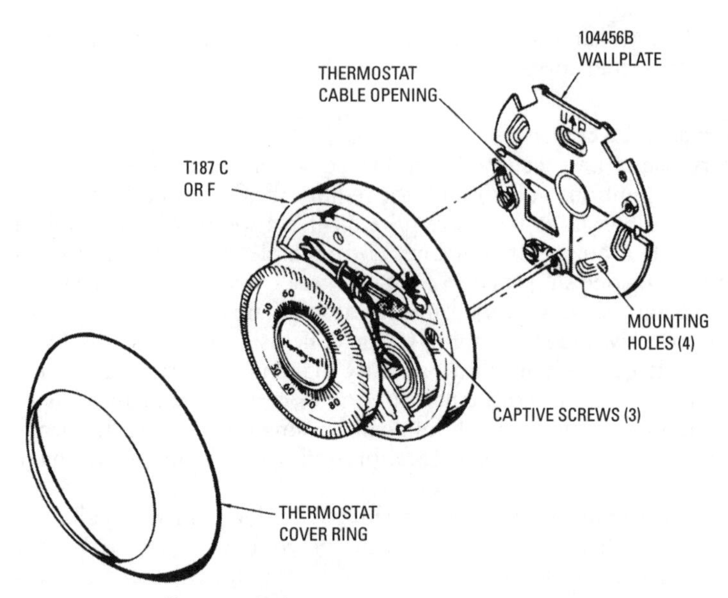

Figure 4-27 Removing the cover ring. *(Courtesy Honeywell Tradeline Controls)*

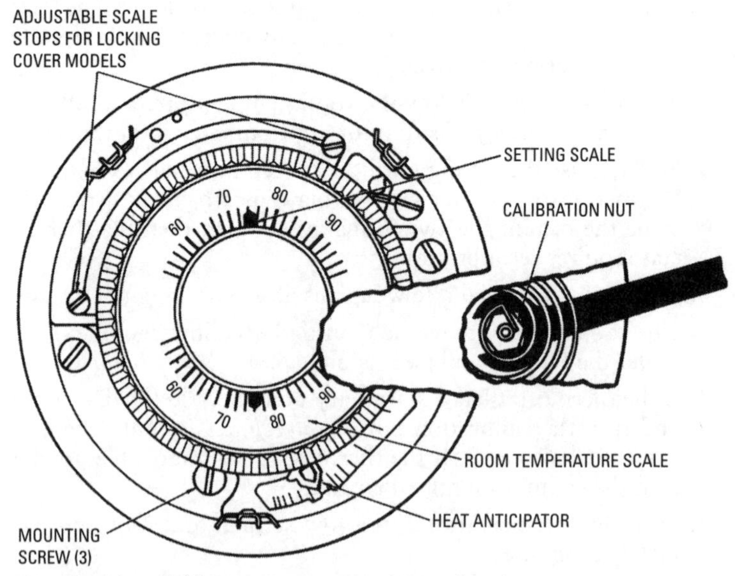

Figure 4-28 Calibration nut under bimetal coil.
(Courtesy Honeywell Tradeline Controls)

9. Hold the dial firmly, and turn the calibration in the opposite direction from the one in step 6 until the mercury switch slips to the heating contact end of the tube.
10. Recheck the calibration, select the desired temperature, and replace the cover.

Programmable Thermostats

Sometimes a room thermostat is combined with a time switch. A time switch is an electrical switching device operated by a clock to provide one or more *on* periods for the space heating or domestic (hot water) heating system. These are called *programmable room thermostats*. There are three basic types of programmable room thermostats: (a) the standard programmer, (b) the full programmer, and (c) the mini-programmer.

The *standard programmer* type controls both the space heating and domestic hot-water heating with the same time settings. A *full programmer* thermostat, on the other hand, provides independent time settings for space heating and domestic hot-water heating. This allows the two to operate independently of one another. Finally, the *mini-programmer* permits the domestic hot-water heating to be on alone (without space heating) or to be on together with the space heating. It does not allow the space heating to be on alone.

Some room thermostats have a night setback feature, which reduces energy use by lowering temperatures at night when the occupants are sleeping. These are called *day-night* (or *twin-type*) *thermostats*. They comprise an assembly of two thermostats mounted on a single base operating in conjunction with a timer or clock. The electric clock can be set to throw the temperature controls from one thermostat for the daytime onto the other for the night (or vice versa) at a predetermined time setting on the clock. This conveniently permits a low temperature at night and normal temperature during the day. Figure 4-29 shows a wiring diagram of a typical twin-type thermostat and illustrates the connections between the clock and primary control.

Day-night thermostats designed to provide automatic temperature-switching control for only a heating system can be modified to provide system and fan switching. The Honeywell T882 Chronotherm clock thermostat provides these functions for a heating and cooling installation when used with the Honeywell Q611A thermostat subbase (see Figures 4-30 and 4-31).

Insertion Thermostats

An *insertion thermostat* is used primarily to measure the air temperature in a duct. The thermostat is mounted on the outside of the

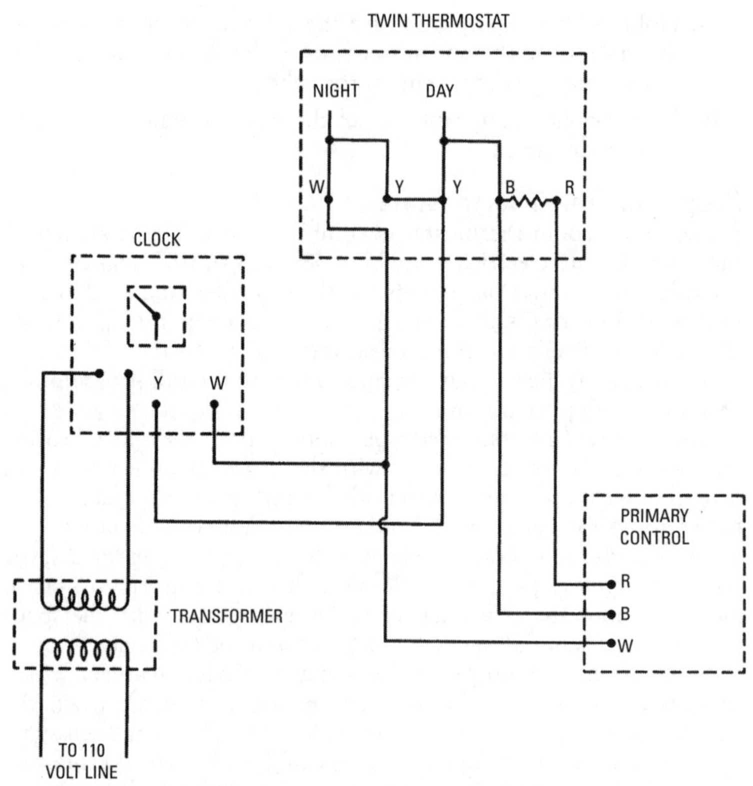

Figure 4-29 Twin-type thermostat wiring diagram.

air duct with the sensing element extending inside. Because of this application, it is sometimes referred to as a *duct thermostat*. Remote-bulb-type thermostats are also used to measure the air temperature in ducts.

Immersion Thermostats

An *immersion thermostat* is commonly used for water temperature control in an automatic gas-fired water heater (see Figure 4-32). This is usually a direct, snap-action, bimetallic thermostat in which contraction of the thermal element immersed in the stored hot water causes the main gas valve to open. This occurs when there is a drop in the temperature of the water in the storage tank. Expansion of this element serves to close the main gas valve when the tank water attains the selected predetermined temperature.

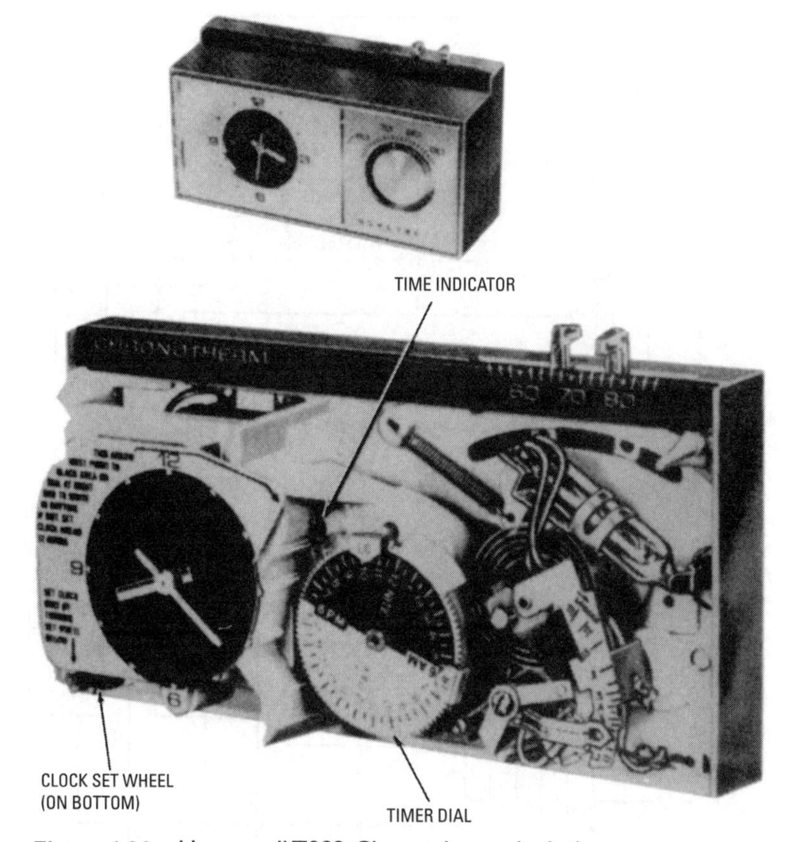

Figure 4-30 Honeywell T882 Chronotherm clock thermostat.
(Courtesy Honeywell Tradeline Controls)

These thermostats normally operate at a temperature differential of approximately 12°F. In other words, if the thermostat is set to shut off the gas to the main burner when the tank water temperature reaches 140°F, it will react to open the valve when the temperature drops to 128°F.

Quite often, an immersion water heater thermostat will be included in a combination control. These combination controls used in water heaters are described in Chapter 4 of Volume 3 ("Water Heaters and Other Appliances").

Cylinder Thermostats

A *cylinder thermostat* is used to control the temperature of the domestic (potable) hot-water tank and to turn on and off the

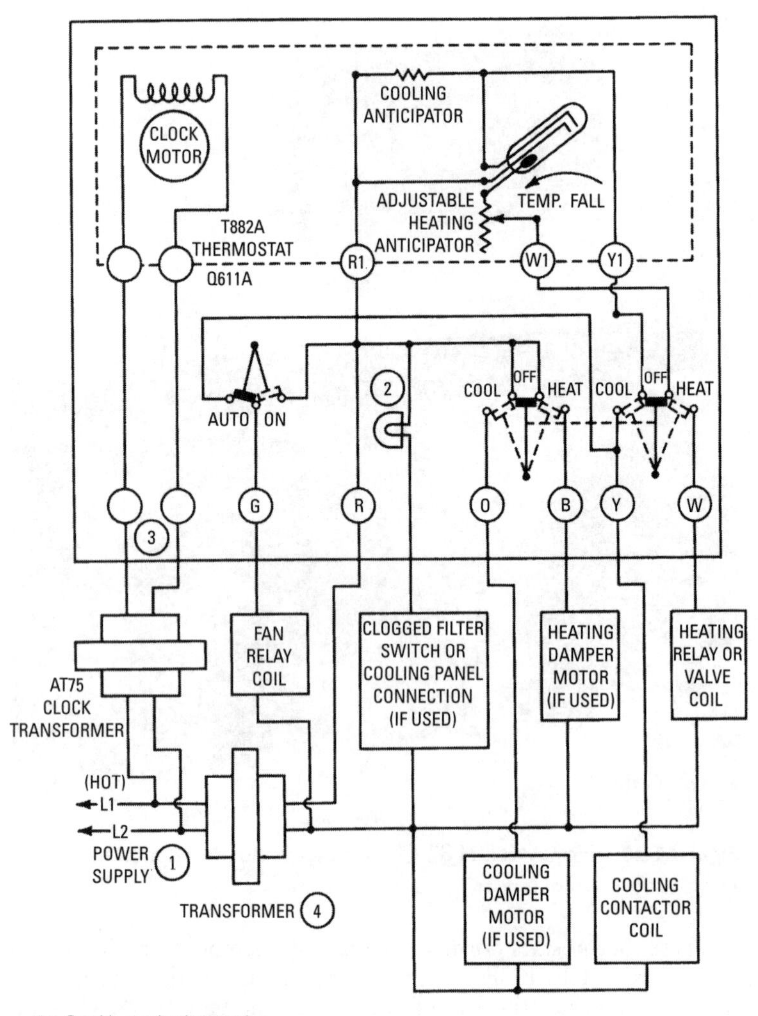

Figure 4-31 Wiring diagram of the T882 thermostat and Q611A subbase. *(Courtesy Honeywell Tradeline Controls)*

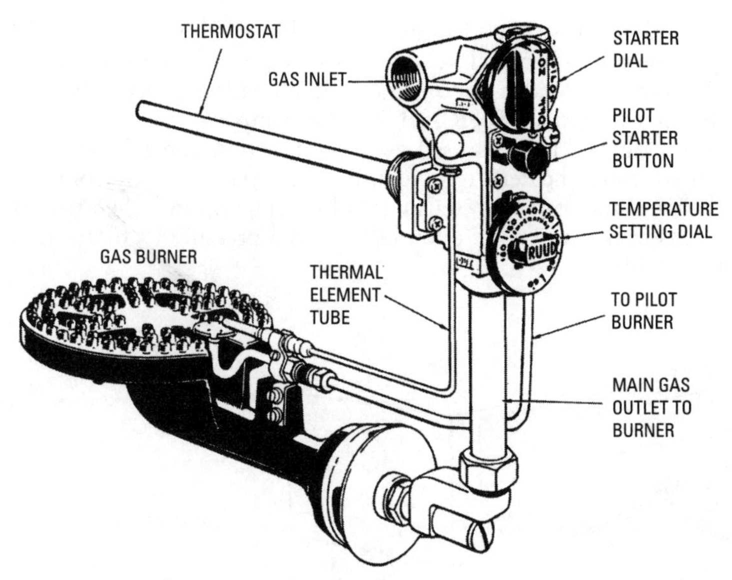

Figure 4-32 Immersion thermostat.

water heater. A single target temperature may be set on this thermostat.

Boiler Thermostats

A *boiler thermostat* is a safety control device installed inside a hydronic boiler. Its function is to limit the temperature of the hot water produced by the boiler. When the water temperature reaches a preset maximum limit, the thermostat switches off the burner and boiler. The target temperature can be permanently fixed or adjusted by the user.

Remote-Bulb Thermostats

A *remote-bulb thermostat* is distinguished from other thermostats by having a sensing element (usually enclosed in a bulb-type device) located at some distance from the thermostat controller body.

In residential heating and/or cooling applications, remote-bulb thermostats are used for space temperature control of room and vented-recess heaters, room air-conditioning units, or radiator valves. Outdoor thermostats also operate on the remote-bulb principle.

Sometimes these thermostats are combined with other controls to provide a number of different control functions in one package. Two examples of this thermostatic combination control are the Robertshaw Unitrol 110SR and 7000SR controls.

The Unitrol 110SR is used as a combination control for small gas-fired room heaters (see Figure 4-33). The control contains a thermostatic valve, a sensing bulb, temperature (thermostat) dial, gas cock, automatic pilot valve, and a pressure regulator (see Figure 4-34).

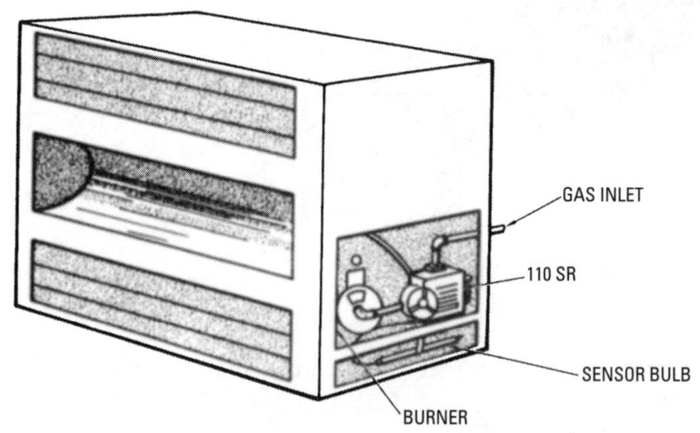

Figure 4-33 Unitrol 110SR combination control on a gas-fired room heater. *(Courtesy Robertshaw Controls Co.)*

The sensing bulb is located in the return air stream at the bottom of the heater. The temperature of the return air is sensed and the valve is actuated to open and close by the hydraulic system in the control. The gas cock and automatic pilot mechanism provide safe lighting of the heater. If the pilot light should go out, the main gas and pilot gas supplies are shut off by the automatic pilot valve. Main burner gas pressure regulation is provided by the pressure regulator incorporated in the Unitrol 110SR control.

Figure 4-35 illustrates the principal components of a Unitrol 7000SR control. This is a diaphragm valve that operates through the center of a thermostatic bleed valve in an internal bleed line. The Unitrol 7000SR combines into a single package a diaphragm valve, thermostat valve, temperature (thermostat) dial, gas cock, automatic pilot valve, and a pressure regulator.

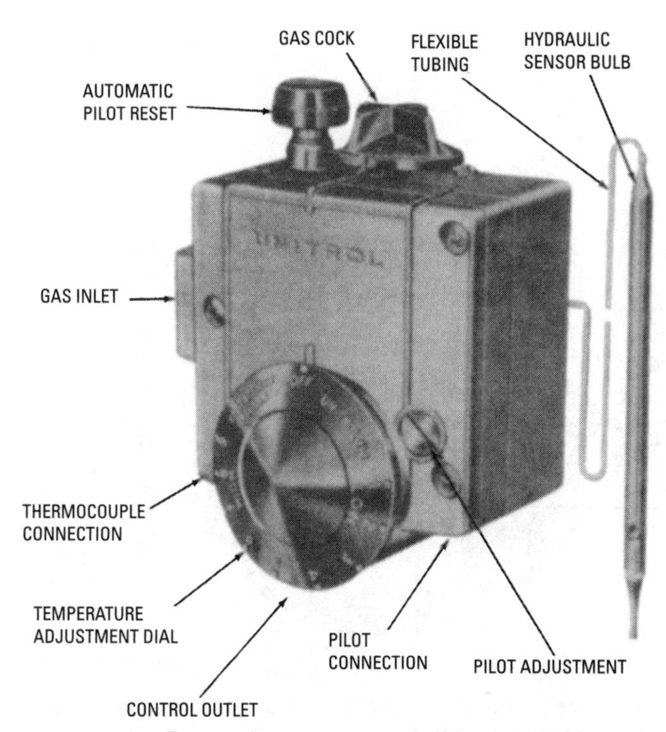

Figure 4-34 Principal components of a Unitrol 110SR combination gas valve. *(Courtesy Robertshaw Controls Co.)*

Application of a Unitrol 7000SR on a gas-fired vented-recess heater is illustrated in Figure 4-36. The sensing bulb is placed in the return air opening at the bottom of the furnace.

Both the Unitrol 110SR and Unitrol 7000SR thermostatic space heater controls use a closed hydraulic sensing and actuating device consisting of a bulb, capillary tube, and a bellows or diastat.

The cross section of a typical hydraulic sensor and actuator is illustrated in Figure 4-37. The bulb, capillary tube, and actuator are filled with a liquid that has a high coefficient of expansion. When the bulb senses a rise in temperature, it causes the volume of the liquid to expand. This expansion is transferred through the capillary tube to expand the hydraulic bellows. The bellows is spring-loaded to operate a snap mechanism to a valve *open* condition (see Figure 4-38). As the temperature rises, the expansion of the liquid opposes the spring-loading to secure a valve *closed* (i.e., off) condition when the required temperature is reached.

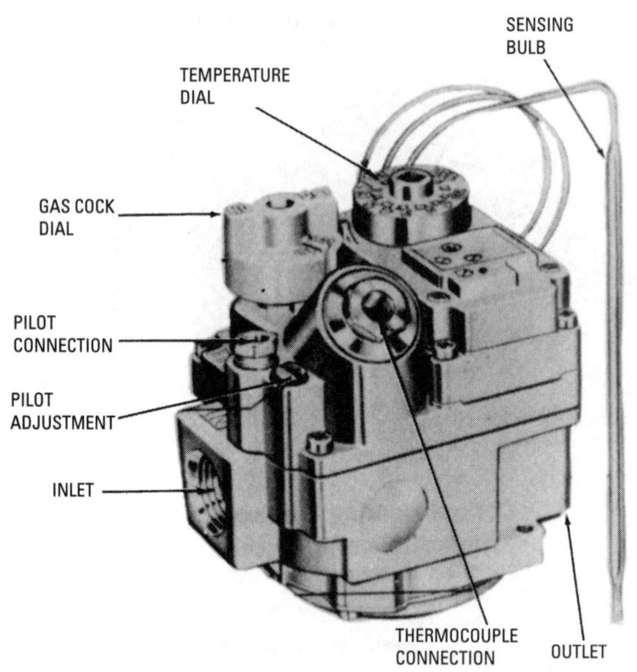

Figure 4-35 Principal components of a Unitrol 7000SR combination gas valve. *(Courtesy Robertshaw Controls Co.)*

Proportional Thermostats

Some thermostats are designed to provide proportional control for valve and damper motors in heating or cooling systems. This type of controller is generally referred to as a *proportional thermostat.*

The Honeywell T92 proportional thermostat, shown in Figure 4-39, contains a bellows that adjusts one or two potentiometers in proportion to temperature changes. These potentiometer adjustments regulate the power supplied to the controlled device. This particular thermostat is designed to provide 24- to 30-volt proportional control for valve and damper motors in the system. Those models equipped with two potentiometers are capable of unison or sequence control.

Outdoor Thermostats

An *outdoor thermostat* is designed to maintain the proper balance between the temperature of the heating medium inside the structure and the outdoor temperature.

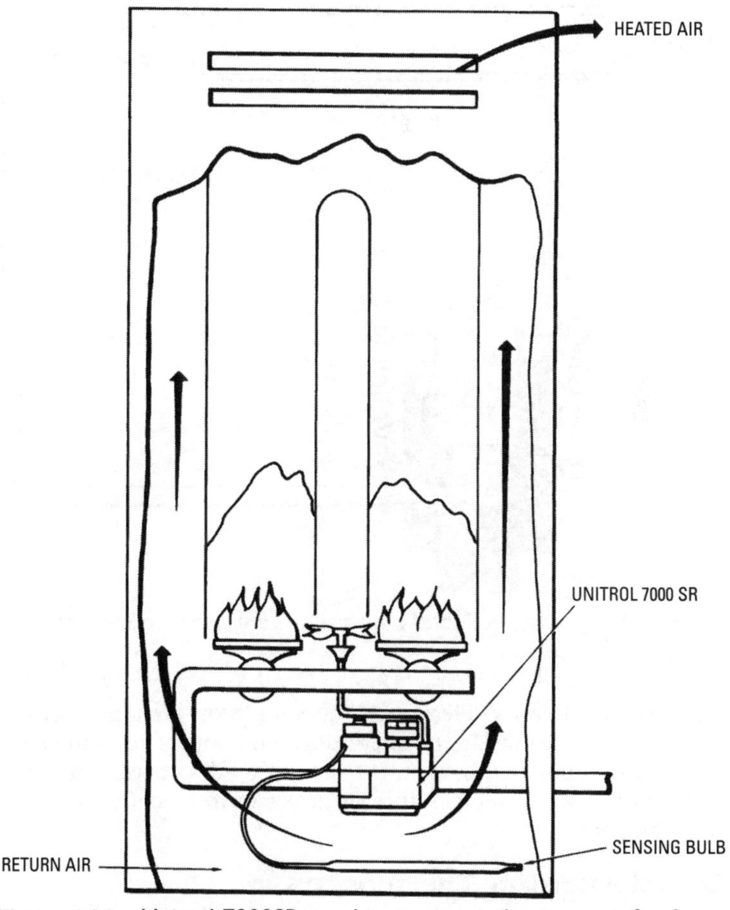

Figure 4-36 Unitrol 7000SR combination gas valve on a gas-fired vented-recess heater.

The outdoor thermostat, illustrated in Figure 4-40, can be used as an operating controller for a hot-water or warm-air heating system. This is a remote-bulb thermostat suitable for line voltage, low-voltage, or millivolt switching. It is designed to automatically raise the heating medium control point as the outdoor temperature falls.

Outdoor thermostats are frequently used as controllers in hot-water heating systems. A simple on-off control is possible, but this usually involves stopping the circulation of the water in the system during those periods when there is no call for heat. Anticipating

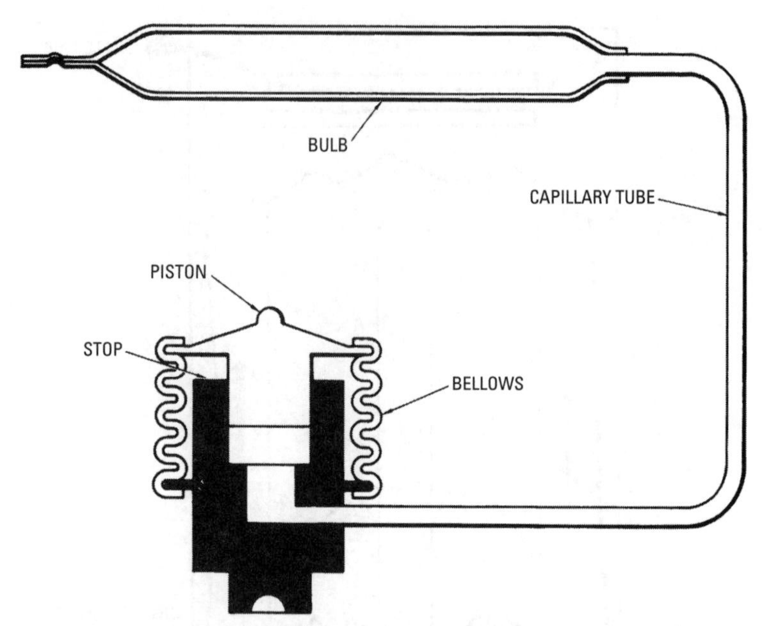

Figure 4-37 Cross section of a typical hydraulic sensor and activator. *(Courtesy Robertshaw Controls Co.)*

control systems are preferred to the simple on-off types because there is no noticeable lag time between cold temperatures and a call for heat. An anticipating control system provides continuous circulation of the water temperature in direct ratio to changes in outdoor temperature.

Troubleshooting Thermostats

Thermostat problems are sometimes the result of poor wiring connections. Check the wiring first. You should also make certain that the fan and system switches and the temperature setpoint are properly set.

The troubleshooting chart in Table 4-3 includes many of the more common symptoms and possible causes of operating problems associated with thermostats.

Humidistats

A *humidistat* is a switching device used to control the level of humidity in a confined space. Standard applications include the basic on-off humidity control of a heating/cooling system or the high-limit safety interlock of a humidifier.

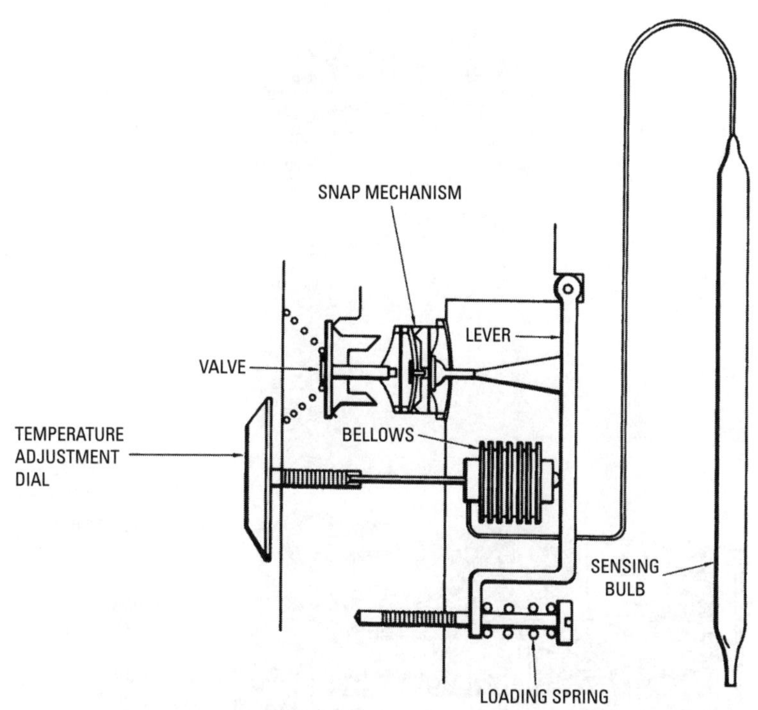

Figure 4-38 Schematic of a hydraulic sensor and actuating system used for snap-action control. *(Courtesy Robertshaw Controls Co.)*

The earliest humidistats used a hygroscopic material in the construction of the sensing or control element (see Figure 4-41). Hygroscopic materials, such as animal hair or certain types of plastics, are those affected by the moisture content of the air. When the material absorbs moisture from the air, it expands. When the surrounding air becomes drier, the hygroscopic material gives up moisture to the air and contracts. This expansion and contraction of the control element in the humidistat opens or closes the electrical circuit controlling the humidifier. These early humidistats, many of which are still in use today, are commonly called *electric humidistats*, *mechanical humidistats*, *electromechanical humidistats*, *humidity controllers*, or *hygrostats*—a lot of different names, but all operating on the same principle of using a hygroscopic material to mechanically switch an electrical current on and off.

Electronic humidistats are being used in most new heating/cooling systems today instead of the older electromechanical types.

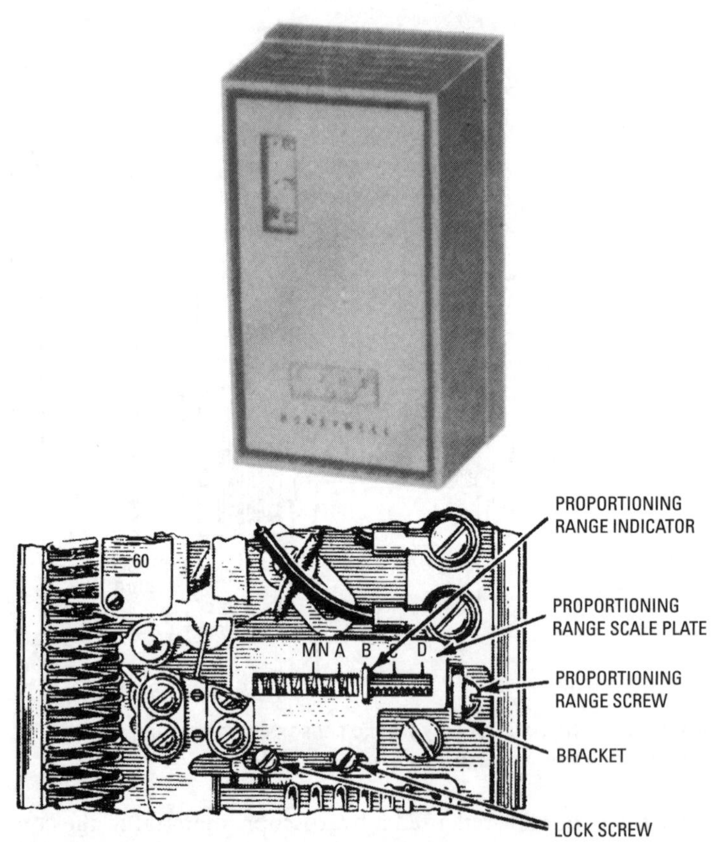

Figure 4-39 Honeywell T92 proportional thermostat.
(Courtesy Honeywell Tradeline Controls)

They are also being used to replace the older units in existing systems. An electronic humidistat uses electronic switching circuitry to create the switching action. In some electronic humidistats, a thin film capacitance is used to sense the moisture content of the surrounding air. Others use various polymer-resistance analog humidity-sensing technologies.

Pneumatic humidistats are also designed to sense changes in ambient relative humidity (see Figures 4-42 and 4-43). As shown in Figure 4-43, the sensing device is a nylon element near the set-point adjustment wheel at the bottom of the humidistat. When the nylon sensing element responds to changes in the ambient relative

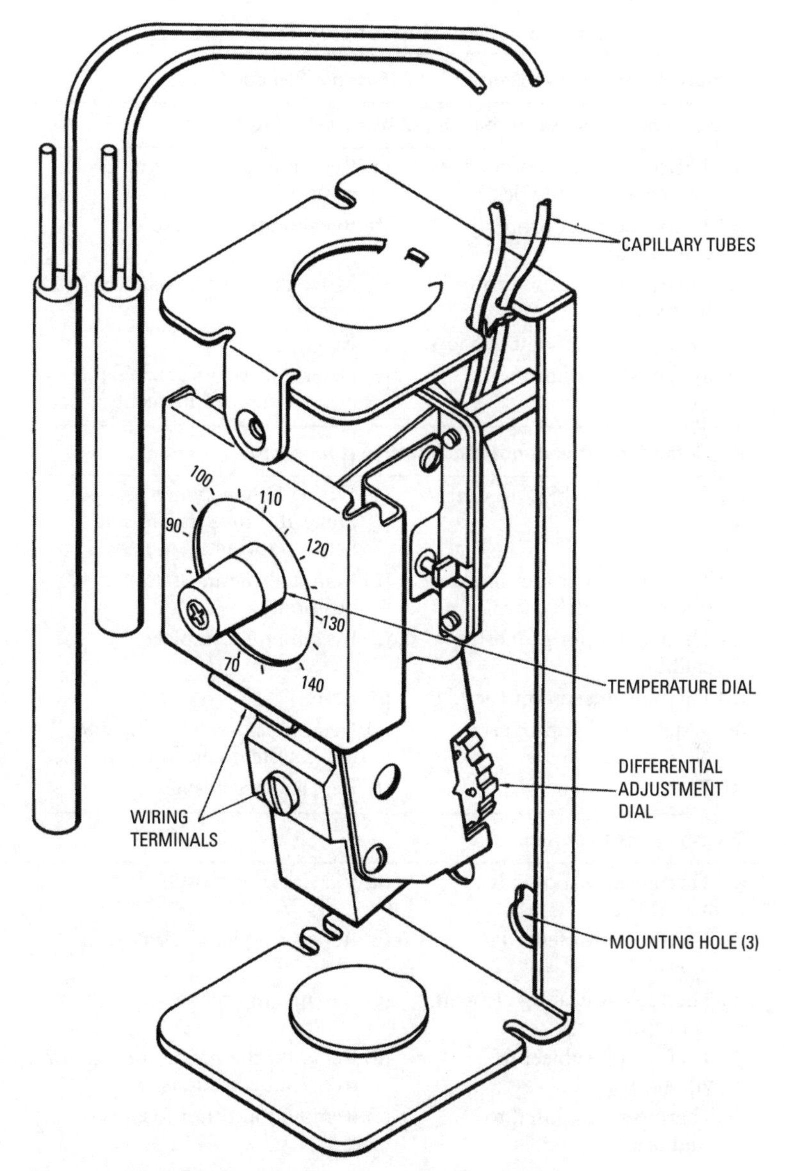

Figure 4-40 Outdoor reset control for hot-water or warm-air heating systems. *(Courtesy Honeywell Tradeline Controls)*

Table 4-3 Troubleshooting Thermostats

Symptom and Possible Cause	*Possible Remedy*
Room temperature overshoots thermostat setting (too cold).	
(a) Thermostat not mounted level (mercury-switch types).	(a) Remount thermostat in level position.
(b) Thermostat not properly calibrated.	(b) Recalibrate or replace.
(c) Thermostat exposed to heat source.	(c) Move thermostat to better location.
(d) Thermostat setpoint too low.	(d) Reset.
(e) System sized improperly.	(e) Determine correct sizing and make system adjustments.
Room thermostat does not reach setting (too warm).	
(a) Thermostat subject to draft.	(a) Wiring hole may not be plugged. Move thermostat to better position.
(b) Thermostat not mounted level (mercury-switch types).	(b) Remount thermostat in level position.
(c) Thermostat not properly calibrated.	(c) Recalibrate or replace.
(d) Thermostat setpoint too high.	(d) Reset.
(e) System sized improperly.	(e) Determine correct sizing and make system adjustments.
(f) Thermostat damaged.	(f) Replace thermostat.
System cycles too often.	
(a) Thermostat exposed to heat source.	(a) Relocate thermostat.
(b) Thermostat differential too small.	(b) Reset or replace thermostat.
(c) Thermostat heating element improperly set.	(c) Reset or replace thermostat.
(d) Thermostat subject to vibrations.	(d) Remount thermostat in location free from vibrations.
(e) Thermostat exposed to cold draft.	(e) Remount in better location.

(continued)

Table 4-3 (continued)

System does not cycle often enough (burner operates too long).	
(a) Thermostat not exposed to circulating air.	(a) Remount in better location.
(b) Contacts dirty.	(b) Clean or replace.
(c) System sized improperly.	(c) Determine correct sizing and make system adjustments.
(d) Thermostat differential too great.	(d) Reset or replace thermostat.
(e) Thermostat heating element improperly set.	(e) Reset or replace thermostat.
(f) Thermostat set too high.	(f) Reset.
Room temperature swings excessively.	
(a) Thermostat not exposed to circulating air.	(a) Remount in better position.
(b) Thermostat exposed to heat source.	(b) Remount in better position.
(c) System sized improperly.	(c) Determine correct sizing and make system adjustments.
Thermostat jumpered (system works).	
(a) Thermostat contacts dirty.	(a) Clean or replace.
(b) Thermostat setpoint too high.	(b) Reset or replace thermostat.
(c) Thermostat damaged.	(c) Replace thermostat.
(d) Break in thermostat circuit.	(d) Locate and correct.
Burner fails to stop.	
(a) Thermostat in cold location.	(a) Relocate in better location.
(b) Thermostat set too high.	(b) Reset.
(c) Defective thermostat.	(c) Replace thermostat.
(d) Thermostat out of adjustment.	(d) Recalibrate or replace thermostat.
(e) Thermostat contacts stuck.	(e) Correct.

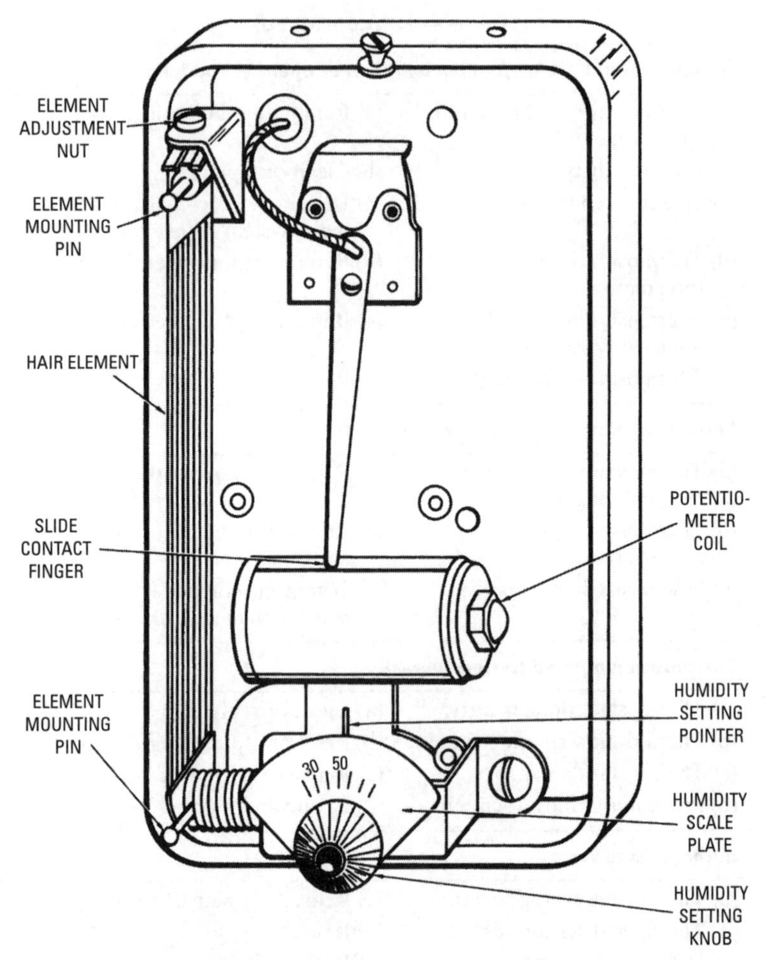

Figure 4-41 Components of a typical humidistat.
(Courtesy Honeywell Tradeline Controls)

humidity, it produces a proportional change in the branch line pressure and changes the control actuator position by the same proportion.

Location of Room Humidistats

The proper location of the room humidistat is important to its efficient operation. The following recommendations are offered as a guide to locating the humidistat:

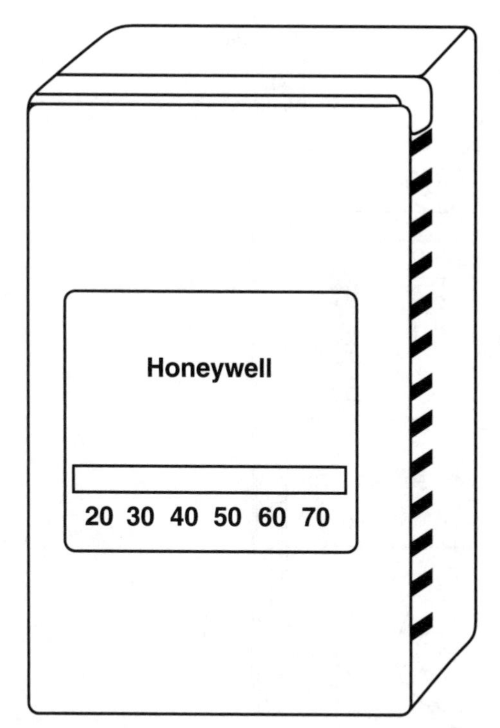

Figure 4-42 Honeywell pneumatic humidistat.
(Courtesy Honeywell, Inc.)

- *Never locate the humidistat in an area where there are heavy concentrations of moisture.* A kitchen, bathroom, or laundry room will frequently have high levels of moisture in the air.
- *Never install the humidistat on the inside surface of an outside wall.* Exterior walls are subject to temperature extremes caused by weather changes.
- *Never locate the humidistat where the air circulation is restricted.*
- *Never locate the humidistat where it can be affected by a nearby heat source.* Heat sources such as sunlight, lamps, television sets, fireplaces, warm-air outlets, or heat-producing appliances will interfere with the efficient operation of the humidistat.

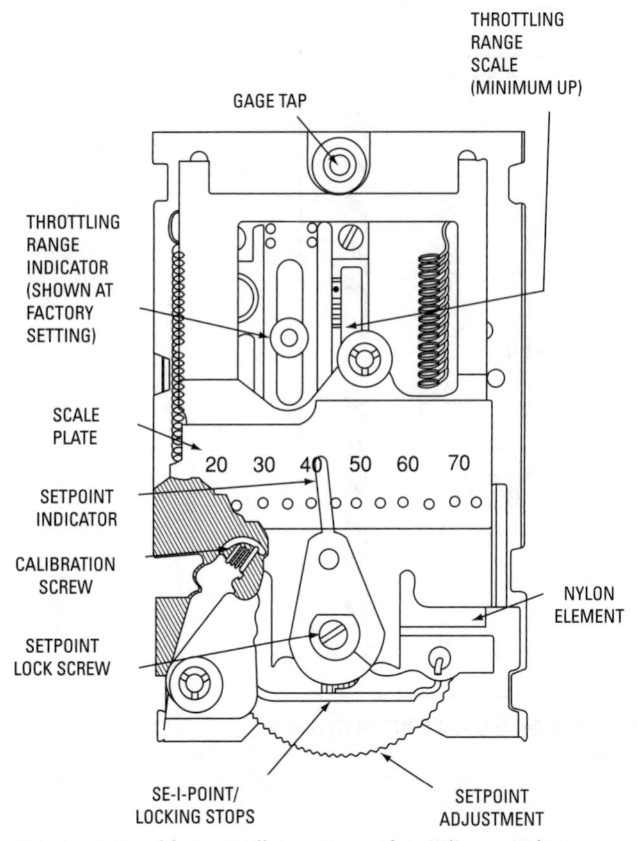

Figure 4-43 Honeywell pneumatic humidistat with cover removed.
(Courtesy Honeywell, Inc.)

Troubleshooting Humidistats

Read the humidistat manufacturer's installation and operating manual for troubleshooting problems, their possible causes, and their suggested remedies. In most cases, they will be specific to the type of humidistat (electromechanical, electronic, or pneumatic) and the model. Some very basic troubleshooting problems that apply uniformly to all humidistats are listed in Table 4-4.

Table 4-4 Troubleshooting Humidistats

Problem and Possible Cause	*Suggested Remedy*
Slow response.	
(a) Humidifier installed in a dead air space.	(a) Relocate to appropriate location.
(b) Inadequate airflow caused by an incorrect cover.	(b) Install a correct cover.
Inaccurate reading.	
(a) Backplate too tight.	(a) Retighten.
(b) Humidistat installed on inside surface of an outside wall.	(b) Move to appropriate location.
(c) Humidistat installed near heat source.	(c) Move nonpermanent heat source (lamp, TV set, etc.) away from humidistat or move humidistat away from permanent heat source (constant sunlight, fireplace, stove, etc.).
Constant readings.	
(a) Defective humidistat.	(a) Replace humidistat.
(b) Incorrectly calibrated humidistat.	(b) Recalibrate humidistat.
(c) Humidistat is undersized.	(c) Replace the humidistat with a correctly sized one.

Chapter 5

Gas and Oil Controls

A variety of different types of controls are used in gas and oil heating systems to ensure the safe, efficient, and automatic operation of the furnace, boiler, or water heater. These controls function together as a control circuit within the heating system.

The most important functions of the control circuit are (1) to start or stop the gas or oil burner in response to a signal from the centrally located room thermostat and (2) to shut down the burner if an unsafe operating condition occurs.

A complete description of all the controls used to govern the operation of a furnace, boiler, or water heater would be too extensive to include in a single chapter. For that reason, thermostats, limit controls, and related safety and control devices are covered in other chapters. This chapter is primarily concerned with a description of those controls that directly govern the flow of gas or oil to the burner. These are primarily the safety, pressure-regulating, and flame-sensing valves and devices. It also includes descriptions of the operating (ignition) systems.

Caution

> Work on gas-fired and oil-fired equipment should be performed only by qualified personnel trained in the proper application, installation, and maintenance of HVAC systems.

Gas Controls

In addition to the room thermostat and the fan and limit control, the basic components of the control system of a gas-fired furnace, boiler, or water heater will generally consist of the following controls:

- Main gas valve
- Pressure regulators
- Pilot gas cock
- Automatic gas control valve
- Automatic pilot valve
- Pilot assembly

The pilot assembly includes the pilot burner and the thermocouple or thermopile (pilot generator) in standing-pilot systems. In more modern systems, the pilot assembly consists of the pilot burner, a spark ignition module, and a flame sensor.

A control system may also include a gas primary control, transformer, or safety pilot relay. Some controls may be eliminated, depending on the design and requirements of the system. For example, a safety pilot relay is not necessary if a nonelectric pilot safety valve is used in the gas line.

Note

The valves and other devices in a control system will vary depending on the type of ignition system.

Gas Control Circuits

The gas control circuits used to operate modern gas-fired heating equipment can be divided into the following three basic types:

1. Low-voltage control circuits.
2. Line voltage control circuits.
3. Millivolt control circuit.

A low-voltage temperature control circuit (see Figure 5-1) uses a step-down transformer to reduce the higher line voltage to approximately 24 to 30 volts. A 24-volt thermostat is used as the controller in most installations.

The line voltage temperature control circuit shown in Figure 5-2 is a 120-volt system. Because the voltage is not reduced, a line voltage thermostat or controller and a line voltage operator must be used in the system.

A millivolt control circuit (see Figure 5-3) operates on the thermocouple principle. A single thermocouple automatically generates approximately 30 millivolts without the aid of an outside source of electricity. A number of thermocouples used together can generate up to 750 millivolts. This combination is variously referred to as a *generator*, *pilot generator*, *thermopile generator*, *thermopile system*, or *powerpile system*.

Each of the three temperature control circuits described in the preceding paragraphs is also wired into a pilot safety shutoff circuit, generally via a switch-type pilot safety shutoff device. An inline pilot safety shutoff device is also located in each safety shutoff circuit, and these provide *complete* gas shutoff.

Gas Burner Primary Control

The primary control shown in Figure 5-4 is a solid-state electronic relay used on gas, oil, or combination gas-oil burners. It is designed

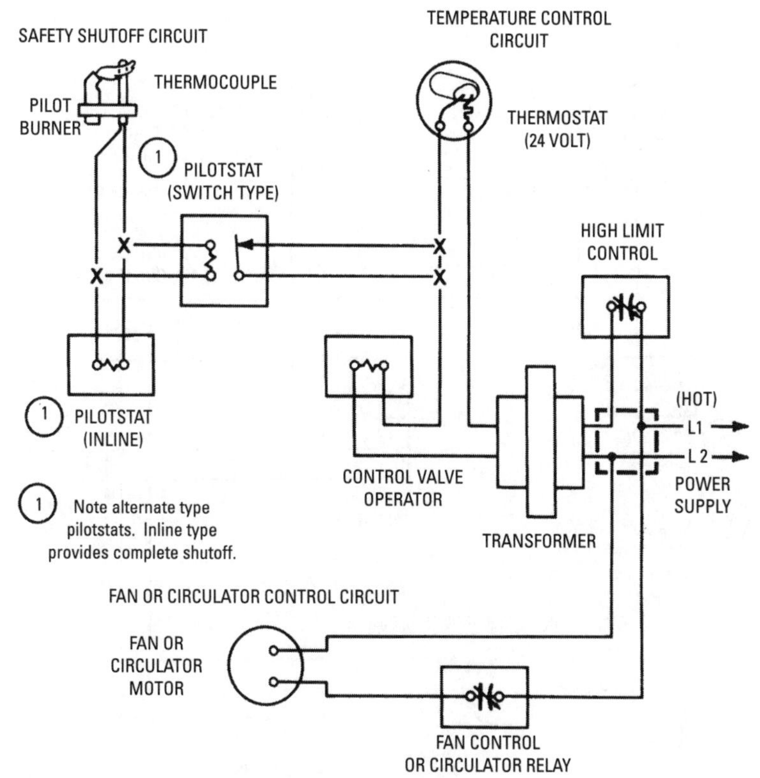

Figure 5-1 Low-voltage control circuit. *(Courtesy Honeywell Tradeline Controls)*

to provide operational control of the burner in response to the room thermostat and limit controls, and to instantly shut off the burner in the event of flame failure. Figure 5-5 illustrates a typical wiring diagram for a Honeywell RA890F Protectorelay primary control used in a control circuit for a gas-fired boiler.

This primary control is used with rectification-type flame detectors to sense the presence or absence of flame. The heart of the flame detector circuit in a gas-fired system is an electrode inserted in the pilot flame. In the wiring diagram, shown in Figure 5-6, the flame rod is connected to terminal *F* of the primary control.

In the event of pilot flame failure, the flame detector circuit responds to control the gas valve in the manifold. When there is an absence of flame, the primary control shuts off the supply of gas by closing the gas valve or by keeping it closed if it is not already open.

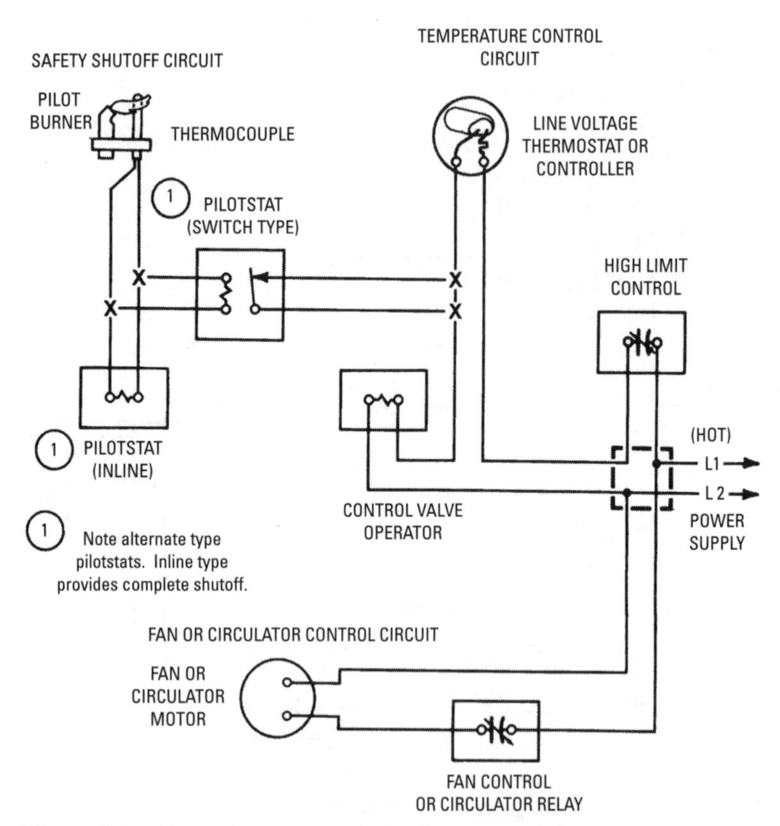

Figure 5-2 Line voltage control circuit. *(Courtesy Honeywell Tradeline Controls)*

These units are designed to be fail-safe. Abnormal conditions in the flame detector circuit, such as an open circuit, short circuit, or current leakage to ground, simulate absence of flame and cause the system to shut down. Safety controls such as temperature or pressure limits or low-water cutoffs are connected *ahead* of the switching terminals of the relay so that shutdown of the burner occurs even in the event of a relay malfunction such as fused contacts.

The main valve circuit is deenergized $8/10$ of a second after flame failure occurs. On starting up or after flame failure, a trial-for-ignition period of approximately 45 seconds maximum occurs. During this ignition period, only pilot gas is allowed to flow to the burner. If the flame circuit is not completed within this time period, safety lockout of the relay occurs, causing a total shutdown of the system. Manual reset is then required to restart.

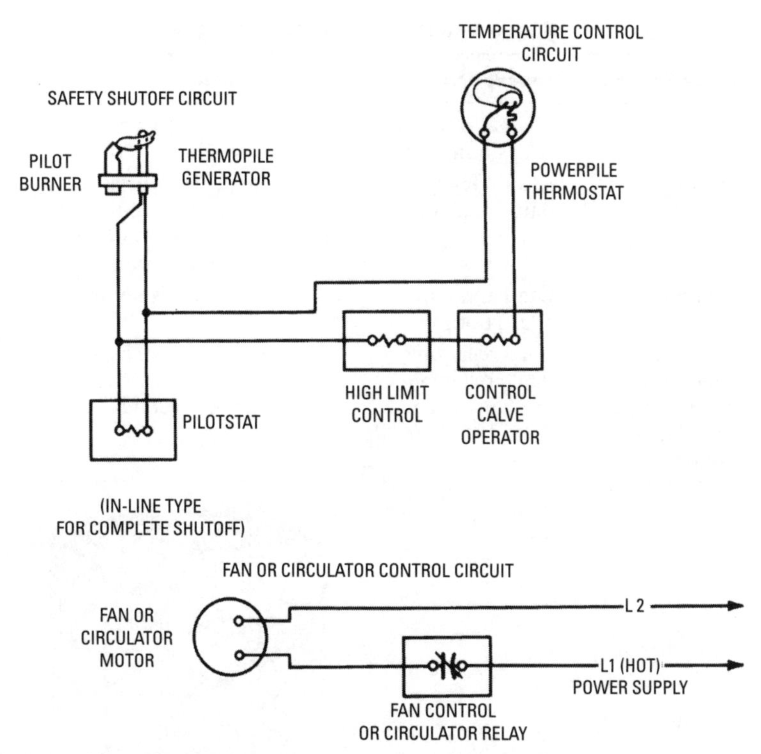

Figure 5-3 **Millivolt control circuit.** *(Courtesy Honeywell Tradeline Controls)*

Two sets of relays are contained in a Honeywell RA890F Protectorelay primary control. The load relay (left hand) supplies current to the No. 3 terminal to control the blower motor. The flame relay (right hand) responds to the load relay but only if allowed by the flame-detecting electronic network of the relay. The flame relay can supply current to the No. 5 terminal controlling the gas valve only if the load relay has also pulled in.

The load relay is responsive to the thermostat or other operating control connected to the T/T terminal provided that safety controls, located in the 1–6 circuit, indicate that safe conditions exist for main burner operation.

On an interruption of power to the No. 1 and No. 2 terminals of the primary control, the relay returns to the standby position. When power is restored, normal operation is resumed except that the starting cycle is maintained longer than usual while the vacuum tube is warming up. Relay positions and their effect on the burner are listed in Table 5-1.

Table 5-1 Relay Position and Effect on Gas Burner

Relay Position	*Description*	*Effect on Burner*
Standby	Load and flame relays both out	Motor and gas valve both deenergized
Starting	Load relay in; flame relay out	Motor energized; gas valve deenergized; trial-for-ignition period; safety lockout occurs after 45 seconds
Running	Load and flame relays both in	Motor and gas valve both energized
Abnormal conditions due to flame simulating failure	Load relay out; flame relay in	Motor and gas valve both deenergized; safety lockout occurs after 45 seconds

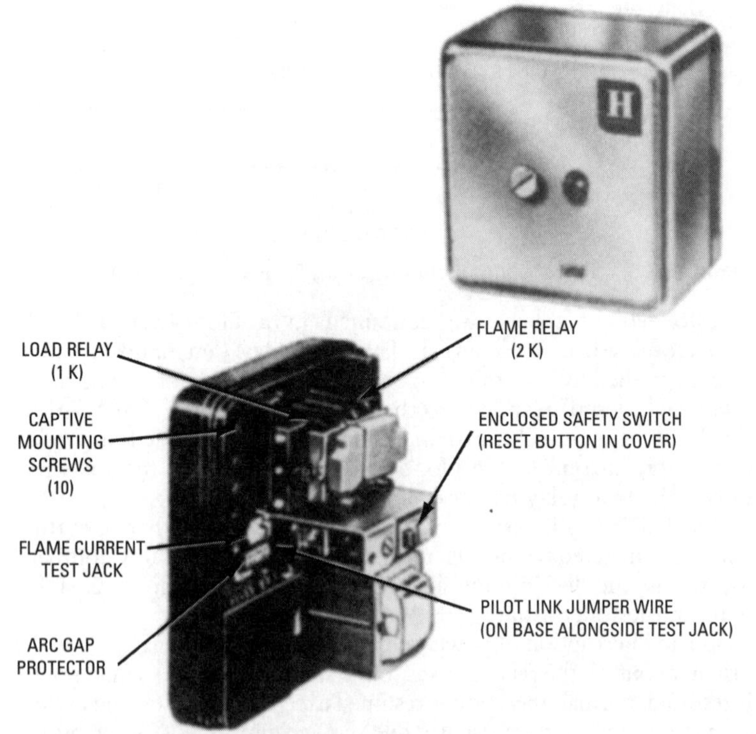

Figure 5-4 Honeywell RA890F Protectorelay primary control.
Courtesy Honeywell Tradeline Controls)

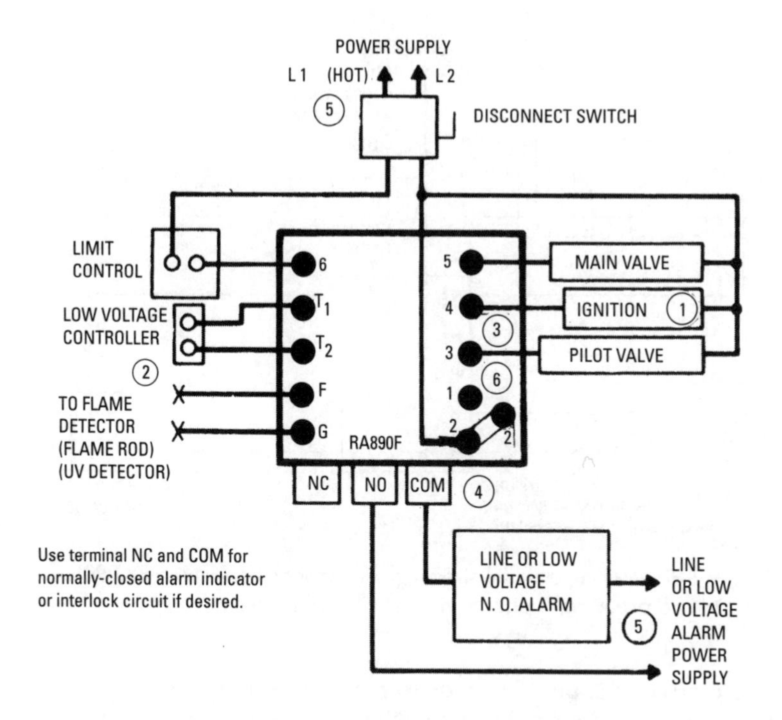

Figure 5-5 Wiring diagram of the RA890F control.

(Courtesy Honeywell Tradeline Controls)

Servicing a Gas Burner Primary Control

Access to the wiring terminals of the primary control illustrated in Figure 5-4 is obtained by loosening the screws that secure the chassis to the base. When remounting the chassis, be sure to tighten all mounting screws because they also serve as electrical connections.

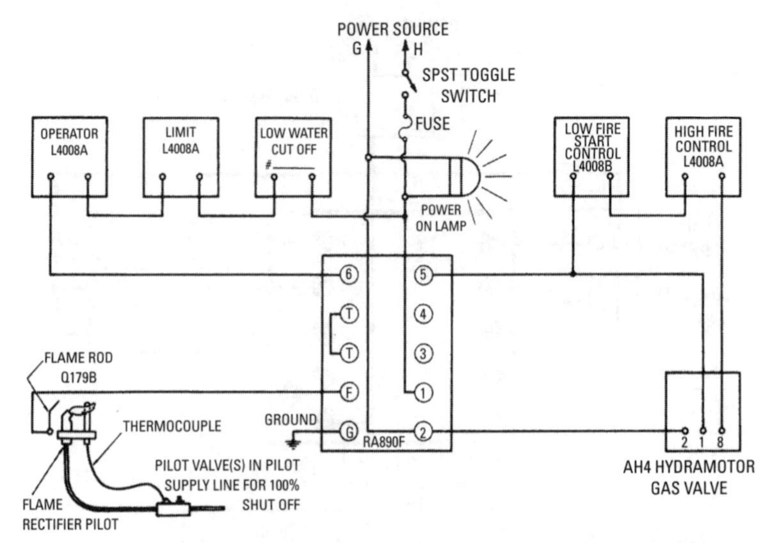

Figure 5-6 Wiring diagram showing connections between a RA890F control and other components in a gas control circuit. *(Courtesy Bryan Steam Corp.)*

No attempt should be made to repair a primary control except for tube replacement. Vacuum tubes are used in Honeywell primary controls. *Never* replace them with radio tubes. If a primary control is defective, the entire chassis should be replaced with a good one.

Operating controls located in the T/T circuit (see Figure 5-6) should be of the low-voltage, two-wire type. A low-voltage transformer for this purpose is built into a Honeywell Protectorelay. Safety controls located in 1–6 terminals must be two-wire, line voltage type. With the exception of the line switch, no controls should ever be placed in the line ahead of the 1–2 terminals of the primary control.

Before assuming that the primary control is defective, be sure to check the pilot, pilot adjustment, flame detector circuit, and all operating and safety controls; proper operation is also dependent on these external factors. The flame circuit can be more accurately checked by the use of a microammeter to read flame current. Normal operation requires a current of 2 microamperes or more.

Never push relays in manually because it can result in accidental opening of the main diaphragm valve. Be sure to turn off the electrical power before removing the primary control chassis from the base.

Gas Valves

The valves used to control the flow of gas through a gas-fired furnace, boiler, or water heater can be divided into two basic categories: (1) manually operated valves and (2) power-operated valves.

The two manually operated valves (gas cocks) used on gas-fired heating equipment provide a backup safety function in case the automatic gas valves fail to operate. One of these valves is located in either the main gas supply riser or the manifold. The other one is located on the pilot gas line.

The manual gas valve installed on the main gas supply line (riser) or manifold is variously referred to as the *main gas shutoff valve*, *manual shutoff plug cock*, or simply the *gas cock* (see Figure 5-7). This valve provides manual control of the gas flow to the main gas burners. It is *not* used to control the gas supply to the pilot burner, the latter being provided with its own separate shutoff valves.

The manual valve located on the pilot gas line is called the *pilot shutoff cock* or the *pilot gas cock*. It is usually the first controlling device on the pilot line (see Figure 5-7). It provides complete gas shutoff whenever it is necessary to remove and service other controls on the pilot line, such as the pilot gas regulator or the pilot solenoid valve.

Power-operated or automatic valves are actuated by some form of auxiliary power such as hydraulic pressure, pneumatic pressure, electricity, or a combination of these sources. The following are the principal types of power-operated valves used on gas-fired heating equipment:

- Solenoid valves
- Direct-acting heat motor valves
- Diaphragm valves

Solenoid Gas Valves

The solenoid gas valve is commonly used on gas-fired heating equipment to provide on-off control of the flow of gas.

The primary function of a solenoid valve is to provide direct valve operation. The power to operate the valve is obtained from the magnetic flux developed in a solenoid coil. A valve disc in the valve body is connected by a rod to the core of an electromagnet. When the room thermostat or power switch directs an electrical current to the solenoid, it pulls the rod (plunger) to the top of the plunger tube and lifts the attached valve disc. Gas then flows through the main valve port until the electrical circuit is interrupted by the controller. This action releases the rod, which falls and shuts off the valve. The weight of the rod and seat assembly and the gas pressure on top of the valve seat ensure a tight shutoff. The ITT

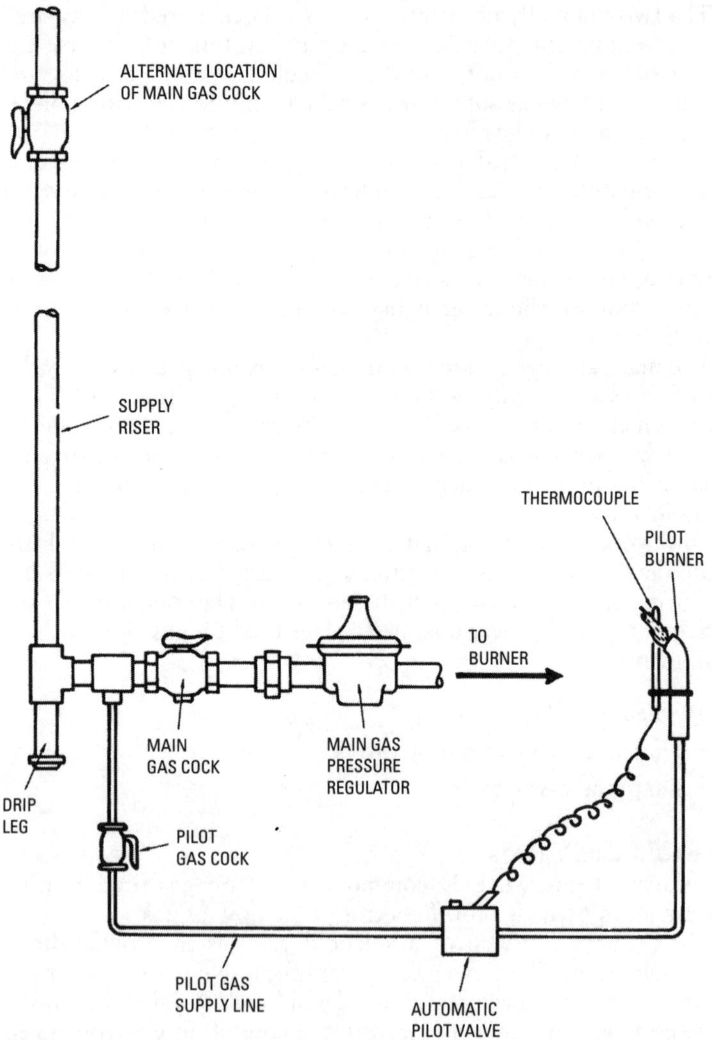

Figure 5-7 Typical arrangement of gas cocks and main gas-pressure regulator.

General Controls K3 Series gas valve shown in Figure 5-8 is an example of a direct-operated solenoid gas valve.

Some solenoid valves use a balanced diaphragm to control the flow of gas (see Figure 5-9). When the solenoid coil is energized, it lifts the rod or plunger just enough to open a bleed valve (or so-called pilot valve). Gas then bleeds from the area above the

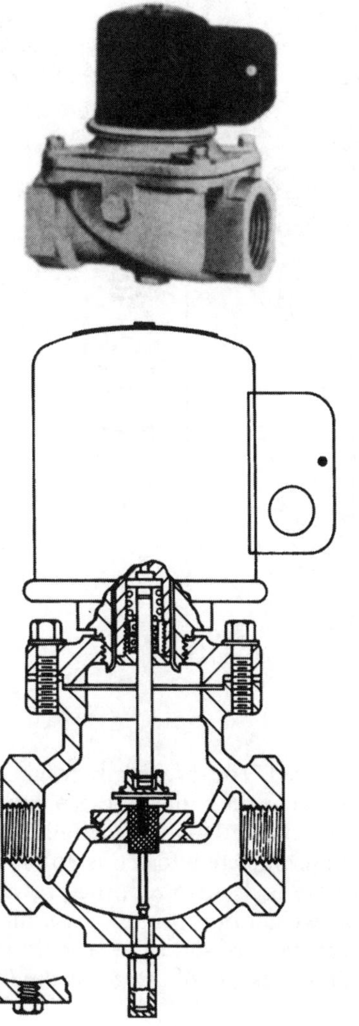

Figure 5-8 Direct-operated solenoid gas valve.

(Courtesy ITT General Controls)

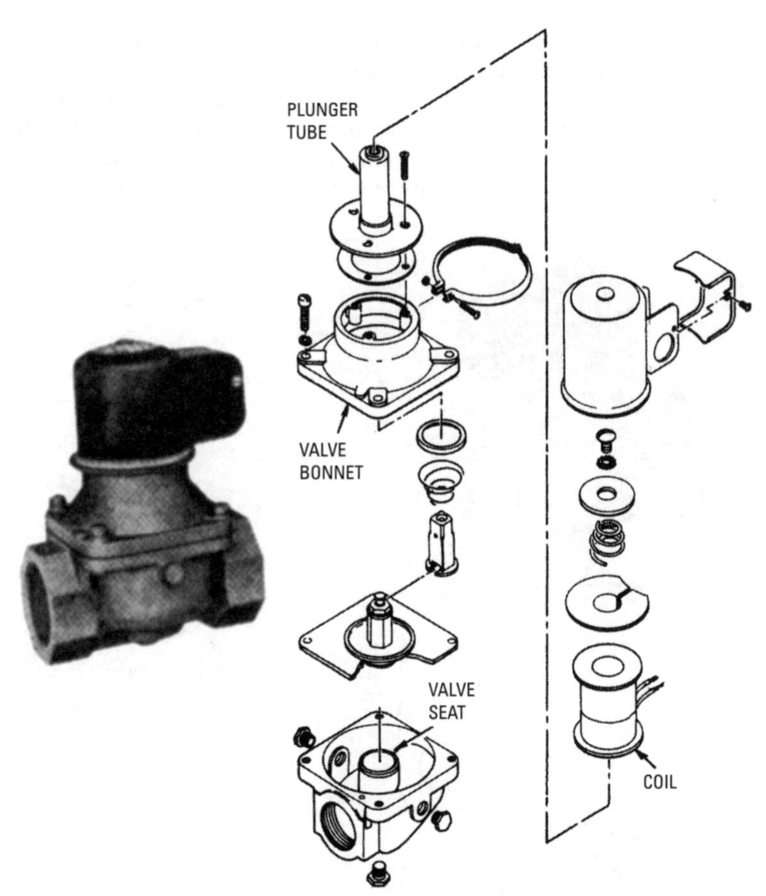

Figure 5-9 Balanced diaphragm solenoid gas valve.
(Courtesy ITT General Controls)

diaphragm faster than it can be replaced. This eventually results in the pressure above the diaphragm being the same as the pressure below the seat disc. This is referred to as a balanced or unloaded condition. The solenoid coil lifts the complete interior assembly to full open position. When the solenoid is deenergized, the pressure recovers above the diaphragm. The weight of the interior assembly and the gas pressure across the seat disc are sufficient to hold the valve closed. In this type of valve, the pressure of the gas is used to control its operation.

A third type of solenoid valve consists of a solenoid-operated (i.e., magnetically operated) puff bleed three-way valve and a diaphragm valve in a single unit (see Figure 5-10). The combined unit provides on-off control of the gas to the gas-fired heating equipment.

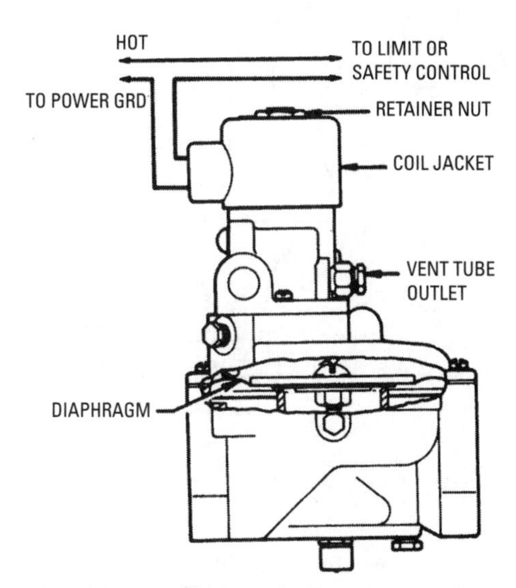

Figure 5-10 Electric diaphragm gas valve.
(Courtesy ITT General Controls)

The three-way valve (also referred to as a *pilot valve*), responding to electrical signals from the limit or safety controls, opens or closes the gas valve by controlling the gas pressure bleed-off above the diaphragm in the main valve body. In the normally closed position, inlet gas pressure above the diaphragm prevents the valve from opening. In the open (energized) position, the three-way or pilot valve closes off the inlet gas pressure and allows the gas pressure above the diaphragm to bleed off so that gas pressure below the diaphragm forces the diaphragm up to open the valve.

Dual-solenoid valves are designed for three-stage control (high-low-off) of the flow of gas (see Figure 5-11). Both a high-fire solenoid and a low-fire solenoid are used to accomplish this purpose. Low-fire adjustments can be made by turning the adjustment screw clockwise (to decrease low fire) or counterclockwise (to increase it).

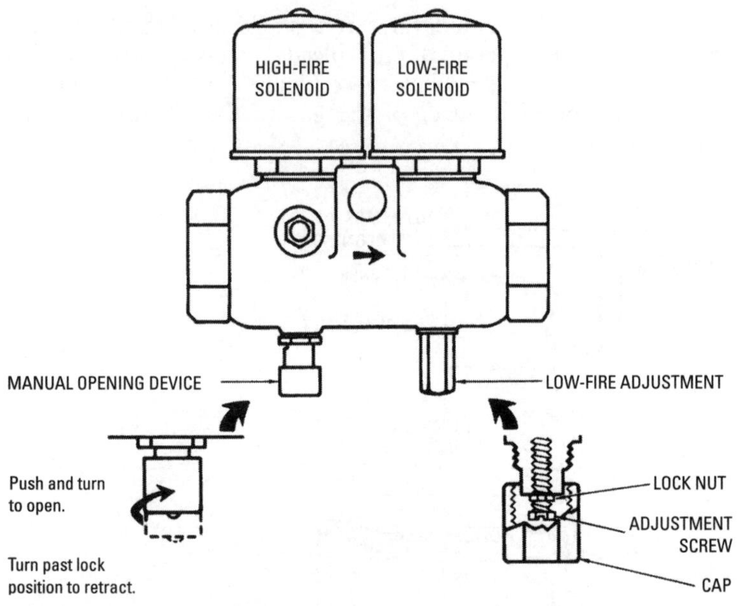

Figure 5-11 Dual-solenoid valve. *(Courtesy ITT General Controls)*

Figure 5-12 shows a schematic wiring diagram of a two-stage control containing a dual-solenoid valve. If both solenoids are to be energized at one time, the circuit requires a 40-volt transformer.

Solenoid Coils

Several different solenoid coils are available from manufacturers, and the type selected for use will depend on the specific application. For example, most standard applications will require a moisture-resistant coil for normal usage of gas or fluid up to 175°F. Special applications include those with especially high ambient and fluid temperatures, high voltage, or high steam pressure. A solenoid coil may also be specifically required for moisture or water applications. Under these circumstances the coil should be both waterproof and fungus proof. Some examples of solenoid coils are shown in Figure 5-13.

A principal cause of coil malfunction is excessive heat. If the valve is subjected to temperatures above the coil rating, it will probably fail. A missing part, a damaged plunger tube or tube sleeve, or improper assembly may also be a cause of excess heat. The applied voltage must be at the coil's rated frequency and voltage.

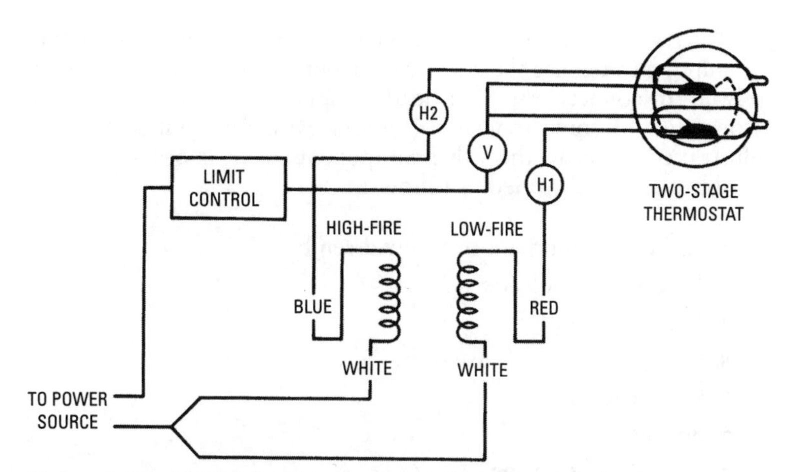

Figure 5-12 Dual-solenoid valve in a two-stage control circuit. *(Courtesy ITT General Controls)*

Figure 5-13 Various types of solenoid coils. *(Courtesy ITT General Controls)*

Always turn off the electrical power to the solenoid valve before attempting to replace the coil. Then, having turned off the electrical power, disconnect the coil leads. Figures 5-14 and 5-15 are schematics of some typical solenoid valves. The numbers in the illustrations refer to the valve components in their order of disassembly and are identified as follows:

1. Jacket retaining nut or screw assembly.
2. Elbow for coil leads.
3. Valve O-ring.
4. Coil jacket or coil assembly.
5. Nut or screw.
6. Spring retainer.
7. Plunger tube spring.
8. Screw assembly spacer.
9. Top washer and/or sleeve assembly.
10. Solenoid coil.
11. Bottom washer and/or sleeve assembly.
12. Plunger tube.

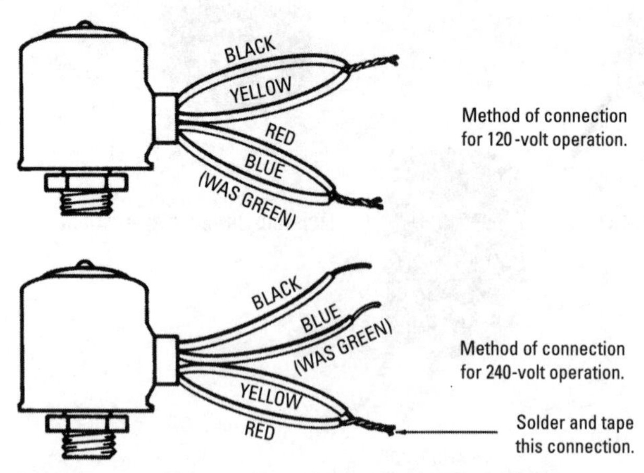

Figure 5-14 Wiring diagrams for dual-voltage coils.
(Courtesy ITT General Controls)

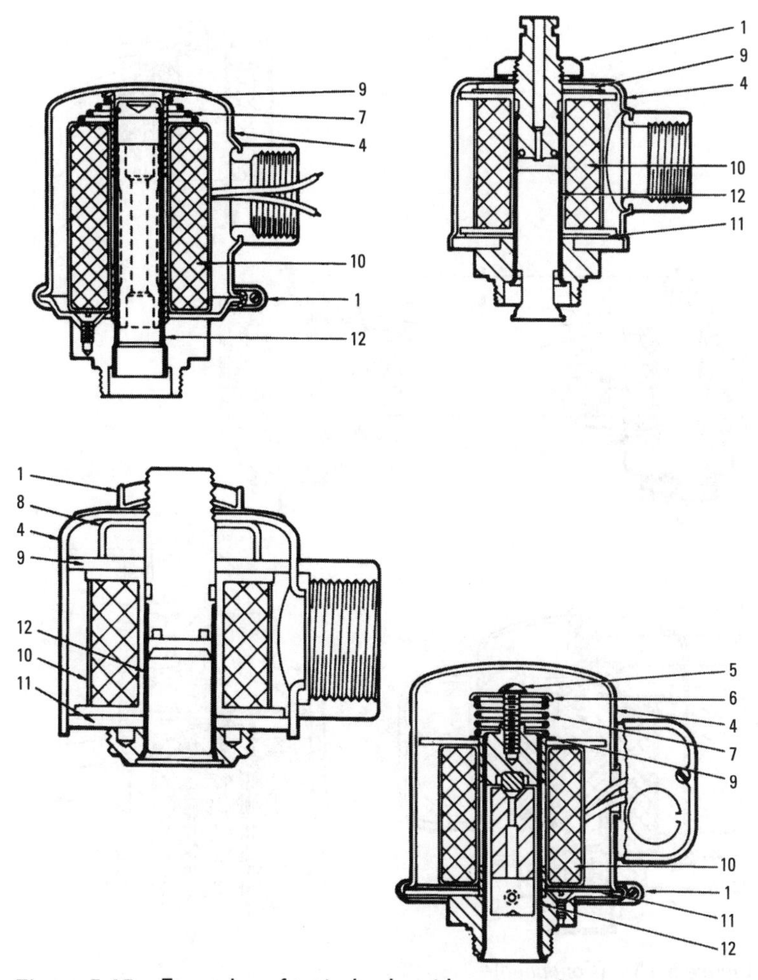

Figure 5-15 Examples of typical solenoid construction.
(Courtesy ITT General Controls)

The solenoid coil should be reassembled in reverse order, but with the following precautions:

1. Be *very* careful to reassemble the top washer and/or sleeve assembly (No. 9 above) exactly as it had been assembled. Improper assembly will cause the solenoid coil to burn out.

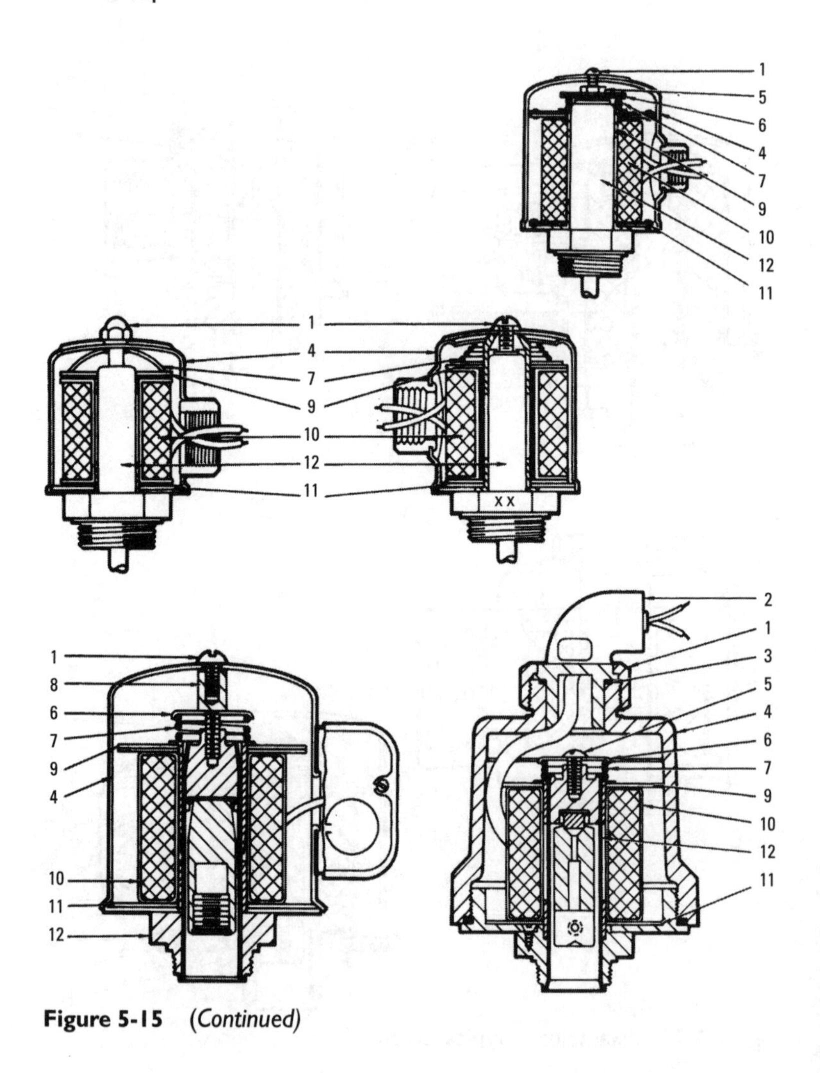

Figure 5-15 *(Continued)*

2. Be sure to align the top washer and/or sleeve assembly so that the coil leads have an unobstructed passage out of the solenoid.
3. Properly align all slots in the bottom washer and/or sleeve assembly.

Direct-Acting Heat Motor Valves

A *direct-acting heat motor valve* depends on the heat-induced expansion and contraction of a rod-type element to provide the movement and force for its operation.

The heat is generated by the passage of an electrical current through a resistance coil wound around a metal rod. One end of the rod is secured in place. The other end rests against a flexible snap mechanism (see Figure 5-16). When the room thermostat calls for heat, the electrical current flows through the coil and heats the metal rod. The heat generated by the resistance of the coil causes the rod to expand against the snap mechanism. When enough force is applied to the rod, the snap mechanism snaps over center and opens the valve. When the rod cools and contracts, the snap mechanism returns to its original position, and the valve closes.

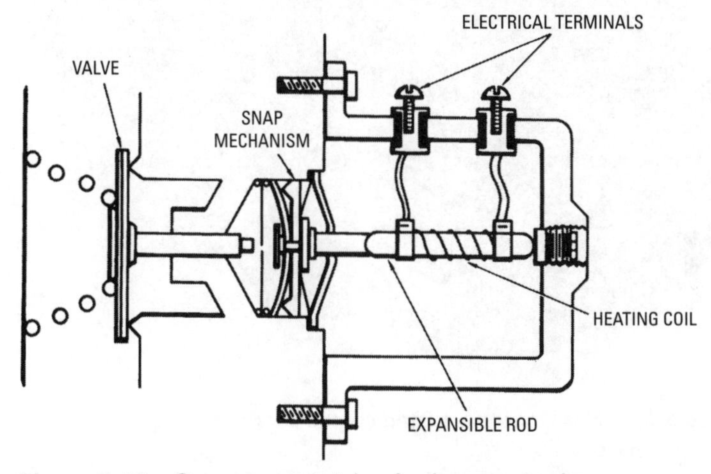

Figure 5-16 Operating principle of a direct-acting heat motor valve. *(Courtesy Robertshaw Controls Co.)*

Some combination gas valves utilize the direct-acting heat motor principle of operation. The Robertshaw Unitrol 1000E is an example of this type of valve (see Figure 5-17). It combines in one unit a gas cock, an automatic pilot, a pilot gas filtration device, and a heat-motor-actuated automatic valve. A main gas-pressure regulator can be added as an option. As shown in Figure 5-17, a manual opener or bypass selector is a common feature on these valves. When the bypass selector is in the *on* position and the room thermostat is in

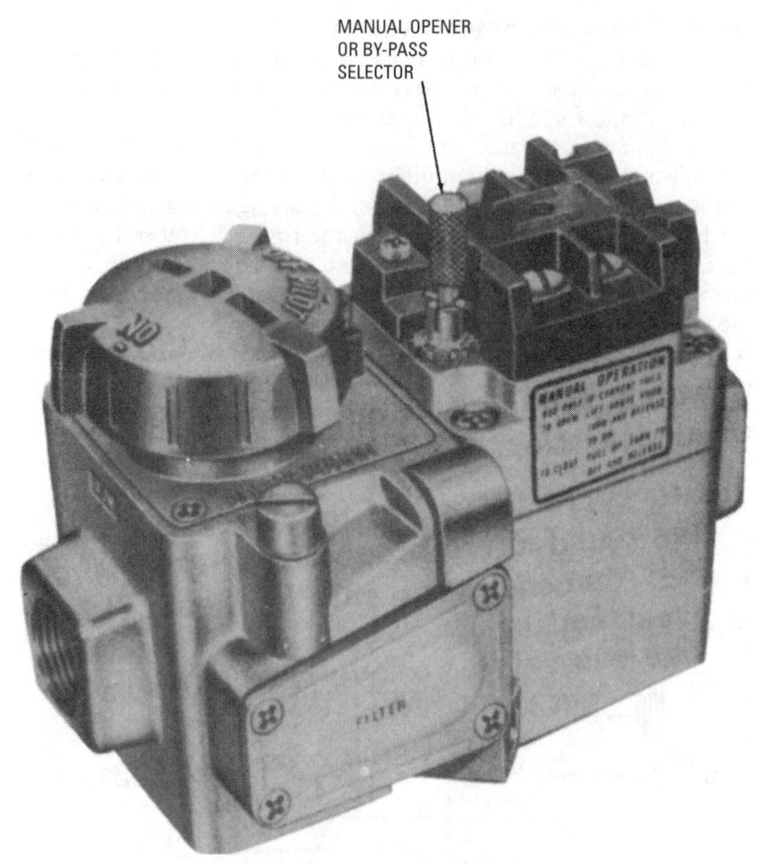

Figure 5-17 Heat-motor-actuated combination gas valve.
(Courtesy Robertshaw Controls Co.)

the *off* position, a bypass rate is provided to the burner for minimum input conditions.

Diaphragm Valves

The following are the three principal diaphragm valves, each distinguished by the kind of power used to actuate them:

- Hydraulic-actuated valves
- Solenoid-actuated valves
- Heat-motor-actuated valves

A hydraulic-actuated valve utilizes a hydraulic element to provide both the thermostatic sensing means and the power for valve operation. The closed hydraulic sensing and actuating device consists of a bulb, capillary tube, and a bellows or diastat (see Figure 5-18). Temperature changes cause the liquid in the remote-bulb sensing device to expand or contract. This expansion or contraction of the liquid operates the valve by controlling the pressure exerted against a bellows in the valve body. Additional information about this type of valve is contained in the section *Remote-Bulb Thermostats* in Chapter 4 ("Thermostats and Humidistats").

Both solenoid-actuated diaphragm valves and heat-motor-actuated diaphragm valves are described elsewhere in this chapter (see *Solenoid Gas Valves*, *Oil Valves*, and *Direct-Acting Heat Motor Valves*).

The term diaphragm valve can be confusing because valves that use a diaphragm are usually referred to by the power used to actuate them (e.g., hydraulic-actuated valves, solenoid-actuated valves) or their specific function (e.g., electric gas valves, oil burner valves).

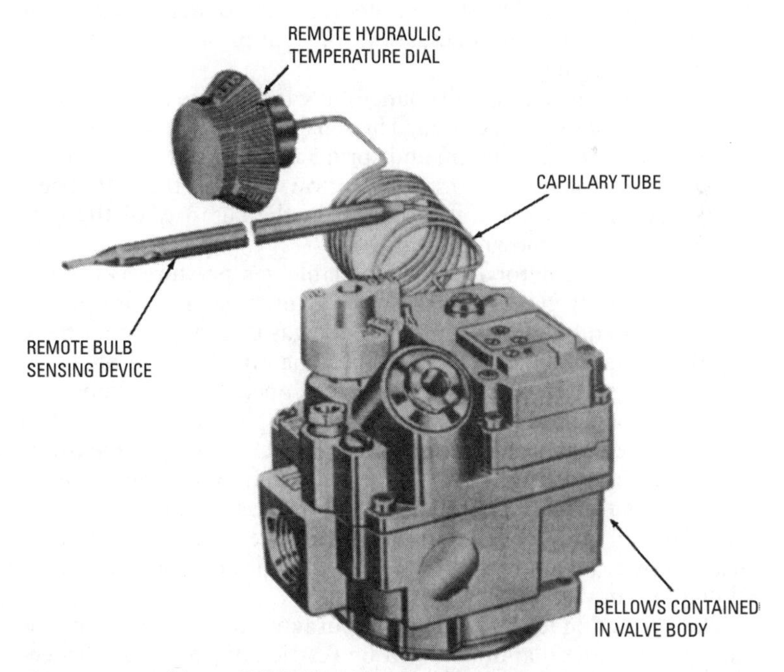

Figure 5-18 Unitrol 7000SR-1H hydraulic-actuated valve.
(Courtesy Robertshaw Controls Co.)

A *diaphragm valve* is *any* valve that contains a diaphragm; its purpose is to respond to pressure variations. Because this is an essential feature of pressure regulators, the operating principles of a diaphragm valve are described in the sections *Gas-Pressure Regulators* and *Combination Gas Controls*.

Pressure Regulators

Natural gas is distributed through the city mains at pressures of 7 inches water gauge or higher. Normally this gas will be at a higher pressure than the heating equipment or appliance can properly use. Furthermore, the gas pressure in the mains (and in the building supply lines) will often fluctuate because of load demand variations. Excessively high gas pressure and gas-pressure variations are detrimental to the operating efficiency and safety of a gas-fired furnace, boiler, or water heater. Hence, they must be brought under control before the gas enters the burners.

A *gas-pressure regulator* (or *manifold pressure regulator* as it is also called) is a regulating device used to control manifold gas pressure. Gas is delivered to the burners from the outlet orifice of the regulator at a single, nonfluctuating constant pressure regardless of inlet pressure changes.

A regulator must sense all changes in gas pressure and be able to adjust the gas flow as required. The sensing device by which this is accomplished is a diaphragm and spring arrangement attached to a valve ball or disc used to restrict gas flow through the seat. These and other components are illustrated in the cutaway of the low-pressure regulator shown in Figure 5-19.

A pressure regulator uses the available gas pressure as the primary force to open or close the valve. Outlet gas pressure presses against the diaphragm and spring. If the gas pressure is too little to overcome the force of the spring, then the attached ball or disc is pushed away from its seat. This enlarged opening allows more gas to flow. If the outlet pressure against the diaphragm is greater than the spring setting, then the valve ball or disc is brought toward its seat, narrowing the opening and restricting flow. As the gas pressure against the diaphragm equals the force exerted by the spring, the valve ball or disc is so positioned from the orifice to maintain a steady downstream pressure. This principle of operation is basic to all diaphragm valves.

The spring-loaded side of the diaphragm must be vented or the movement of the diaphragm will be restricted. The most elementary form of venting is shown in Figure 5-20. This is simply an orifice installed in a vent hole on the spring-loaded side of the

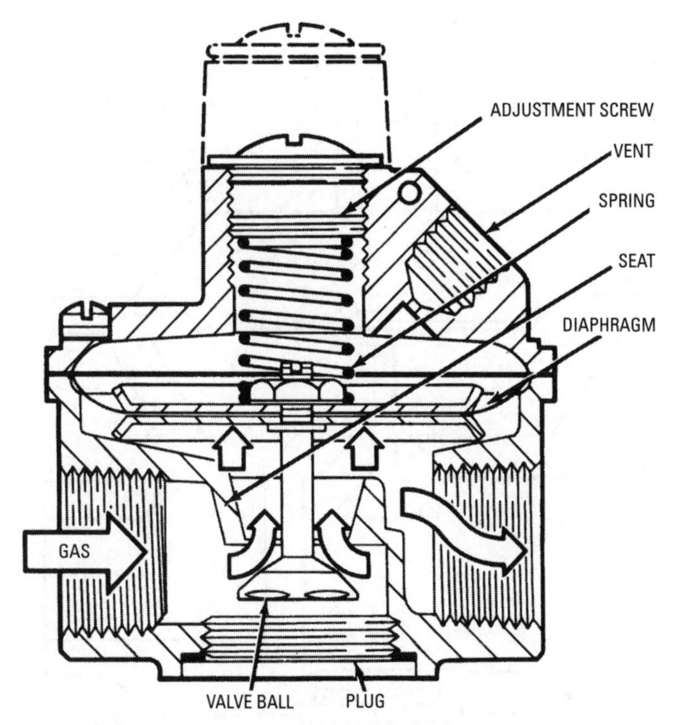

Figure 5-19 Principal components of a low-pressure regulator. *(Courtesy ITT General Controls)*

diaphragm. A more complex method of venting involves connecting a tube to a tapping in the vent. This represents either the internal or external bleed system of venting gas. The principal difference between the two systems lies in how and where the vented gas is disposed.

Proper venting allows the valve diaphragm to move freely in either direction. Installing an orifice in the vent hole slows the diaphragm action, thereby providing a smoother operating control (see Figure 5-20). The use of a vent hole orifice also prevents the rapid and potentially dangerous escape of gas in the event of a diaphragm rupture.

Both internal and external bleed systems are used to vent the gas from the spring-loaded side of the diaphragm. In an *internal bleed system*, the bleed gas is routed to the burner or pilot where it is burned, or to the burner manifold where it is mixed with the main

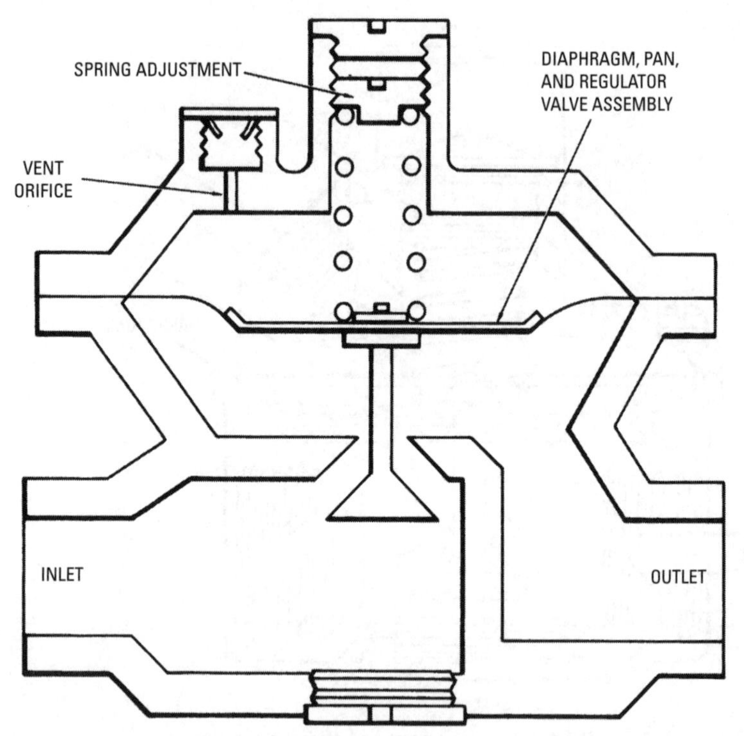

Figure 5-20 Pressure regulator with vent orifice. *(Courtesy Robertshaw Controls Co.)*

gas supply and eventually burned. In an *external bleed system*, the bleed gas is vented by a tube extending to the outdoors. A variation of the external bleed system is to place the outlet end of a tube in the heat exchanger to vent the gas outside.

A simple diaphragm valve functions only to open or close the valve and is generally of the external bleed type (see Figure 5-21). The operator mechanism for a restricting internal bleed orifice diaphragm valve is basically as shown in Figure 5-22.

A gas-pressure regulator can be either an independent control on a gas manifold or a part of a combination gas control. The obvious advantage to using a combination control is the simplification of appliance assembly and the saving of space obtained when compared to the use of separate components. Less obvious are the operational advantages.

When the pressure regulator is *not* an integral part of the combination gas control, it either precedes or follows the location of the

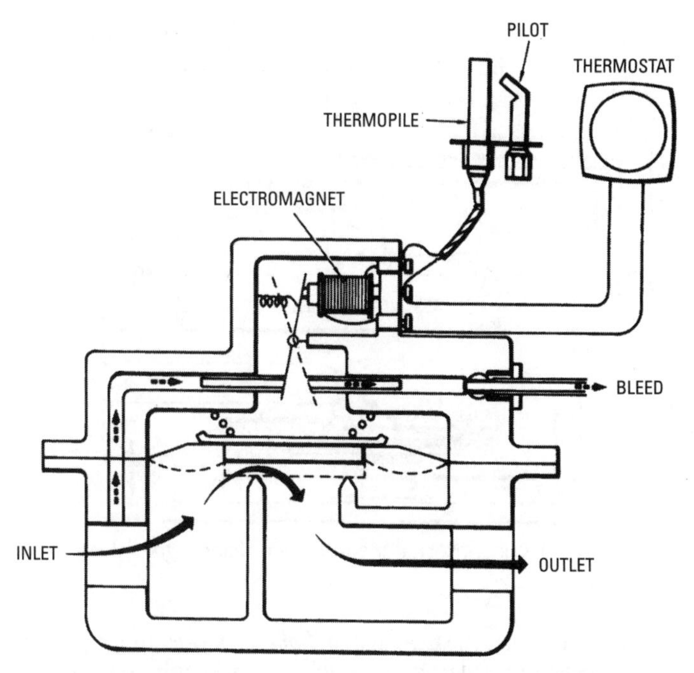

Figure 5-21 Schematic drawing of a millivolt diaphragm bleed valve.
(Courtesy Robertshaw Controls Co.)

latter in the manifold. If the separate pressure regulator precedes the combination control, both the main burner gas and the pilot gas are regulated by the same regulator in most installations. If this is the case, a problem will sometimes occur with pilot outage. As the main gas valve opens to provide gas to the main burners, a temporary starving of the pilot gas can occur due to regulator response delay. This condition can be avoided by installing a separate pilot gas regulator for pilot gas only or by using a regulator-equipped combination gas control with the proper sequencing of operation.

Installing a separate pressure regulator *after* the combination gas control (i.e., downstream from it) can sometimes result in overgassing the main burners. This occurs because the regulator remains in a wide-open position when the main gas valve remains closed. When the gas valve opens, overgassing can result from the delay of the regulator valve in resuming regulation. A combination gas control can eliminate this problem.

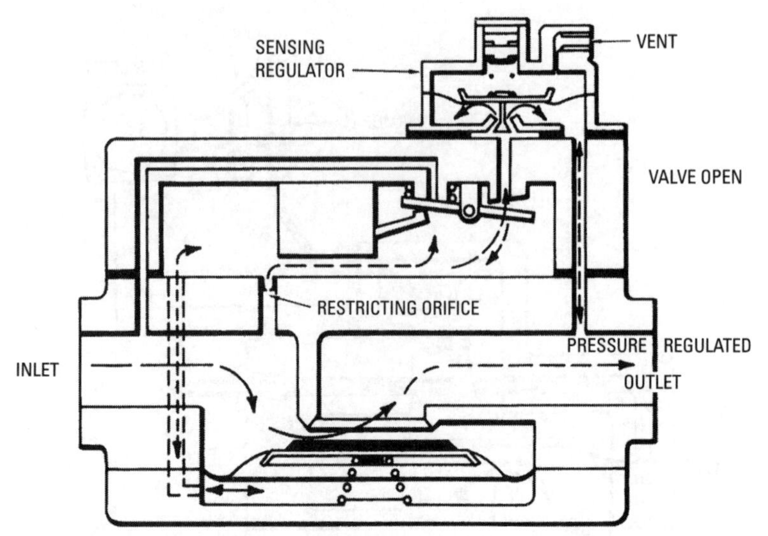

Figure 5-22 Restricting internal bleed orifice diaphragm valve.
(Courtesy Robertshaw Controls Co.)

Some gas-fired water heaters use a *balanced* pressure regulator. This type of pressure regulator uses two internal diaphragms to control gas pressure. The operating principle is quite different from that described for the regulators used on gas-fired heating equipment. For additional information, read the section *Balanced Pressure Regulators* in Chapter 4 of Volume 3 ("Water Heaters and Other Appliances").

Pressure Switches

A *pressure switch* is a safety device used in positive-pressure or differential-pressure systems to sense gas- or air-pressure changes. A typical arrangement of gas-pressure switches on a gas manifold is shown in Figure 5-23. The wiring diagram in Figure 5-24 illustrates the connections between the high and low gas-pressure switches on a gas-fired boiler manifold and the primary control.

Gas-pressure switches are available in two basic types:

- Falling-pressure switches
- Rising-pressure switches

In the falling-pressure switch, decreased pressure on the diaphragm actuates the device. The switch is designed to lock out when the pressure falls to the setpoint (minus differential). As the pressure

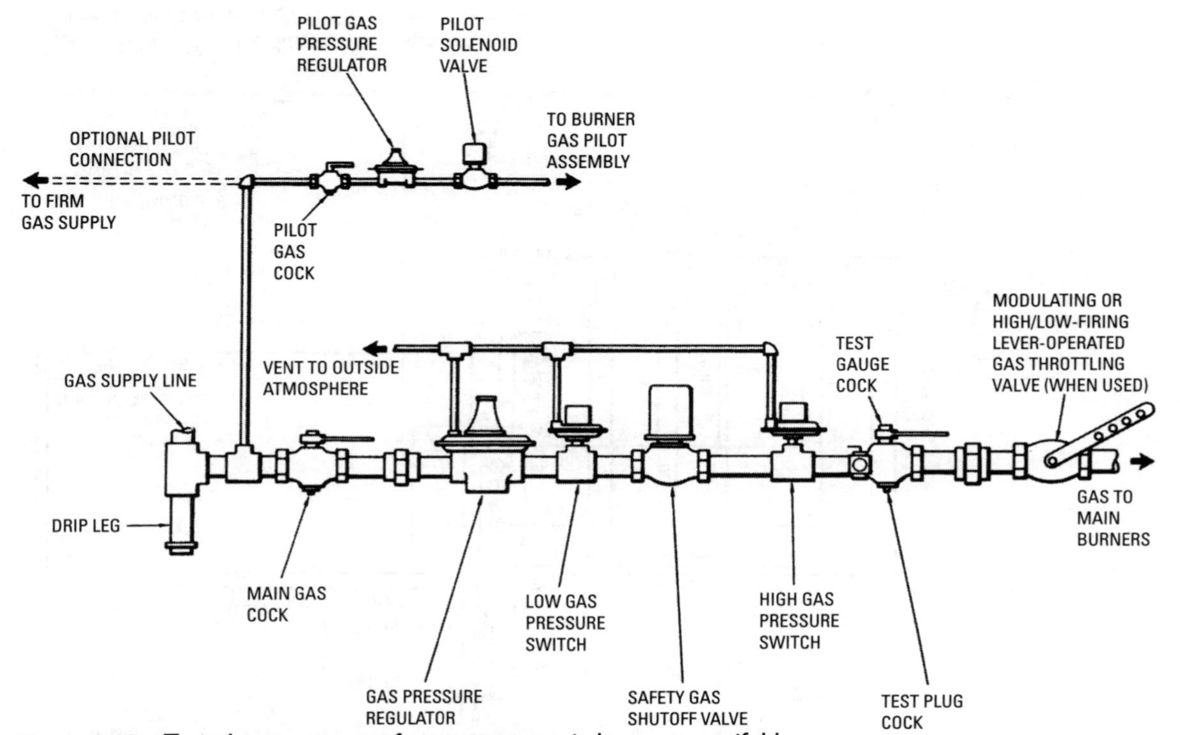

Figure 5-23 Typical arrangement of gas-pressure switches on a manifold.

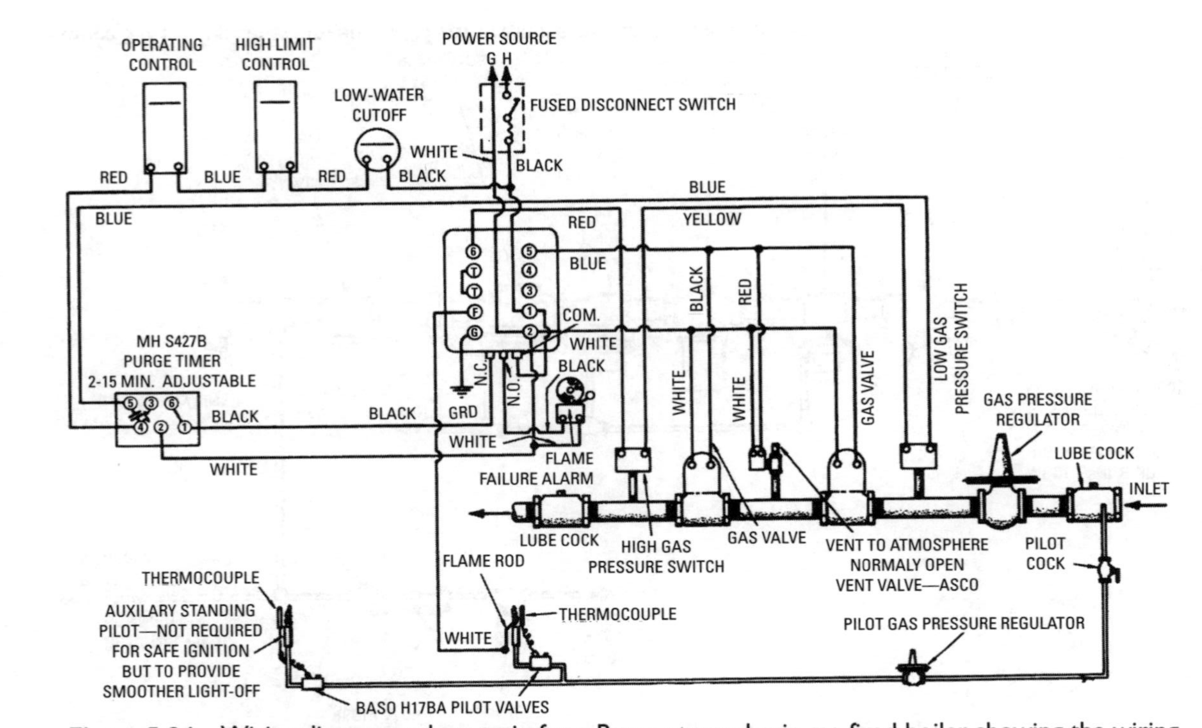

Figure 5-24 Wiring diagram and gas train for a Bryan atmospheric gas-fired boiler showing the wiring connections for the high and low gas-pressure switches. *(Courtesy Bryan Steam Corp.)*

increases, the diaphragm rises and the switch is deactuated (except on manual reset models). An adjustable spring-loaded diaphragm determines the amount of pressure required to actuate the switch.

In a rising-pressure switch, the switch is actuated by increased pressure on the diaphragm. As the pressure falls, the diaphragm lowers and the switch is deactuated (again, except on manual reset models). An adjustable spring-loaded diaphragm determines the amount of pressure required to actuate the switch.

The pressure required to move the diaphragm in these switches is adjustable within the pressure range stamped on the switch nameplate. The pressure switch shown in Figure 5-25 can be adjusted by removing the cover and turning the adjustment screws clockwise. This action raises the actuation point of the switch. Turning the screws counterclockwise lowers the actuation point. The range scale plate in the switch is marked for four relative pressure settings. Setting *A* corresponds to a minimum range, *D* to a maximum range, and both *B* and *C* to intermediate ranges.

A typical wiring diagram for a single-pole double-throw SPDT switch is shown in Figure 5-26. An SPDT switch may be wired to open or close the circuit on pressure rise.

On switches equipped with manual reset (see Figure 5-25), the switch contacts open on pressure rise or pressure drop (depending on which model is used) and remain open regardless of pressure change. The reset button is pushed to close the switch *after* the pressure has returned to an acceptable level.

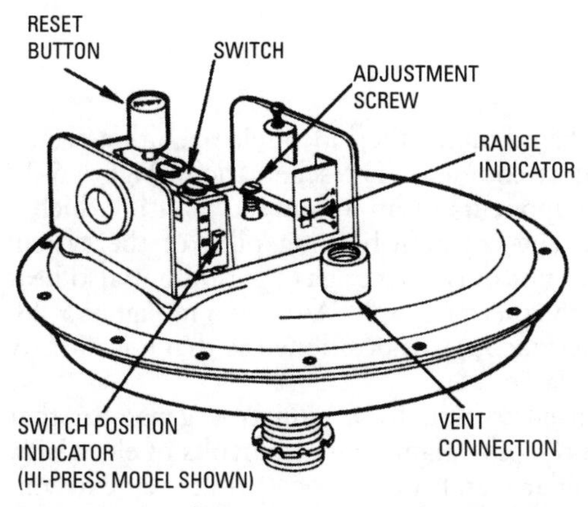

Figure 5-25 Gas-pressure switch with manual reset. *(Courtesy ITT General Controls)*

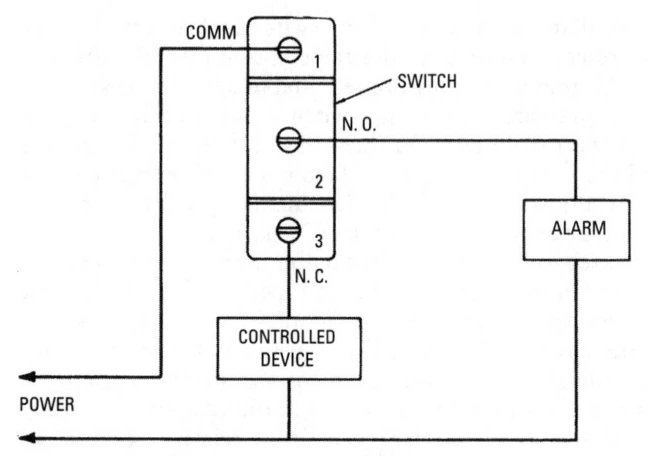

Figure 5-26 Wiring diagram of SPDT gas-pressure switch.
(Courtesy ITT General Controls)

Automatic Pilot Safety Valve

The automatic pilot safety valve is a device used to shut off the gas supply when the pilot flame is extinguished or fails to light.

There are a variety of different pilot safety controls available on the market, but all are based on one of the following three operating principles:

- Thermocouple
- Metal expansion
- Liquid pressure

Pilot safety controls based on the thermocouple operating principle are probably the most common. The schematic in Figure 5-27 shows the principal components of an automatic pilot in which a thermocouple is used. The constant-burning pilot of the system illustrated here provides not only burner gas ignition but also heat for the hot junction of the thermocouple. As shown in Figure 5-28, the hot junction of a thermocouple is positioned so that it is directly in the path of the pilot flame.

The thermocouple itself is actually a miniature generator that can convert heat (from the pilot flame) into millivolts of electricity. It consists of two dissimilar metals joined together at their extremities. When one of the junctions (the hot junction) remains cold, electrical energy is generated. The amount of energy generated by

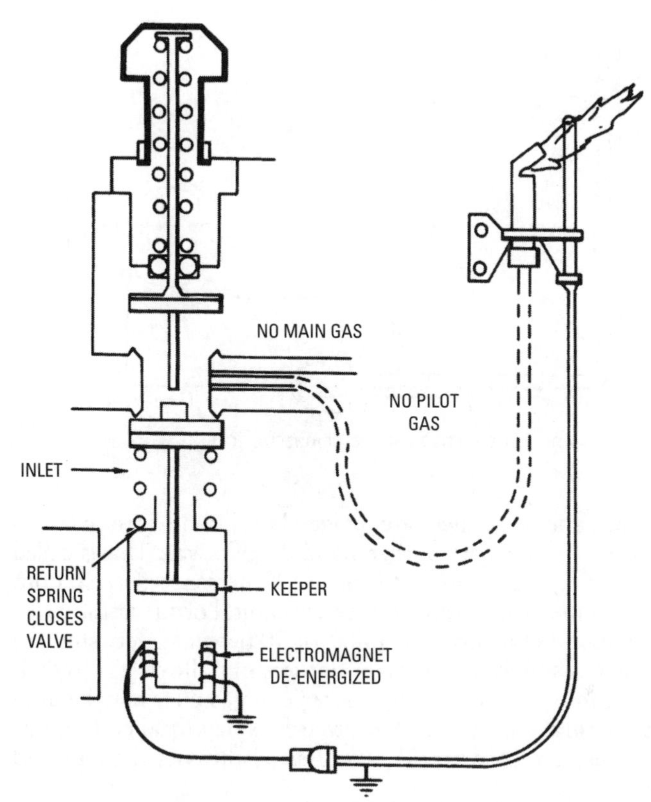

Figure 5-27 Thermocouple used in conjunction with an automatic pilot valve. *(Courtesy Robertshaw Controls Co.)*

the thermocouple is directly proportional to the temperature difference between the hot and cold junctions (see Figure 5-28). One thermocouple or junction will deliver approximately 25 millivolts. Several thermocouples or junctions can be wired in series to produce a higher voltage (see *Pilot Generators*). The electrical energy generated by the thermocouple energizes an electromagnet that operates the automatic pilot valve.

The automatic pilot safety valve may be an individual control but is more commonly a part of a combination control, which usually combines manual valve and thermostatic or automatic valve functions. These valves are described in this chapter (see *Combination Gas Valves*).

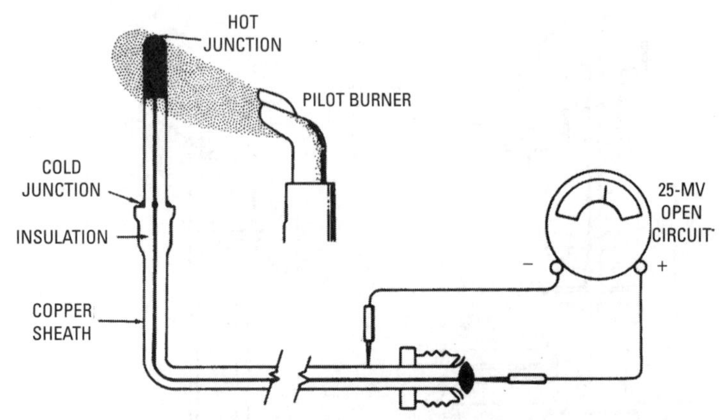

Figure 5-28 Common thermocouple construction. *(Courtesy Trane Co.)*

Some automatic pilot valves are designed to shut off both the main gas (i.e., the gas to the main burners) and the pilot gas. This is called 100 percent safety automatic pilot or a 100 percent safety shutoff. They are recommended for use with either natural or LP gases.

A 90 percent safety automatic pilot (or 90 percent safety shutoff) shuts off the gas supply to the main burners but allows gas to continue flowing to the pilot burner. This type of gas pilot may be used safely with natural gas (and is permitted by some local codes), but it should *never* be used with LP gas. LP gas is heavier than air and will not vent.

The procedure for lighting the pilot in a system containing a thermocouple (or pilot generator) and an automatic pilot safety valve is fairly simple. The reset button on the automatic pilot valve must be pushed down while the pilot is being lit (see Figures 5-29 and 5-30). While the reset button is depressed, an auxiliary valve causes the main gas porting to close. At the same time, the spring-loaded automatic pilot valve is opened, allowing gas to flow only to the pilot burner, which can now be lit.

The same action that opened the spring-loaded automatic valve has also forced the keeper (see Figure 5-29) against the pole faces of the electromagnet. Eventually the heat of the pilot flame causes the thermocouple to generate an electrical current, which causes the electromagnet to become magnetized. As a result, the keeper is held against the magnet and the automatic pilot valve is maintained in an open position. The pilot should be allowed to burn for at least 60 seconds before releasing the button. This permits enough current to

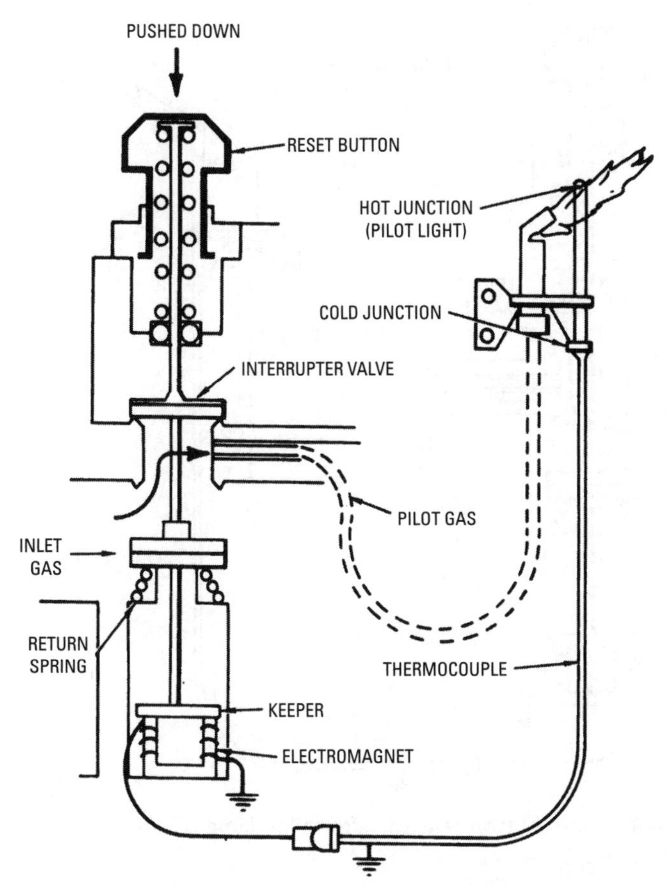

Figure 5-29 Reset button depressed while pilot is being lit.
(Courtesy Robertshaw Controls Co.)

be built up by the thermocouple or pilot generator to hold the valve open.

Once the pilot has been established, the reset button can be released, and the main gas and pilot gas will be free to move past the automatic pilot valve (see Figure 5-30). When the pilot flame is extinguished, the thermocouple hot junction cools and breaks the electrical current. The electromagnet is demagnetized as a result of this loss of current, and the keeper is released, causing the automatic pilot valve to close and shut off the flow of gas. Before relighting the pilot, allow at least 5 minutes for all the gas to clear the system.

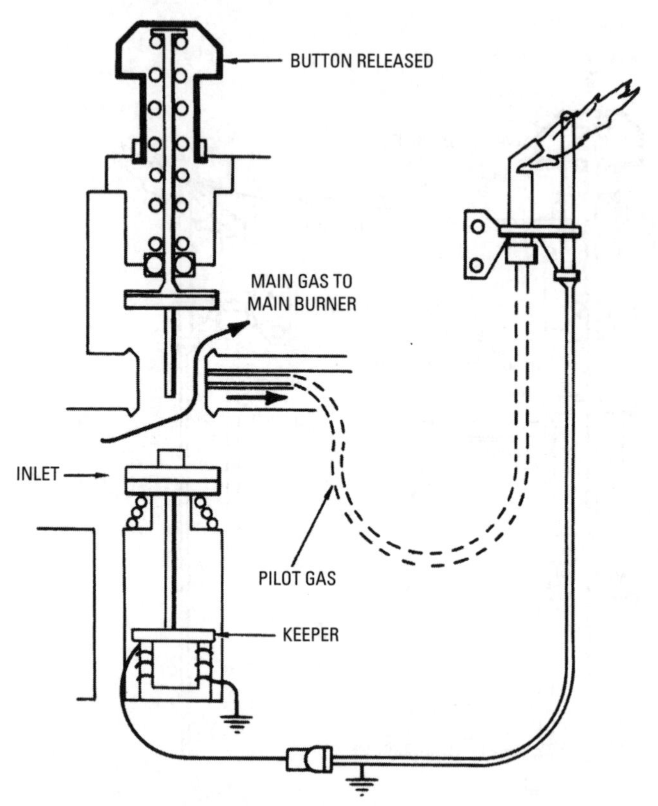

Figure 5-30 Reset button released after pilot flame is established. *(Courtesy Robertshaw Controls Co.)*

Thermopilot Valves

A *thermopilot valve* is a 100 percent safety shutoff control for gas-fired heating equipment and appliances. It operates on current supplied by a thermocouple or pilot generator. This current energizes a thermomagnet, which holds the valve open after manual reset. When there is an unstable or low pilot flame or no flame at all, the current to the thermomagnet is lost. As a result, the valve closes, shutting off the gas flow.

An example of a thermopilot valve operated by a thermocouple is the ITT General Controls MR-2G valve. Using it in conjunction with a solenoid gas valve provides 100 percent safety shutoff control (see Figure 5-31).

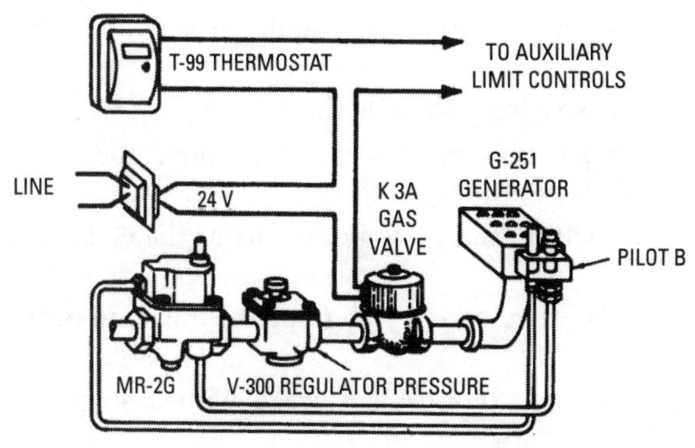

Figure 5-31 ITT control valve model MR-2G.
(Courtesy ITT General Controls)

The thermomagnet of an ITT General Controls MR-2YA valve is energized by a pilot generator. It must be used with a pilot-operated diaphragm valve to provide 100 percent safety shutoff control (see Figure 5-32).

The following suggestions should be carefully observed when installing ITT thermopilot valves in order to ensure efficient valve operation:

- Locate the valve so that it is easily accessible and where the ambient temperature is below 200°F.

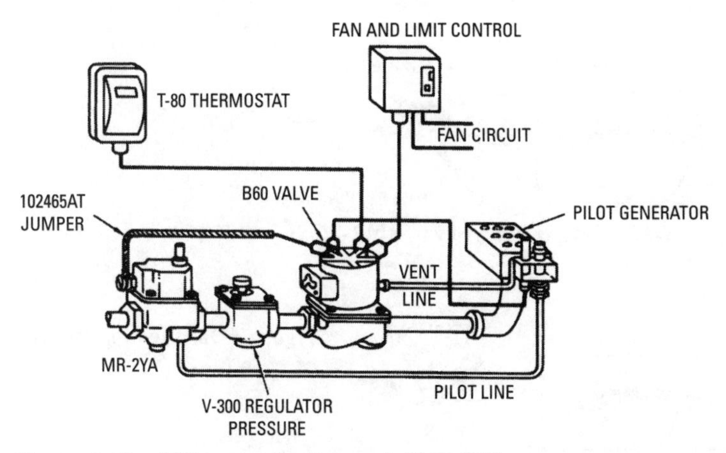

Figure 5-32 ITT control valve model MR-2YA. *(Courtesy ITT General Controls)*

- Make sure the piping is clean (blow out all particles and other impurities).
- Apply thread seal sparingly to male threads only.
- Install valve with gas flow in the same direction as the arrow in the valve body is pointing.
- Use the pipe wrench on the valve body flats at the end being connected.
- Avoid stress on the valve body by aligning the inlet and outlet pipe connections.
- Check all pipe connections for gas leaks with a soap and water solution. *Never* use a flame.
- Check the valve with a millivolt meter (see next paragraph).
- All wiring connections should be clean and tight.
- Pilot burner should be installed on main burner so that the ignition flame will light the main burner with the pilot turned down low.

Testing an ITT General Controls MR-2G thermopilot valve with a millivolt meter requires the use of a special adapter. The adapter and thermocouple are connected to the valve as shown in Figure 5-33. Attach the meter clips as shown. Figure 5-34 illustrates the testing of an ITT MR-2YA valve. In either case, the valve should be replaced if the meter reading is above that shown on the applicable scale *and* the valve fails to open after following the lighting instructions.

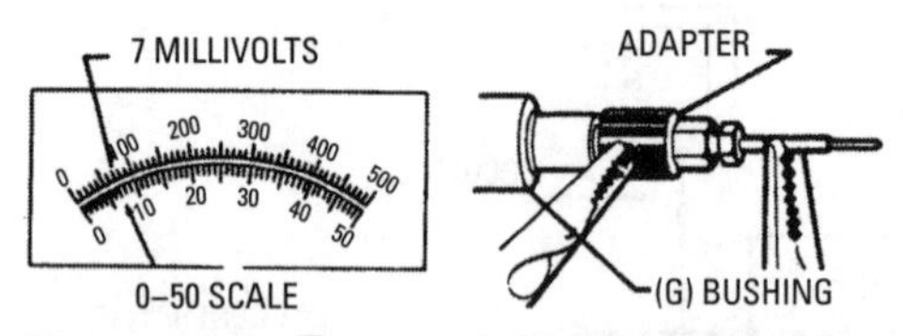

Figure 5-33 Testing control valve MR-2G.
(Courtesy ITT General Controls)

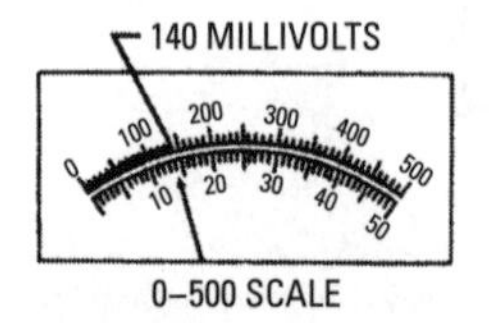

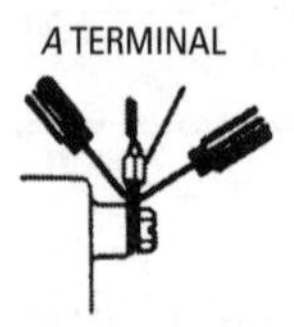

Figure 5-34 Testing control of valve MR-2YA. *(Courtesy ITT General Controls)*

Thermocouples

The function of the thermocouple is to detect (prove) the existence of the pilot flame. If the pilot light is lit and the thermocouple detects the pilot flame, it opens the main gas valve and allows natural gas to flow to the burner(s). If there is no pilot flame, the thermocouple shuts off the gas to the pilot and closes the main gas valve so that no gas can flow to the burner(s).

Thermocouples are made from flame-resistant metal alloys and are available in standard lengths of 13.5 to 48 inches (see Figure 5-35). Industry-standard G bushing or R bushing threads are used for connecting the thermocouple to the valve. Thermocouples from the various manufacturers can also be interchanged. The thermocouples manufactured by ITT General Controls, Robertshaw-Grayson, Baso, Honeywell and others are designed for universal installation when used with the proper adapter and retainer (where required) (see Figures 5-36 and 5-37).

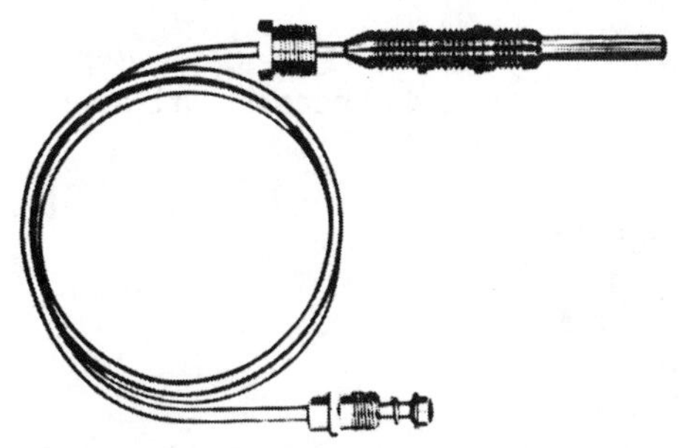

Figure 5-35 Typical thermocouple.
(Courtesy ITT General Controls)

Always test the thermocouple before replacing it. The problem with the pilot flame or safety control may be caused by something other than a malfunctioning thermocouple.

Manufacturers of thermocouples generally provide a special adapter for testing thermomagnet valves. The adapter is screwed into the gas valve, the thermocouple (or generator) leads into the adapter, and the test is run as follows:

1. Attach the meter clips as shown in Figure 5-38. Reverse the meter clips if the needle moves to the left of zero on the millivoltmeter.
2. A meter reading of less than 7 millivolts for thermocouples or less than 140 millivolts for pilot generators indicates the orifice and primary air holes on the pilot burner need to be cleaned.

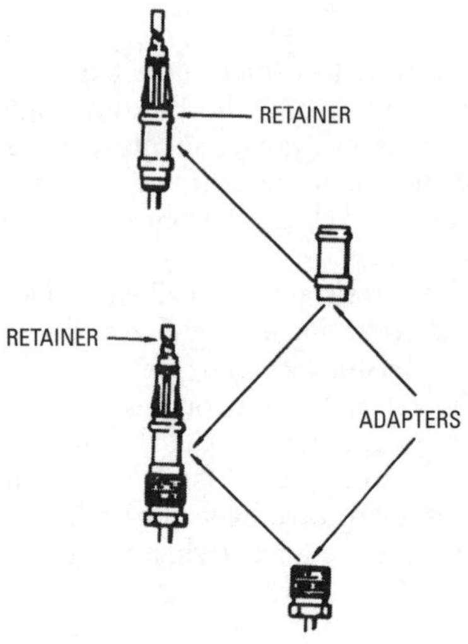

Figure 5-36 Thermocouple adapters and retainers. *(Courtesy Robertshaw Controls Co.)*

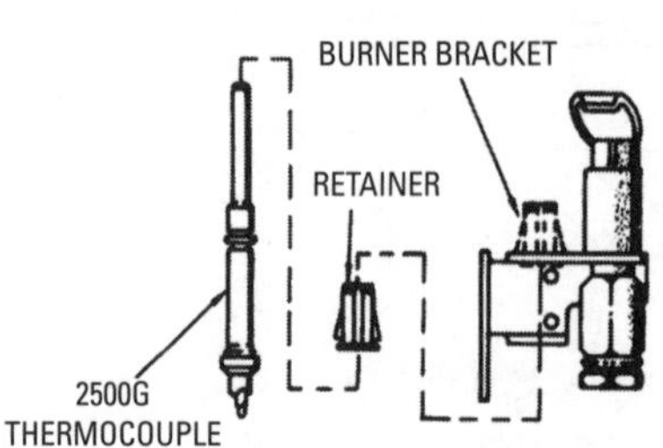

Figure 5-37 Retainer used to hold thermocouple in burner bracket.

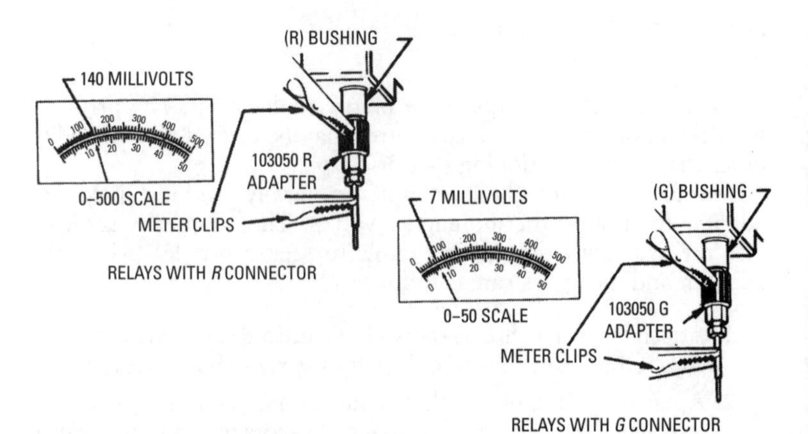

Figure 5-38 Relay bushing connections. *(Courtesy ITT General Controls)*

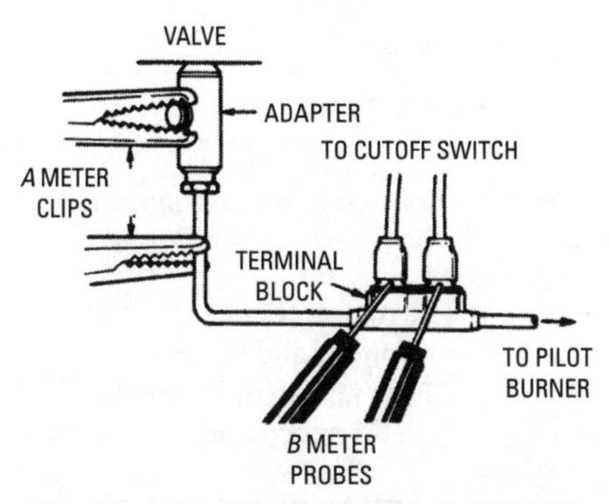

Figure 5-39 Checking a terminal block connected to a high-energy cutoff switch.
(Courtesy ITT General Controls)

Sometimes a thermocouple will have a terminal block connected to a high-energy cutoff switch. The millivoltage should be checked with meter probes as shown in Figure 5-39. The various readings suggest the following courses of action:

- Replace the high-energy cutoff switch if the reading exceeds 4 millivolts.
- Replace the thermocouple if the reading is still less than 7 millivolts.
- Replace (or repair) the valve if it fails to hold open with a meter reading of more than 7 millivolts.

Troubleshooting Thermocouples

Table 5-2 lists possible remedies for a number of different operating problems associated with thermocouples.

Table 5-2 Troubleshooting Thermocouples

Symptom and Possible Cause	*Possible Remedy*
Pilot flame lit but safety control fails to function.	
(a) Thermocouple not hot enough to generate current.	(a) Wait at least 1 minute for thermocouple to become hot enough.
(b) Drafts deflecting flame away from thermocouple.	(b) Eliminate source of draft.

(continued)

Table 5-2 (continued)

Symptom and Possible Cause	*Possible Remedy*
(c) Pilot flame too small or yellow in color due to restricted pilot line or dirt in primary air opening or burner head.	(c) Disconnect, clean thoroughly, and reconnect. Change orifice if necessary.
(d) Loose or dirty electrical connections.	(d) Disconnect, clean, reconnect, and tighten.
(e) Thermocouple tip too low in pilot flame mounted in bracket.	(e) Check installation to make sure thermocouple is properly.
Safety control operates but fails when main burner has been on a short time.	
(a) Restriction in pilot or main gas tubing.	(a) Eliminate restriction. Provide normal pressure.
(b) Draft-deflecting flame couple.	(b) Eliminate draft or baffle.

Thermopiles (Pilot Generators)

The amount of electrical energy generated by a thermocouple that has a single hot junction and a single cold junction (i.e., approximately 25 millivolts) is considered adequate for most residential heating equipment. However, a few residential furnaces and boilers and most commercial types require a higher voltage. An increase in voltage can be obtained by using a number of thermocouples wired in series (see Figure 5-40). A series of thermocouples in one unit is called a *thermopile*, *pilot generator*, or *thermopile generator.* A thermopile forms a part of a millivolt or self-energizing control circuit.

An example of a thermopile used in gas-fired heating equipment and appliances is shown in Figure 5-41. This particular unit produces

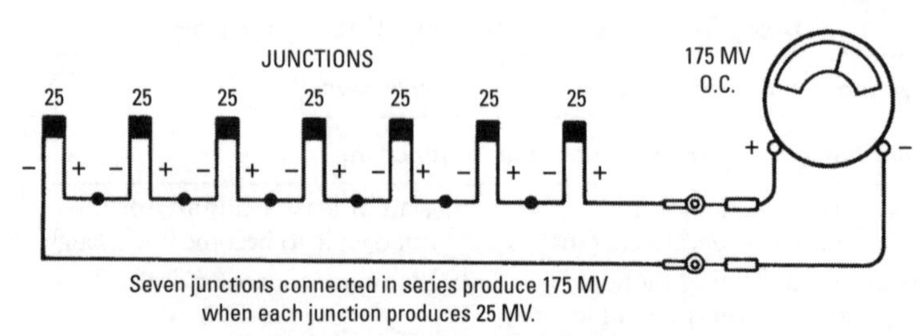

Figure 5-40 Thermocouple construction and series wiring. *(Courtesy Trane Co.)*

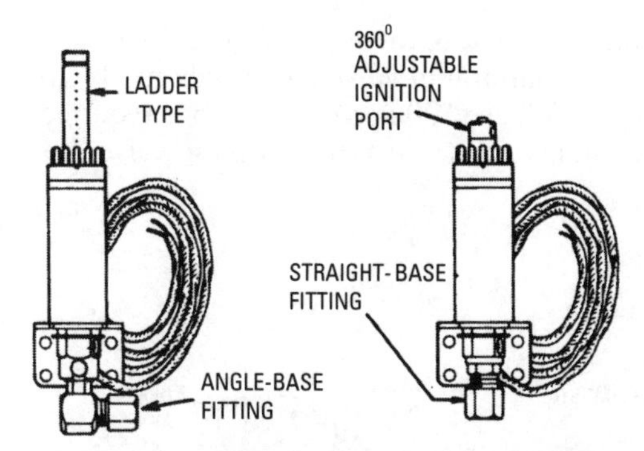

Figure 5-41 ITT pilot generator model PG-1.
(Courtesy ITT General Controls)

approximately 320 millivolts in an open circuit for gas valve control. It is available with an adjustable ignition port and interchangeable orifices for use with any gas.

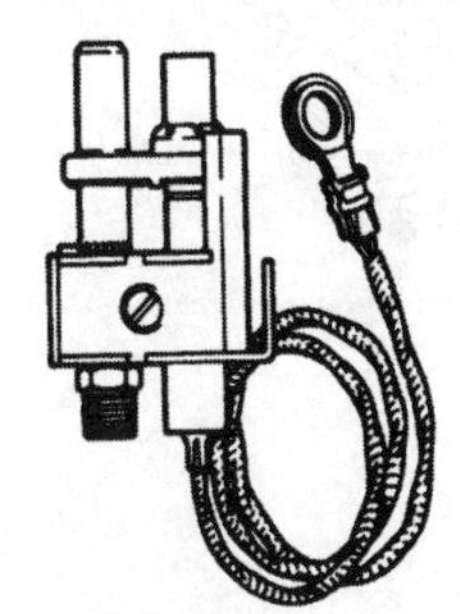

Figure 5-42 ITT pilot generator model PG-9A with cartridge.
(Courtesy ITT General Controls)

A thermopile that produces an even higher voltage is the ITT General Controls PG-9A (see Figure 5-42). This thermopile provides a pilot flame for gas burner ignition and generates electricity from the heat of the pilot flame to operate millivolt gas valves and relays. The replaceable cartridge in the thermopile contains many thermocouples connected in series. The top ⅜ to ½ inch of the cartridge is heated by the pilot flame, which produces approximately 500 to 750 millivolts (½ to ¾ of 1 volt) open circuit.

A millivoltmeter must be used to test a thermopile. The meter leads are attached to the valve or relay terminals to which the wires of the pilot generator are also attached. The thermostat must be calling for heat and the pilot burning during a millivoltmeter test.

Pilot-Operated Diaphragm Valves

A pilot-operated diaphragm valve is used to provide automatic shut-off of main line gas when there is an unstable pilot flame or no flame at all. These valves are operated by electrical energy (millivoltage) produced by the pilot generator. Their operation is controlled by the room thermostat, limit devices, and other operating controls.

The ITT General Controls B60 gas valve (see Figure 5-43) is an example of a pilot-operated diaphragm valve commonly used with gas-fired heating equipment. Where 100 percent shutoff is required, it should be used in conjunction with a thermopilot valve (see Figure 5-44).

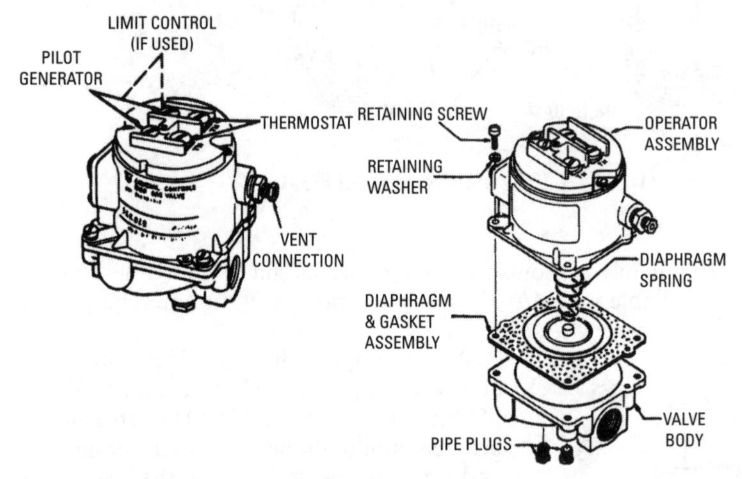

Figure 5-43 Pilot-operated diaphragm valve. *(Courtesy ITT General Controls)*

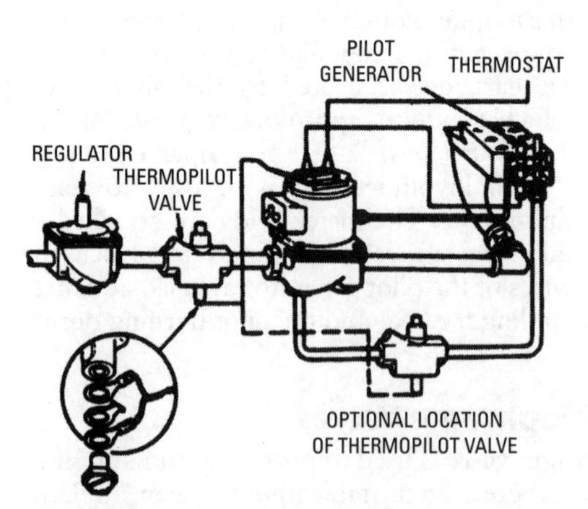

Figure 5-44 ITT valve model B60Y used in conjunction with a thermopilot valve.
(Courtesy ITT General Controls)

Combination Gas Valves

A *combination gas valve* (or *combination gas control*) combines in a single unit all manual and automatic control functions required for the operation of gas-fired heating equipment. In other words, a single valve replaces the various individual pilot line and main line gas controls. A gas-pressure regulator is usually optional.

Many manufacturers of gas controls offer a complete line of combination gas valves; each valve is designed for a different kind of installation or application. Usually these valves will differ on the basis of the controller voltage or voltage source, valve application or function, required Btu capacity for the installation, and type of gas used.

Honeywell manufactures a line of standardized and interchangeable gas control components. A complete preassembled combination gas control can be ordered from the factory or one can be assembled in the field from a variety of different standardized components in order to meet the needs of a particular installation. This add-on feature also allows field replacement of a defective component without removing the complete valve from the installation. A number of possible combinations are illustrated in Figures 5-45 and 5-46.

Standing Pilot Combination Gas Valves

A typical Honeywell combination gas control used with a standing pilot consists of the following basic components (see Figure 5-47):

1. Main valve body.
2. Valve operator.
3. Pressure regulator.

As shown in the schematic diagram of the control (see Figure 5-48), the main valve body (or manifold control) contains a valve diaphragm (5) and disc (3). This portion of the combination gas control operates as a conventional diaphragm valve. The valve opens and closes in response to the presence or absence of gas in the pressure chamber (4). This gas is called the working gas because it provides the lifting force necessary to raise the valve disc off its seat.

The valve operator controls the flow of working gas by means of an electrically actuated lever (1). This lever is opened or closed by the temperature control circuit. When the burner is off and there is no call for heat from the room thermostat, the lever is in the position shown in Figure 5-48. Note that while the lever is in this position, it blocks the admission of gas into the valve. Furthermore,

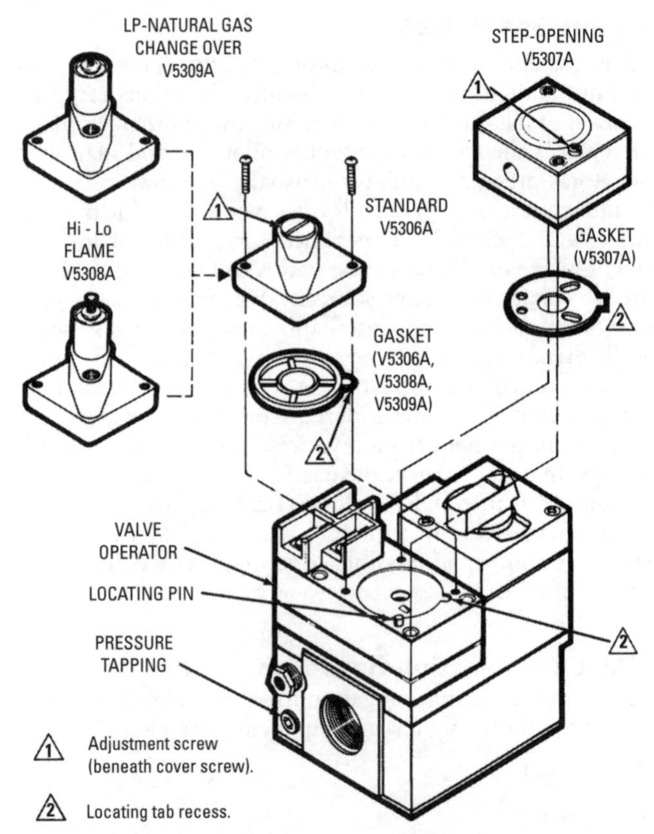

Figure 5-45 Various gas-pressure regulators available for use on a combination gas control equipped with a valve operator. *(Courtesy Honeywell Tradeline Controls)*

the working gas can escape through the working channel (2), resulting in a reduction of the gas pressure in the pressure chamber. The main gas valve closes with this loss of working gas pressure.

A call for heat from the room thermostat energizes the valve operator and causes the lever to open the inlet port (see Figure 5-49). The working gas then flows into the pressure chamber (4) and pushes the diaphragm (5) up against the valve disc assembly (3) to allow the flow of gas through the valve to the burner. At the same time, gas also flows into the pressure regulator chamber (8) and through the evacuation gas channel (6) into the combination gas control outlet.

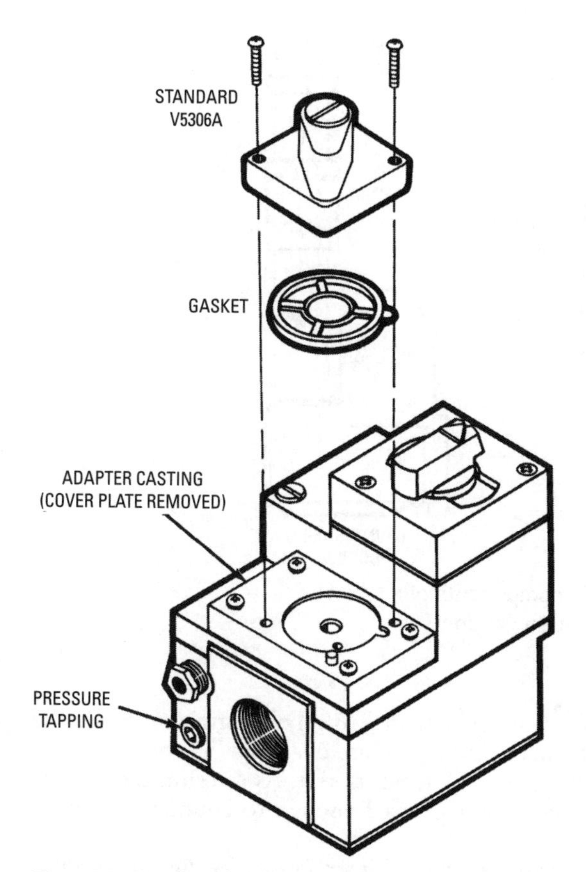

Figure 5-46 Standard gas-pressure regulators installed on manual manifold control not equipped with valve operator. *(Courtesy Honeywell Tradeline Controls)*

Changes in the outlet pressure of the gas cause changes in the position of the regulator diaphragm (9). If the outlet pressure rises, the regulator valve (7) opens slightly to allow more working gas into the evacuation gas channel (6). This discharge of working gas causes the main valve diaphragm (5) to drop and allows the main valve disc (3) to move downward on its seat. This action reduces the flow of main burner gas through the control to correct the rise in outlet pressure.

A drop in the outlet pressure has the opposite effect. The regulator valve (7) closes slightly and reduces the amount of working gas

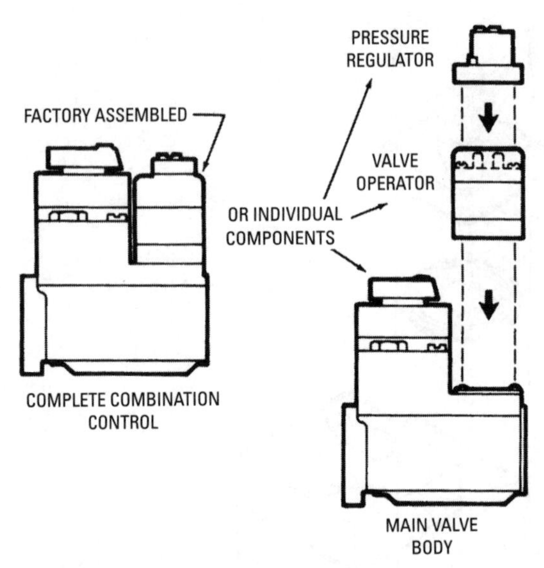

Figure 5-47 Basic components of a typical Honeywell combination gas control.
(Courtesy Honeywell Tradeline Controls)

entering the evacuation gas channel (6). This action increases the gas pressure in the main valve pressure chamber (4) and forces the main valve disc upward away from its seat. As a result, the flow of gas through the control is increased enough to counter the fall in outlet pressure.

Manufacturers provide detailed instructions for lighting a pilot in an installation equipped with a combination gas valve. These instructions should be read carefully before attempting to light the pilot burner. If there are no instructions with the equipment, the following procedure is suggested:

1. Turn the wall thermostat to *off* or its lowest setting.
2. Depress the gas cock and rotate it to the *off* position (see Figure 5-50).
3. Allow 5 minutes for any gas in the burner compartment to escape.
4. Turn the gas cock dial to the pilot position.
5. Depress and hold the gas cock dial down while lighting the gas pilot.

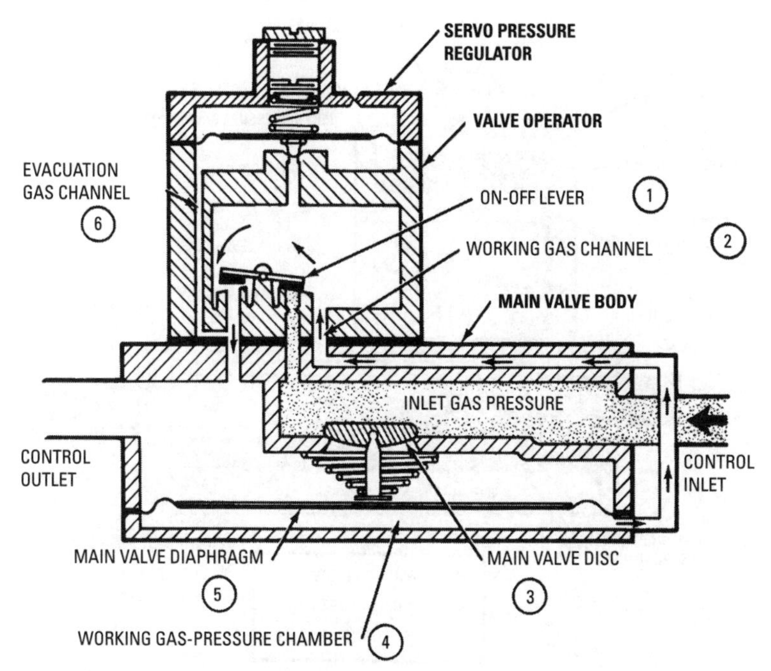

Figure 5-48 Schematic diagram of combination gas control in the burner off position. *(Courtesy Honeywell Tradeline Controls)*

6. Allow the pilot to burn 60 seconds before releasing the gas cock dial.
7. Turn the gas cock dial to *on* position after the pilot burner flame has been established.
8. Set the room thermostat to the desired temperature position.

The pilot position on a combination gas valve is used for temporary or seasonal shutdown. The *off* position is used when complete shutdown is necessary.

Continuous Pilot Dual Automatic Gas Valve

The Honeywell VR4205 direct-spark ignition (DSI) combination gas valve shown in Figure 5-51 contains a safety shutoff, a manual valve, two automatic operators, a pressure regulator, and a pilot adjustment device. Some models use a single electrode for spark ignition and flame sensing; others use separate electrodes for spark ignition and flame sensing. The Honeywell VR4205 gas control is

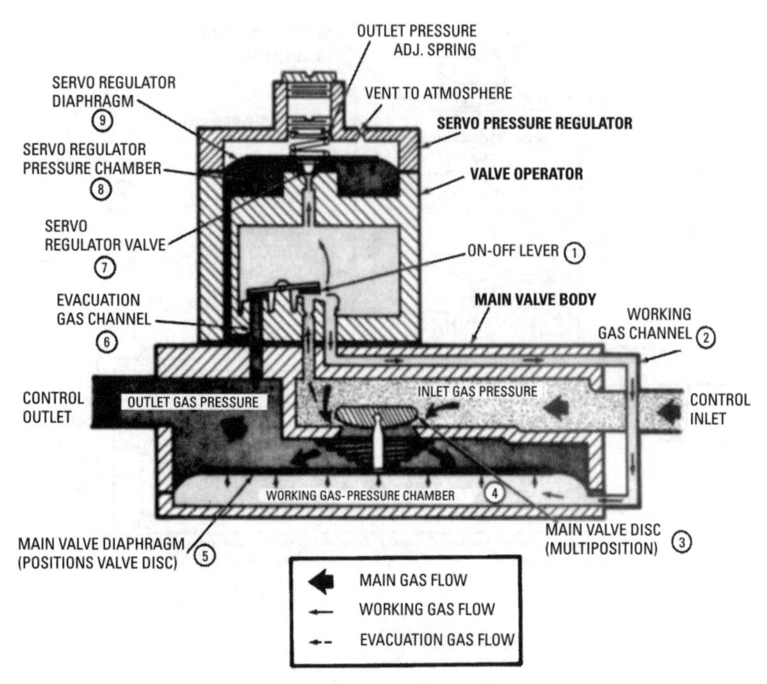

Figure 5-49 Wiring diagram of a combination gas control in the burner on position. *(Courtesy Honeywell Tradeline Controls)*

also designed to provide on-off manual control of gas flow. In the off position, gas flow to the main burner is mechanically blocked. In the on position, gas flows to the main burner under control of the thermostat, the direct-spark ignition (DSI) module, and the two automatic main valves.

The operation of a Honeywell VR4205 valve gas control is illustrated in Figures 5-52 and 5-53. When the thermostat calls for heat, the DSI module is energized. The module activates the first and second automatic valves of the gas control, which allow gas to flow to the main burner. At the same time, the DSI module generates a spark at the igniter-sensor to light the main burner. The second automatic valve diaphragm, controlled by the servo pressure regulator, opens and adjusts gas flow as long as the system is powered. The servo pressure regulator monitors outlet pressure to provide an even flow of gas to the main burner.

Loss of power (thermostat satisfied) deenergizes the DSI module and closes the automatic valves. The first automatic valve and the

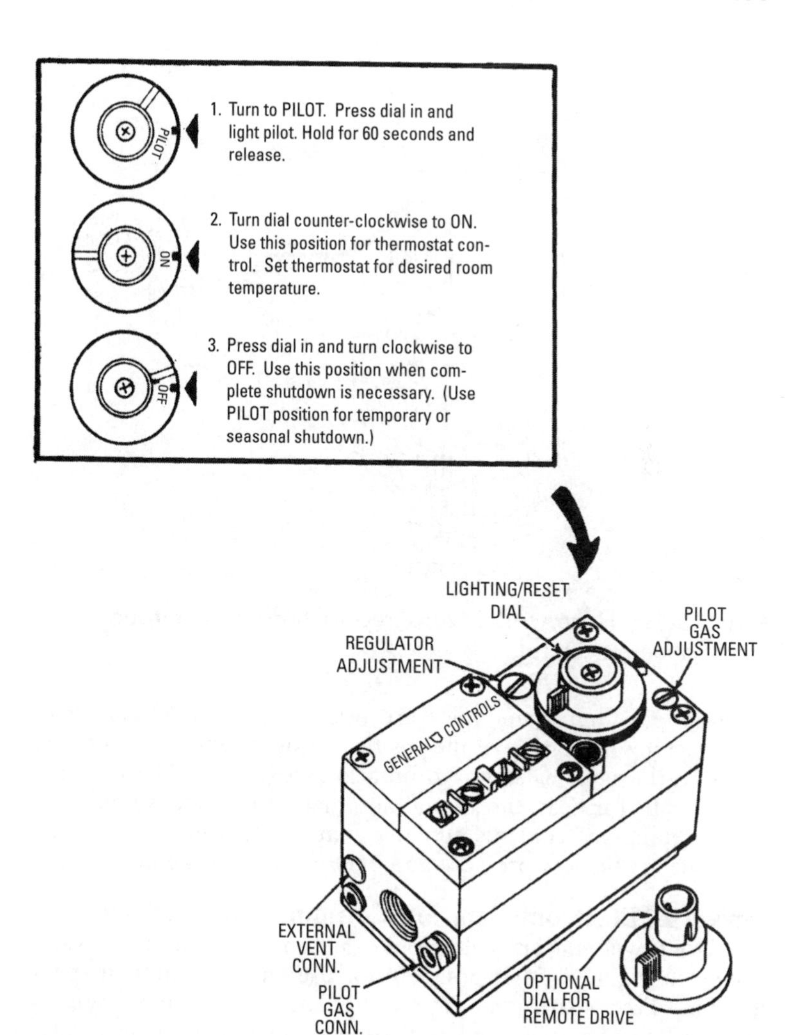

Figure 5-50 Lighting procedure for a combination gas control.
(Courtesy ITT General Controls)

second automatic valve operator close, bypassing the regulator(s) and shutting off the main burner and the pilot. As pressure inside the gas control and underneath the automatic valve diaphragm equalizes, spring pressure closes the second automatic valve to provide a second barrier to gas flow. The system is now ready to return to normal service when the thermostat again calls for heat and power is restored.

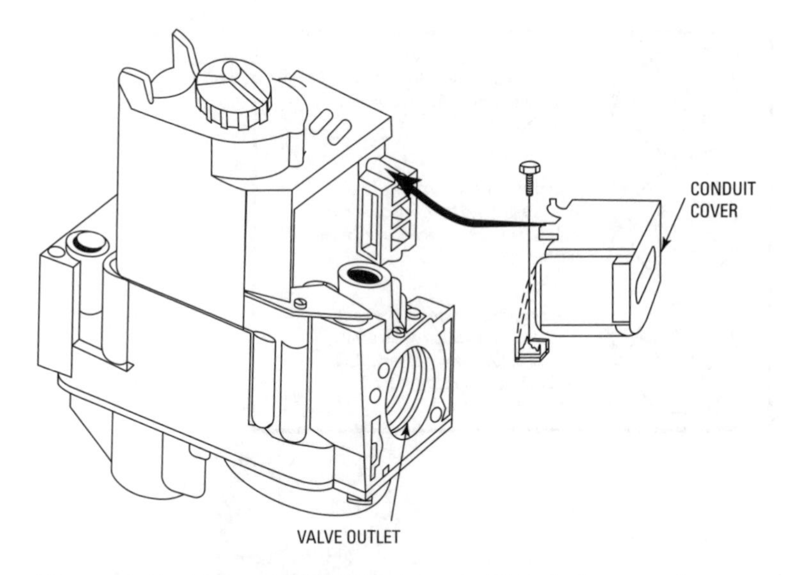

Figure 5-51 Honeywell VR4205 direct ignition dual automatic combination gas control. *(Courtesy Honeywell, Inc.)*

Some gas control modules are offered with slow-opening or step-opening regulation of the gas flow. Slow-opening gas controls function the same as standard models except that when the thermostat calls for heat, the second automatic valve opens gradually. Step-opening gas controls actually combine two pressure regulators, one for the low pressure and one for the full-rate pressure.

Universal Electronic Ignition Combination Gas Valve

The Honeywell universal electronic ignition combination gas valve shown in Figure 5-54 is designed for use with intermittent-spark ignition, direct-spark ignition, and hot-surface ignition systems in 24 VAC gas furnaces and boilers. This combination gas valve includes a manual valve, two automatic operators, pressure regulator, pilot adjustment, pilot plug, and ignition adapter.

Pilot Burners

A *pilot burner* is a device used in a gas-fired appliance to light the main gas burners and generate sufficient millivoltage to operate a thermocouple or thermopile pilot safety shutoff device (see *Automatic Pilot Valves* in this chapter). Modern pilot burners are designed to burn continually in order to provide an automatic ignition of the burners when the main gas supply is turned on.

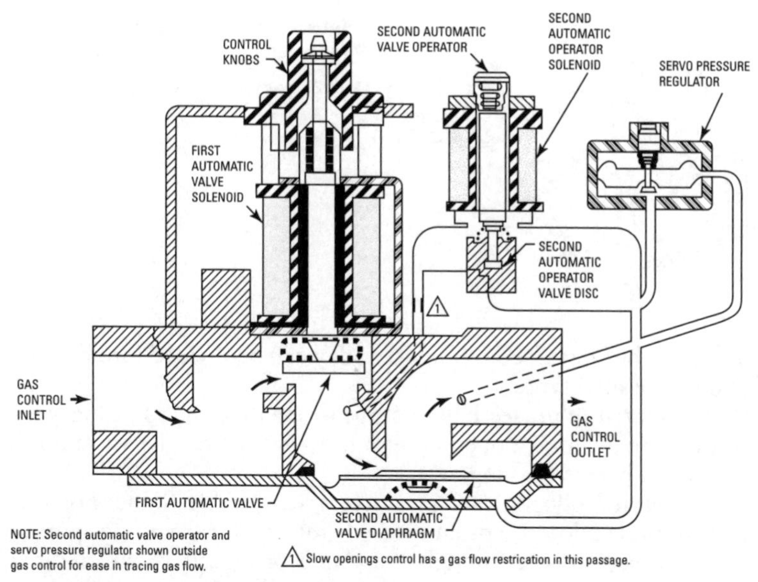

Figure 5-52 Continuous pilot dual automatic gas valve operation during burner on cycle. *(Courtesy Honeywell, Inc.)*

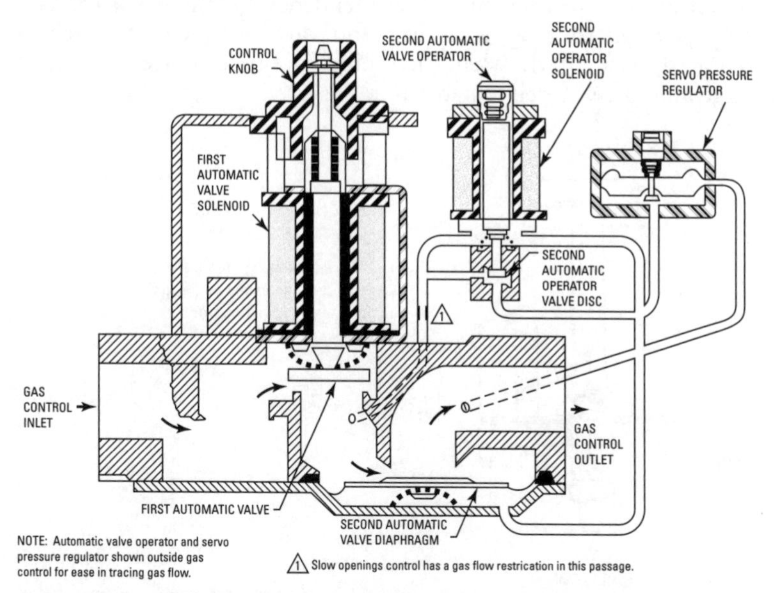

Figure 5-53 Continuous pilot dual automatic gas valve operation during burner off cycle. *(Courtesy Honeywell, Inc.)*

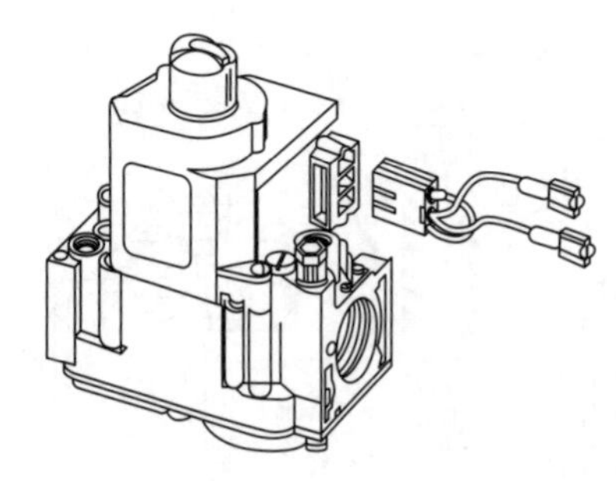

Figure 5-54 Honeywell universal ignition combination valve. *(Courtesy Honeywell, Inc.)*

The pilot burners in common use today can be divided into aerated and nonaerated types.

An *aerated pilot* (see Figure 5-55) is one that injects primary air through an air intake opening into the gas stream. The air and gas are mixed before burning.

An aerated pilot burner produces a very stable flame. For this reason, these pilots are often used where the pilot location is particularly inaccessible. Although the flame produced by an aerated pilot burner is more stable than one produced by a nonaerated type, an aerated pilot burner does have some disadvantages. An important one to remember is the tendency for the small primary air openings to clog with lint and dirt. Frequent cleaning is required, particularly when these pilots are used in areas having a large amount of foreign material in the air.

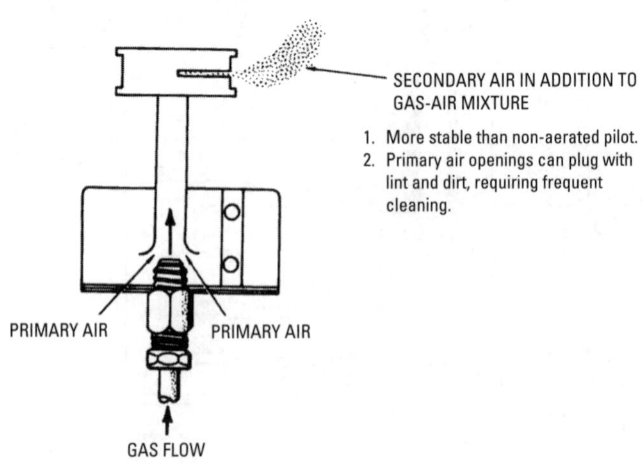

Figure 5-55 Aerated pilot. *(Courtesy Trane Co.)*

A *nonaerated pilot* (see Figure 5-56) does not inject primary air. As a result, the air and gas are not premixed, and the combustion process must be completed with secondary air only. This results in a less-stable flame than the one produced by an aerated pilot. On the credit side, a nonaerated pilot requires less maintenance than an aerated pilot. For this reason, the nonaerated pilot burner is usually preferred by most utility companies.

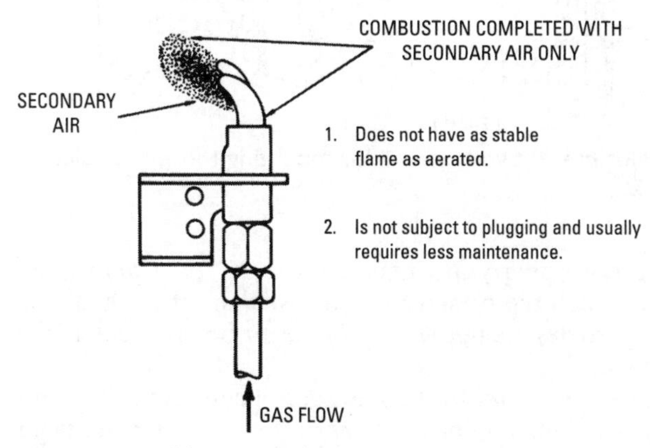

Figure 5-56 Nonaerated pilot. *(Courtesy Trane Co.)*

A pilot burner assembly consists of the pilot bracket, pilot orifice, primary air intake, lint screen, mixing chamber, pilot ports, and pilot hood.

The *pilot bracket* is a device used to mount the pilot in a fixed relationship to the burner. Some pilot brackets also contain means for mounting the thermocouple or pilot generator so that the hot junction is located directly in the path of the pilot flame (see Figure 5-57).

The pilot ports are the openings through which the gas (in nonaerated pilot burners) or the gas and air mixture (in aerated pilot burners) passes before burning. The gas and air are premixed in the *mixing* chamber of an aerated pilot. The air is injected into the mixing chamber of an aerated pilot through a hole or opening called the *primary air intake*. The amount of primary air can be controlled by adjusting an *air shutter* that covers the primary air intake opening.

A *lint screen* is generally used to remove lint, dirt, and other contaminants from the primary air before it enters the primary air opening. In some pilots, the lint and other particles are burned

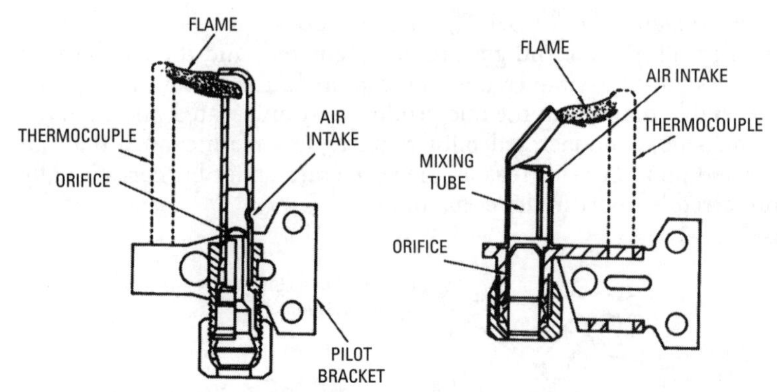

Figure 5-57 Pilot bracket with means for mounting thermocouple.
(Courtesy Robertshaw Controls Co.)

before entering the primary air intake. An *incinerated pilot* is an aerated pilot in which the primary air passes adjacent to the flame area where the particles are burned out of the air before it enters the primary air intake.

Impurities are also found in the pilot gas. These impurities can be removed by installing a pilot gas *filter* in the line upstream from the pilot adjustment means in the control. The pilot gas filter is expected to operate several years without service. Clogging of the filter will be indicated by shortened pilot flames, which will result in improper pilot operation. Shortened pilot flames can also be caused by pilot tube stoppage or a dirty pilot orifice. These possibilities should be checked before removing the filter. If the filter should become clogged, replace the *entire* filter rather than just the filtering medium.

The *pilot orifice* is a removable component in the pilot that contains precisely sized openings that control the admission of gas to the pilot. Pilot orifices are either of the spud or insert type. A *spud orifice* screws into the bottom of the pilot burner. It is both an orifice and a fitting combined into a single unit with threads at either end (see Figure 5-58). An *insert orifice* must be held in position by a separate fitting (see Figure 5-59). The pilot gas line (tubing) is connected to the bottom of the pilot orifice.

Installing a Pilot Burner

Always follow the manufacturer's instructions when installing a new pilot burner. Read these instructions very carefully before beginning any work.

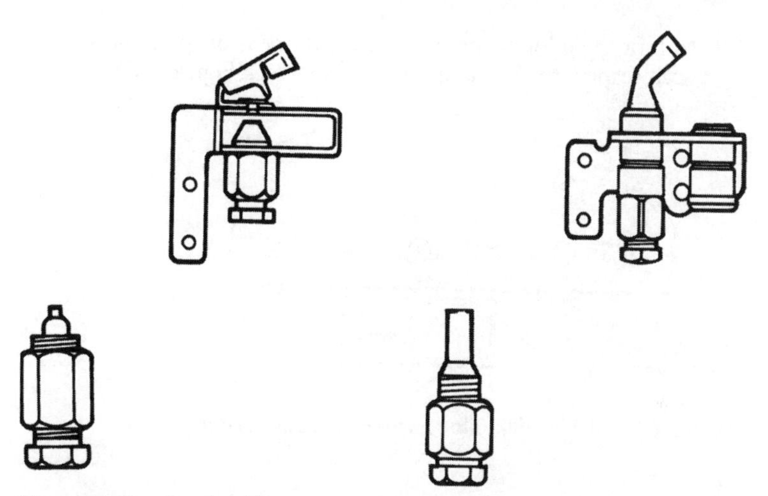

Figure 5-58 Spud orifice. *(Courtesy Honeywell Tradeline Controls)*

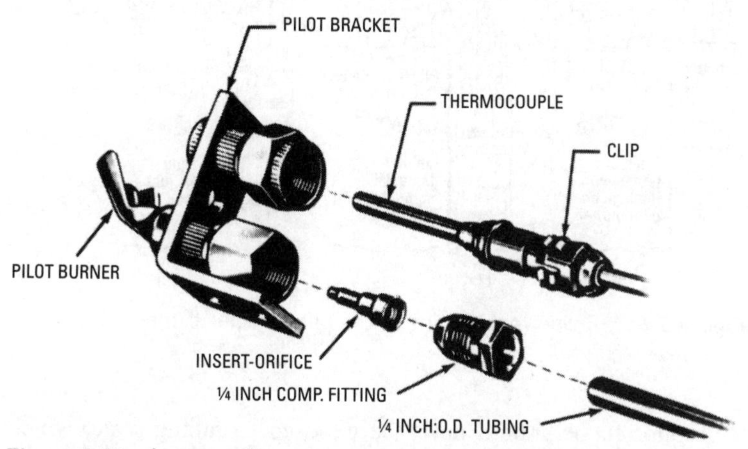

Figure 5-59 Insert orifice. *(Courtesy Honeywell Tradeline Controls)*

If no installation instructions are available, the following location and mounting requirements should be carefully observed:

1. Choose a location for the pilot burner that provides easy access, observation, and lighting.
2. Rigidly affix the pilot burner to the main burner. Other mounting surfaces should not be used (see Figure 5-60).

3. Mount the pilot burner so that the flame is properly positioned with respect to the main burner flame (see Figure 5-61).

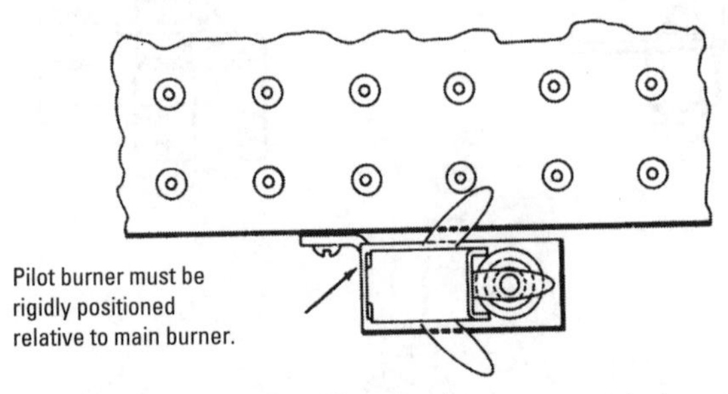

Figure 5-60 Rigidly affix pilot burner to main burner.

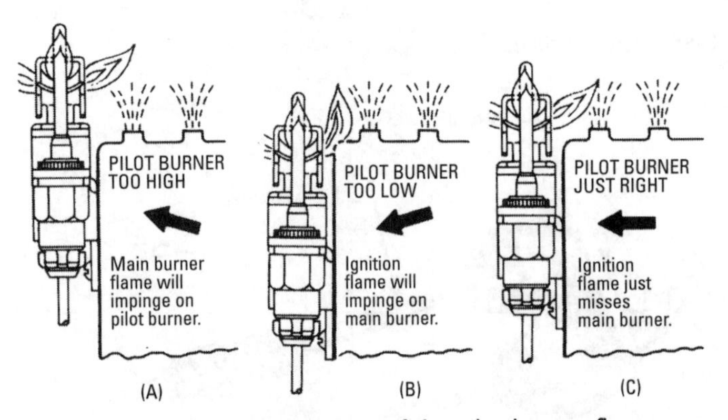

Figure 5-61 Proper positioning of the pilot burner flame.
(Courtesy Honeywell Tradeline Controls)

A pilot flame should never be exposed to falling scale, which could impair the ignition of the main burners. Furthermore, the flame should not be exposed to draft conditions or to sudden puffs of air sometimes caused by igniting or extinguishing the main burner. Always provide an ample air supply free of contaminating products of combustion.

Replacing the Pilot Burner Orifice

The procedure for replacing a pilot burner orifice depends on whether it is a spud or an insert orifice. For a Honeywell spud orifice, the procedure is as follows:

1. Disconnect the gas supply tubing from the pilot burner.
2. Unscrew the spud orifice and throw it away.
3. Remove any burrs from the tubing and square off the ends.
4. Remove the one-piece nut and ferrule from the new assembly and slip them over the tubing.
5. Install the new assembly in the pilot burner and tighten securely.
6. Push the tubing (along with the nut and ferrule) into the burner as far as it will go and engage the nut.
7. Tighten the nut by hand until it will turn no further. Use a wrench to make one final turn.

The connection in step 7 should be tight enough to prevent any gas leakage. *Do not tighten it too much* or you will run the risk of stripping the threads. Try not to bend the tubing near the fitting after the nut has been tightened.

As shown in Figure 5-59 the procedure for replacing an insert orifice presents no serious problems. The gas supply tubing must first be disconnected from the pilot burner by unscrewing the compression fitting. The small insert orifice can then be removed. Sometimes a light tap on the pilot burner bracket will be required to dislodge the orifice.

Place the new orifice on the end of the gas tubing and insert both the orifice and tubing into the pilot burner. Tighten the compression nut until it is secure. Use the same procedure described for tightening a spud orifice.

Lighting the Pilot

Read the appliance manufacturer's lighting instructions before attempting to light the pilot burner. The basic procedure for lighting a pilot is as follows:

1. Turn the room thermostat to its lowest setting.
2. Shut off the main gas supply to the main burner and the pilot burner.
3. Allow at least 5 minutes for the unburned gas to vent.
4. Light the pilot burner in accordance with the appliance manufacturer's lighting instructions.

Venting the unburned gas (step 3) is very important. This is especially true for LP gas because it is heavier than air and will not vent upward naturally. *Every* precaution should be taken to ensure that the appliance is properly venting any unburned gas.

Appliance and pilot burner manufacturers provide very detailed lighting instructions for their equipment. Moreover, the development of various types of combination gas controls has simplified the lighting procedure and increased the safety factor (see *Combination Gas Valves* in this chapter).

Pilot Flame Adjustment

The pilot flame must be adjusted for proper color and size. Appliance manufacturers refer to this procedure as pilot *flame* adjustment or pilot *gas* adjustment.

The appliance manufacturer will provide instructions for making pilot flame adjustments. On combination gas valves, this involves the removal of a pilot adjustment cap and turning the adjustment screw until the desired flame characteristics are obtained. The best pilot flame is a steady, nonblowing blue flame that envelops the upper ⅜ to ½ inch of the thermocouple or generator (see Figure 5-62).

Main Burner Ignition

The pilot burner should ignite the main burner quietly and reliably under all operating conditions, including low gas supply pressure. The ignition of the main burner gas should occur within 4 seconds from the time that gas is admitted to the main burner. This should occur when the pilot gas supply is reduced to an amount just above the point of pilot flame extinction.

A main burner ignition test should be performed *after* the main burner gas input and primary air adjustments have been made. The type of test used to check main burner ignition will depend on the type of pilot used in the gas-fired appliance. For example, in a pilot

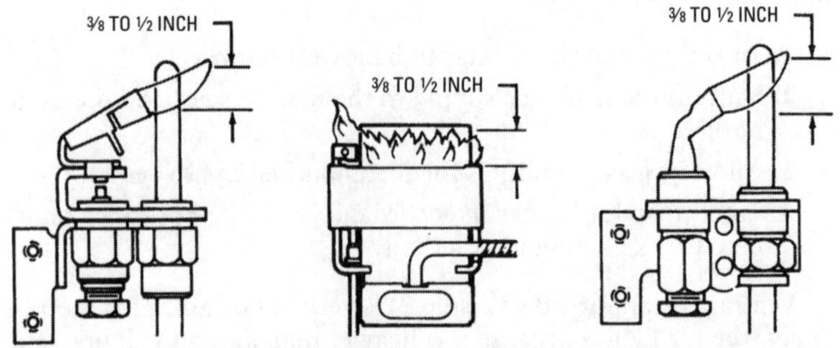

Figure 5-62 **Pilot flame should generally envelop ⅜ to ½ inch of the thermocouple or thermopile tip.** *(Courtesy Honeywell Tradeline Controls)*

generator system, main burner ignition is checked with the pilot flame adjusted to the minimum millivoltage required to open the main valve. The manufacturer's installation literature for the appliance will probably include instructions for testing main burner ignition. If no literature is available, check with the local gas company.

Pilot-Pressure Switch

A *pilot-pressure switch* operates on the same principle as the gas-pressure switch used in the manifold of the gas-fired furnace or boiler. Its function is to prevent the premature failure of the spark igniter or glow coil should a prolonged gas interruption occur while the thermostat is calling for heat.

The pilot-pressure switch is installed in the pilot gas line between the pilot and pilot regulator. Installing a pilot switch requires that the pilot gas line be disconnected downstream from the pilot gas regulator. The gas line is then connected to the pilot-pressure switch by cutting the existing tubing to the regulator and connecting into the tee provided for switch mounting. A typical wiring of a pilot-pressure switch is shown in Figure 5-63.

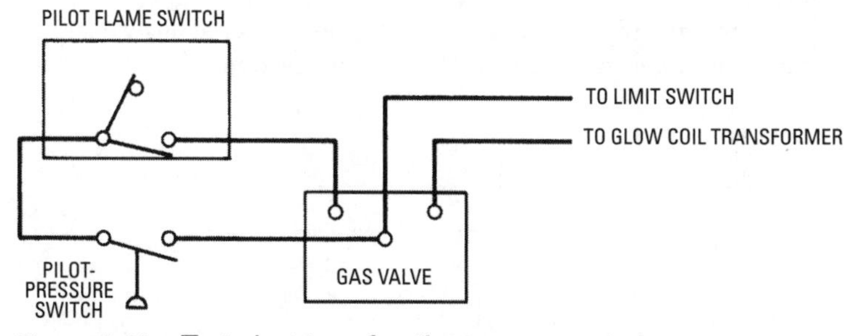

Figure 5-63 Typical wiring of a pilot-pressure switch. *(Courtesy Janitrol)*

Electronic Ignition Modules

Electronic (solid-state) ignition systems have been developed to improve the energy efficiency of gas furnaces, boilers, and water heaters. These electronic ignition systems replace the less energy-efficient standing pilot used in traditional gas-fired appliances. There are three different types of controls used in electronic ignition systems:

- Intermittent pilot ignition module
- Direct ignition gas module
- Hot-surface ignition module

Note

Ignition modules do not have replaceable parts. If defective, the entire unit must be replaced. A replacement module must be of the exact same model and type as the defective one.

Manufacturers of electronic ignition modules offer a wide range of models designed to fit a variety of different applications (see Figure 5-64). Always make certain that the make, model, and operating specifications of the original gas valve control module match those of the replacement one. Specification sheets and installation instructions for electronic ignition modules are available online at the manufacturer's Web site or by writing directly to the manufacturer (see Appendix B for contact information).

The following three sections briefly describe the three types of electronic ignition modules used in gas-fired furnaces, boilers, and water heaters. For more detailed and model-specific information, consult the owner's manual, installation manual, and/or the specification/data sheets provided by the manufacturer.

Intermittent Pilot Ignition Module

An intermittent pilot ignition module is a solid-state ignition device used to automatically light a pilot burner and simultaneously energize (operate) the main gas valve of the heating system when the room

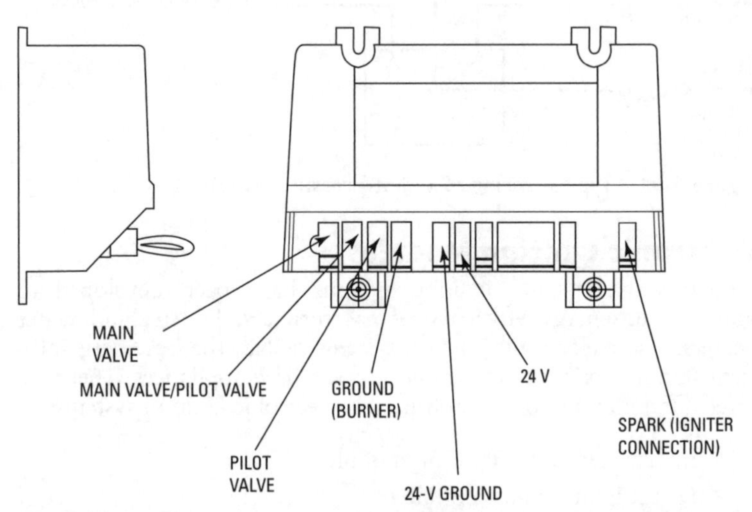

Figure 5-64 Typical intermittent pilot ignition module. *(Courtesy Honeywell, Inc.)*

thermostat calls for heat (see Figure 5-65). Figure 5-66 illustrates the wiring connections between an intermittent pilot module, the dual-valve combination valve, and the combined pilot burner and igniter-sensor unit.

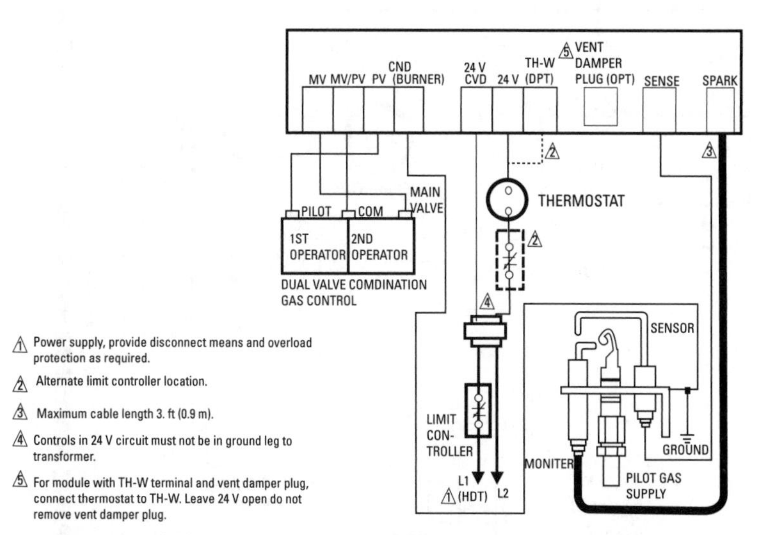

Figure 5-65 Intermittent pilot ignition module. *(Courtesy Honeywell, Inc.)*

The operating sequence for a gas burner operated by an intermittent pilot ignition control is as follows:

1. Room thermostat calls for heat, and the intermittent pilot ignition module simultaneously opens the pilot valve and supplies a continuous spark to the electrode in the pilot burner.
2. Pilot burner gas ignites and produces a flame.
3. Pilot flame sensor detects the pilot flame and signals the intermittent pilot ignition control to discontinue the spark and energize (open) the main gas valve.

Note

The main gas valve will not be energized until the flame sensor detects the presence of the pilot flame. As long as the main gas valve remains closed, no gas from the supply line can flow through the burners. Should a loss of flame occur, the main gas valve closes and the spark recurs within 0.5 second.

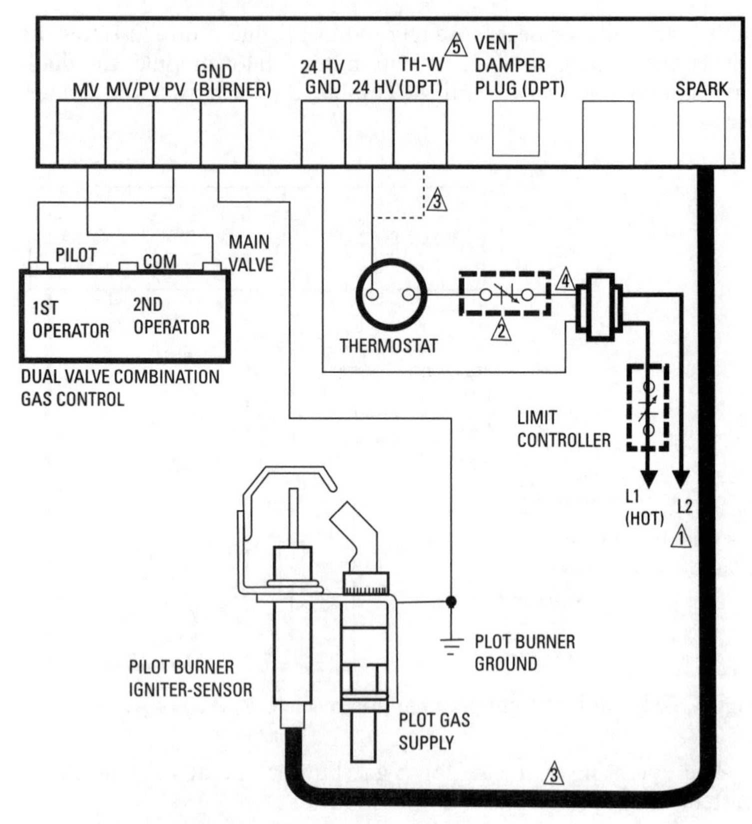

1. Power supply, provide disconnect means and overload protection as required.
2. Alternate limit controller location.
3. Maximum cable length 3 ft (0.9 m).
4. Controls in 24 HV circuit must not be in ground leg to transformer.
5. For module with TH-W terminal and vent damper plug, connect thermostat to TH-W. Leave 24 HV open. Do not remove vent damper plug.

Figure 5-66 Intermittent pilot module with a combination pilot burner and igniter–flame sensor in a heating system with an atmospheric burner. *(Courtesy Honeywell, Inc.)*

4. Gas from the gas supply line flows through the now-open main gas valve to the burners and ignites. This is the burn or *on* cycle. When the heat has reached the level required by the thermostat setting, the main gas closes and the burner or burners shut down. This is the *off* cycle in an intermittent pilot ignition control system.

Some Useful Definitions

- **Run**—The period during which the main gas valve remains energized and the spark is turned off after the successful ignition.
- **Trial for ignition**—The period during which the pilot valve and spark are activated, attempting to ignite gas at the main gas burner.
- **Flameout**—The loss of proven flame.
- **Proven flame**—A pilot flame detected by a flame sensory device.
- **On cycle**—Period of time during which the main gas valve is open and the burners are operating.
- **Off cycle**—Period of time during which the main gas valve is closed and the burners are not operating.

If the pilot flame is extinguished, even though the room thermostat is still calling for heat, the intermittent pilot ignition control immediately deenergizes the main gas valve, causing it to close its open supply port and stop the flow of gas to the burners. A spark at the pilot burner electrode will recur within 0.8 second.

As soon as the pilot flame is reignited and detected by the pilot flame sensor, the main gas valve is energized, the valve port is opened, and the spark is extinguished. The intermittent pilot ignition control then deenergizes the pilot gas and main burner gas valve when the thermostat stops calling for heat.

Direct-Spark Ignition Module

The direct-spark ignition (DSI) module illustrated in Figure 5-67 is a low-voltage, solid-state unit that controls the gas valve, monitors the burner flame, and generates a high-voltage spark for ignition. DSI modules are available with or without a purge timer and with separate or combined igniters and flame sensors. Typical wiring connections for a direct-spark ignition system are illustrated in Figure 5-68.

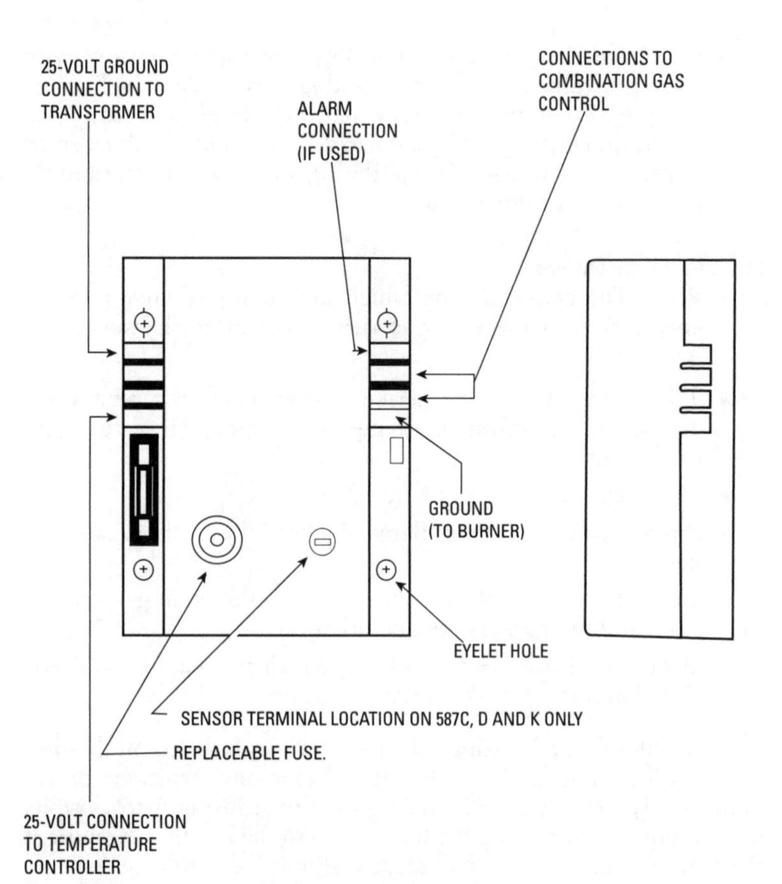

Figure 5-67 Typical direct-spark ignition module. *(Courtesy Honeywell, Inc.)*

Hot-Surface Ignition Module

The principal components of a hot-surface ignition system are the hot-surface ignition module, a line voltage silicon carbide igniter (also sometimes called a *glow stick* or *glow plug*), a remote flame sensor, a 24-volt AC ignition-detection control, and a 24-volt (AC) redundant gas valve (see Figure 5-69). The flame sensor is designed to detect the presence of a flame. It can be mounted remotely on multiple burners or next to the gas burner.

The hot-surface ignition module, similar to the one shown in Figure 5-70, is a microprocessor-based gas ignition control designed for direct ignition gas-fired appliances. It provides direct

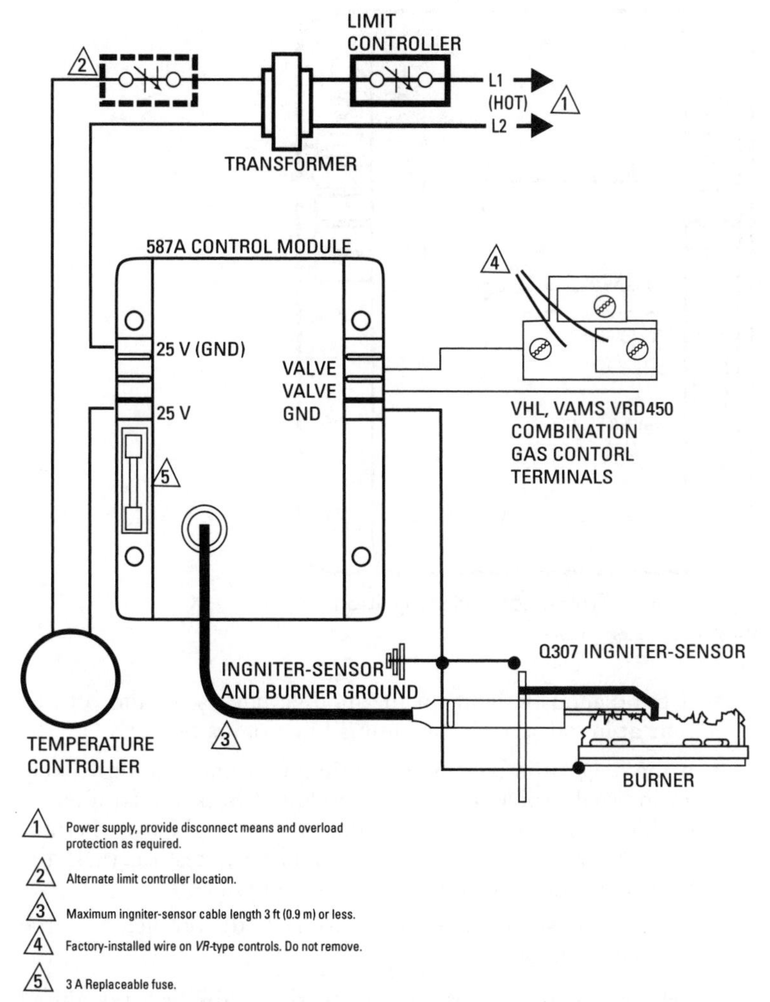

Figure 5-68 Typical wiring connections for a direct-spark ignition system. *(Courtesy Honeywell, Inc.)*

main gas burner ignition, remote sensing, and prepurging. It will retry for ignition and has a fixed time for flame lockout.

Some hot-surface ignition modules have self-diagnostic capabilities. A diagnostic light on the HSI module provides the following information:

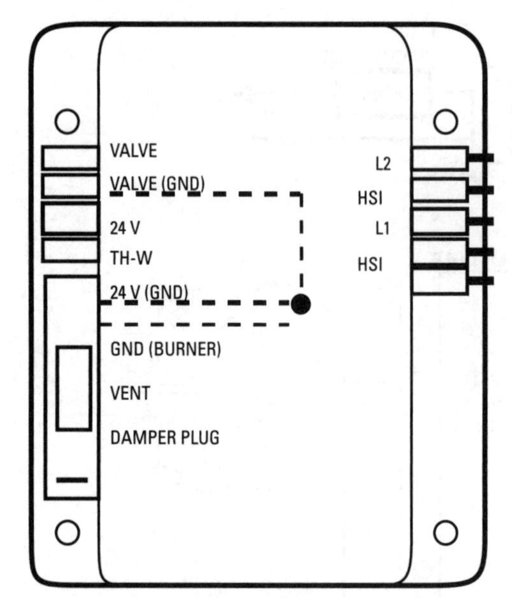

Figure 5-69 Typical hot-surface ignition module. *(Courtesy Honeywell, Inc.)*

- If the diagnostic light on the module flashes on and off one time at initial startup, the unit is functioning properly.
- If the diagnostic light is lit continuously, there is most likely an internal problem with the module. Check for an internal problem by interrupting the line power or 25-volt thermostatic power for a few seconds and then restore it. If the burner still fails to ignite, replace the module.
- If the diagnostic light continues to flash, the problem is in the external components or wiring.

For HIS modules without self-diagnostic capabilities, a qualified HVAC technician or electrician should troubleshoot the system with the appropriate test equipment. The test equipment should include the following:

- A volt-ohm meter for checking both the voltage and the resistance.
- A precision microammeter for checking the flame sensor output and location.
- A pressure gauge (low scale) for checking gas pressure.

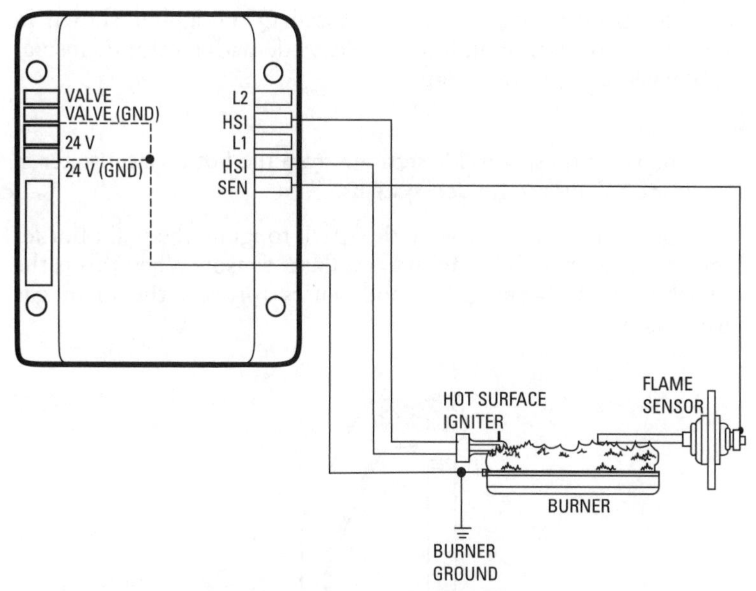

Figure 5-70 Honeywell hot-surface ignition module with wiring connections to the flame sensor and the hot-surface igniter.

(Courtesy Honeywell, Inc.)

Warning

> Extreme caution must be taken when working on a hot-surface ignition system. Because of the high voltage present, there is always the potential for serious electrical shock.

If the unit is not equipped with a self-diagnostic light, closely follow the troubleshooting suggestions provided by the manufacturer. These will be specific to the make and model. Some simple things to look for include the following:

- Checking to make certain the manual knob on the gas valve is in the on position and gas is available at the inlet piping
- Checking the outlet gas pressure to make sure it matches the nameplate rating
- Checking the wire leads to the gas valve for proper connection or damage

Igniters

An igniter produces the spark for direct ignition of the main gas burner in various heating applications (gas furnaces, gas boilers, gas water heaters, etc.). The Honeywell Q347 igniter shown in Figure 5-71 consists of an internal electrode with a ceramic insulator, bracket, and ground strap.

Note

> The flame-sensing rod is separate from the hot-surface igniter in most hot-surface ignition systems.

An igniter is used to provide the spark to ignite the main burner flame. Some igniters have an integral flame sensor. When this is the case, the igniter both ignites and senses (proves) the main gas burner flame.

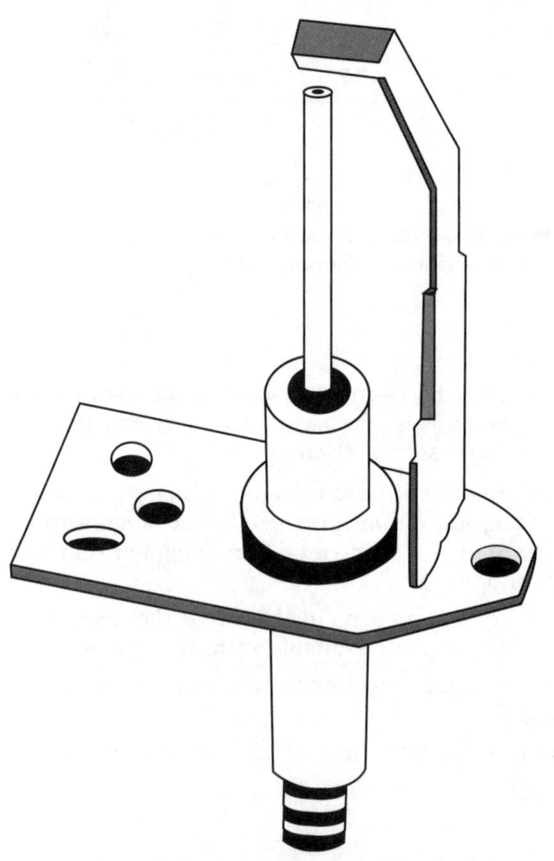

Figure 5-71 Honeywell Q347 igniter. *(Courtesy Honeywell, Inc.)*

The operating sequence of a system in which an igniter is used may be summarized as follows:

1. Room thermostat calls for heat.
2. Gas valve opens and gas flows to the burner(s).
3. Burner ignites when the gas reaches the main burner.
4. Spark igniter shuts off.

The duration of the spark operation must be within the igniter manufacturer's specified lockout timing period. The igniter manufacturer will provide a chart of the ignition control lockout times in the service literature for the igniter. The example shown in Table 5-3 is for Honeywell's Q347 igniter.

Table 5-3 Ignition Control Lockout Times

Specified Lockout Time (Stamped on Ignition Module)	*Maximum Safety Lockout Time*
4.0 seconds	5.0 seconds
6.0 second	7.0 seconds
11.0 seconds	15.0 seconds
15.0 seconds	20.0 seconds
21.0 seconds	35.0 seconds

(Courtesy Honeywell, Inc.)

The electrode spark gap in the igniter must be within the specified maximum (see Figure 5-72). If the gap is not within specifications, it will have to be adjusted for optimum performance.

The flame rod of a combined igniter–flame sensor unit must be immersed 1 inch in the burner flame to produce the best flame signal

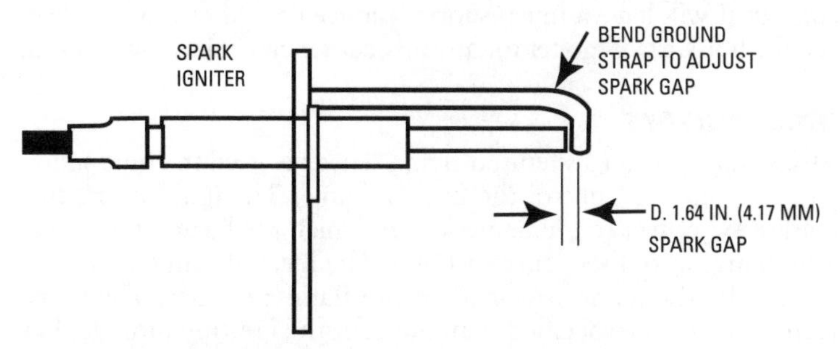

Figure 5-72 Igniter gap adjustment. *(Courtesy Honeywell, Inc.)*

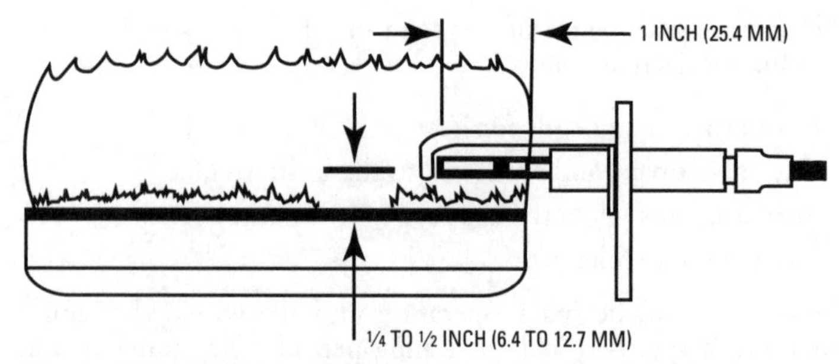

Figure 5-73 Flame rod immersed 1 inch in burner flame.
(Courtesy Honeywell, Inc.)

(see Figure 5-73). Examples of poor flame conditions and their probable causes are illustrated in Figure 5-74.

The flame signal can also be adversely affected by a bent bracket, bent rod, or cracked ceramic insulator. Sometimes the bracket can be bent back into shape. If the rod is bent or the ceramic insulator is cracked, the igniter should be replaced.

Note

> Always check the specifications of the replacement hot-surface igniter before installing it. Not all igniters have the same voltage or warm-up time as the original design.

The igniter used in a hot-surface ignition operating system differs in design from the type used in intermittent pilot or direct-spark ignition systems described in the preceding paragraphs (see Figure 5-75).

Be careful when replacing a hot-surface igniter because they are fragile and easily damaged. Sometimes a crack in the igniter surface is so small that it is not visible. A cracked hot-surface igniter may still work, but it will have a much shorter service life. After it is installed, check the hot-surface igniter for any inconsistencies in its glow pattern.

Flame Sensors

In electronic ignition systems, a flame sensor is used in conjunction with the igniter to control the burner flame. The ignition module provides AC power to the flame sensor, which the burner flame rectifies (changes) to direct current (DC). The level of flame current is measured by the flame sensor to ensure flame presence. The flame current must be the specified minimum for the ignition module. For example, the flame current for a Honeywell S87C direct-spark

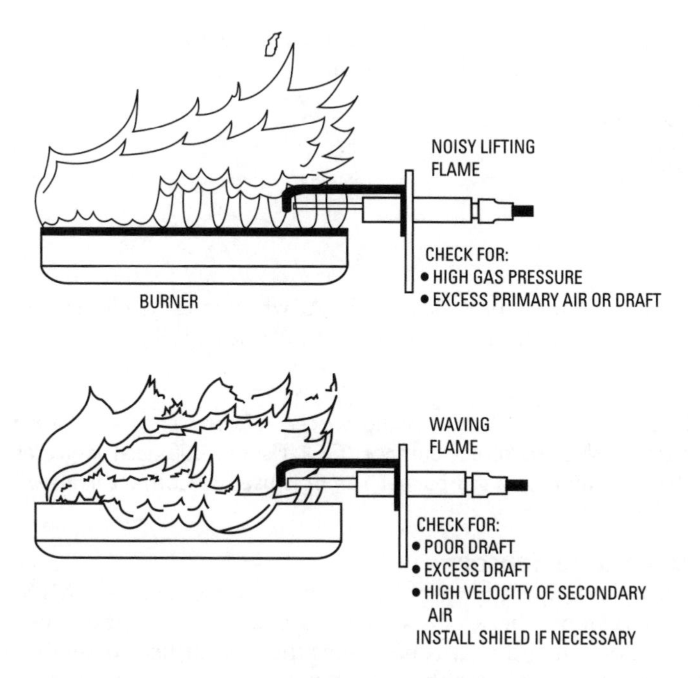

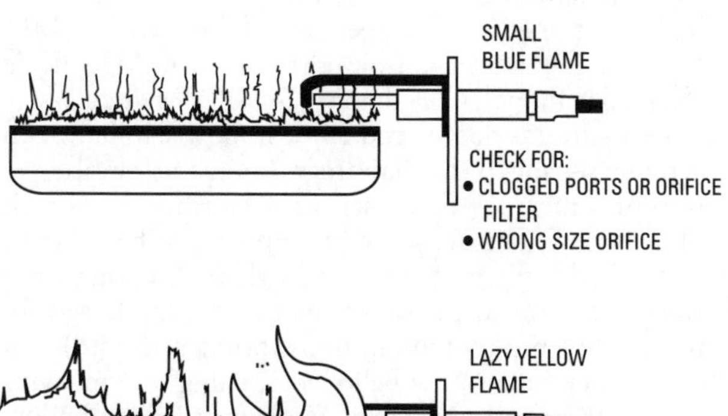

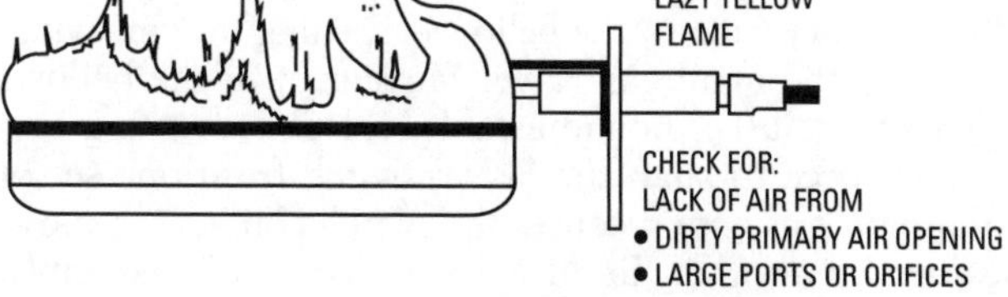

Figure 5-74 Examples of poor flame conditions. *(Courtesy Honeywell, Inc.)*

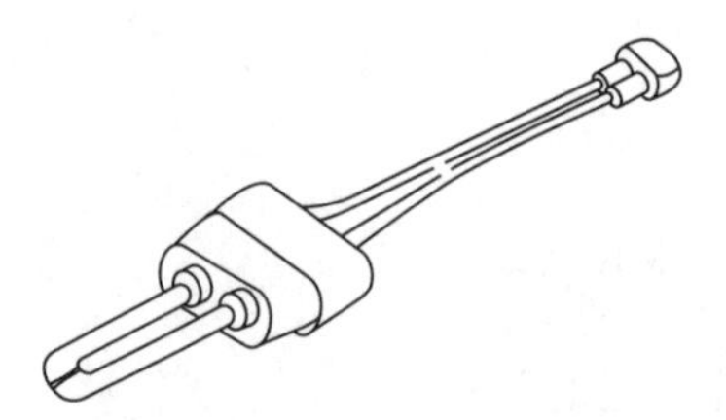

Figure 5-75 Typical hot-surface igniter.

ignition module must be at least 1.5 μA, whereas for a Honeywell S89 hot-surface ignition module it must be at least 0.8 μA.

Note

In electronics, *rectification* is the process of converting alternating current (AC) to direct current (DC). *Flame rectification* indicates that a flame (e.g., a gas burner flame) is used to convert from AC to DC current.

Mercury Flame Sensors

Many people frequently confuse mercury flame sensors (MFS) with thermocouples because each has a similar sensor that extends into the pilot flame and a tube connecting the device to the gas valve. Moreover, they have the same function in the control system of a gas-fired appliance, but there are significant differences.

The principal components of a mercury flame sensor (MFS) device are (1) a pilot flame sensor (the portion of the MFS device extending into the pilot flame), (2) a diaphragm/SPDT switch assembly located at the main gas valve, and (3) a hollow capillary tube connecting the flame sensor to the diaphragm/switch assembly.

The operation of a mercury flame sensor device depends on the evaporation of mercury. The sensor end, capillary tube, and the SPDT switch are filled with mercury. When there is enough heat produced by the pilot flame at the sensor end of the capillary tube, it vaporizes and pushes the remaining nonvaporized (liquid) mercury down the capillary tube to the bellows-type diaphragm/switch assembly located at the main gas valve. Movement of the bellows diaphragm presses against a nonadjustable, calculated spring tension with enough force to snap the SPDT switch from one set of contacts to another. This action causes the switch contacts to move from one position to another. In an MFS device switch assembly, the normally closed contact opens and the normally open contact closes. This action deactivates the igniter (after the pilot flame is proven) and opens the main gas valve to allow raw gas to flow to the burners.

Note

Mercury flame sensors are no longer used in gas-fired furnaces and boilers, especially those equipped with solid-state control modules. However, manufacturers still produce replacement MFS units along with their compatible main gas valves.

Oil Controls

The principal functions of the oil controls are (1) to turn the oil burner on and off in response to temperature changes in the space or spaces being heated and (2) to stop the system if an unsafe condition develops. The following controls are necessary to perform these functions:

- Thermostat
- Limit controls
- Primary control
- Oil valves
- Time-delay controls
- Circulator or fan control
- Other auxiliary controls

This chapter is concerned with a description of the oil burner primary control, oil valves, and time-delay controls. The remaining controls found in an oil burner control system are described in Chapter 4 ("Thermostats and Humidistats") and Chapter 6 ("Other Automatic Controls").

Oil Valves

Oil valves are used to provide on-off control of the flow of oil to the oil burner. These are normally closed solenoid valves that open when energized and close immediately when deenergized. They are variously referred to as *solenoid oil valves*, *magnetic oil valves*, or *oil burner valves* and are available in either immediate-discharge or delayed-discharge models.

An immediate-discharge oil valve discharges oil as soon as it is energized. A delayed-discharge valve is equipped with an integral thermistor to delay the valve opening for about 3 to 15 seconds (the length of time will vary depending on the manufacturer). This delay allows the burner fan to reach operating speed and establish sufficient draft before the oil is discharged.

A solenoid oil valve will make an audible click when it is opening and closing properly. If the valve fails to open after the room thermostat calls for heat, the following conditions may be responsible:

- Inadequate fuel pressure available at the valve
- An obstructed bleed line
- No voltage indicated at valve

Check the voltage at the coil lead terminals against the voltage shown in the nameplate. Also check the inlet pressure against the rating on the nameplate. If none of these conditions is causing the problem, the failure of the valve to open is probably due to a malfunctioning solenoid coil. The position of the coil is shown in the exploded view of the valve in Figure 5-76. The steps for replacing the solenoid are as follows:

1. Remove the nut on top of the valve by turning it counterclockwise.

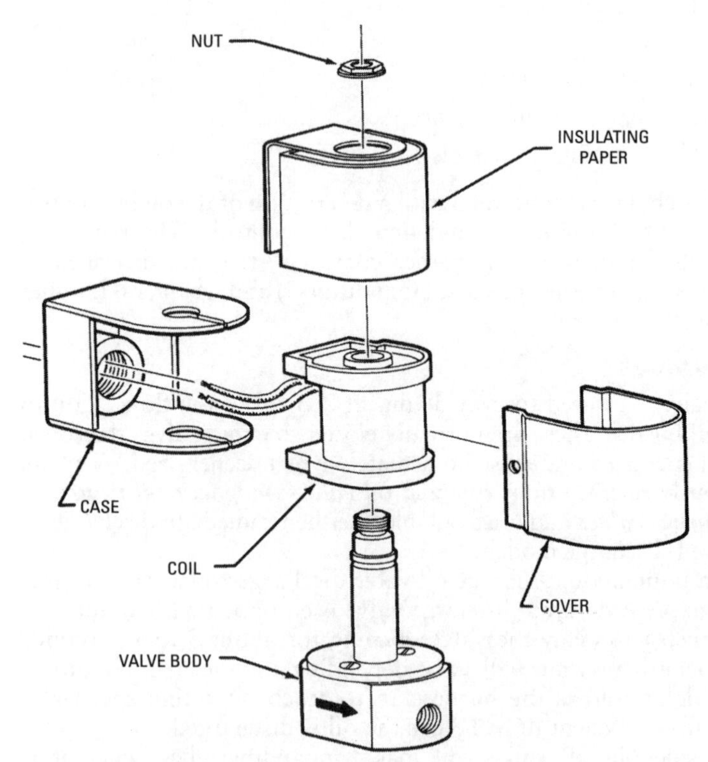

Figure 5-76 Position of the coil in a solenoid oil valve.

(Courtesy Honeywell Tradeline Controls)

2. Remove the powerhead assembly from the spindle.
3. Disconnect and remove the solenoid coil.
4. Connect the replacement coil and reassemble.

Examples of delayed-discharge valves are shown in Figures 5-77 and 5-78. In both valves, the timing delay is governed by a thermistor attached to the solenoid coil. In these valves, the timing delay will vary with ambient temperature, voltage level, and other factors during normal operation. If the timing is *significantly* off, it may be necessary to replace the thermistor. Because the thermistor is attached to the solenoid coil, the coil must also be replaced in order to replace the thermistor.

Figure 5-77
Magnetic valve used in controlling oil flow to the oil burner.
(Courtesy Honeywell Tradeline Controls)

Delayed valve opening can also be obtained by using an electronic time delay wired in series with the oil valve (see Figure 5-79). Unlike the thermistor, the timing of this device is not affected by ambient temperature. On a call for heat, the valve opening is delayed for approximately 5 seconds.

Figure 5-78
Magnetic oil valve.
(Courtesy Honeywell Tradeline Controls)

Oil Burner Primary Control

The oil burner *primary control* is an automatic safety device designed to turn off the oil burner motor should ignition or flame failure occur.

Each primary control operates in conjunction with a sensor by which the burner flame is monitored throughout the burner *on* cycle. The method used to sense the burner flame determines the type of primary control used and its location in the heating system.

The two types of primary controls commonly used in oil burner control systems are (1) the cadmium cell primary control and (2) the stack detector primary control. The cadmium cell primary control is burner mounted and uses a light-sensitive cad cell flame detector (sensor). The stack detector primary control relies on a thermal sensor to detect flame or ignition failure. This type of primary control assembly is available with the thermal sensor mounted in the stack and the

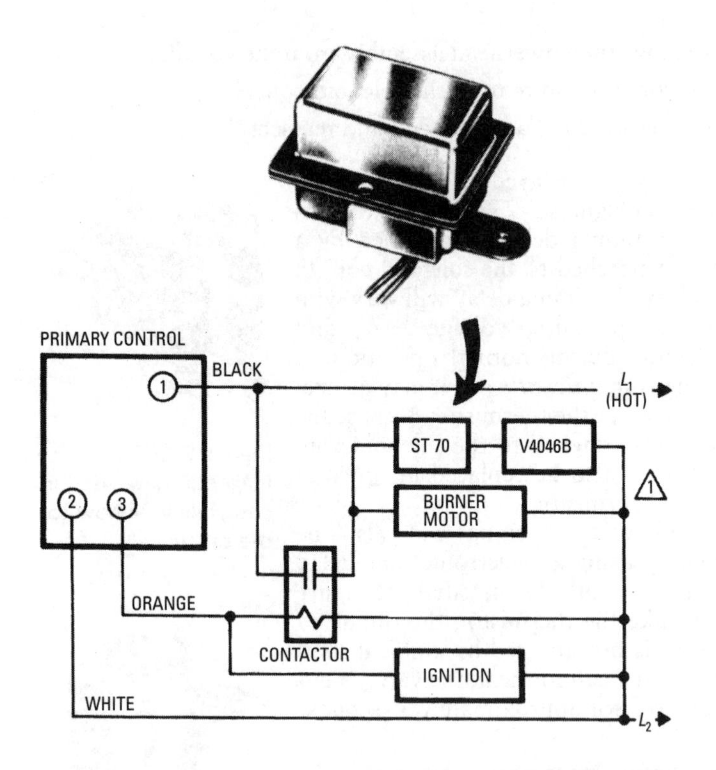

Figure 5-79 Electronic time delay wired in series with the oil valve. *(Courtesy Honeywell Tradeline Controls)*

primary control mounted on the burner, or with both the primary control and thermal sensor mounted in the stack as a single unit.

Cadmium Cell Primary Controls

A cadmium cell primary control consists of a primary control assembly operating in conjunction with a cadmium detection cell.

The cadmium detection cell is considered the most effective sensor used to monitor the burner flame. It consists of a light-sensitive photocell, a holder, and a cord assembly (see Figure 5-80). The surface of the detection cell is coated with cadmium sulfide and overlaid with a conductive grid. Electrodes attached to the detection cell are used to transmit an electrical signal to the primary control.

Figure 5-80 Cadmium detector cell. *(Courtesy Honeywell Tradeline Controls)*

The variable resistance of this surface to the presence of light (i.e., the burner flame) is used to actuate the flame detection circuit. When light is present (in the form of the burner flame), the resistance of the cadmium sulfide surface to the passage of electrical current is very low. Consequently, as long as the burner flame lasts, an electrical current will pass between the cadmium detection cell and the primary control unit, and the burner motor *on* cycle will continue operating. If the burner flame should fail or if ignition should fail to occur, the resistance of the cadmium sulfide surface to the passage of electrical current will be very high. This will interrupt the passage of the electrical current to the primary control and will cause the latter to shut off the burner motor.

The detection cell is mounted inside the burner air tube so that it faces the burner flame (see Figure 5-81). The exact location of

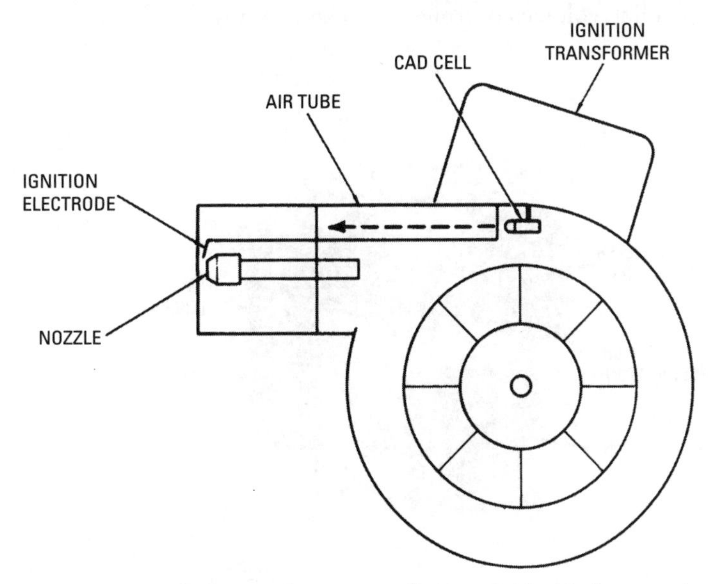

Figure 5-81 Cadmium detection cell mounted inside burner air tube and facing burner flame.

the detection cell is determined by the oil burner manufacturer and dictated by the design of the oil burner. In any event, the detection cell must be placed so that it views the burner flame directly. The fact that the detection cell responds to *any* light source means that it must be located where its surface will be shielded from any form of direct or reflected external light. Moreover, the ambient temperature should be kept below 140°F because excessive temperatures can also cause the detection cell to malfunction.

Sometimes a malfunctioning oil burner will cause a heavy layer of soot to accumulate on the cell surface. The cell surface should be carefully wiped to remove the soot and restore full view of the oil flame. A damaged detection cell should be replaced.

The *type* of primary control used with a cadmium detection cell will depend on the type of controller voltage, the type of ignition system, and the length of safety switch timing required by the installation.

The Honeywell R8184G Protectorelay primary control shown in Figure 5-82 has a transformer included in the unit to supply 24-volt power to the control circuit. This is a low-voltage primary, and it requires a 24-volt thermostat. Other models are available that require a line voltage controller (see Figure 5-83).

Figure 5-82 Honeywell model R8184G Protectorelay primary control. *(Courtesy Honeywell Tradeline Controls)*

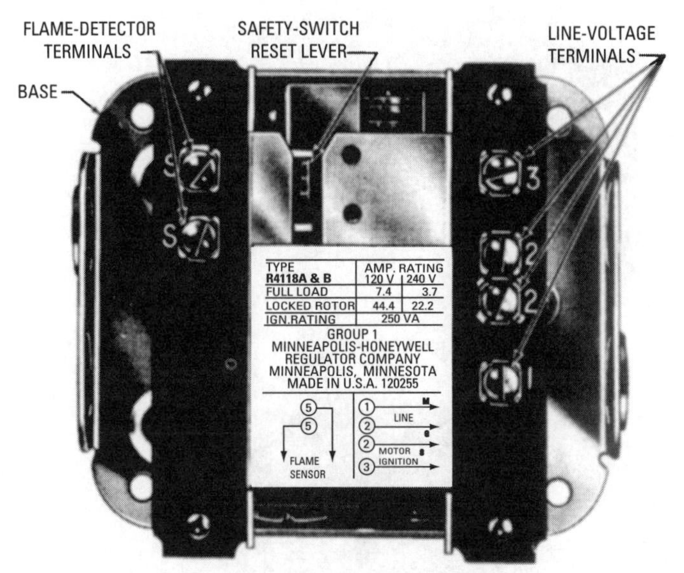

Figure 5-83 Primary control used with a line voltage thermostat.
(Courtesy Honeywell Tradeline Controls)

The primaries illustrated in Figures 5-82 and 5-83 are designed for use with nonrecycling constant-ignition oil burners. Automatic recycling control of an *intermittent*-ignition oil burner can be obtained by using the primary shown in Figure 5-84. The basic differences between the constant-ignition and intermittent-ignition systems can best be illustrated by the wiring diagrams shown in Figures 5-85 and 5-86. An intermittent-ignition system contains the same components as a constant-ignition system *plus* the following:

1. A interlock contact in the ignition circuit (T_1).
2. An ignition timer heater (T).
3. An interlock contact (T_2) in the circuit between the safety switch heater (SS) and the ignition timer heater (T).

Safety switch timing can be 15, 30, 45, or 80 seconds depending on the manufacturer and model.

Stack Detector Primary Control

Stack-mounted oil burner primary controls employ thermal sensors to detect ignition or flame failure. A typical stack detector thermal sensor (see Figure 5-87) consists of a bimetal element

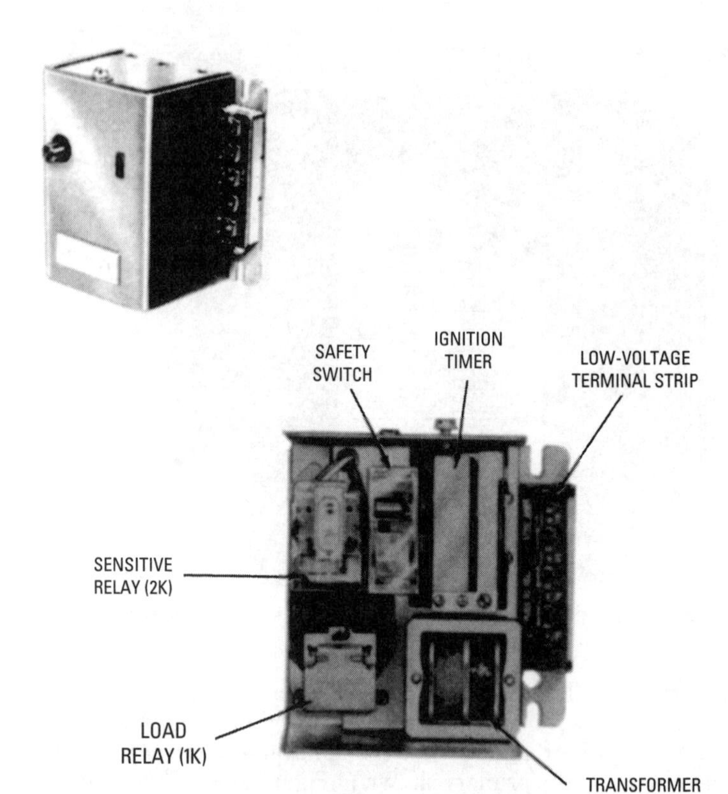

Figure 5-84 Honeywell R8185E Protectorelay primary control. *(Courtesy Honeywell Tradeline Controls)*

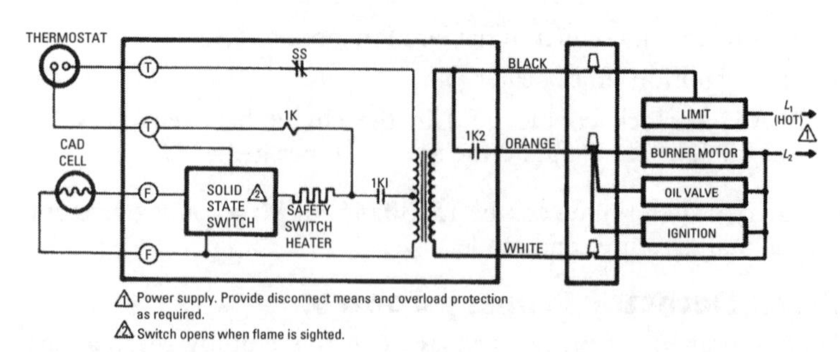

Figure 5-85 Schematic diagram of Honeywell model R8184G primary control. *(Courtesy Honeywell Tradeline Controls)*

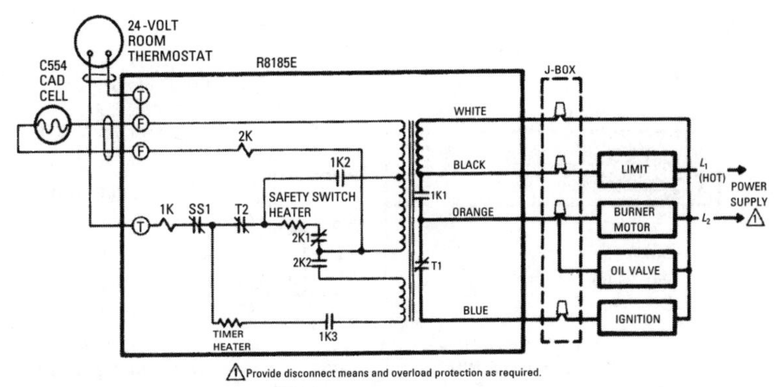

Figure 5-86 Schematic diagram of Honeywell model R8185E primary control. *(Courtesy Honeywell Tradeline Controls)*

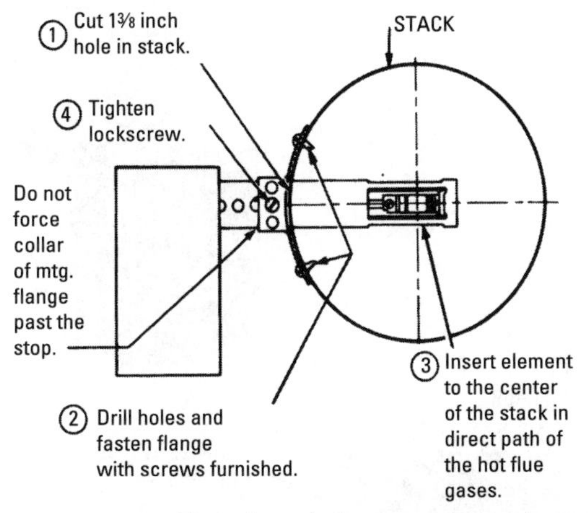

Figure 5-87 Typical stack detector thermal sensor.
(Courtesy Honeywell Tradeline Controls)

inserted into the stack (see Figure 5-88). The thermal sensor (combustion thermostat) is usually located on the stack where the element will be exposed to the most rapid temperature changes. The thermal sensor should always be mounted ahead of any draft regulator. If installed on an elbow, it should be mounted on the outside curve of the elbow.

The stack-mounted primary control illustrated in Figure 5-89 combines a Honeywell RA117A Protectorelay control for burner

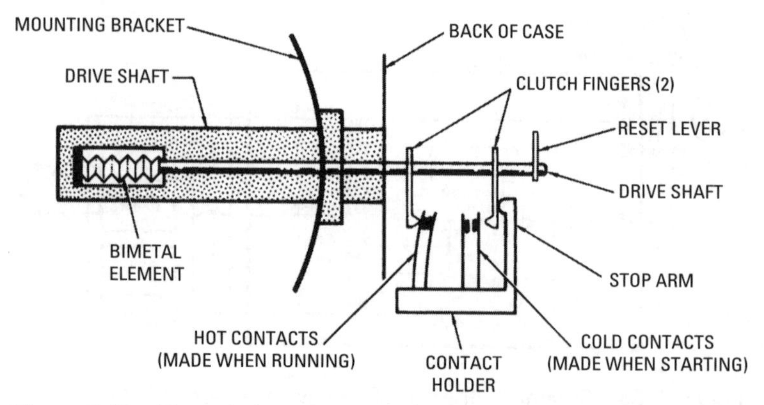

Figure 5-88 Bimetal element inserted into the stack.
(Courtesy Honeywell Tradeline Controls)

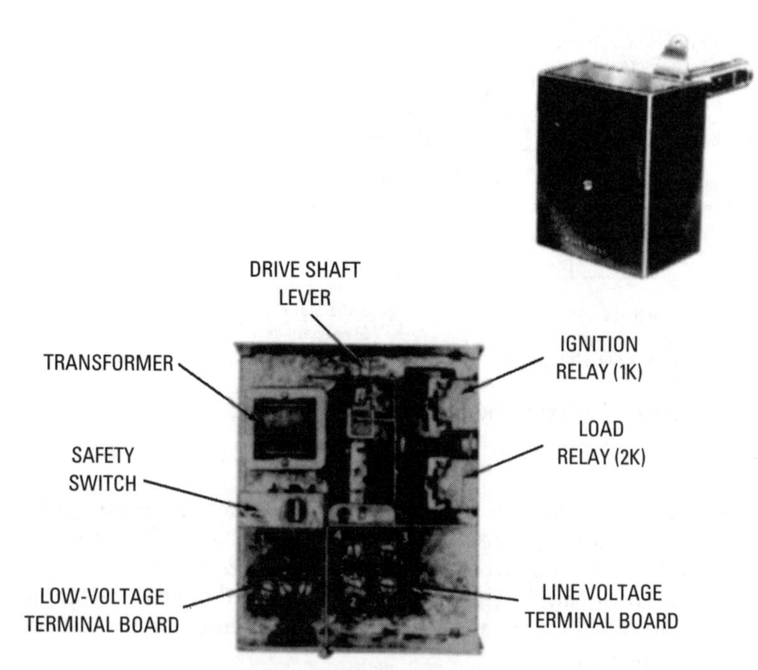

Figure 5-89 Honeywell RA117A Protectorelay primary control for stack-mounted installation. *(Courtesy Honeywell Tradeline Controls)*

cycling control and a thermal detector for sensing temperature changes of the flue gases (as high as 1000°F maximum temperature). The safety switch shown on the center-left of the unit is designed to lock out if the flame is not properly established. If the flame goes out during the burner *on* cycle, the primary control will make one attempt to restart. If the attempt is unsuccessful, the safety switch will lock out. A manual reset is then required in order to restart the burner. The primary control shown in Figure 5-89 is used with a two-wire or three-wire primary controller.

A stack-mounted combination line voltage primary control and flame detector is shown in Figure 5-90. This type of primary control is used with constant-ignition oil burners and is designed for flange-mounting on a stack, flue pipe, or combustion chamber door of a furnace or boiler. It must be used with a line voltage thermostat or controller.

Combination Primary Control and Aquastat

The combination primary control and aquastat is designed for use with a *constant*-ignition oil burner in a hydronic heating system. The purpose of this unit is to supervise the operation of the oil burner and provide both water temperature and circulator control. A remote sensor (cadmium detection cell) is used to detect any irregularities in the oil burner flame.

Figures 5-91 and 5-92 illustrate a number of different combination primary control and aquastat units used in hydronic heating systems. In operation, the high-limit switch (SPST) will automatically turn off the burner if the boiler overheats. The low-limit circulator switch (SPDT) is used to maintain water temperature for the domestic hot-water supply. It will also prevent the circulator from operating if the water temperature is too low (i.e., below the setpoint).

On the units shown in Figures 5-91 and 5-92, a call for heat from the room thermostat pulls in relays 1K and 2K to turn on the oil burner and start heating the safety switch. Under normal operating conditions, the burner should ignite within safety-switch timing. If such is the case, the cadmium cell detects the flame, and relay 3K pulls in to deenergize the safety-switch heater. The oil burner then continues to operate until the call for heat is satisfied.

The circulator (pump) in the heating system operates when relay 1K pulls in *only* if the *R* to *W* contact on the aquastat control is made (see Figure 5-92). A drop in water temperature will cause the *R* to *B* (low limit) contact to be made. This acts as a call for heat, pulling in relay 2K to turn on the oil burner.

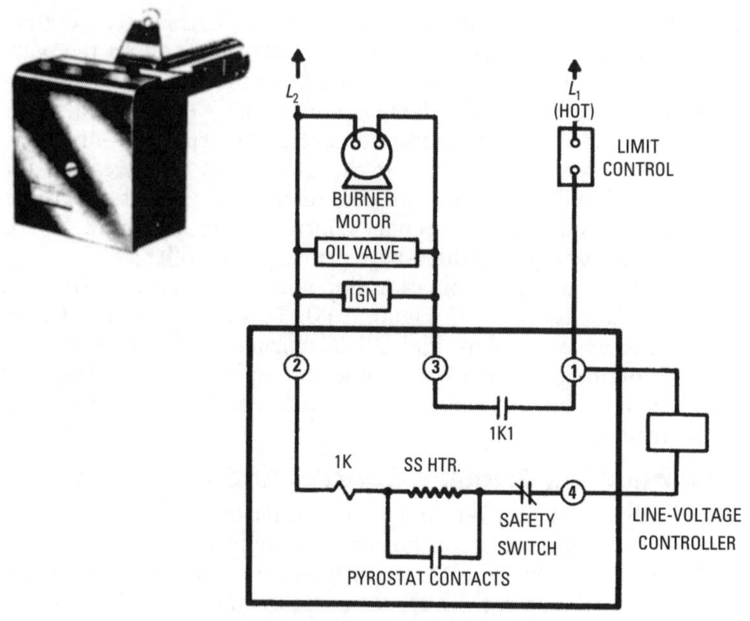

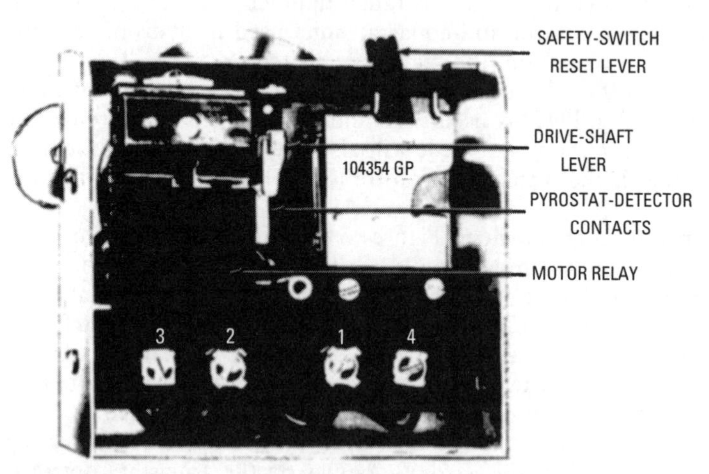

Figure 5-90 Stack-mounted combination line voltage primary control and flame detector. *(Courtesy Honeywell Tradeline Controls)*

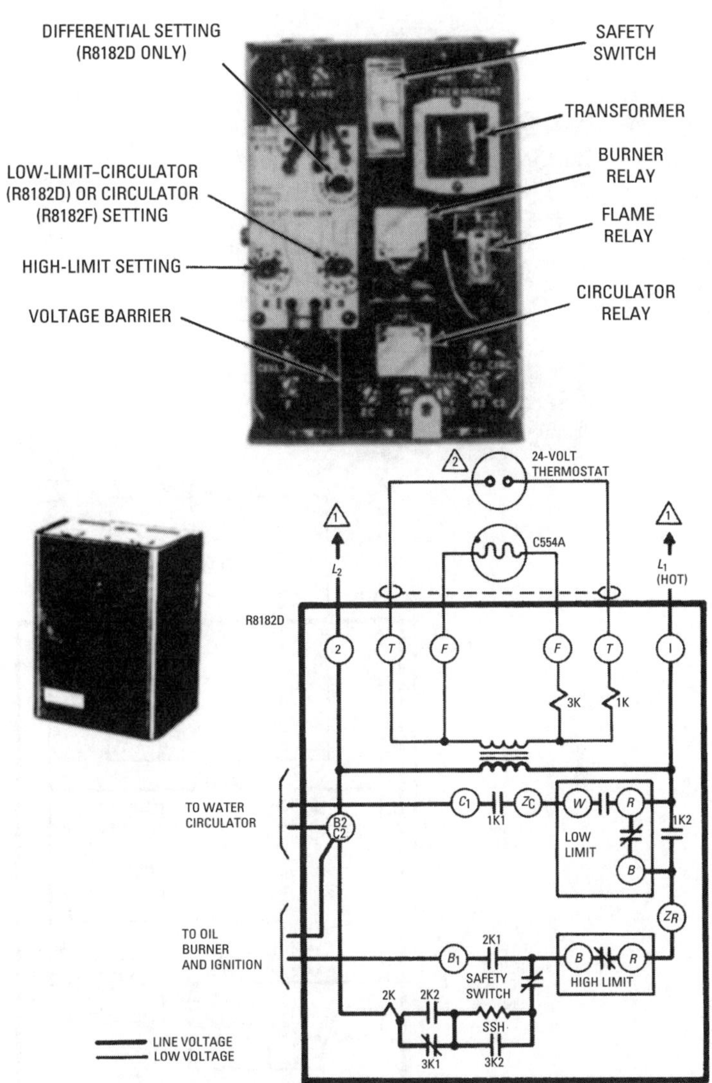

1 Power supply, 120 volts AC. Provide disconnect means and overload protection as required.

2 Control wires can be run with line-voltage wires in conduit but then must have NEC class 1 insulation.

Figure 5-91 Model R8182H Protectorelay primary control. High-limit/ low-limit aquastat switching with remote-bulb sensor.

(Courtesy Honeywell Tradeline Controls)

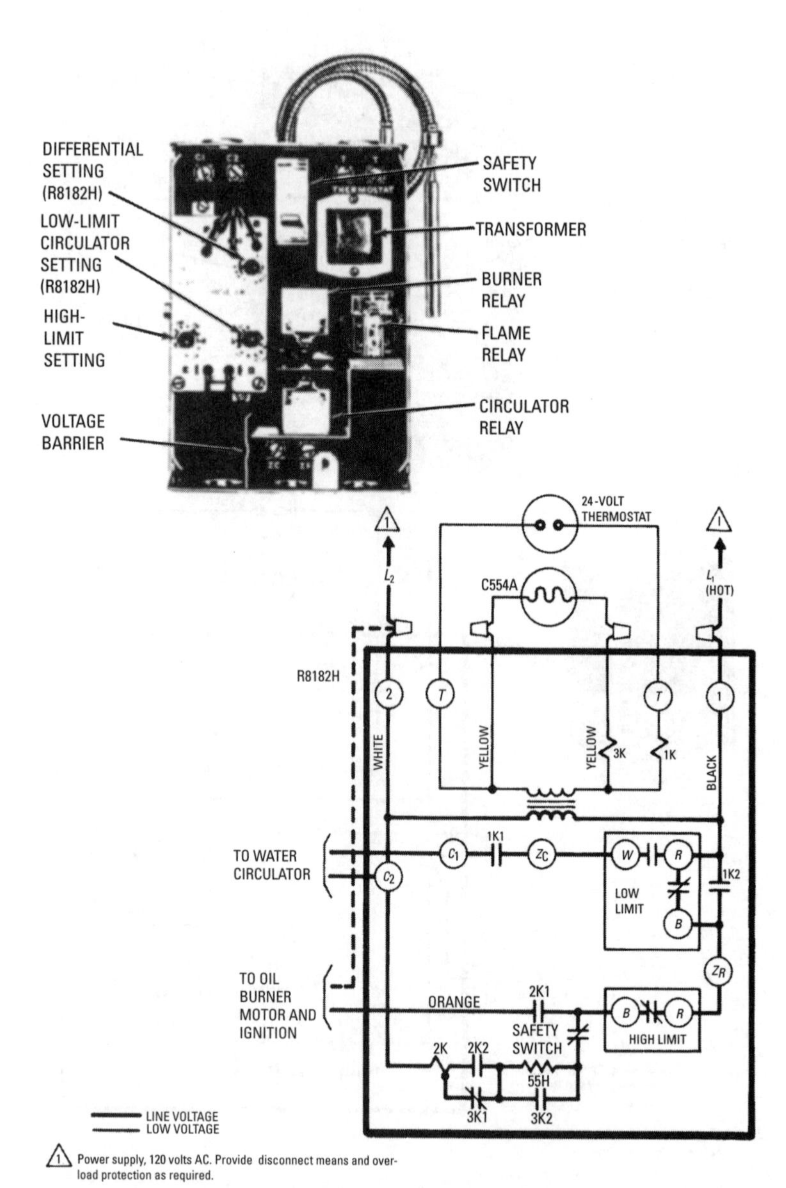

Figure 5-92 Model R8182D Protectorelay primary control. High-limit/low-limit circulator aquastat switching. *(Courtesy Honeywell Tradeline Controls)*

Combination primary control and aquastat units are also used on water heaters (see Chapter 4 of Volume 3, "Water Heaters and Other Appliances").

Troubleshooting the Oil Burner Primary Control

Table 5-4 lists possible remedies to a number of different operating problems associated with oil burner primary controls. Before checking the primary control, examine the following parts of the oil burner and ignition systems:

- Main power supply and burner motor fuse
- Ignition transformer
- Electrode gap and position
- Contacts between ignition transformer and electrode

Other system components that should be checked before examining the primary control are the oil piping to the tank, the oil filter, oil pump pressure, oil nozzle, and oil supply.

Table 5-4 Troubleshooting Oil Burner Primary Control

Symptom and Possible Cause	*Possible Remedy*
Repeated safety shutdown.	
(a) Slow combustion thermostat response.	(a) Move combustion thermostat to better location. Adjust for more-efficient burner flame. Clean surface of cad cell.
(b) Low line voltage.	(b) Check wiring and rewire if necessary. Contact local power company.
(c) High resistance in combustion thermostat circuit.	(c) Replace combustion thermostat.
(d) High resistance in thermostat or operating control circuit.	(d) Check circuit and correct cause.
(e) Short cycling of burner.	(e) Clean filters. Reset or replace differential of auxiliary controls. Repair or replace faulty auxiliary control. Set thermostat heat anticipation at higher amp value. Clean holding circuit contacts.

(continued)

Table 5-4 (continued)

Symptom and Possible Cause	*Possible Remedy*
(f) Short circuit in combustion thermostat cable.	(f) Repair cable or replace combustion thermostat.
Relay will not pull in.	
(a) No power; open power circuit.	(a) Repair, replace, or reset fuses, line switch, limit control, auxiliary controls.
(b) Open thermostat circuit.	(b) With power to relay, momentarily short thermostat terminals on relay. If burner starts, check wiring.
(c) Combustion thermostat open.	(c) Repair or replace combustion thermostat.
(d) Ignition timer contacts open.	(d) Clean magnet.
(e) Open circuit in relay coil.	(e) Replace relay.

Chapter 6

Other Automatic Controls

Modern heating and cooling systems contain electrical control circuits that are interconnected and interlocked with the various system components by a series of switches and relays. Most of these components, particularly the heating and cooling equipment (furnaces, boilers, compressors, condensers, and so on) and most system controls have been described in other chapters of the book. This chapter is reserved for a description of the fan and limit controls; the various electrical control circuit switches and relays; transformers; and a number of different control devices used in cooling systems.

Fan Controls

A number of different devices are available for controlling the operation of fans in heating and/or cooling installations. Most of these devices function as fan safety controls; a few of them serve as fan primary controllers. The following are fan controls described in this chapter:

- Fan control
- Air switch
- Fan relays
- Fan center
- Fan manager
- Fan timer switch
- Fan safety cutoff switch

Caution

> Always disconnect the power supply before installing, servicing, or repairing any of the electrically operated devices described in this chapter. Failure to do so may result in damage to equipment and/or electric shock.

Fan Control

A *fan control* is a device used to turn the system fan on and off in response to air temperature changes in the furnace plenum. This fan controller is frequently combined with a limit controller in one unit (see *Combination Fan and Limit Control* in this chapter).

In the operation of the furnace, the burner or burner assembly starts first and heats the air, which rises through the heat exchanger to the furnace plenum. The fan control is located in the plenum and is present for a specific cut-in temperature. When the temperature of the rising air reaches the cut-in temperature setting on the fan controller, the fan is automatically turned on and warm air is moved through the distribution ducts. After a period of time, the room thermostat will no longer call for heat and will shut off the burner. The air in the plenum then begins to cool. When the air temperature drops below the cut-in temperature of the fan controller, the fan is automatically shut off.

In most forced warm-air heating systems, the fan control is usually a line voltage device wired in the *hot* lead (L1) of the power supply to the fan motor (see Figure 6-1). If a step-down transformer is used to provide a low-voltage control circuit for the room thermostat, the fan motor and fan controller will connect at the line side of the transformer (see Figure 6-2).

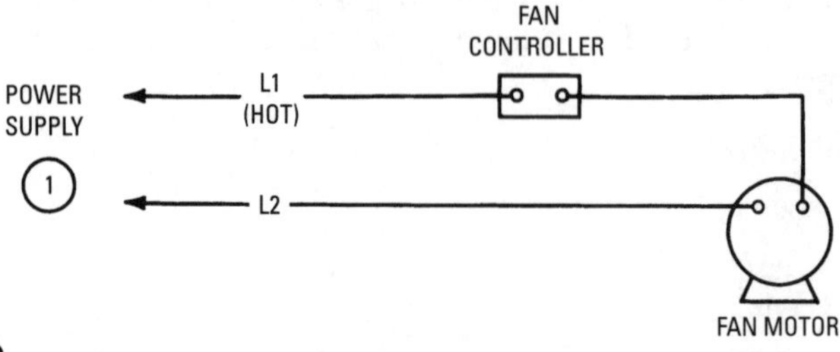

Figure 6-1 Fan controller wired in the hot lead.
(Courtesy Honeywell Tradeline Controls)

The following procedure may be used for setting a fan control:

1. Allow the burner or burner assembly to operate for a normal running period.
2. Lower the thermostat setting so that the burner(s) will not operate during the fan control setting procedure.
3. Place a thermometer in the furnace plenum or bonnet or in one of the warm-air ducts near the furnace.

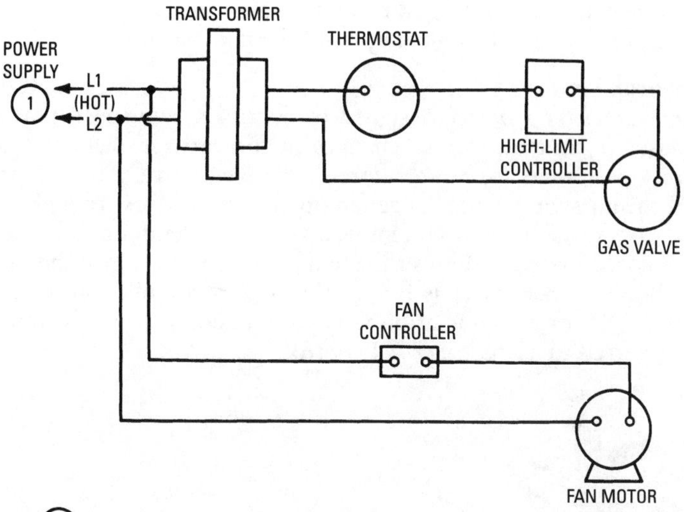

Figure 6-2 Typical warm-air fan control circuit.
(Courtesy Honeywell Tradeline Controls)

4. Set the fan adjustment lever and the fan differential adjustment (when used) to their lowest or coldest position so that the fan will run continuously.
5. Watch the thermometer until the temperature drops to about 5°F (3°C) above the temperature normally maintained in the rooms being heated.
6. As soon as the temperature on the thermometer has reached the appropriate level (see step 5), slowly move the fan temperature adjustment lever up to a point where it will stop the fan.

When setting a fan according to the aforementioned procedure, the speed of the fan must be set so that the *average* temperature rise through the furnace is about 90°F (50°C).

In some installations, a drafty condition may result from the location or types of warm-air outlets. This condition can be corrected by the following adjustments to the procedure described previously:

1. Place the thermometer in front of the return air grille in the room or space that is most difficult to heat.

2. When the air *leaving* the room begins to feel cool, slowly move the fan adjustment up until the fan just stops.

Air Switch

An *air switch* is a device designed to control a two-speed fan in response to air temperature changes in the furnace plenum (see Figure 6-3). When the temperature in the plenum rises to the set-point, the air switch will change fan operation from low to high. In other words, the air will be removed from the plenum at a higher rate of speed in order to lower the temperature of the air in the furnace. Because this device is frequently used in conjunction with a fan controller or a combination fan and limit controller, it is sometimes referred to as an *upper fan control.*

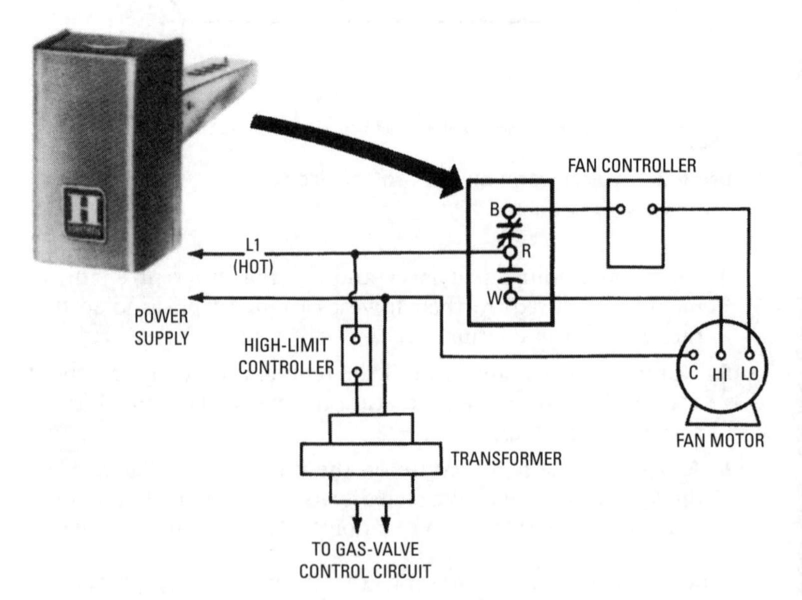

Figure 6-3 Model L6068A air switch and wiring diagram.

(Courtesy Honeywell Tradeline Controls)

Air switches are generally available with fixed temperature settings (e.g., 125°F, 135°F, 165°F, or 200°F) or with an adjustable temperature range (125°F to 165°F or 160°F to 200°F). Temperature setting adjustments can be made as shown in Figure 6-4.

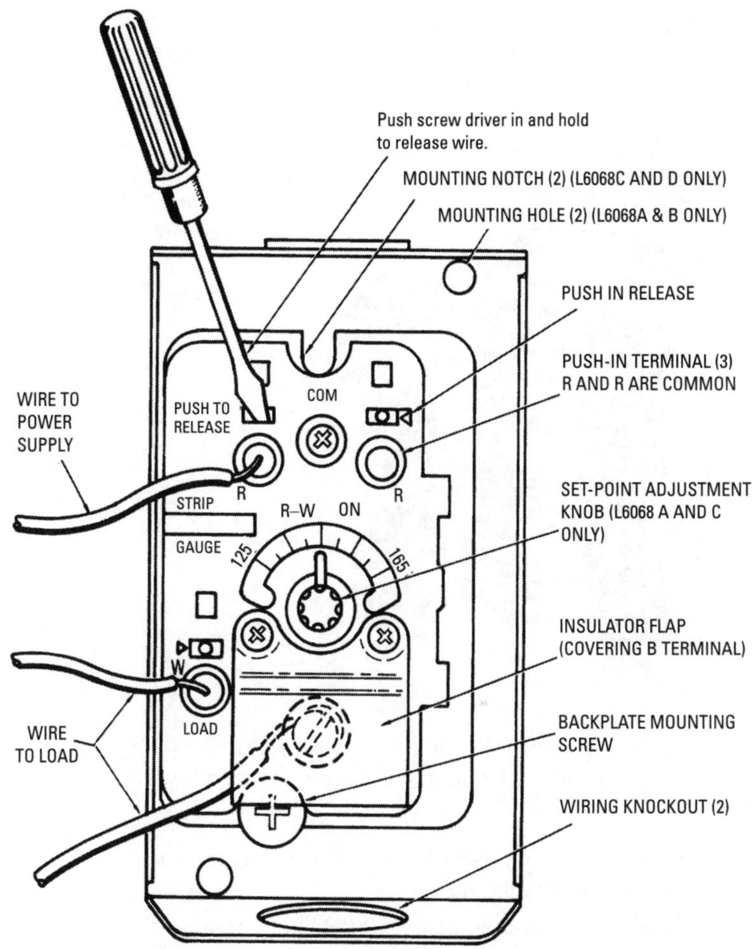

Figure 6-4 Adjustment and connection points on a model L6068 air switch. *(Courtesy Honeywell Tradeline Controls)*

An air switch can also be used to shut off the burner and turn on the fan when the temperature between the filter and heat exchanger rises to the setpoint. In this manner, it functions as a limit control (see *Secondary High-Limit Switch* in this chapter).

Fan Relays

A *fan relay* is a primary controller designed to provide 24-volt circuit control of line voltage fan motors and auxiliary circuits in heating and/or cooling systems (see Figures 6-5 and 6-6). It also provides manual fan operation at any time by using the manual fan switch of the thermostat base.

Figure 6-5 Model R851 fan relay (contactor) provides 24-volt control of single- or two-speed fan motors up to ½ hp.

(Courtesy Honeywell Tradeline Controls)

Fan relays are available in a variety of different models based on the switching element contact position. The following are some examples:

- Single-pole, single-throw (SPST) switching element—both contacts normally open.
- Single-pole, double-throw (SPDT) switching element—one contact normally open and one normally closed.
- Double-pole, single-throw (DPST) switching element—one normally open main contact and one normally open auxiliary pole.

Fan relays are often used with multiple-speed fans to provide low-speed fan operation during the heating cycle and high-speed operation during the cooling cycle.

A 24-volt room thermostat is generally used to switch the fan relay controlling the indoor fan (120- or 240-volt AC power). In cooling systems, a second fan relay must be added if switching control of the condenser fan motor is desired.

Figure 6-6 Fan relay used for low-voltage control of line voltage fan motors and auxiliary circuits.
(Courtesy Honeywell Tradeline Controls)

The operation of a fan relay can be checked by applying power to the coil and listening for the click of the contacts closing or by testing for electrical continuity. If the fan relay does not operate, check the voltage to the coil.

Sometimes a fan will operate at low speed but not at high speed. Check the fan relay first. If it is not defective and is receiving proper voltage, the failure of the fan to operate at high speed may be caused by loose wiring or dirty contacts. The method used for cleaning relay contacts is described elsewhere in this chapter (see *Cleaning Contactors*).

A defective fan relay is also the *occasional* cause of compressor short cycling; however, this operating problem is more commonly traced to dirty air filters and other air movement restrictions on the low side of the compressor. These possible causes should be checked first.

Fan Center

A *fan center* is a primary controller designed to provide automatic low-voltage control of line voltage fan motors and auxiliary circuits in heating, cooling, and heating/cooling circuits. A typical wiring hookup with a two-speed fan motor, electronic air cleaner, and humidifier is illustrated in Figure 6-7.

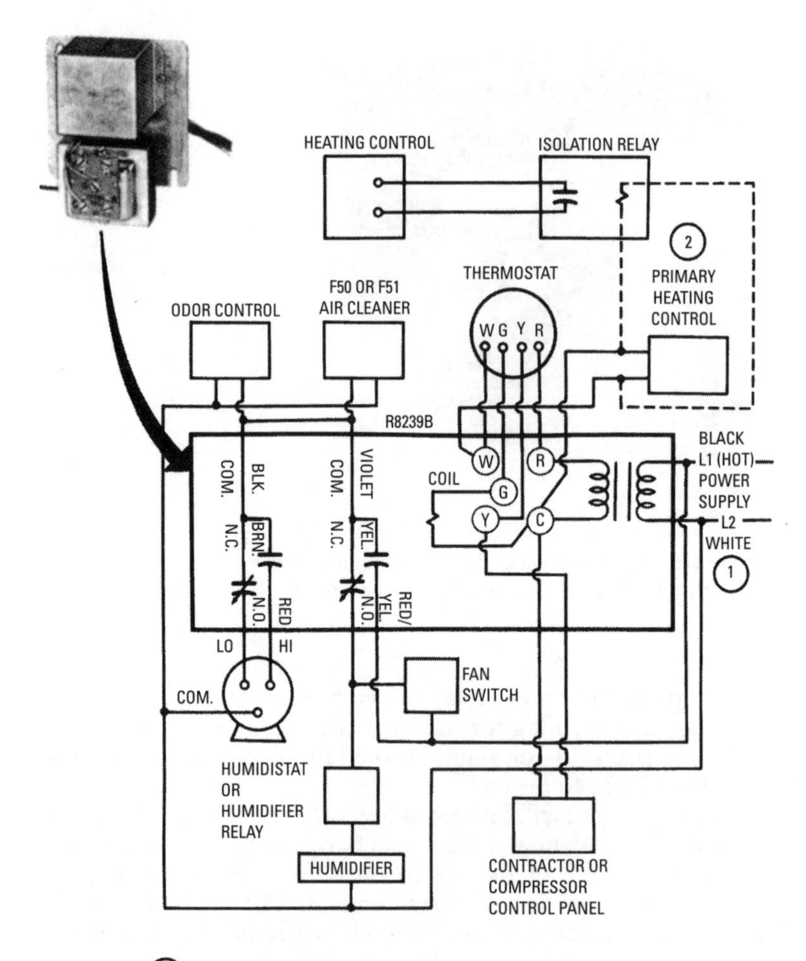

(1) Provide disconnect means and overload protection as required.

(2) Use optional hook up with isolating relay (dashed line) if heating control has a separate power supply. Isolation of the power supplies may also be accomplished by using special thermostat subbase combinations with isolated circuits.

Figure 6-7 Fan control center and wiring diagram.

(Courtesy Honeywell Tradeline Controls)

In addition to providing the same switching functions as a fan relay (see previous section), a fan center includes an integral low-voltage transformer and a terminal board for low-voltage system wiring. In air-conditioning systems, a thermal delay relay is often added to a fan center to prevent short cycling of the compressor motor.

Fan Manager

A *fan manager* is used in compressor-operated air-conditioning systems and heat-pump systems to enable the blower to continue running for a short time after the compressor has shut off (see Figure 6-8). This short delay increases the cooling efficiency of the system by allowing the blower enough time to force the residual cooled air into the living spaces. Heat-pump systems can be wired so that the blower shutoff delay also occurs in the heating cycle.

Fan Timer Switch

A *fan timer switch* is a device that provides *timed* fan operation for forced warm-air furnaces and unit heaters when they are wired in parallel with a furnace or heater controller. The switch operation is

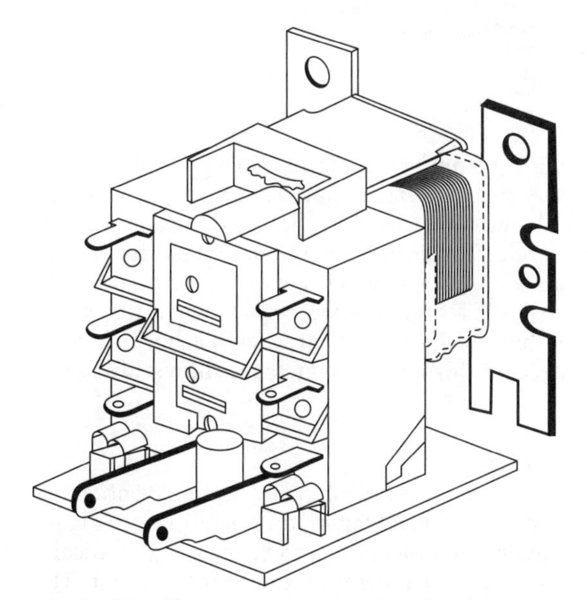

Figure 6-8 Fan manager and wiring diagram.

(Courtesy Honeywell Inc.)

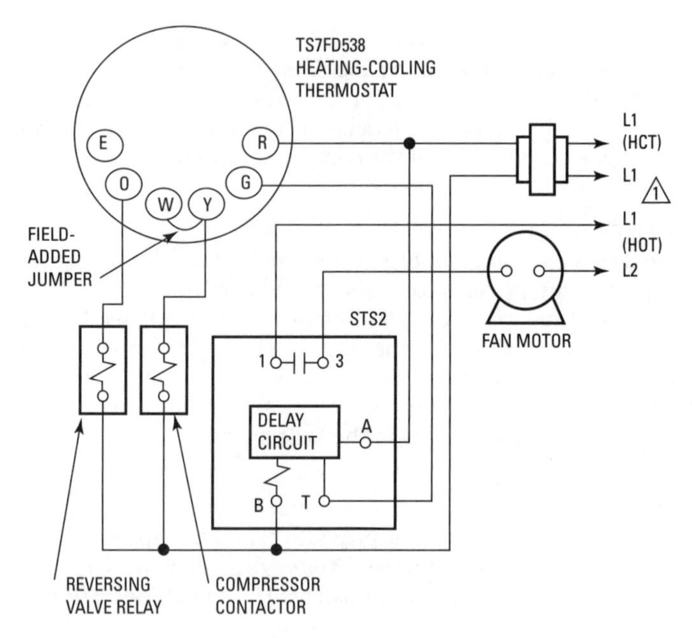

Figure 6-8 *(continued)*

independent of furnace plenum temperature changes. This factor ensures fan operation and eliminates unnecessary recycling at the beginning and end of burner operation. The use of a fan switch is particularly recommended for horizontal and downflow furnaces.

The Honeywell S876 fan timer switch shown in Figure 6-9 contains a heater-actuated SPST bimetal snap-action switch that turns the fan on after the burner starts and off after the burner stops. Typical wiring connections for a timer switch are shown in Figures 6-10 and 6-11.

Fan Safety Cutoff Switch

A *fan safety cutoff switch* can be installed in any heating, ventilating, or air-conditioning system to control fan motor operation (see Figure 6-12). These are manually reset mercury switches that automatically break the fan motor circuit on temperature rise to the setpoint. The setpoint is established by adjusting the temperature-setting screw on the unit. This switch is designed to lock out to prevent the return of fan operation until manually reset.

Figure 6-9 Model S876 fan timer switch. *(Courtesy Honeywell Tradeline Controls)*

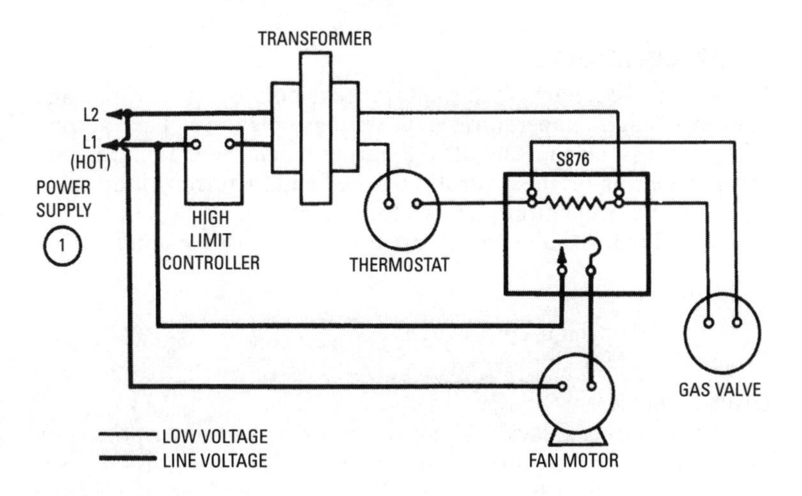

Figure 6-10 Fan motor controlled by a time switch.
(Courtesy Honeywell Tradeline Controls)

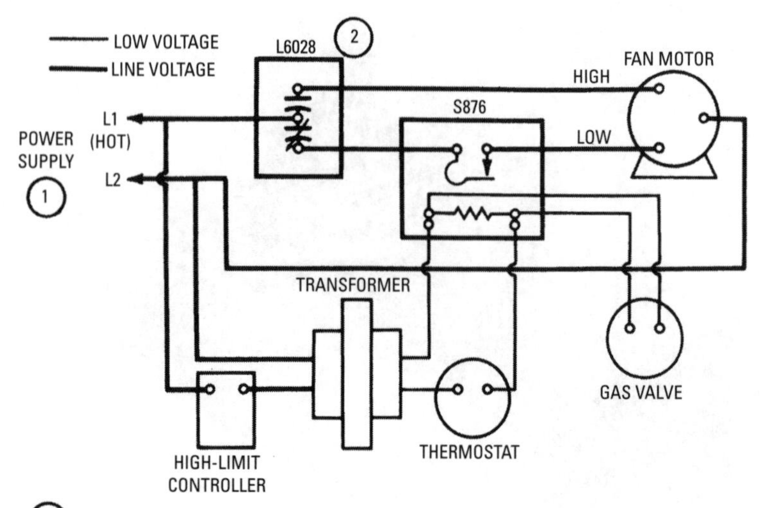

Figure 6-11 Two-speed fan control circuit with low fan speed controlled by a fan timer switch. *(Courtesy Honeywell Tradeline Controls)*

Limit Controls

Limit controls are also used to prevent the buildup of excessive and dangerous high temperatures in the furnace plenums. They accomplish this task by shutting off the burner when the maximum temperature setting on the control is reached and by turning it on again when the air temperature returns to normal.

The following limit controls are described in this chapter:

- Limit control
- Secondary high-limit switch

Limit Control

A *limit control* is a device designed to provide high-limit protection for a forced warm-air furnace (see Figure 6-13). It controls the operation of a burner or burner assembly in response to air temperature changes in the furnace plenum. If the air temperature in the plenum becomes excessively high, the limit controller shuts off the burner or burner assembly until the air temperature returns to normal.

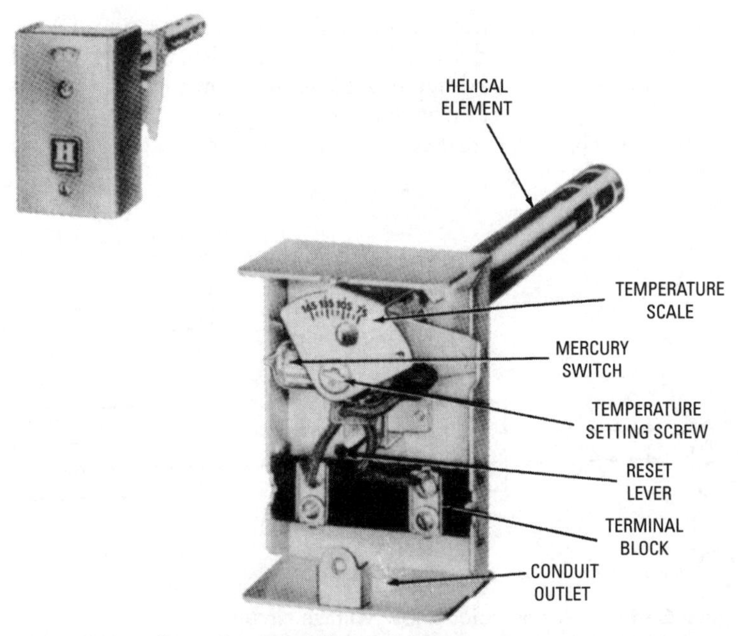

Figure 6-12 Fan safety shutoff switch. *(Courtesy Honeywell Tradeline Controls)*

Figure 6-13 Limit control. *(Courtesy Honeywell Tradeline Controls)*

Limit controls are available in models suitable for use in low-voltage, line voltage, and self-energizing (millivolt) systems. Typical wiring hookups for these different systems are shown in Figures 6-14 and 6-15. These limit controls have so-called universal contacts in the limit switch, which makes them suitable for all voltages from millivolt to line voltage.

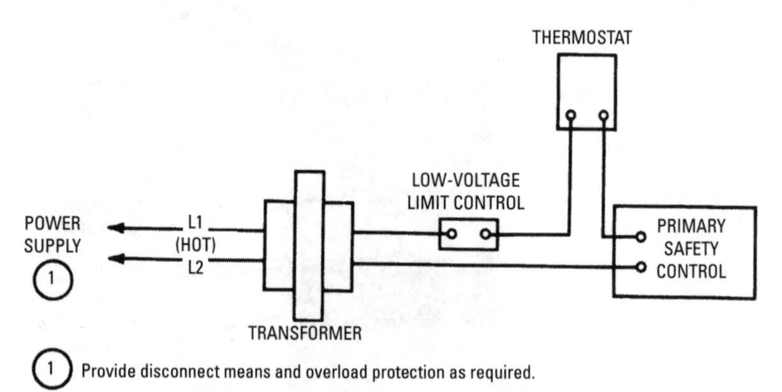

Figure 6-14 Limit control in a low-voltage circuit.
(Courtesy Honeywell Tradeline Controls)

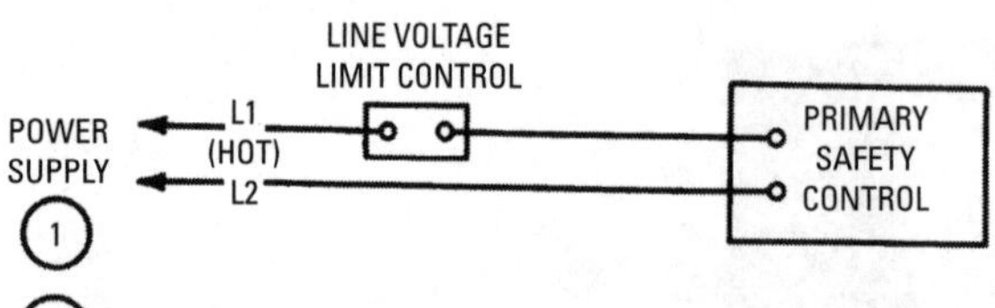

Figure 6-15 Limit control in a line voltage circuit.
(Courtesy Honeywell Tradeline Controls)

A limit control contains a snap-action switch operated by either a fluid-filled or bimetallic sensing element.

Fluid-filled sensing elements are connected to the control by a length of capillary tube, which is available in lengths up to 72 inches. The tube is filled with a temperature-sensitive liquid. A temperature change causes the liquid to expand against a diaphragm that operates a snap-action switch (see Figure 6-16).

Bimetal sensing elements are available in helical, flat-blade, and spiral types (see Figures 6-17, 6-18, and 6-19). The bimetal sensing element is connected directly to the switch operator.

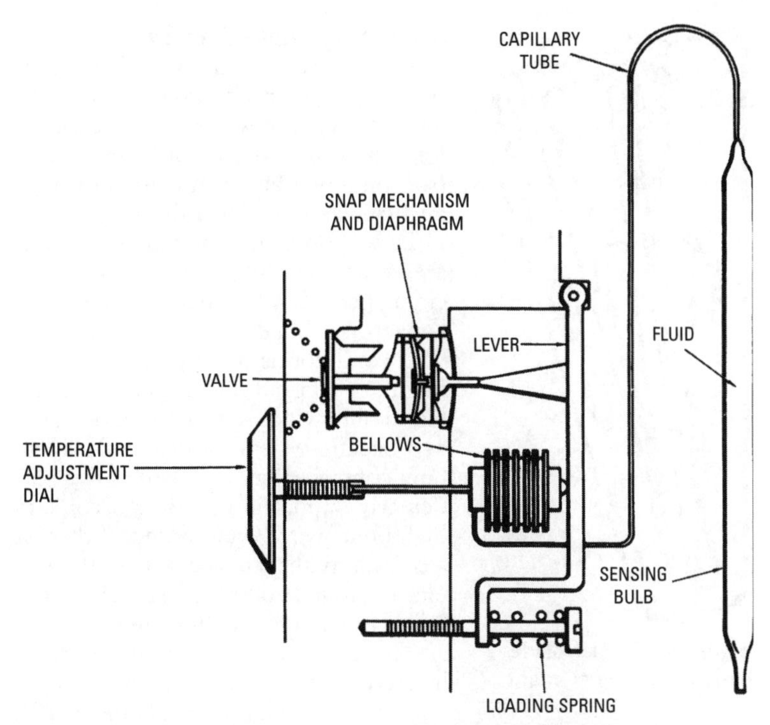

Figure 6-16 Fluid-filled sensing bulb, diaphragm, and snap mechanism.
(Courtesy Robertshaw Controls Co.)

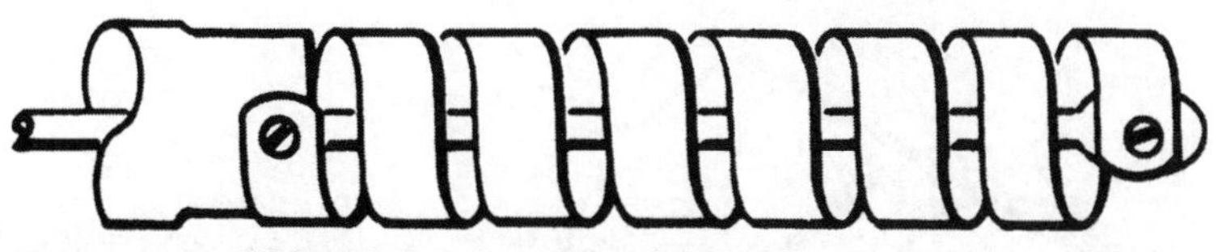

Figure 6-17 Helical bimetal sensing element.

The temperature setting of the limit control should be high enough not to interfere with the normal operation of the furnace, but low enough to shut off the burner or burner assembly before air temperatures in the furnace plenum reach the danger point. After the temperature in the furnace plenum has cooled and dropped below the setting on the limit control, the limit switch closes and starts the burner or burner assembly again.

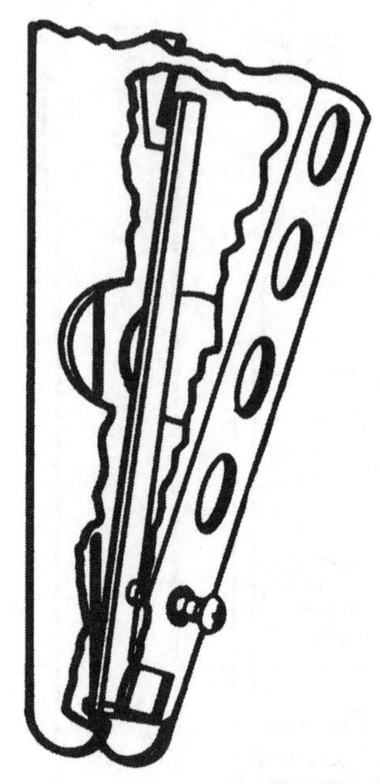

Figure 6-18 Flat-blade bimetal sensing element.

Secondary High-Limit Switch

The same type of air switch used to provide two-speed control of fan motors can also serve as a *secondary high-limit switch* (or *upper-limit control*) on downflow or horizontal warm-air furnaces (see Figure 6-20).

Downflow and horizontal furnaces are sometimes subject to a reverse air circulation condition that can result in a dangerous buildup of temperatures. This condition is usually caused by fan failure or clogged filters. The secondary high-limit switch is a safety device used as a backup system for the regular high-limit controller (see Figure 6-3). It is particularly important to have a secondary high-limit switch on a furnace if there is a possibility that the location of the regular high-limit controller may cause it to fail to detect a fan malfunction.

The secondary high-limit switch is located between the filter and the furnace fan (see Figures 6-21 and 6-22). When the air temperature exceeds a certain setting, the switch opens and shuts off the burner or burner assembly and turns on the fan. Secondary high-limit switches are either automatic or manually

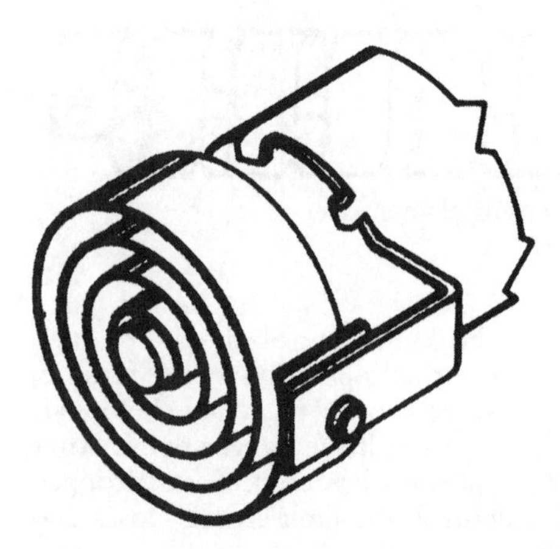

Figure 6-19 Spiral bimetal sensing element.

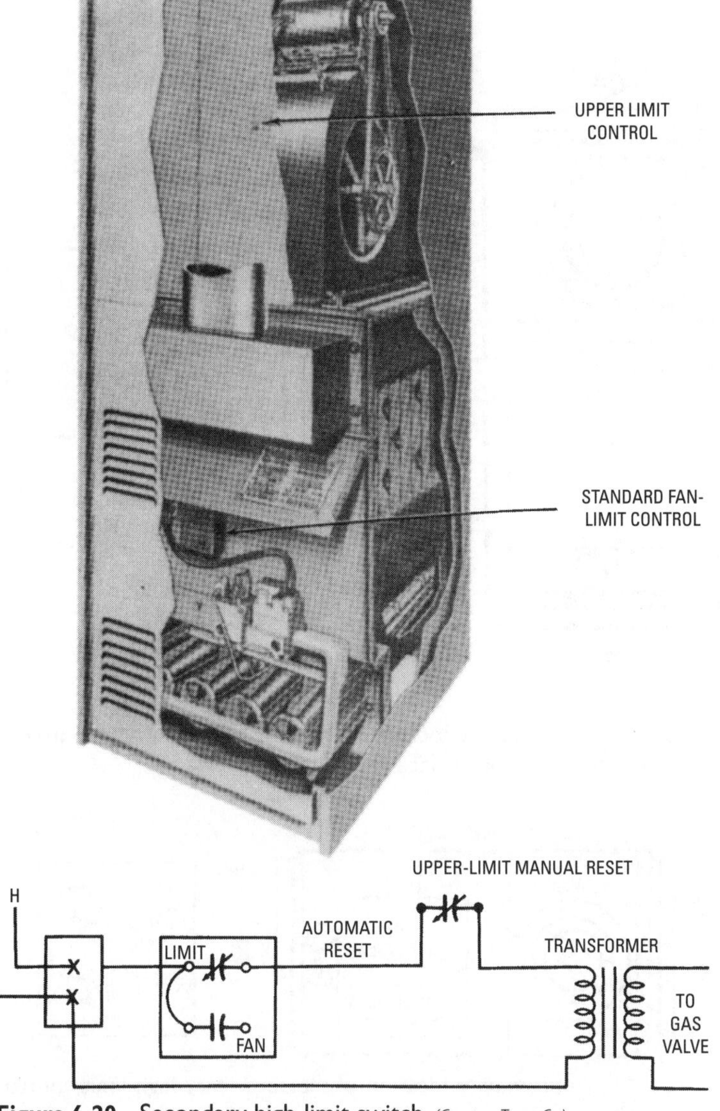

Figure 6-20 Secondary high-limit switch. *(Courtesy Trane Co.)*

reset types. The automatic type is a single-pole, double-throw (SPDT) switch that turns on the burner when the limit cuts out. The manually reset switch must be reset before the burner will operate again.

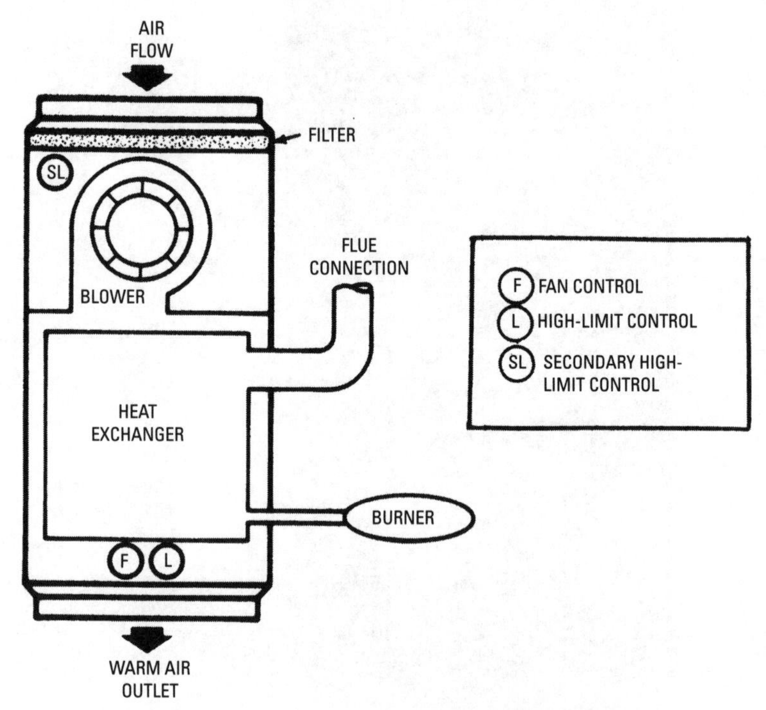

Figure 6-21 Approximate location of a secondary high-limit control on a downflow warm-air furnace. *(Courtesy Honeywell Tradeline Controls)*

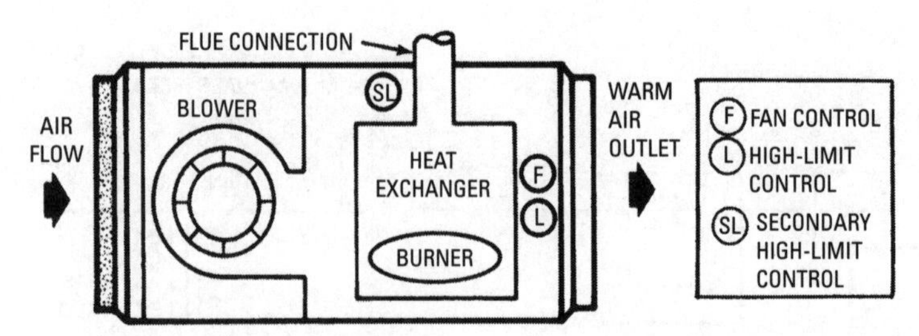

Figure 6-22 Approximate location of the secondary high-limit control in a horizontal warm-air furnace. *(Courtesy Honeywell Tradeline Controls)*

When the air temperature between the filter and the heat exchanger reaches the setting (i.e., setpoint) in the air switch, one internal switch closes (*R* to *W*) and another opens (*R* to *B*). This operation shuts off the burner and starts the fan.

Combination Fan and Limit Control

A *combination fan and limit control* combines the functions of a fan controller and a limit controller in a single unit. One sensing element (either bimetal or fluid-filled) is used for both controls (see Figures 6-23 and 6-24).

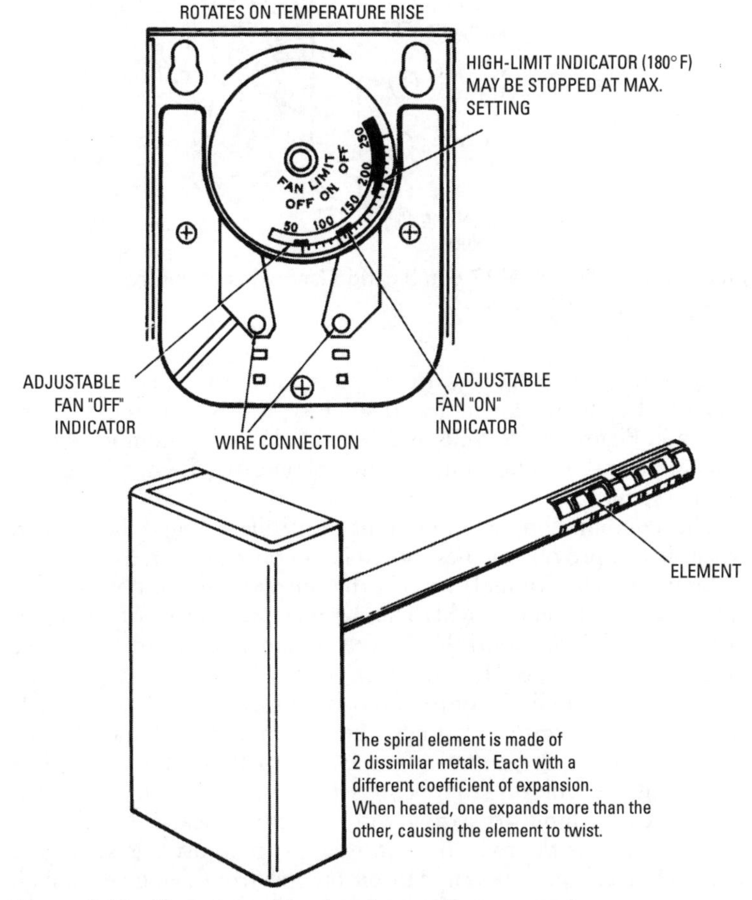

Figure 6-23 Typical combination fan and limit control. *(Courtesy Trane Co.)*

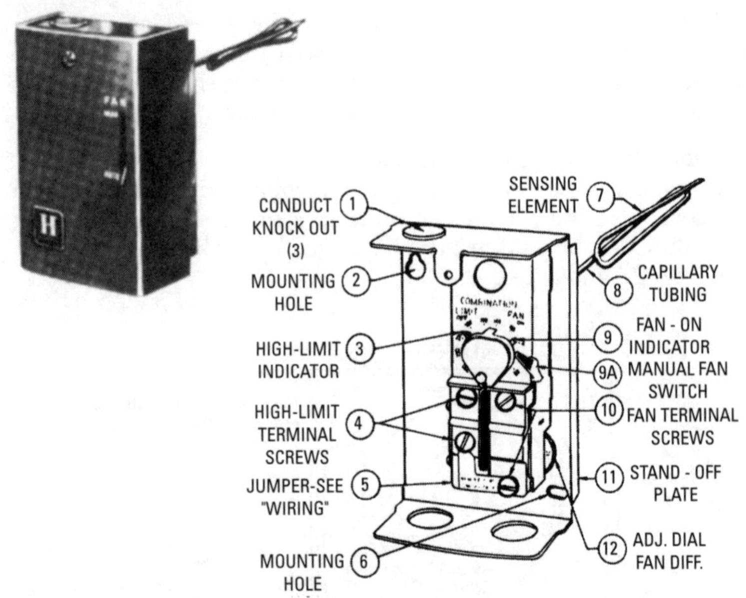

Figure 6-24 Model L4017 combination fan and limit control.

(Courtesy Honeywell Tradeline Controls)

Combination controllers are wired in much the same way as the individual controls. Examples of some typical wiring hookups are shown in Figures 6-25, 6-26, 6-27, and 6-28. These combined controls can be used in line voltage, low-voltage, or self-energizing millivolt systems.

The combination fan and limit controller should be located where it will provide the best possible operating characteristics.

Limit switch terminals are on the left side of the control, fan switch terminals on the right. This arrangement is true of combination fan and limit controls as well as single-purpose types (see Figures 6-29, 6-30, 6-31, and 6-32).

On the fan and limit controls shown in Figures 6-29, 6-30, 6-31, and 6-32, temperature settings can be changed by moving the temperature-setting pointers. Temperature settings are interlocked to prevent the limit *off* from being set as low as the fan *on* pointer. Sometimes the limit *off* setting will be factory-locked to a specific setting. If this is the case, do *not* attempt to adjust this setting. A safety interlock prevents the fan *on* pointer from being set as high as the limit *off* pointer.

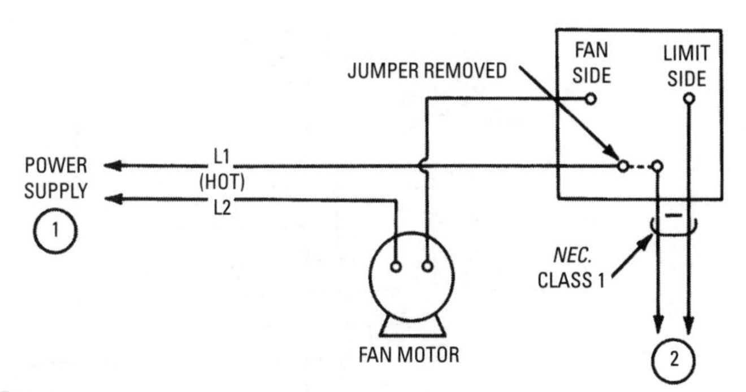

(1) Add disconnecting means and overload protection as required.

(2) To controlled low-voltage equipment.

Figure 6-25 Combination fan and limit control wiring diagram with limit controller in the low-voltage circuit. *(Courtesy Honeywell Tradeline Controls)*

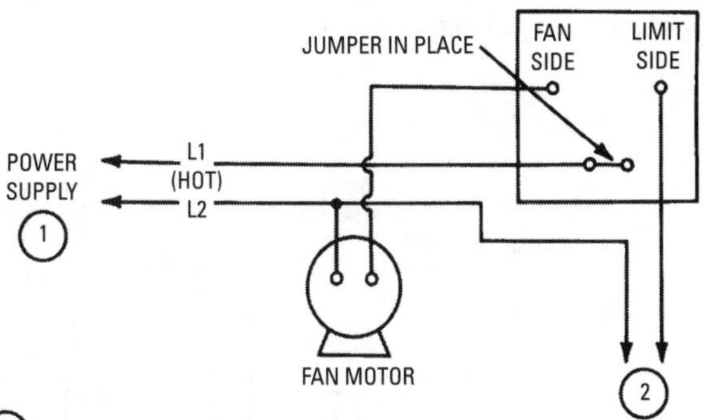

(1) Add disconnecting means and overload protection as required.

(2) To controlled low-voltage equipment.

Figure 6-26 Combination fan and limit control wiring diagram with limit controller in the line voltage circuit. *(Courtesy Honeywell Tradeline Controls)*

Some combination fan and limit controls are equipped with a manual summer fan switch to provide continuous fan operation for summer ventilation. To operate the blower during summer weather *without* the burner in operation, move the switch lever on the manual summer fan switch from *auto* to *on* position. This will provide continuous fan operation until the lever is moved to the *auto* position.

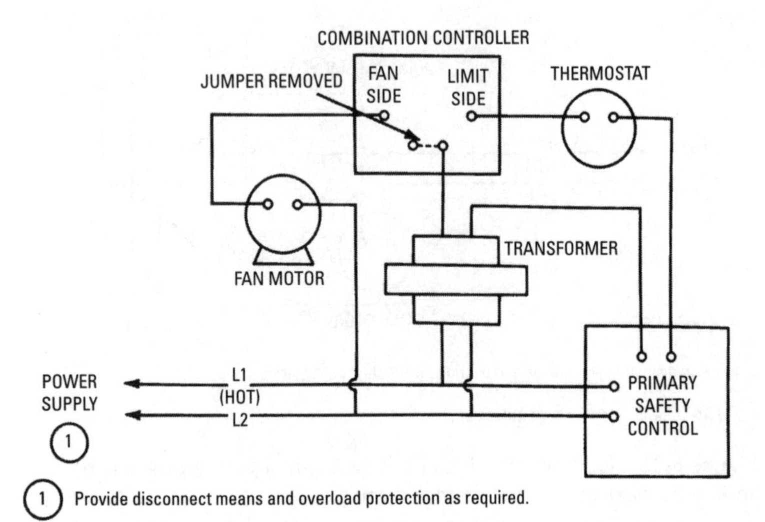

Figure 6-27 Combination fan and limit control in warm-air heating control circuit with low-voltage limit. *(Courtesy Honeywell Tradeline Controls)*

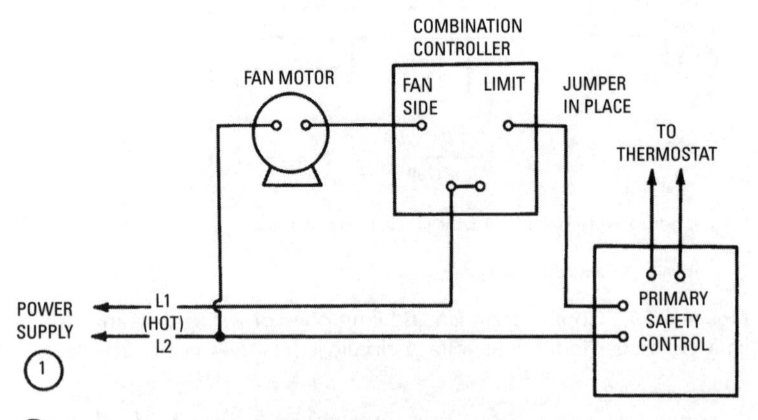

Figure 6-28 Combination fan and limit control in warm-air heating control circuit with line voltage limit. *(Courtesy Honeywell Tradeline Controls)*

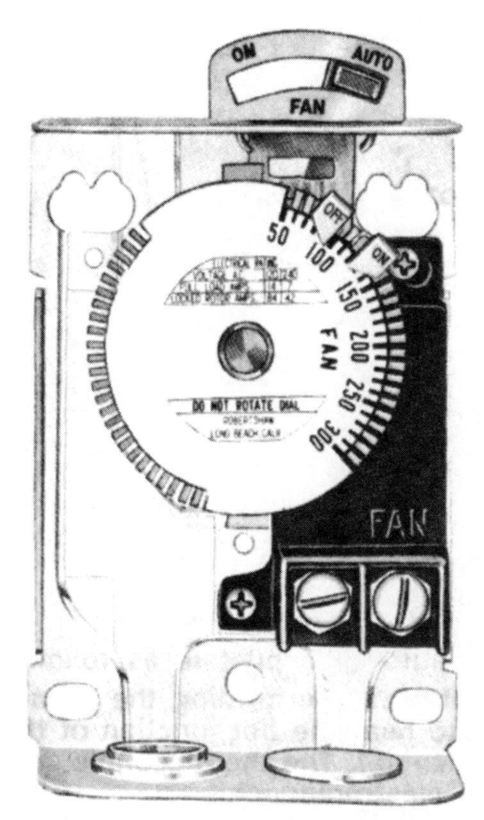

Figure 6-29 Fan control with adjustable fan differential and summer fan switch.

(Courtesy Robertshaw Controls Co.)

As shown in Figure 6-33, the metal tabs on the temperature dial of the Robertshaw combination fan and limit controller can be bent back to serve as stops for the fan and limit setting pointers. The dial is held to prevent rotation, and the pointer is pressed in and rotates to the proper temperature setting. The first tab *above* the pointer is bent back 90° toward the sensing element.

A certain amount of caution must be exercised when installing a combination fan and limit control. These controls can be either located in the furnace plenum or mounted directly on a panel, *but* the sensing element *must* be located in the path of free-flowing air. Never mount the control near the cool-air intake, and keep the

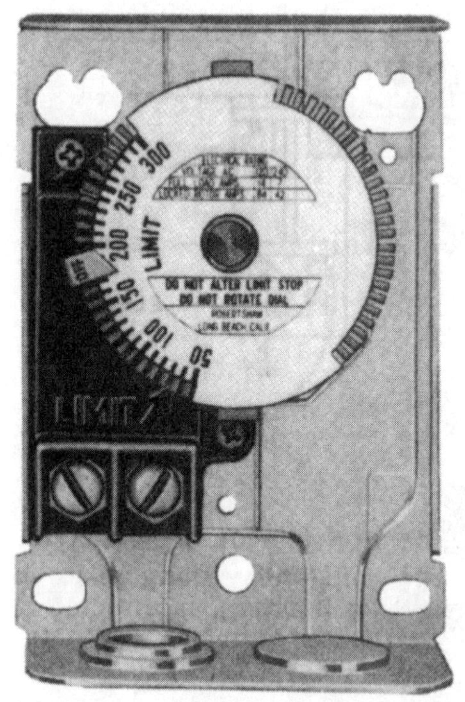

Figure 6-30 Limit control without summer fan switch. *(Courtesy Robertshaw Controls Co.)*

sensing element away from any hot metal surfaces. Furthermore, make sure you mount the control where it is accessible for making temperature adjustments.

Switching Relays

A *switching relay* is a device used to increase system switching capabilities, to isolate electrical circuits, and to provide electrical interlocks in a heating and cooling system. These devices are especially useful in systems where the heating and cooling equipment have separate supplies.

A typical switching relay contains an integral transformer and a magnetic relay with contacts designed to make or break an electrical circuit. These contacts will be either normally open or normally closed, depending on the design of the relay and its purpose in the heating and/or cooling system.

Figure 6-31 Combination fan and limit control with nonadjustable fan differential and summer fan switch. *(Courtesy Robertshaw Controls Co.)*

The 24-volt switching relay illustrated by the wiring diagram in Figure 6-34 is designed to control 115-volt and 24-volt or millivolt circuits. It incorporates a 20 VA 115 V/24 V transformer and a 24 V/60 Hz 0.2-ampere magnetic relay with two normally open contacts. One set of the relay contacts is line voltage rated for the switching of a circulator or other device. The other set of contacts is used for the switching of a self-energized (millivolt) or 24-volt circuit. A terminal board is located on top of the relay cover with screw terminals for connecting a thermostat (terminals T_1 and T_2) and a gas valve or oil burner control (terminals X_1 and X_2).

The switching relay illustrated in Figure 6-34 is shown as used in a gas-fired, forced hot-water heating system. In operation, a

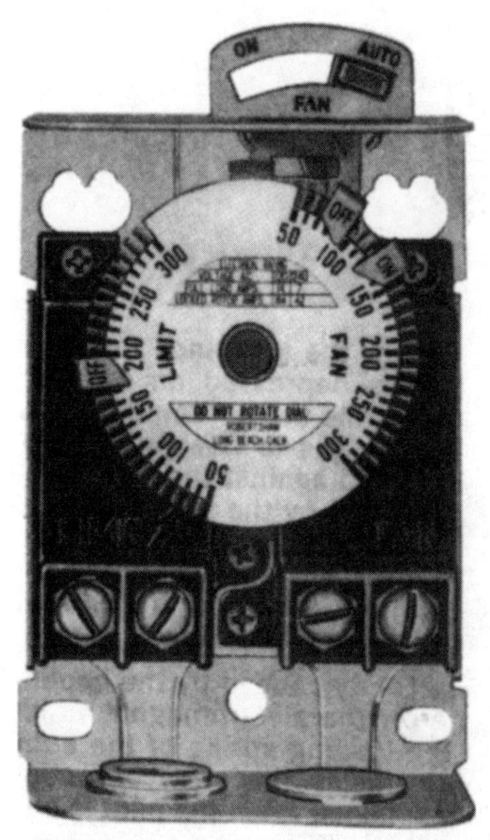

Figure 6-32 Combination fan and limit control with adjustable fan differential and summer switch. *(Courtesy Robertshaw Controls Co.)*

thermostat or some other switching accessory (e.g., an aquastat or zone valve) connected to the T_1 and T_2 terminals starts the circulator and boiler simultaneously by energizing and closing the two-pole, normally open relay. When the relay is activated by the thermostat, it closes a circuit from the integral transformer to a magnetic switch, which causes high voltage to be fed to the circulator to start the system pump. At the same time, the second pole of the relay switches 24-volt power to the gas valve or oil burner control, thereby starting the boiler.

A switching relay can also be used as a pilot duty relay to power a contactor and control a crankcase heater for the compressor motor. This type of switching relay has normally closed contacts

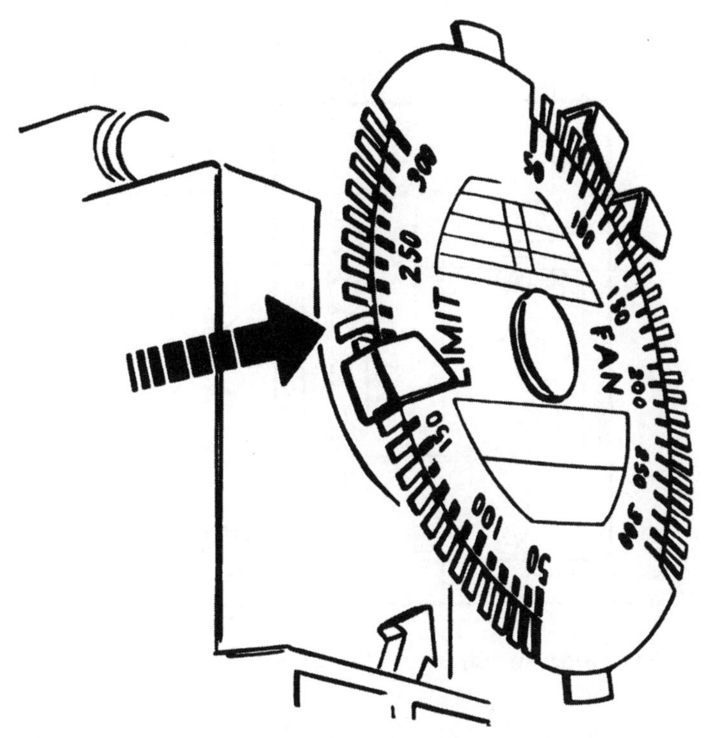

Figure 6-33 Temperature dial metal tabs bent to form a stop.
(Courtesy Robertshaw Controls Co.)

that complete an electrical circuit to the heater until the thermostat calls for cooling. When this occurs, the relay switches to break the heater circuit and power the compressor motor circuit.

In installations where a cooling system has been added to a self-energizing (millivolt) heating system, a switching relay may be used to isolate the cooling and heating power supplies (see Figure 6-35). When the room thermostat calls for heat, an isolating relay is used to switch the heating equipment directly.

Heavy-duty switching relays are used for control of high-current loads such as cooling compressors or electric heating where sudden high-current demands are not unusual. The relay is wired to break both sides of the circuit with DPST switching (see Figure 6-36).

Impedance Relays

An *impedance relay* (see Figure 6-37) is used to provide lockout and remote reset in refrigeration, air-conditioning, and other

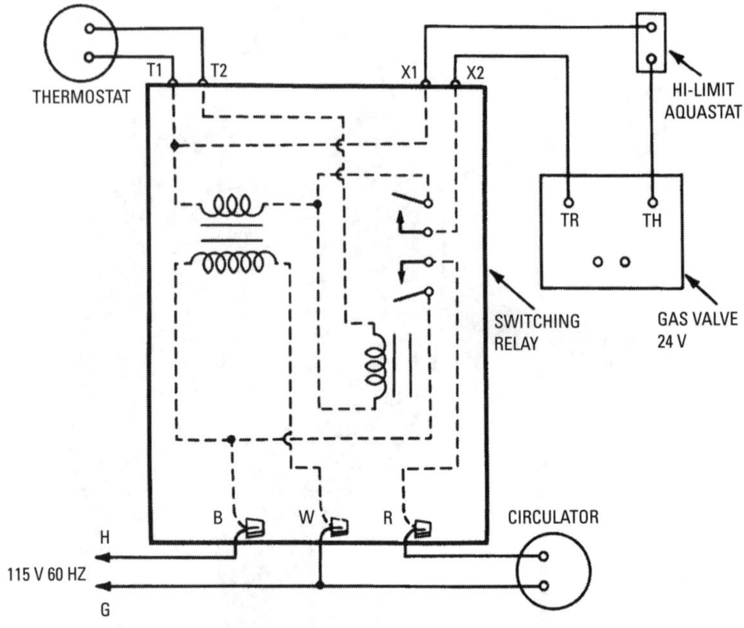

Figure 6-34 Switching relay used with 24-volt power supply.
(Courtesy Hydrotherm, Inc.)

systems. A typical wiring diagram for a low-voltage impedance relay is shown in Figure 6-38. A low-voltage relay requires the use of a transformer with an open circuit secondary of between 24 and 27 volts AC.

As shown in Figure 6-38, one pair of contacts is normally open and the other pair is normally closed. During normal operation, the normally closed contacts of the pressure controls (i.e., the low-pressure and high-pressure cutout switches) and the motor overloads short out the impedance relay coil so that the compressor contactor pulls in. If one of the pressure controls or overloads opens, the impedance relay coil is energized in series with the contactor coil and most of the available voltage is used by the high impedance of the relay coil. Because insufficient voltage remains to operate the contactor coil, the contactor drops out and compressor operation stops.

As the impedance relay pulls in, its normally closed contacts open to keep the contactor out, even though the pressure control or

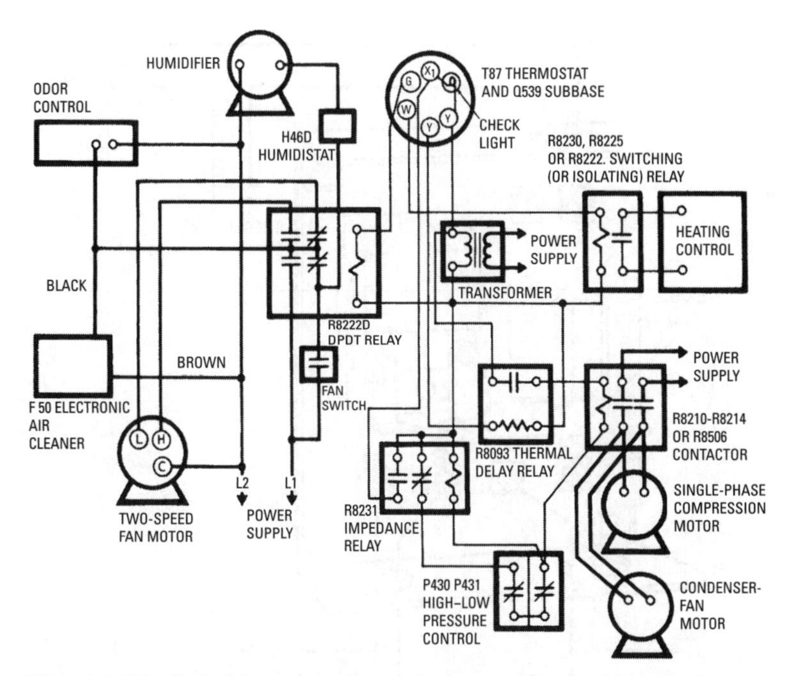

Figure 6-35 Switching relay used to isolate cooling and heating power supply. *(Courtesy Honeywell Tradeline Controls)*

overload (automatic reset) remakes. The system can be reset by breaking the contactor circuit to allow the impedance relay to drop out. In most systems, this is accomplished by moving the thermostat subbase switch to *off* and back to *cool* again.

Heating Relays/Time-Delay Relays

In air-conditioning installations, a *time-delay relay* is often installed in the control circuit to provide protection for the compressor and contactor. This device is activated or deactivated by the room thermostat. In operation, it causes a time delay between turning down the thermostat setting and the start of the compressor unit of approximately 20 to 45 seconds (depending on the manufacturer). On the shutoff cycle at the thermostat, the same delay occurs. This control prevents rapid short cycles from occurring.

A time-delay relay is essentially a switching relay that contains a small heater wound around a bimetal element. The relay heater is

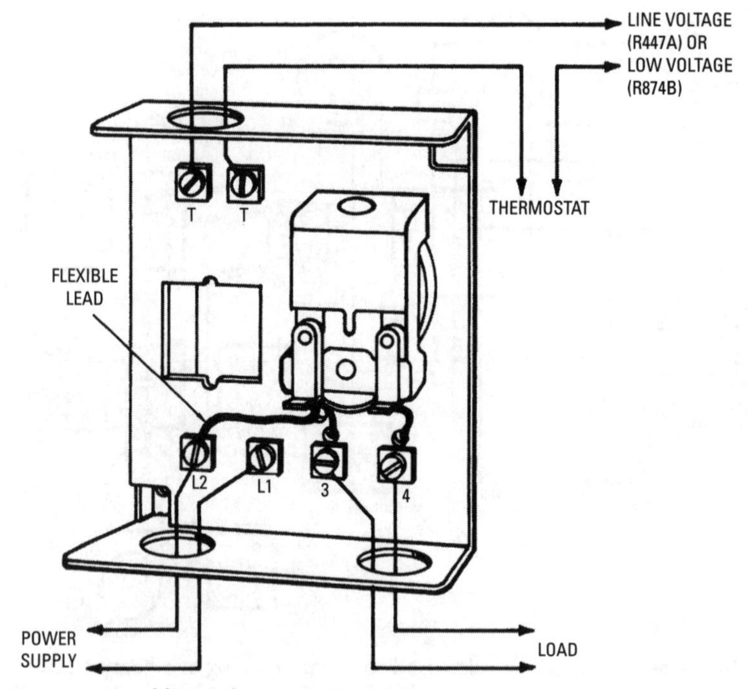

Figure 6-36 Heavy-duty switching relay. *(Courtesy Honeywell Tradeline Controls)*

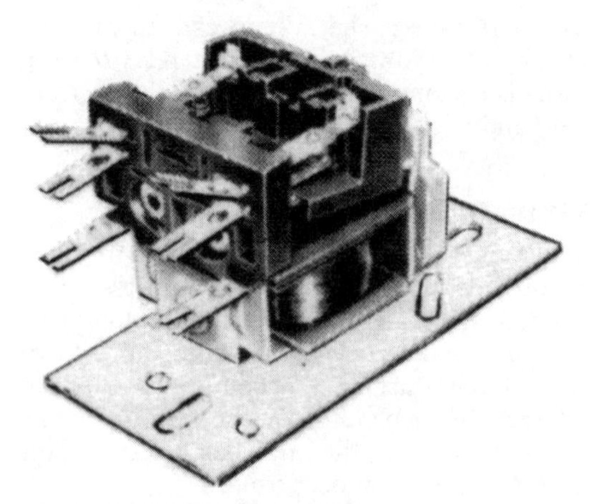

Figure 6-37 Impedance relay.
(Courtesy Honeywell Tradeline Controls)

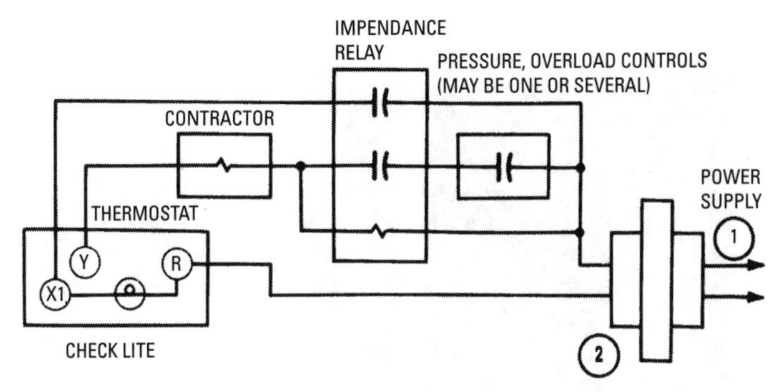

Figure 6-38 Impedance relay in a low-voltage circuit.
(Courtesy Honeywell Tradeline Controls)

energized through the cooling contacts of the thermostat and heats the bimetal element. After a time delay of approximately 20 to 45 seconds, the heated bimetal element bends to provide the switching force. In other words, it closes a set of snap-action contacts, thereby completing the circuit through the starter or contactor coil required to start the compressor. The relay may be wired to provide a delay in breaking the circuit after the thermostat is satisfied.

The time-delay relay shown in Figure 6-39 is used with a two-wire low-voltage thermostat (and remote-mounted thermostat) to provide a time delay between stages for electric heaters in furnace ducts. Each relay can control up to 6000 watts at 240 volts AC. One relay is required per stage or time increment. This particular model provides a delay of approximately 75 seconds between the *on* cycles of consecutive stages. The sequencing of heating loads is permitted by auxiliary contacts.

Because of the dual nature of their function, time-delay relays are also referred to as *time-delay switches*, *thermal switching relays*, *thermal time-delay relays*, *thermal relays*, and *heating relays*. Other examples of time-delay/heating relays are shown in Figures 6-40 and 6-41.

Potential Relay

In air-conditioning installations, the *potential (start) relay* serves as a switch to disconnect the starting capacitors when the compressor motor has overcome the initial starting torque.

Figure 6-39 Thermal switching relay used for control of electric furnaces or electric duct heaters.
(Courtesy Honeywell Tradeline Controls)

Figure 6-40 Electric heating relay used with a two-wire thermostat for control of electric boilers, duct heaters, fan coils, or electric furnaces.
(Courtesy Honeywell Tradeline Controls)

An important fact to remember is that each potential relay is specifically designed for the compressor to which it is attached. Should this relay fail or become erratic in its operation, no attempt should be made to repair it. The relay must be replaced with an *identical* component.

Figure 6-41 Time-delay relay used in electric baseboard heating.
(Courtesy Singer Controls Co. of America)

Pressure Switches

A *pressure switch* is a safety device designed to stop or start heating or air-conditioning equipment in response to gas- or air-pressure changes. These switches are used in either positive-pressure or differential-pressure systems.

Gas-pressure switches used in the control of gas-fired furnaces and boilers are described in the section *Pressure Switches* in Chapter 5 ("Gas and Oil Controls"). Pressure switches used as refrigerant controllers in a cooling system are described in this chapter (see *Low-Pressure Cutout Switch and High-Pressure Cutout Switch*).

The *National Electrical Code* requires that a duct heater be interlocked with the system fan so that the heater cannot be energized unless the fan is also energized. This can be accomplished by using either a fan interlock relay or a built-in air-pressure switch.

An air-pressure switch designed to provide fan interlock control consists of an internal diaphragm that is actuated by positive air pressure. The switch sensor is mounted so that it extends into the air stream (see Figure 6-42). Movement of the diaphragm closes an electrical switch, which permits the duct heater to turn on. When there is no airflow or low airflow, the switch will open and turn off the duct heater.

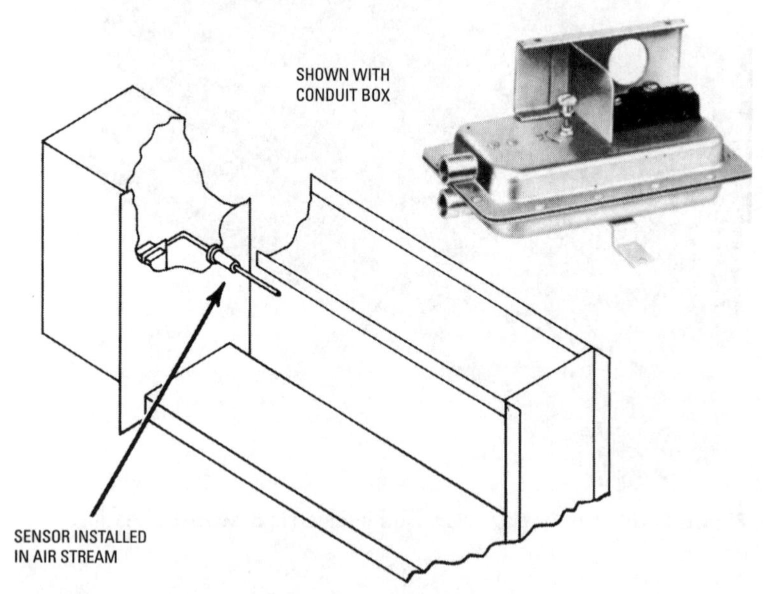

Figure 6-42 Air-pressure switch. *(Courtesy Vulcan Radiator Co.)*

Sail Switches

A *sail switch* consists of a steel or polyester film sail mounted on a switching device (see Figure 6-43). The combined unit is mounted so that the sail is located in an air duct. When the air velocity increases, the switch makes an electrical circuit. Figure 6-44 illustrates the location of the sail switch in a gas control circuit.

Sail switches are used in forced warm-air heating systems, in air-conditioning systems, and with gas-fired unit heaters. Some sail switches are designed to provide on-off control of electronic air cleaners, odor-control systems, humidifiers, and other equipment that is energized when the fan is operating. In these applications,

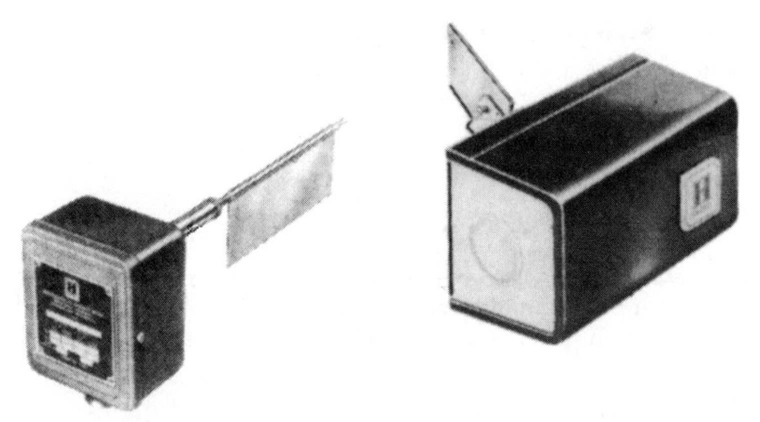

Figure 6-43 Sail switches. *(Courtesy Honeywell Tradeline Controls)*

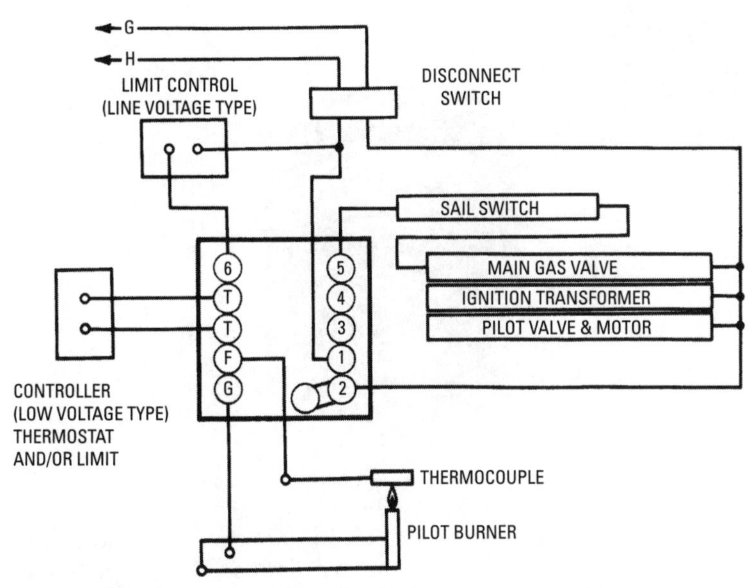

Figure 6-44 Location of the sail switch in a gas control circuit.

the sail switch completes a power circuit to auxiliary equipment, which can be wired independently of the blower motor. Sail switches are also used in electric heating systems to provide minimum airflow.

Other Switches and Relays

Other switches and relays used in heating and/or cooling systems include the following:

- Balancing relays
- Manual switches
- Auxiliary switches

A *balancing relay* (see Figure 6-45) is used with an electric motor that does not have an integral balancing relay. The relay is mounted separately from the motor so that vibrations will not affect it.

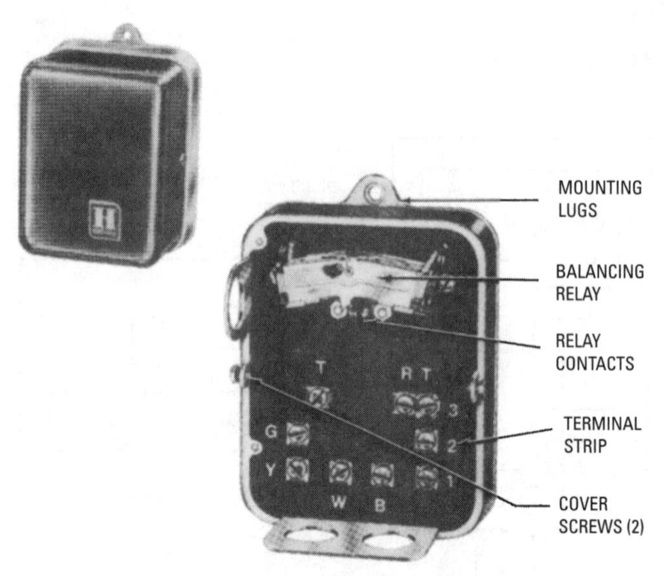

Figure 6-45 Balancing relay. *(Courtesy Honeywell Tradeline Controls)*

A *manual switch* is used to manually perform one or more operations in a heating and/or cooling installation. These switches are generally two-position types (on-off), although multiple-position switches are also available. An example of the latter would be the heat-off-cool switch on a heating and cooling thermostat. Examples of some special switching hookups in which manual switches are used are shown in Figures 6-46, 6-47, and 6-48. In each of these examples, a relay could have been used instead of the manual switch.

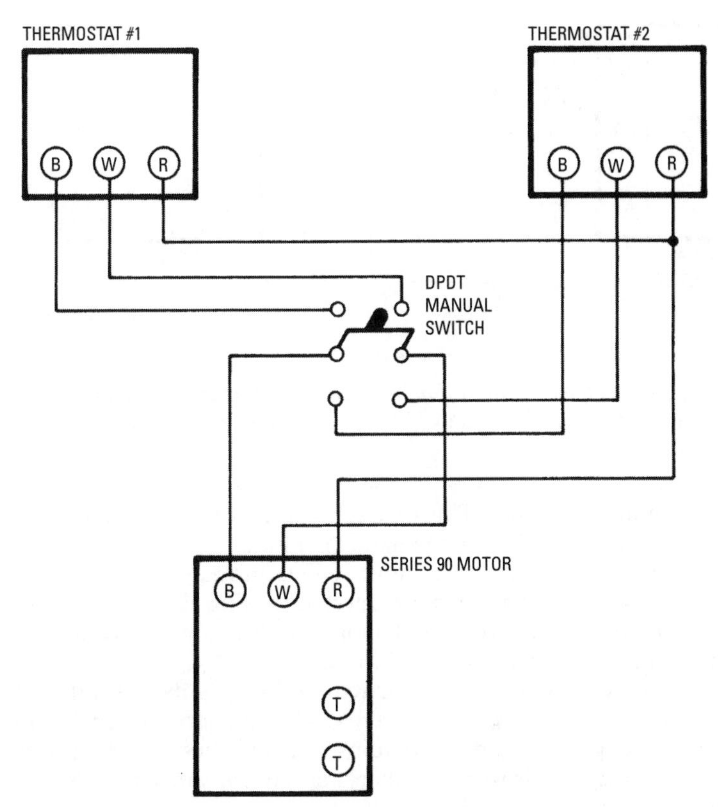

Figure 6-46 Manual switch (DPDT) used to transfer control of an electric motor from one thermostat to another.

(Courtesy Honeywell Tradeline Controls)

An *auxiliary switch* (see Figure 6-49) is used in conjunction with an electric motor to provide control of auxiliary equipment. This control functions as a direct extension of motor operation. The auxiliary switch may be an integral part of the motor or an external unit fitted to the motor and adjusted to open or close at the desired point in the motor stroke (see Figure 6-50). As shown in Figure 6-51, the internal auxiliary switches are in a single-pole, double-throw (SPDT) configuration.

Sequence Controllers

A *sequence controller* (also referred to as a *sequencer* or *step controller*) is a device used to operate two or more electric switches in

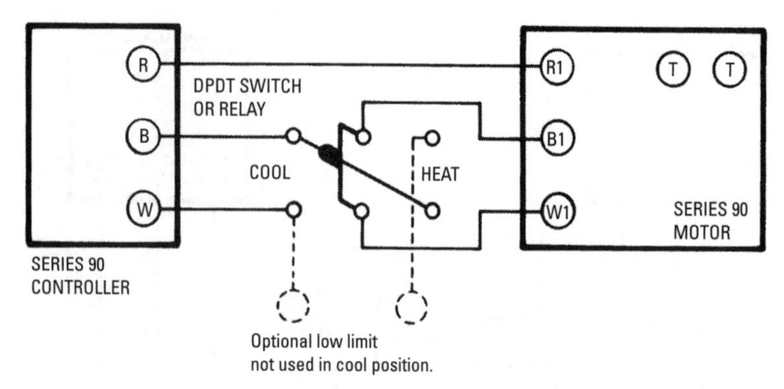

Figure 6-47 Manual switch (DPDT) used for reversing control so that the same thermostat can be used for both heating and cooling.

(Courtesy Honeywell Tradeline Controls)

predetermined sequence. This function is accomplished by means of a proportional electric or pneumatic operator.

Sequence controllers are most commonly used to provide sequenced starting of a number of electric heating elements or compressor motors. This prevents the massive drawing on the current that simultaneous starting would cause.

Sequence controllers are manufactured in a number of different sizes and capacities. The Honeywell S984 step controller, shown in Figure 6-52, provides up to 10 adjustable switches. It can also be used to control at least two other step controllers when greater switching capacity is required.

Each switch in the Honeywell step controller is operated by a cam mounted on the main shaft (see Figure 6-53). Adjustments of the step controller can be made by setting each of the switches to make or break a circuit at the correct time or angle in the stroke. The procedure for setting the switches in the Honeywell step controller is as follows:

1. Loosen the setscrew so that the cam assembly will turn freely on the motor shaft.
2. Run the step controller motor until the desired switch make point is reached. This will be determined by the time or degrees of camshaft rotation.
3. Turn the cam until the switch just makes and tighten the setscrew.
4. Run the step controller *back* to the desired switch break point.

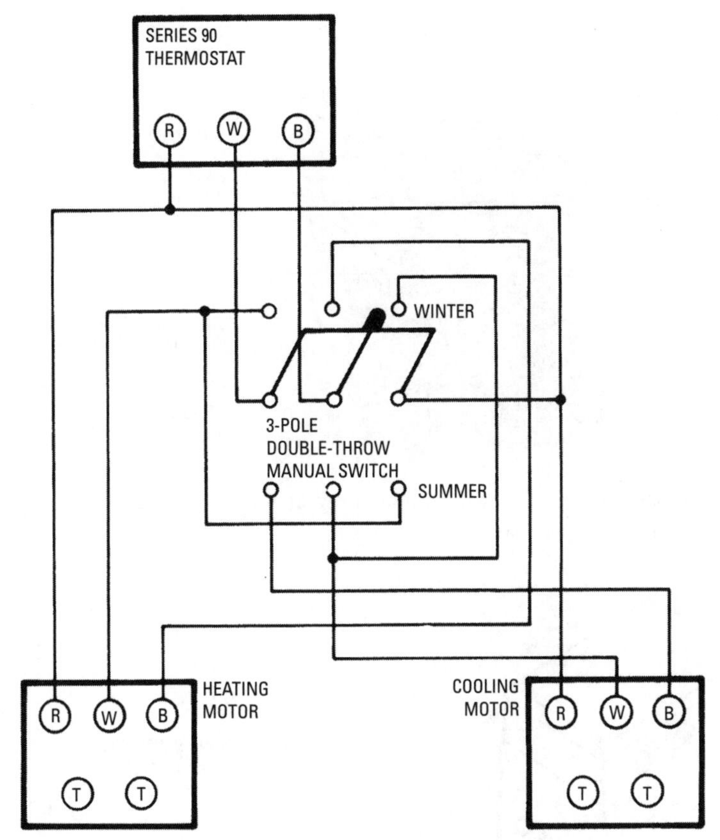

Figure 6-48 Manual switch (three-pole, double-throw) used to transfer thermostat control to one or the other of two motors.
(Courtesy Honeywell Tradeline Controls)

5. Loosen the lockscrew and turn the differential cam so that the switch will break at this point.

The instructions for adjusting the Honeywell step controller recommended setting the make points of *all* the switches first and then setting all the break points. This procedure avoids unnecessary cycling of stages.

The sequence controller, shown in Figure 6-54, provides time-delay switching of up to eight electric heater banks and a fan or pump. When the thermostat calls for heat, it starts the low-voltage

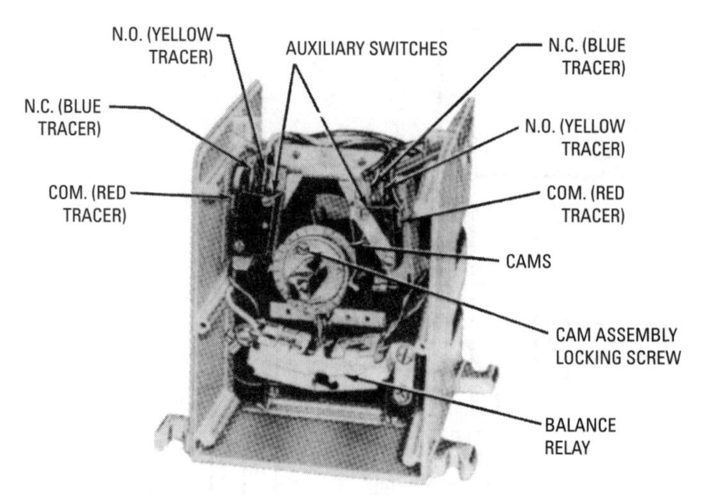

Figure 6-49 Electric motor with an integral auxiliary switch.
(Courtesy Honeywell Tradeline Controls)

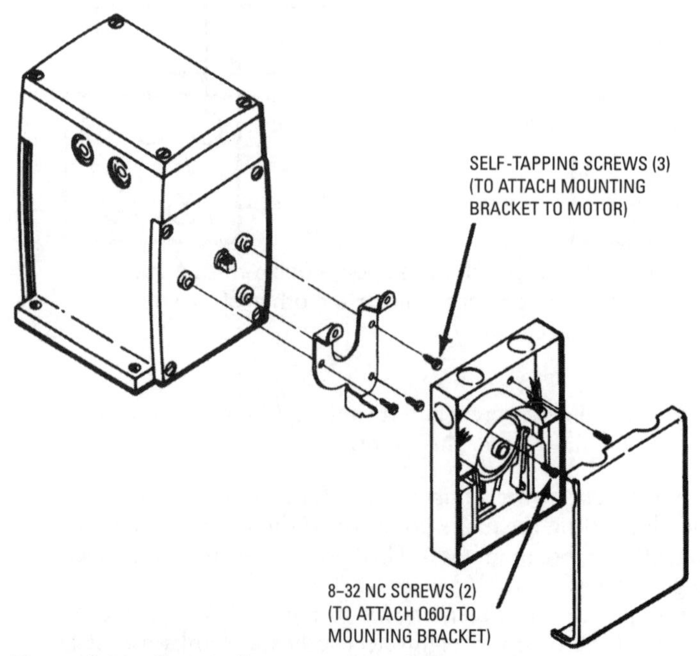

Figure 6-50 Externally mounted auxiliary switch.
(Courtesy Honeywell Tradeline Controls)

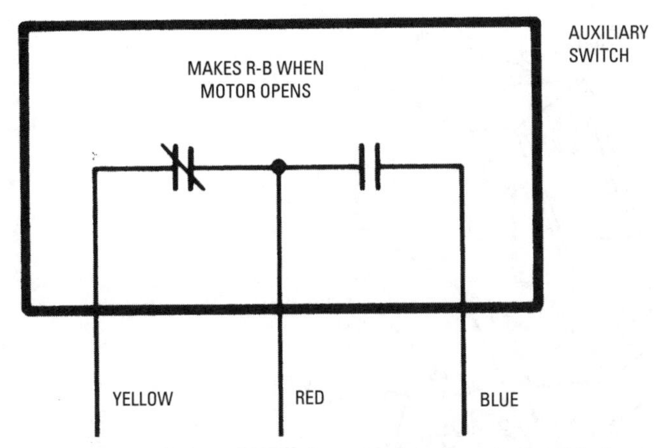

Figure 6-51 Internal auxiliary switches in a single-pole, double-throw (SPDT) configuration. *(Courtesy Honeywell Tradeline Controls)*

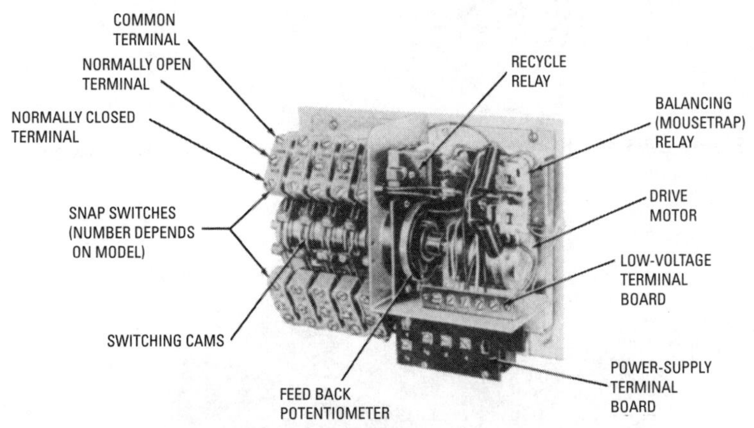

Figure 6-52 Model S984 step controller. *(Courtesy Honeywell Tradeline Controls)*

sequence motor, which operates the rotating cams. The first cam locks in the motor circuit so that if the thermostat should break immediately, the motor will continue running, rotating all cams back into the starting position.

The second cam starts the fan. The remaining cams switch on the heater banks with a specific time delay between the energizing of each bank. The motor will stop running after 180° rotation as long as the thermostat calls for heat. The fan and all heater banks will

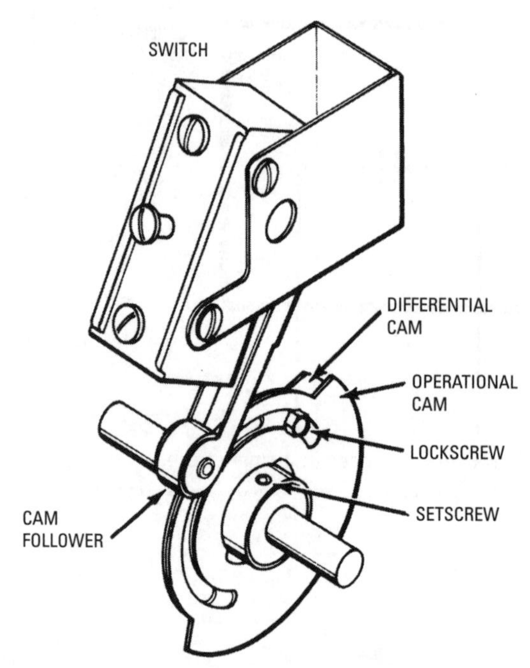

Figure 6-53 Adjusting the step controller. *(Courtesy Honeywell Tradeline Controls)*

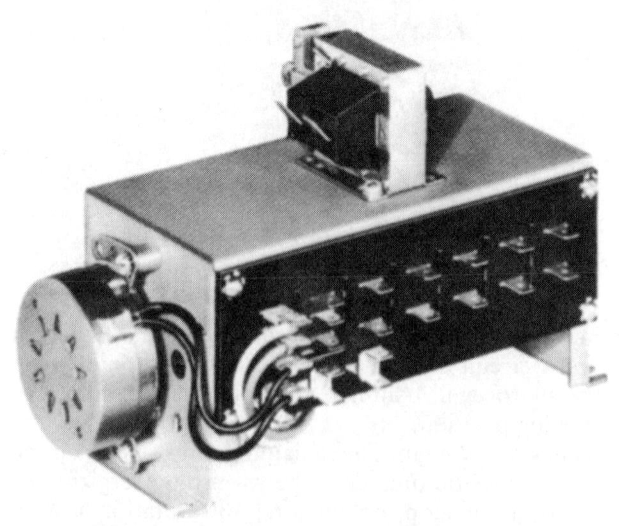

Figure 6-54 Model 38 sequence control. *(Courtesy Singer Control Co. of America)*

continue to be energized. When the thermostat is satisfied, the motor starts again and switches off the heater banks at the predetermined time interval. The fan is deenergized, and the motor stops when the cams are back in the starting position.

Contactors

A *contactor* is essentially a switching relay device that functions as a primary control in a cooling system. Its operation is based on either magnetic or mercury-to-mercury contact action.

A magnetic contactor (see Figure 6-55) is similar in design and operating principle to a relay but is larger in size. In a mercury contactor (see Figure 6-56), the contacts are made and broken between two pools of mercury separated by a ceramic insulator. Mercury-to-mercury contact action results in a quieter operation than magnetic contactors can provide, but a mercury contactor has the disadvantage of being position sensitive. It *must* be mounted in an upright position.

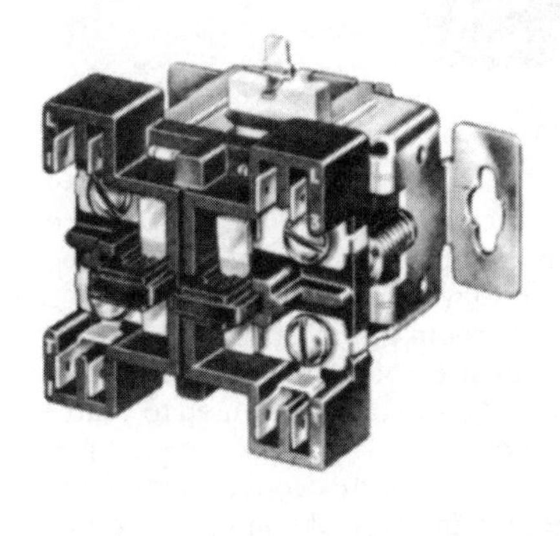

Figure 6-55 Magnetic contactors. *(Courtesy Honeywell Tradeline Controls)*

A contactor is used for applications requiring heavy current, high voltage, or a large number of poles—in other words, applications where the capacity of a relay would be inadequate.

In a cooling system, an electric-driven compressor motor may be cycled by a thermostat (either a low-voltage or line voltage type) or a low-pressure control. Because these controllers are usually unable to handle the high current drawn by the compressor motor,

Figure 6-56 Mercury contactors. *(Courtesy Honeywell Tradeline Controls)*

a contactor is installed between the thermostat or pressure control and the motor where it functions as a primary control. This contactor is the electrical contact switch to which the main electric power is supplied. When activated by the room thermostat, the contactor causes the compressor and condenser blower motor in an air-conditioning system to begin operating. Contactors are used to control all but the smallest compressor motors.

The size of the contactor selected for use in a cooling system will depend on such variables as the size and type of compressor motor and the auxiliary loads.

Both single-phase and three-phase motors are used in compressors. As shown in Table 6-1, the motor current (expressed in amperes) will vary according to the size of the compressor. Contactors are rated in amperes and should be selected to match the rating of the compressor motor. The number of pole contacts is also an important consideration in selecting a contactor. A one- or two-pole contactor is required for a single-phase compressor motor, and a three-pole contactor for a three-phase motor. Auxiliary poles may be used for interlock switching, fan loads, or crankcase heaters (see Figure 6-57).

Table 6-1 Compressor Size and Motor Current Ratings

Single-Phase Motor		*Three-Phase Motor*	
Compressor Size (tons)	***Motor Current (amperes)***	***Compressor Size (tons)***	***Motor Current (amperes)***
2	18	3	18
3	25–30	4	25–30
4	30–40	5	30–40
5	35–50	7½	35–50

(Courtesy Honeywell Tradeline Controls)

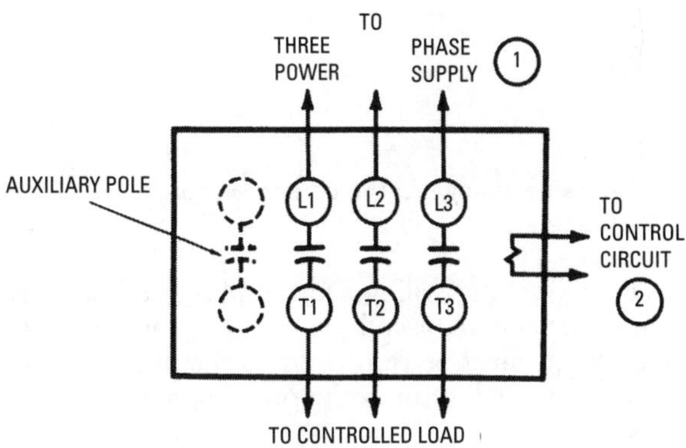

Figure 6-57 Typical hookup of a three-pole contactor.
(Courtesy Honeywell Tradeline Controls)

Troubleshooting Contactors

Always check the fuse box (or circuit breakers) first to make certain that the fuses are good and the switch is in the *on* position. If this is not the source of the problem, then set the thermostat and base to cooling position and check for continuity across contacts. If the circuit is open, repair or replace the thermostat.

If the system fails to start and the contactor is open and buzzing, check the voltage at the contactor coil (see Figure 6-58). Normal voltage will be within plus or minus 10 percent of the rated coil voltage. Subnormal voltage may be caused by an undersized transformer or low voltage at the supply side of the transformer. Thermostat wiring that is too long can also cause a subnormal voltage reading. A normal voltage reading under these circumstances (that is, an open, buzzing contactor) is generally caused by a tight or fouled contactor armature. The armature should be cleaned or the contactor replaced.

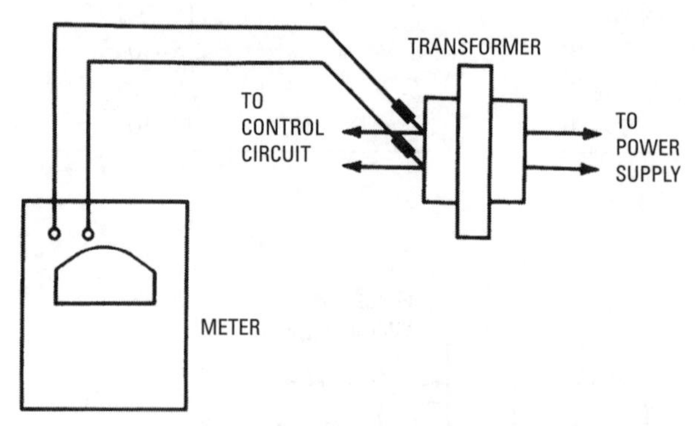

Figure 6-58 Checking voltage to the contactor coil.
(Courtesy Honeywell Tradeline Controls)

If the system compressor will not start and the contactor is open but *not* buzzing, the contactor coil may not be powered. A voltage and continuity check must be run to locate the problem. This is a more complicated procedure than the one described for an open, buzzing contactor.

The voltage to the contactor coil should be checked first (see Figure 6-58). If the voltage is normal (within plus or minus 10 percent of the rated voltage), the contactor should be replaced. If a zero voltage reading is obtained, check the voltage at the transformer secondary (see Figure 6-59). If a zero voltage reading is obtained, check the line voltage side of the transformer (see Figure 6-60). A normal supply voltage reading here indicates that the transformer is defective and should be replaced. A zero voltage reading, on the other hand, indicates a problem with the power supply. Check the fuses, circuit breakers, line disconnect switch, and the power at the service entrance.

If the voltage checked at the transformer secondary (as shown in Figure 6-59) is *normal*, the control circuit to the contactor coil is open. With a continuity checker, jump across the terminals of the following controls:

- Low-pressure switch
- High-pressure switch

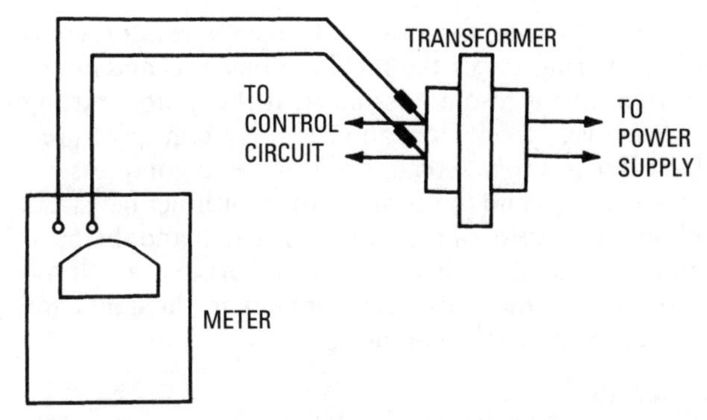

Figure 6-59 Checking voltage at the transformer secondary.
(Courtesy Honeywell Tradeline Controls)

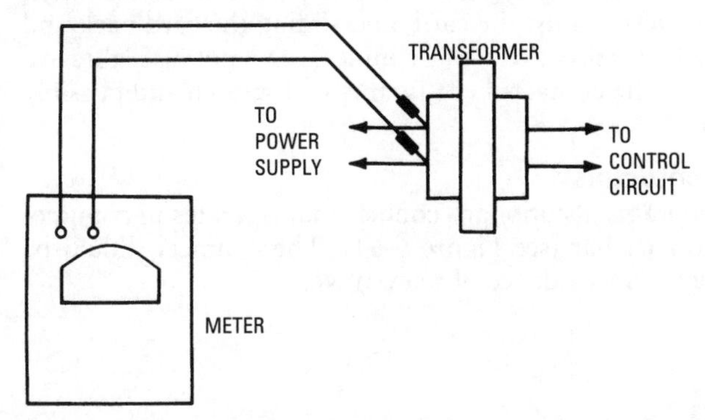

Figure 6-60 Checking voltage at the line side of the transformer.
(Courtesy Honeywell Tradeline Controls)

- Room thermostat
- Heating/cooling interlock switch
- Time-delay switch
- Lockout relay

If the system starts with a particular control out of the circuit, the control is probably defective and needs to be replaced. If none of these controls is defective, check the circuits for broken wires or loose connections. An open internal thermostat or overload relay switch in the compressor is another possible cause.

If the compressor hums but will not start, and the contactor remains closed, the problem may be in the motor starting circuit.

Check the motor circuit wiring for loose or broken wires. If there is no problem with the wiring, check the starting capacitors and motor starter. A defective motor is also a possibility. If the motor starting circuit is not defective, check for abnormal system pressures. Another possible cause is a tight, stuck, or burned-out compressor.

If the contactor is closed and the compressor motor neither starts nor hums, check the continuity of the overload switch and the open compressor motor windings. Check also for broken or loose wiring. If none of these is the cause, the contacts in the contactor are probably burned. Replace the contacts.

Cleaning Contactors

Sometimes a contactor will fail to operate because a layer of dust and lint has accumulated on the electrical contacts. This dust and lint can be removed by placing a file card between the contacts, closing the contacts against the card, and sliding the card back and forth. This will usually clean the contacts. Do not use abrasive material to clean the contacts because this will scratch and possibly ruin the surface.

Replacing Contactors

A contactor contains a stationary contact that operates in conjunction with a contact bar (see Figure 6-61). The contacts should be replaced if they show evidence of uneven wear.

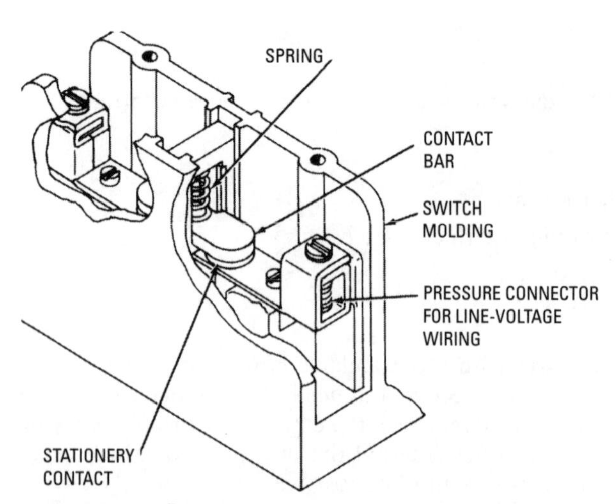

Figure 6-61 Principal components of a contactor.
(Courtesy Honeywell Tradeline Controls)

Always compare the ratings of the replacement contactor with the old one. They should at least be equal in rating. To be on the safe side, it is better to overrate a contactor and replace the old one with a contactor of slightly higher rating.

When replacing a contactor, make certain the terminal connections of the new one fit the installation. The mounting holes and dimensions should also be compatible.

Some manufacturers provide replacement coils for their contactors. *Never* replace a contactor coil until you have located and corrected the cause of the original coil failure. If you fail to do this, the replacement coil will probably burn out, too.

Always disconnect the power supply before attempting to remove a contactor. Be sure to tag the wiring connections to the contact terminals as soon as you have removed the contactor. This will minimize the possibility of confusion when the replacement is installed.

Motor Starter

When an electric-driven compressor motor stalls or is overloaded, it draws current many times its full load rating. If the condition lasts any length of time, the motor windings overheat and a fire may start in the insulation. This will result in very expensive damage to the motor. One method of guarding against the occurrence of an overload condition is by installing a motor starter in the control circuit (see Figure 6-62).

A *motor starter* consists of a contactor plus one or more overload relays. Each overload relay consists of a bimetal contact in series with the motor contact coil and a heater in series with the compressor motor. The overload relay (or relays) disconnects the motor from the power supply when the motor temperature and/or the current drawn by the motor become excessive.

Overload Relay Heater

As shown in Figure 6-62, an overload relay consists of a bimetal contact in series with the motor coil, and a heater in series with the compressor motor. An *overload relay heater* is a small electric heating device designed to work in conjunction with the bimetal contact. When the compressor motor becomes overloaded or stalled, the heavy continuous current through the heater causes the bimetal contact to bend until it opens the motor-starter coil circuit. This action stops the flow of current through the motor starter and results, in turn, in the opening of the load contacts to stop the flow of current to the compressor motor.

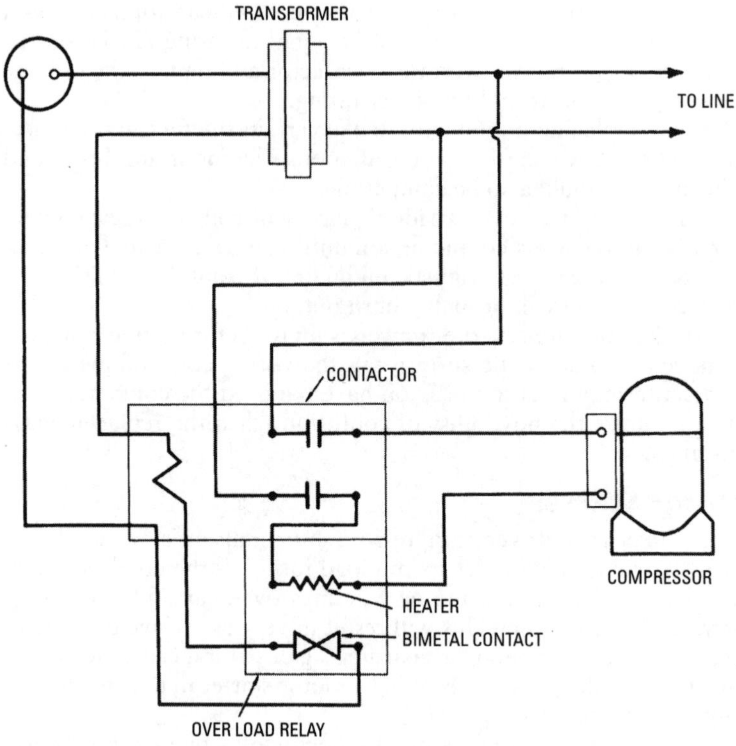

Figure 6-62 Motor starter in the control circuit.
(Courtesy Honeywell Tradeline Controls)

An overload relay heater must be accurately sized for the installation. Generally the manufacturer of the cooling equipment will provide instructions for sizing overload relay heaters.

Inherent Protector

An *inherent protector* is another safety device used to protect an electric-driven compressor motor from overload damage. It accomplishes this purpose by disconnecting the motor from the power supply when the motor temperature or current becomes excessive. Its function is similar to that of the overload relay.

An inherent protector is essentially a thermostat operated by the snap action of a bimetal disc. As shown in Figure 6-63, it consists of a heater, thermostatic disc, and contacts.

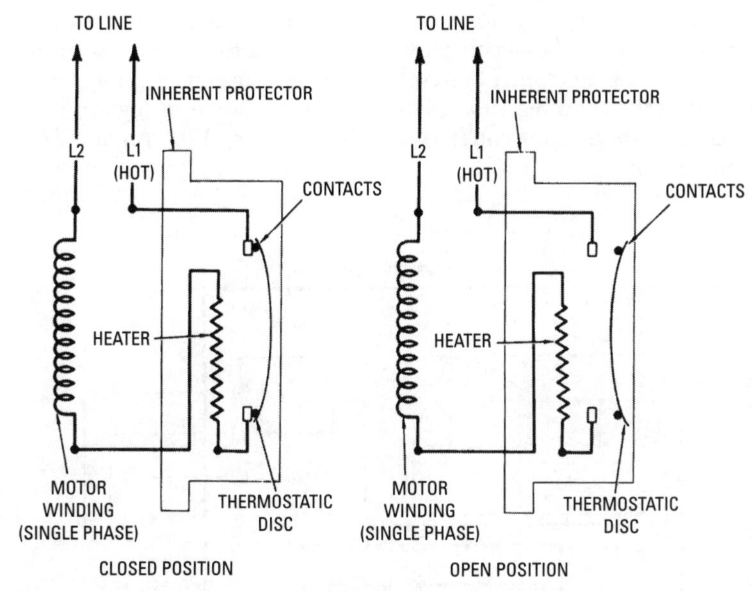

Figure 6-63 Inherent protector. *(Courtesy Honeywell Tradeline Controls)*

In operation, the motor current flows through both the motor winding and the heater. When an overload condition occurs, the critical temperature level is reached in the motor winding at exactly the same time that the protector reaches its tripping point. When the temperature rises to the rating of the bimetal disc, the disc snaps open, reversing its curvature, and cuts off the flow of current to the motor. When the temperature drops, the action is reversed.

Inherent protectors are available for all sizes of single-phase compressor motors. For three-phase motors, they are available for motor sizes up to 1½ hp.

Pilot Duty Motor Protector

Some manufacturers will install a *pilot duty motor protector* (or *pilot duty thermostat*) on the compressor motor to protect the motor from overcurrent damage. This device is a small temperature-sensitive thermostat mounted inside the compressor on the motor windings. If the motor windings become overheated, the pilot duty thermostat breaks the circuit to the contactor relay and shuts off the compressor motor.

The pilot duty thermostat is usually wired in the 24-volt control circuit. This can be accomplished by interrupting the transformer secondary of the control circuit. A wiring diagram of a pilot duty thermostat connected in a 24-volt circuit is shown in Figure 6-64. Pilot duty thermostat contacts can also interrupt 120-volt and 240-volt circuits.

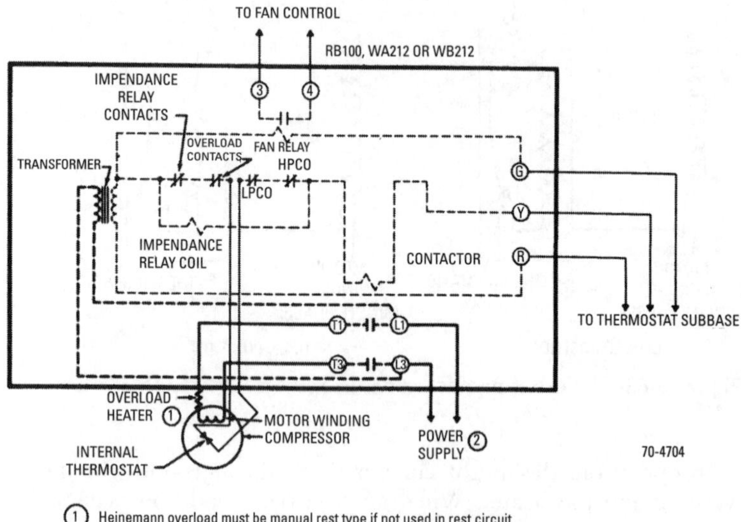

Figure 6-64 Pilot duty thermostat. *(Courtesy Honeywell Tradeline Controls)*

Capacitors

A *capacitor* is a device that provides the phase shift in the running and starting windings of an electric motor in order to increase the torque and efficiency of the motor-compressor assembly.

Capacitors are used in the following single-phase induction motors:

- Capacitor-start motors
- Permanent-split capacitor motors
- Capacitor-start, capacitor-run motors

A *capacitor-start motor* (see Figure 6-65) is used to power fans, blowers, and centrifugal pumps where constant-speed drive is

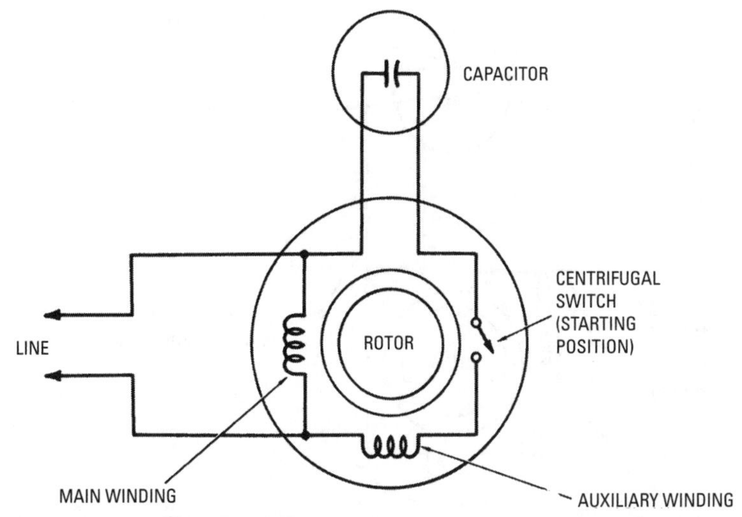

Figure 6-65 Capacitor-start motor. *(Courtesy Honeywell Tradeline Controls)*

necessary. This motor develops high starting torque on the frictional horsepower ratings and moderate starting torque in the lower ratings.

In operation, an auxiliary winding is connected in series with a capacitor in the motor circuit. When the motor approaches running speed, a centrifugal switch cuts the capacitor and auxiliary winding out of the circuit.

A *capacitor-start, capacitor-run motor* (see Figure 6-66) also develops high starting torque. This is accomplished by employing a starting capacitor and a running capacitor. The starting capacitor gives good starting ability but is suited for short-time operation only. The starting capacitor is cut out of the circuit during the running period. The running capacitor provides high efficiency at running speed. Capacitor-start, capacitor-run motors are used to power compressors, reciprocating pumps, and similar types of equipment.

On some capacitor-start, capacitor-run motors, the starting capacitor may be cut out of the circuit with a centrifugal switch (see Figure 6-66). An alternative method of cutting out the starting capacitor is by using a back-emf relay (see Figure 6-67). Back-emf relays are usually used on hermetic (sealed) compressors where it is impractical to install a centrifugal switch. The coil of the back-emf relay is connected in parallel with the auxiliary winding. When the

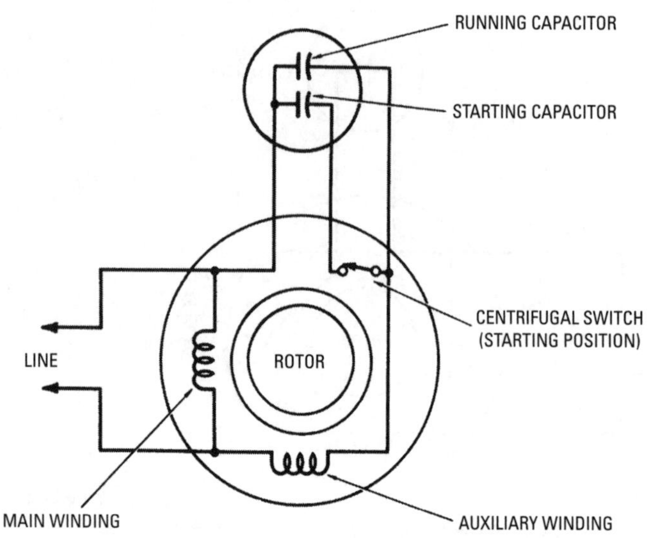

Figure 6-66 Capacitor-start, capacitor-run motor.
(Courtesy Honeywell Tradeline Controls)

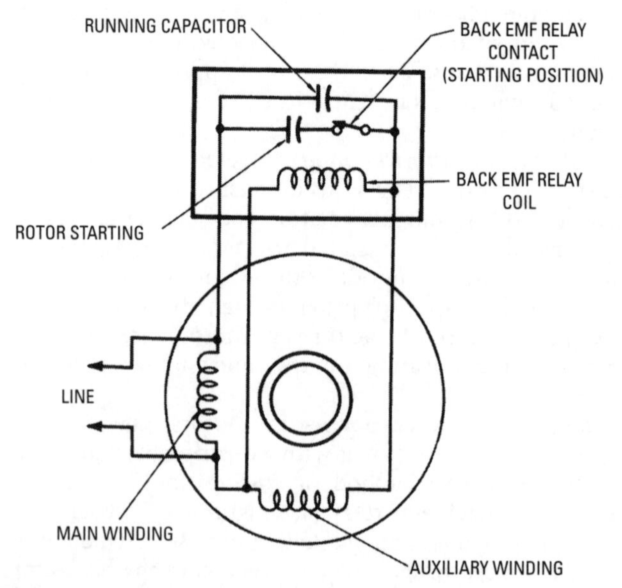

Figure 6-67 Back-emf relay. *(Courtesy Honeywell Tradeline Controls)*

motor approaches running speed, the relay pulls in, opening its contacts and taking the capacitor out of the circuit.

Troubleshooting Capacitors

A defective capacitor may be the cause of the following operating problems:

- Condenser fan will not run.
- Condenser fan will run, but compressor will not start.
- Compressor hums but will not start.

A defective running (or run) capacitor is often the direct cause of the compressor motor cycling on overload. A defective capacitor can be removed and replaced in the field, but first it should be tested to make certain that the capacitor is actually the source of the problem. Before testing the capacitor, inspect it for leaking or swelling. Corrosion around the terminals and bulging caps are caused by leaking electrolytics. These symptoms are an indication of a defective capacitor that can be removed and replaced. There is obviously no need to test it.

Warning

> Always completely discharge a capacitor before testing it. This must be done to avoid serious shock injury as well as to protect the testing equipment from potential damage.

Discharge the capacitor by connecting it to a 15,000-ohm and 2-watt resistor, and then disconnect the wires. Check for a ground in a motor-run capacitor by connecting an ohmmeter or test neon lamp in series with the capacitor and the metal part of the case for each capacitor. A ground is indicated if a continuity of circuit exists. Both terminals should be tested to ground, and this must be done with the capacitor disconnected from the compressor.

Replacing Capacitors

When putting a running (or run) capacitor back into a unit, check the capacitor terminals for a marked or identified terminal. The marked terminal may be indicated by a dab of solder, a paint mark, or a stamping on the case. This terminal must always be connected with the wire leading directly back to the contactor. If this capacitor should become defective by a ground, a fuse in the power circuit will blow. If the capacitor is not properly connected, a defect by ground will cause a flow of power through the start-and-run windings before reaching a fuse and cause compressor damage.

High-Pressure Cutout Switch

A *high-pressure cutout switch* is a pressure-actuated refrigerant controller connected to the pilot (low) voltage circuit of the main contactor in a refrigeration system (see Figure 6-68). These switches are installed in all refrigerating units over 1 hp to provide protection against dangerously high head pressures. These excessive and unsafe pressures develop as a result of a number of different abnormal operating conditions, including (1) air in the refrigerant lines, (2) excessive refrigerant charge, (3) dirty condenser, (4) faulty or inoperative condenser fan, and (5) insufficient water in a water-cooled condenser.

Figure 6-68 High-pressure cutout switch.
(Courtesy Honeywell Tradeline Controls)

As shown in Figure 6-69, the high-pressure cutout switch is connected to the so-called high side of the system. When the head pressure exceeds the control setting, the high-pressure cutout switch opens the compressor power circuit and prevents excessive head pressures from building in the condenser by shutting off the compressor motor.

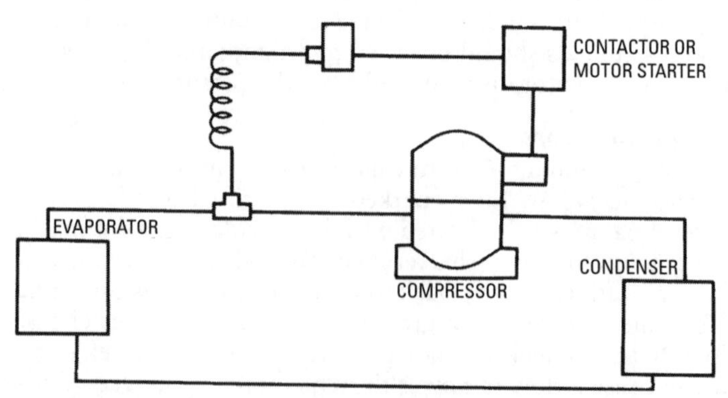

Figure 6-69 High-pressure cutout switch. *(Courtesy Honeywell Tradeline Controls)*

If automatic reset is used, the compressor motor will start again when the pressure returns to normal. More specifically, the contacts of the high-pressure cutout switch close automatically and reestablish the compressor power circuit when the head pressure drops the amount of control differential.

Automatic reset allows the equipment to cycle off the high-pressure control and continue to provide some degree of cooling while the high-pressure condition is being corrected. However, if the high-pressure condition is not corrected and the equipment continues to cycle over an extended period of time, damage may be caused to the motor or compressor. For obvious reasons, this cannot occur when a manual reset high-pressure cutout switch is used.

Low-Pressure Cutout Switch

A *low-pressure cutout switch* is similar in design to a high-pressure switch (see previous section) except that it provides low-pressure cutout protection.

As shown in Figure 6-70, this switch is connected to the so-called low side of the compressor and is designed to open the compressor power circuit if the low-side pressure drops below a desired level. These excessively low pressures are usually caused by dirty filters, evaporator fan failure, a stuck damper, damper motor failure, or some other interruption in the air supply.

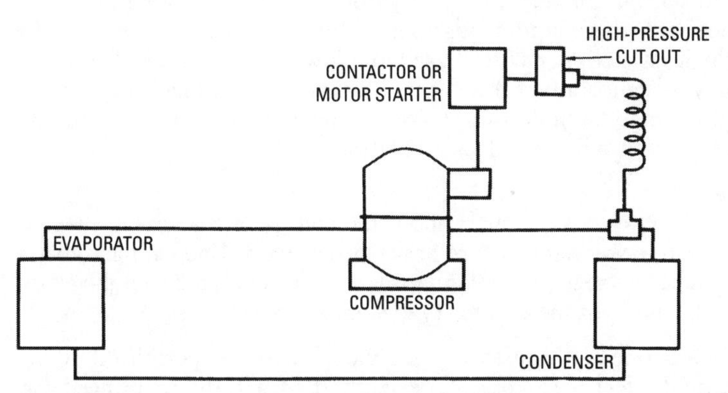

Figure 6-70 Low-pressure cutout switch. *(Courtesy Honeywell Tradeline Controls)*

The cutout action of the switch prevents the temperatures in the evaporator from falling below the temperature at which frost would form on the coils. It also prevents the feeding of an excessive amount of liquid to the compressor. Low-pressure cutout switches may be designed to provide either manual or automatic reset.

Transformers

A transformer is an inductive stationary device designed to transfer electrical energy from one circuit to another. Each transformer contains a primary and secondary winding. A changing voltage applied to one of the windings induces a current to flow in the other winding. In this manner, electrical energy is transferred from one circuit to another. Usually the changing voltage is applied to the primary winding and a current is induced in the secondary. The electrical energy may be transferred at the same voltage (a *coupling* transformer), at a higher voltage (a *step-up* transformer), or at a lower voltage (a *step-down* transformer).

The transformers used in heating and cooling systems are step-down transformers. They are designed to reduce (step down) the higher line voltage power to the 24 to 30 volts required by low-voltage control circuits. Some models are designed to power thermostats, gas valves, and relays in HVAC 24-volt systems. Others are designed primarily for powering air-conditioning systems, although they can be used in other applications *if they do not exceed the listed ratings*. In most installations, the room thermostat is operated by a low-voltage (24-volt AC) circuit. Some gas valves are also operated by a low-voltage circuit.

All wiring connections to transformers must be done in accordance with the recommendations of the *National Electrical Code*. A single thin copper wire is used in a low-voltage circuit. It is commonly (but not always) identified by its red and black insulation. High-voltage circuits use a larger-diameter wire that is commonly covered with white or black insulation.

Note

> If you have any doubts about the voltage of the circuit, check it with a voltmeter before beginning any work. Keep in mind that a small number of transformer models are connected to high-voltage circuits, not the customary low-voltage ones.

Interconnected transformer secondaries are not permitted by the *National Electrical Code*. One method of avoiding the need for interconnecting transformers is by using a single transformer rated to carry both the heating and cooling load. Using a thermostat and subbase combination with isolated heating and cooling circuits is also an acceptable method. Still another successful method utilizes an isolating relay to isolate the heating power supply from the cooling power supply.

Sizing Transformers

Transformers are not 100 percent efficient. There will generally be some loss of energy between the primary and secondary coils. In any event, the secondary coil must have enough remaining energy to drive the load connected to it.

When the equipment in a heating and/or cooling system is not adequately powered, check the transformer primary and secondary voltages. If these readings are within plus or minus 10 percent of the rated voltage and there is no problem with the wiring, the transformer may not be large enough for the system. A transformer too small for the system can be a very serious matter because it will supply abnormally low voltage to the control circuit. As a result, contactors or motor starters will not operate properly, and eventually the compressor may suffer damage.

When replacing a transformer, always select one that is the same size or larger than the one being replaced. For new installations, follow the equipment manufacturer's recommendations.

The capabilities of a transformer are described by its electrical rating. This information will include the primary voltage and frequency, the open-circuit secondary voltage, and the load rating in volt-amperes (VA).

The Class 2 transformers used in low-voltage control circuits have a maximum load rating of 100 VA and a maximum open-circuit secondary voltage of 30 volts. The secondary current must also be limited. This can be accomplished by using an energy-limiting transformer or by adding a 3.2-ampere (or less) fuse in the secondary. In the latter case, the maximum load rating of a typical 4-volt Class 2 transformer is 77 VA (24 volts × 3.2 amperes = 77 VA).

Installing Transformers

Always closely follow the transformer manufacturer's installation instructions, because they will vary depending on the model and the specific application. The following guidelines apply to most transformers:

Caution

> Never attempt to remove or install a transformer unless you are a qualified electrician or have the required training and experience. Improper removal or installation can result in damaged equipment and/or serious electric shock.

Note

> The following guidelines cannot replace the specific step-by-step instructions in the transformer manufacturer's installation instructions.

1. Disconnect the power supply to prevent equipment damage or electric shock before attempting to remove or install a transformer.
2. Separate and tape each exposed, unused lead wire. **Note:** *All wiring must comply with local electrical codes and ordinances.*
3. Do NOT short the transformer secondary terminals or you may burn out the overload protection.
4. Check the specification section of the transformer for lead wire color-coding (see Figure 6-71).
5. Connect the primary lead wires to the line voltage power supply.
6. Connect the transformer secondary to the 24-VAC control system. If the transformer model has a primary or secondary conduit spud, connect the wires first and then screw the conduit onto the spud.

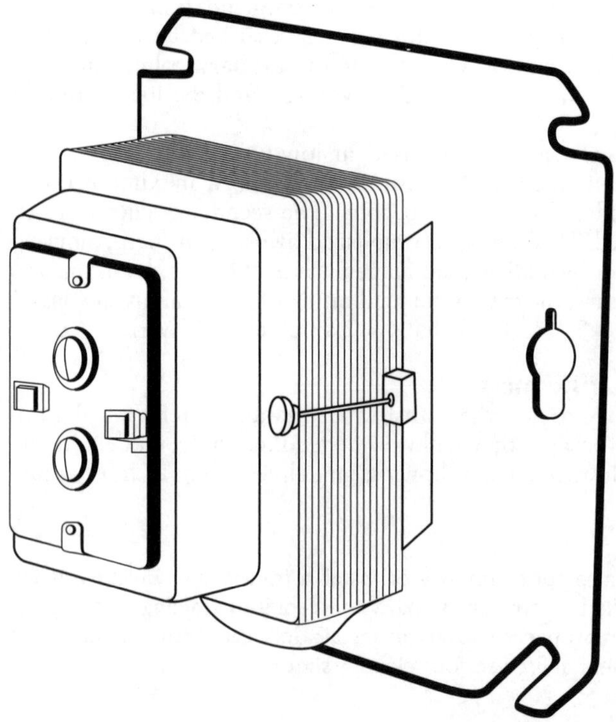

Figure 6-71 Honeywell transformer for a 24-volt system.
(Courtesy Honeywell Inc.)

7. Check the secondary voltage before connecting the transformer to the power supply.
8. Turn on the power supply and operate the system for one or two complete cycles.

Control Panels

A *control panel* combines many of the heating and cooling controls into a single, unified preassembled package. As a result, field wiring and control troubleshooting are greatly simplified.

A complete line of standard control panels is available for a variety of different functions. Depending on the requirements of the installation, it is possible to obtain a variety of different combinations of the following control components in a panel:

1. Transformer.
2. Line voltage and low-voltage wiring terminals.
3. High- or low-pressure cutout.
4. Two-stage cooling time-delay circuit.
5. Compressor contactor or motor starter.
6. Auxiliary relays.

The internal view of the Honeywell heating-cooling panel that is illustrated in Figure 6-72 shows the typical arrangement of these control components.

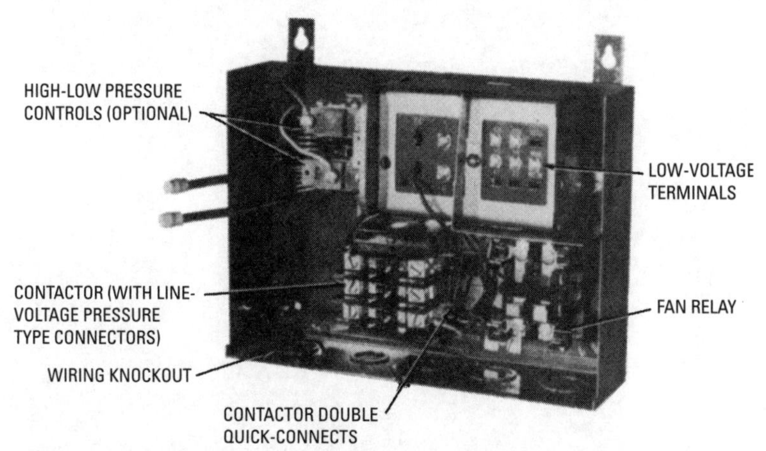

Figure 6-72 Heating-cooling control. *(Courtesy Honeywell Tradeline Controls)*

Chapter 7

Ducts and Duct Systems

Air distribution systems based on the forced-air principle of delivery utilize a system of ducts to deliver the heated or cooled air to the various rooms and spaces within the structure. These air ducts are generally rectangular or round pipes made from a variety of different materials. When these ducts are accurately sized and the duct system correctly designed, the air will be delivered to the rooms and spaces with a minimum of resistance, the result being a more efficient operation with reduced operating costs. The purpose of this chapter is to suggest methods for sizing ducts and designing an efficient duct system.

Methods for sizing fans are described in Chapter 7 of Volume 3 ("Ventilation and Exhaust Fans"). Chapter 4 of Volume 1 ("Sizing Residential Heating and Air-Conditioning Systems") and Chapter 8 of Volume 3 ("Air-Conditioning") provide information and methods for sizing the heating and cooling units.

Codes and Standards

Always consult local codes and standards first before designing and installing a duct system. Any aspect of a duct system that does not comply with these codes and standards will have to be changed, and these changes could be expensive.

Information about duct sizing, installation methods, air distribution, and air duct design methods is contained in the latest edition of the *ASHRAE Handbook of Fundamentals* and publications of the Air Conditioning Contractors of America (ACCA). Detailed information about ducts and duct fittings is available from the Commodity Standards Division of the U.S. Department of Commerce.

Types of Duct Systems

The two duct systems most commonly used in forced-warm-air heating are (1) the perimeter duct system and (2) the extended plenum duct system. Both are available in several design modifications and are described in the sections that follow.

Details about the piping arrangements used with gravity warm-air furnaces are included in the section describing these furnaces in Chapter 10 of Volume 1 ("Furnace Fundamentals").

An excellent source of information for designing a warm-air heating system are the publications and software from the ACCA. The recommended publications are *Manual D—Residential Duct Systems* and *Manual T—Air Distribution Basics for Residential and Small Commercial Buildings*. The latter manual provides step-by-step procedures for selecting, sizing, and locating the supply air diffusers, grilles, registers, and return grilles. The *Ductsize* software available from the ACCA describes how to calculate duct sizes for both supply and return duct systems using either the equal-friction or constant-velocity method.

Perimeter Duct Systems

A *perimeter duct system* is one in which the supply outlets are located around the perimeter (that is, outer edge) of the structure close to the floor of the outside wall, or on the floor itself. The return grilles are generally placed near the ceiling on the inside wall.

The two basic perimeter duct systems used in warm-air heating are (1) the perimeter-loop duct system, and (2) the radial perimeter duct system.

The *perimeter-loop duct system* (see Figure 7-1) is characterized by feeder supply ducts that extend outward from the furnace plenum to a loop duct running around the perimeter. Warm-air supply outlets are located in the loop duct.

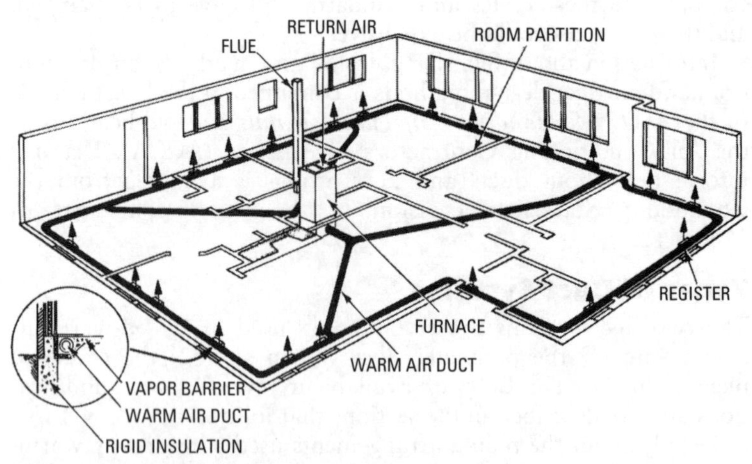

Figure 7-1 Typical perimeter-loop system.
(Courtesy U.S. Department of Agriculture)

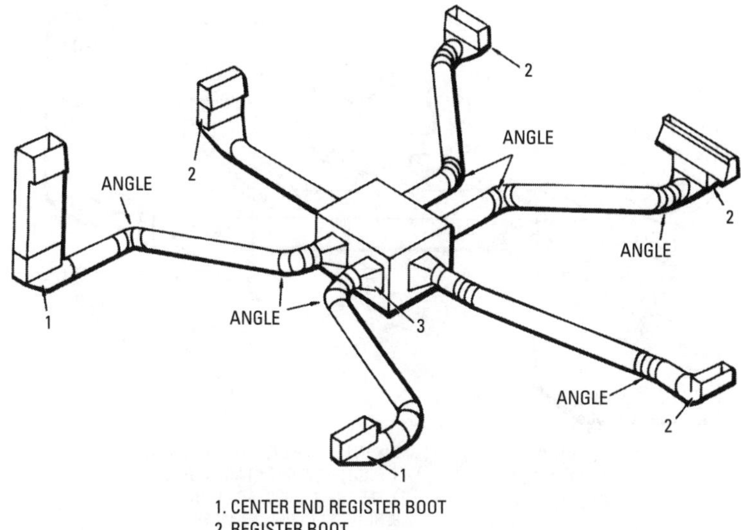

Figure 7-2 Radial perimeter duct system.

There is no loop duct in the *radial perimeter duct system* (see Figure 7-2). The feeder supply ducts extend from the furnace plenum to the warm-air supply outlets located on the outside walls or on the floor next to the outside walls.

Extended Plenum Systems

In the *extended plenum system* (see Figure 7-3), a large rectangular duct extends straight out from the furnace plenum (hence the term extended plenum) and generally in a straight line down the center of the basement, attic, or ceiling. Round or rectangular supply ducts extend as branches from the plenum extension to the warm-air supply outlets. The large extension to the plenum permits a better airflow rate with reduced resistance because of its large duct diameter. The branching ducts are usually located between joists and can be easily covered with a ceiling.

Crawl-Space Plenum Systems

It is possible to incorporate the entire crawl space into a heating system if the crawl-space walls are tight and well insulated. The heated air is forced down into the crawl space and enters the rooms through perimeter outlets, usually located beneath windows.

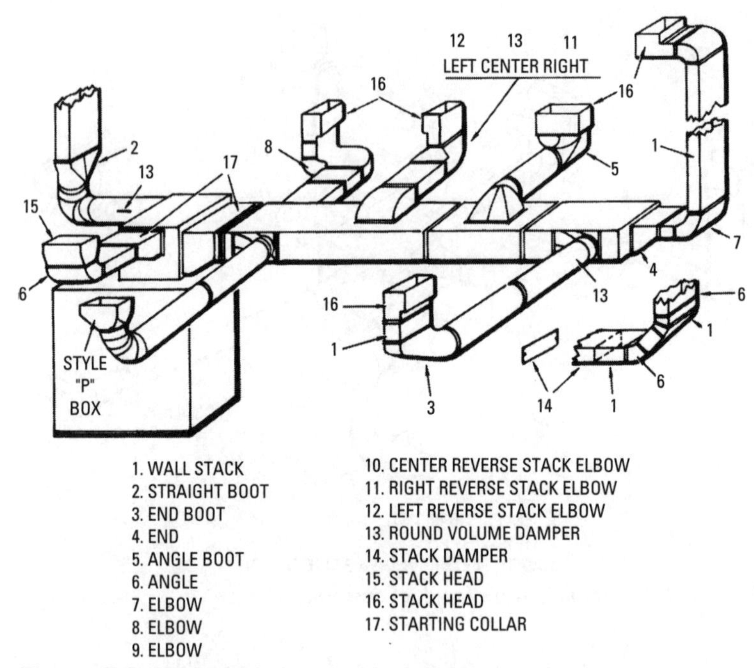

Figure 7-3 Extended plenum duct system.

This type of duct arrangement may be referred to as a *crawl-space plenum system* and represents a modification of the extended plenum system. Because the entire crawl space is filled with warm air, this system provides relatively uniform temperatures throughout the structure.

Duct Materials

Ducts are manufactured from a variety of different materials. The material selected will depend on the use for which it is intended. It is very important that the duct material be taken into consideration when designing a duct system because not every material is suitable for all conditions in which ducts are used.

It is possible to purchase ducts manufactured from the following materials:

- Steel
- Galvanized sheet steel
- Aluminum
- Copper

- Glass fiber
- Paper fiber
- Vitrified clay tile

Plain steel and *galvanized sheet steel* ducts are available in thicknesses ranging from 0.0163 to 0.1419 inch. Ducts manufactured from this material are preferred for use in warm-air gravity and forced-circulation warm-air heating systems. Table 7-1 indicates the thicknesses, gauges, and weights in which plain steel and galvanized sheet steel ducts are available.

Aluminum ducts are available in thicknesses ranging from 0.012 to 0.064 inch (see Table 7-2) and are used in the same types of heating systems as steel ducts. Although aluminum ducts are lighter than steel ones, they generally cost more. Aluminum ducts are frequently used in duct systems located on the outside of buildings.

Copper ducts are available in sizes and gauges matching the aluminum ones and are frequently used in outside ductwork.

Round *glass-fiber ducts* can be purchased in a number of different sizes ranging up to 14 inches in diameter with duct walls up to 1 inch in thickness. Square or rectangular glass-fiber ducts can also be made from flat glass-fiber board. Because of their composition, glass-fiber ducts dampen sound.

Paper-fiber ducts are laid in concrete and used in warm-air heating systems. They are accordingly not recommended for use in attics, basements, or other exposed areas.

Vitrified clay tile ducts represent another duct material suitable for installation under a concrete slab. These ducts range in outside diameter from 5⅛ to 42¼ inches.

Duct System Components

The components of a typical duct system are illustrated in Figure 7-4. In a forced warm-air heating system, the warm air collects in an area at the top of the furnace called the *furnace hood* or *plenum*. An extended plenum duct system will have a large rectangular duct connected to the plenum by a *starting collar* and extending out along the ceiling in a straight line. *Round* (see Figure 7-3) or *square supply ducts* are connected to the plenum (or plenum extension) usually by *adjustable side takeoffs* and extend to either a *register boot* or an *elbow*. Changes of direction in the round duct are accomplished with flexible *angle ducts*. A *nonflexible elbow* provides the same function in rectangular ducts. A vertical duct or warm-air riser is sometimes referred to as a *stack*. A warm-air duct that carries the warm air horizontally in a straight line from the furnace plenum to

Table 7-1 Thicknesses, Gauges, and Weights of Plain (Black) and Galvanized Sheet Metal

	U.S. Std. Gauge	Approximate Thickness (in)		Weight per Square Foot	
		Steel	Iron	Ounces	Pounds
Black sheets	30	0.0123	0.0125	8	0.500
	28	0.0153	0.0156	10	0.625
	26	0.0184	0.0188	12	0.750
	24	0.0245	0.0250	16	1.000
	22	0.0306	0.0313	20	1.250
	20	0.0368	0.0375	24	1.500
	18	0.0490	0.0500	32	2.000
	16	0.0613	0.0625	40	2.500
	14	0.0766	0.0781	50	3.125
	12	0.1072	0.1094	70	4.375
	11	0.1225	0.1250	80	5.000
	10	0.1379	0.1406	90	5.625
Galvanized sheets*	30	0.0163	0.0165	10.5	0.656
	28	0.0193	0.0196	12.5	0.781
	26	0.0224	0.0228	14.5	0.906
	24	0.0285	0.0290	18.5	1.156
	22	0.0346	0.0353	22.5	1.406
	20	0.0408	0.0415	26.5	1.656
	18	0.0530	0.0540	34.5	2.156
	16	0.0653	0.0665	42.5	2.656
	14	0.0806	0.0821	52.5	3.281
	12	0.1112	0.1134	72.5	4.531
	11	0.1265	0.1290	82.5	5.156
	10	0.1419	0.1446	92.5	5.781

** Galvanized sheets are gauged before galvanizing and are therefore approximately 0.004 inch thicker.*
(Courtesy ASHRAE 1960 Guide)

the stack is often referred to as a *leader. Dampers* are located in the duct so that the quantity of warm air can be regulated manually or automatically by thermostatic control. Ducts that carry the warm air to the rooms are called *supply ducts.* All ducts that carry the return air back to the furnace are referred to as *return ducts.*

Table 7-2 Thicknesses, Gauges, and Weights of 2S Aluminum (Density 0.098 lb/in³)

B. & S.	Thickness (in)		Weight per Square Foot	
Gauge	Decimal	Nearest Fraction	Pounds	Ounces
28	0.012	1/64	2.7	0.169
26	0.016	1/64	3.6	0.226
24	0.020	1/64	4.5	0.282
22	0.025	1/32	5.4	0.353
20	0.032	1/32	7.2	0.452
18	0.040	3/64	9.0	0.563
16	0.051	3/64	11.5	0.720
14	0.064	1/16	14.4	0.903

(Courtesy ASHRAE 1960 Guide)

Supply Air Registers, Grilles, and Diffusers

The three basic types of supply air outlets used in an air distribution system are (1) grilles, (2) registers, and (3) diffusers.

Grilles (see Figure 7-5) are used not only to admit the airflow but also to deflect it up or down, or to one side or the other, depending on the direction in which the hand-operated bar moves. They are used primarily on high or low wall locations. Floor grilles are used extensively in gravity warm-air heating systems.

A *register* (see Figure 7-6) is similar in design and function to the grille but with the added feature of being able to regulate the *volume* of the air with a damper. They may be located on walls (high or low) or floors. Floor registers are often used when installing a new heating system in an old house. The major objection to floor registers is that they tend to collect dust and trash.

A *diffuser* (see Figure 7-7) is also used to deflect the airflow, but it differs fundamentally in design from the grille. Diffusers manufactured in the form of concentric cones or pyramids are usually mounted on ceilings or walls. Baseboard diffusers are used in perimeter forced warm-air heating systems. The major objection to ceiling diffusers is that they cause drafts when the air is discharged downward and dirt smudges when the air is discharged horizontally across the ceiling.

If the duct system is designed primarily for cooling, outlets are sometimes located on the ceiling or high on the wall; however, satisfactory heating and cooling can be achieved with baseboard outlets placed low on walls by increasing the air volume and velocity and by properly directing the airflow.

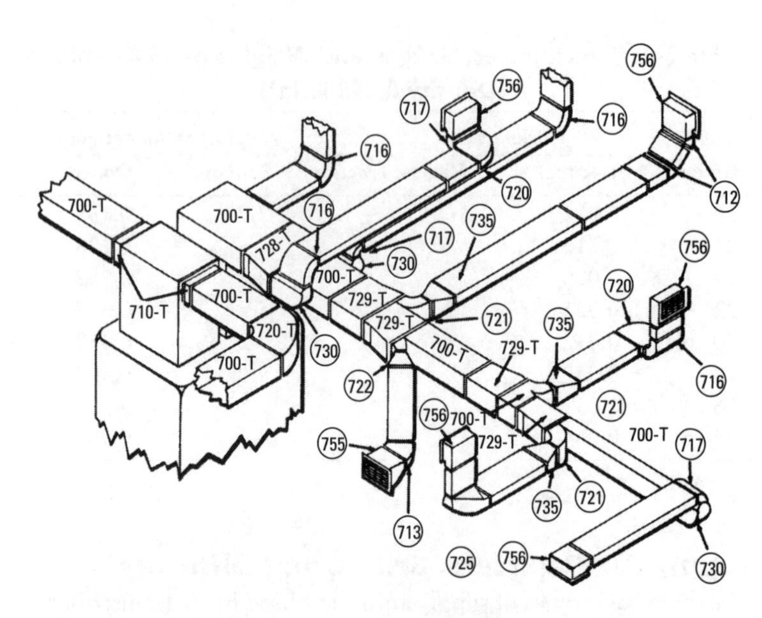

700-T DUCT SIZES 4×8 TO 36×8
710-T STARTING COLLAR, SIZED SAME/ AS/DUCT
728-T & 729-T INCREASER-REDUCER SECTIONS, SIZES SAME AS DUCT (MAX. ING. 10" FOR 728-T AND 5" FOR 729-T)
712-T MAIN TRUNK ANGLES, SIZES SAME AS DUCT
713-T MAIN TRUNK ANGLES, SIZES SAME AS DUCT
717-T MAIN TRUNK ELLS, SIZES SAME AS DUCT
720-T MAIN TRUNK ELLS, SIZES SAME AS DUCT
721 SIDE TAKE OFFS 4×8 , 5×8, 6×8.
722 SIDE TAKE OFFS 4×8, 5×8, 6×8.
224 - 725 - 726 REVERSE STACK ELBOWS $10 \times 3\frac{1}{4}$, $14 \times 3\frac{1}{4}$
712 ANGLE $10 \times 3\frac{1}{4}$, $12 \times 3\frac{1}{4}$, $14 \times 3\frac{1}{4}$
713 ANGLE $10 \times 3\frac{1}{4}$, $12 \times 3\frac{1}{4}$, $14 \times 3\frac{1}{4}$
716 ELBOW $10 \times 3\frac{1}{4}$, $12 \times 3\frac{1}{4}$, $14 \times 3\frac{1}{4}$
717 ELBOW $10 \times 3\frac{1}{4}$, $12 \times 3\frac{1}{4}$, $14 \times 3\frac{1}{4}$
720 ELBOW $10 \times 3\frac{1}{4}$, $12 \times 3\frac{1}{4}$, $12 \times 3\frac{1}{4}$, $14 \times 3\frac{1}{4}$
700 WALL STACK $10 \times 3\frac{1}{4}$, $12 \times 3\frac{1}{4}$ $14 \times 3\frac{1}{4}$
755 STACK HEAD 4, 5, 6, 8×10, 4, 5, 6, 8×12, 4, 5, 6, 8×14
756 STACK HEAD 4, 5, 61 8×10, 4, 5, 6,

Figure 7-4 Principal components of a warm-air duct system.

(Courtesy Clayton and Lambert Mfg. Co.)

Return Air and Exhaust Air Inlets

Grilles and registers are the two principal types of air inlets used to exhaust the air from a space or to return the air to the centrally located heating or cooling unit. The grilles are generally fixed-angle types because there is no need to direct air circulation when return air is involved.

Figure 7-5 Examples of various grilles. *(Courtesy A-J Mfg. Co.)*

Duct Run Fittings

Round and rectangular duct run fittings are available in a variety of different shapes and sizes, depending on the requirements of the air distribution system. Some duct run fittings are used only for cooling systems; others are designed for use in both heating and cooling systems. Duct run fittings can be purchased from manufacturers

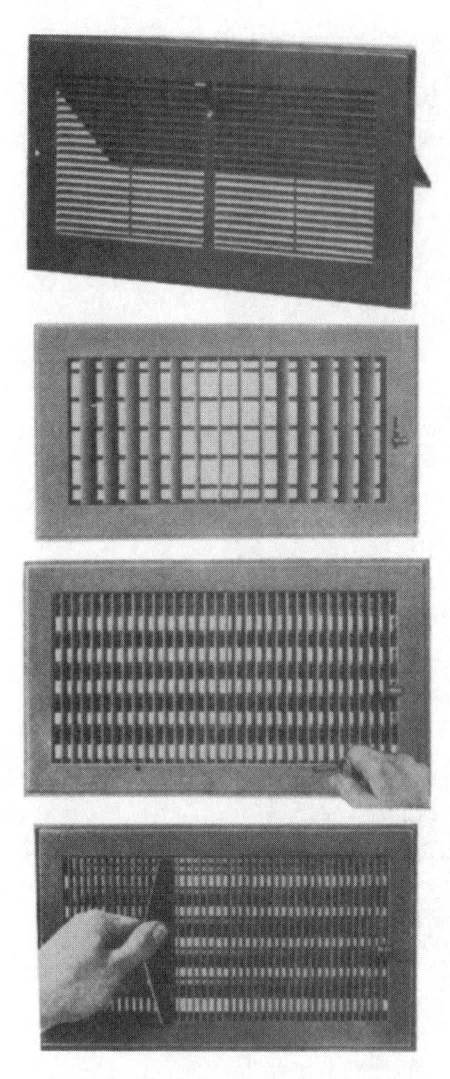

Figure 7-6 Examples of various registers.
(Courtesy United States Register Co.)

and local supply houses, or they can be made locally. Making your own fittings requires knowledge of sheet-metal work. Components of a typical duct system are shown in Figure 7-4. Based on their function, these duct run fittings can be divided into the following principal categories:

- Supply-air and return-air bonnet or plenum (see Figure 7-8)
- Plenum and extended plenum takeoffs (see Figure 7-9)

Figure 7-7 Examples of diffusers. *(Courtesy United States Register Co.)*

- Trunk duct angles and elbows (see Figure 7-10)
- Stack angles and elbows (see Figure 7-11)
- Boot fittings (see Figures 7-12 and 7-13)
- Wall sections (see Figure 7-14)

Air Supply and Venting

Any boiler or furnace fired with a combustible fuel (for example, coal, oil, gas) must be equipped with a piping system to remove smoke and other low-temperature flue gases and to provide sufficient air for combustion. These air supply and venting systems are composed of round pipes and fittings made from sheet metal and resemble ducts in design and construction.

The design and installation of air supply and venting systems are described in the several chapters dealing with furnaces and boilers. See, for example, the appropriate sections of Chapter 11 of Volume 1 ("Gas-Fired Furnaces").

Duct Dampers

A *duct damper* is a device used for controlling the direction or volume of air flowing through a duct.

The ASHRAE defines a damper as being "a device used to vary the volume of air passing through a confined cross-section by varying

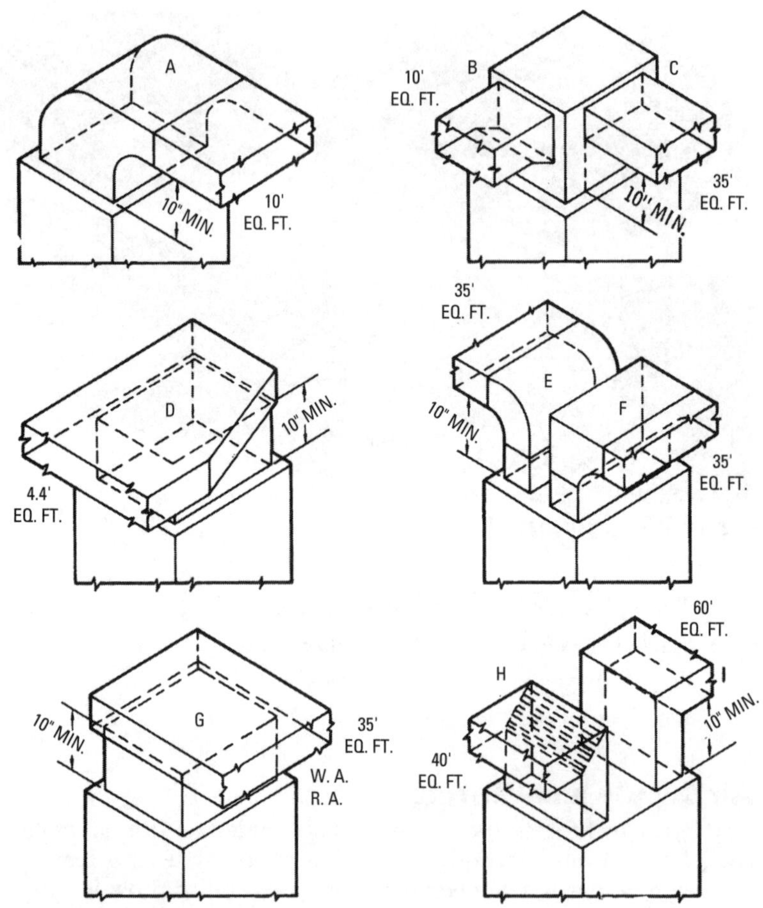

Figure 7-8 Warm-air and return-air bonnet or plenum.
(Courtesy ASHRAE 1952 Guide)

the cross-sectional area." In other words, a damper functions as an obstruction to airflow through the duct; however, it is a *movable* obstruction that can be adjusted to give various-size openings for the passage of air.

The air volume in an air distribution system may have to be changed between the heating operation and the cooling operation because it generally requires *more* air at 50°F to cool a house to 75°F than warm air at 160°–170°F to heat a house to 75°F. Bearing this in mind, the ductwork and fan should be sized for whichever

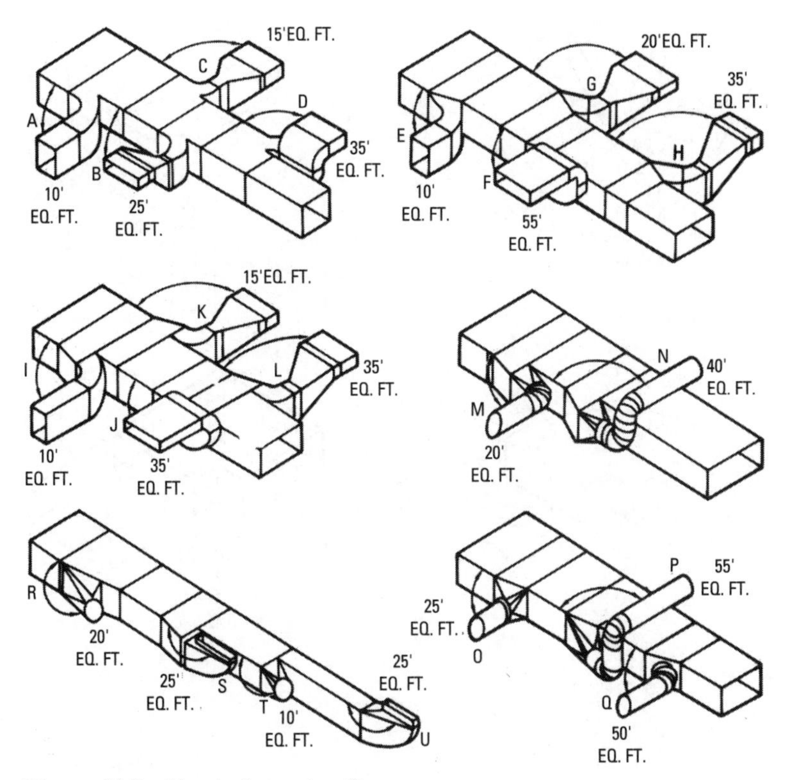

Figure 7-9 Trunk duct takeoff. *(Courtesy ASHRAE 1952 Guide)*

operation (that is, heating or cooling) requires the greater air volume. *Volume control dampers* may then be installed in the duct system to reduce the air volume during the season that requires the smaller volume.

A damper is usually made in the form of a round or rectangular blade and can be either manually operated or motorized. The ASHRAE lists the following three basic types of dampers:

- Volume dampers
- Splitter dampers
- Squeeze dampers

A *volume damper* or *volume-control damper* (see Figure 7-15) is installed in a duct to either completely cut off or regulate the flow

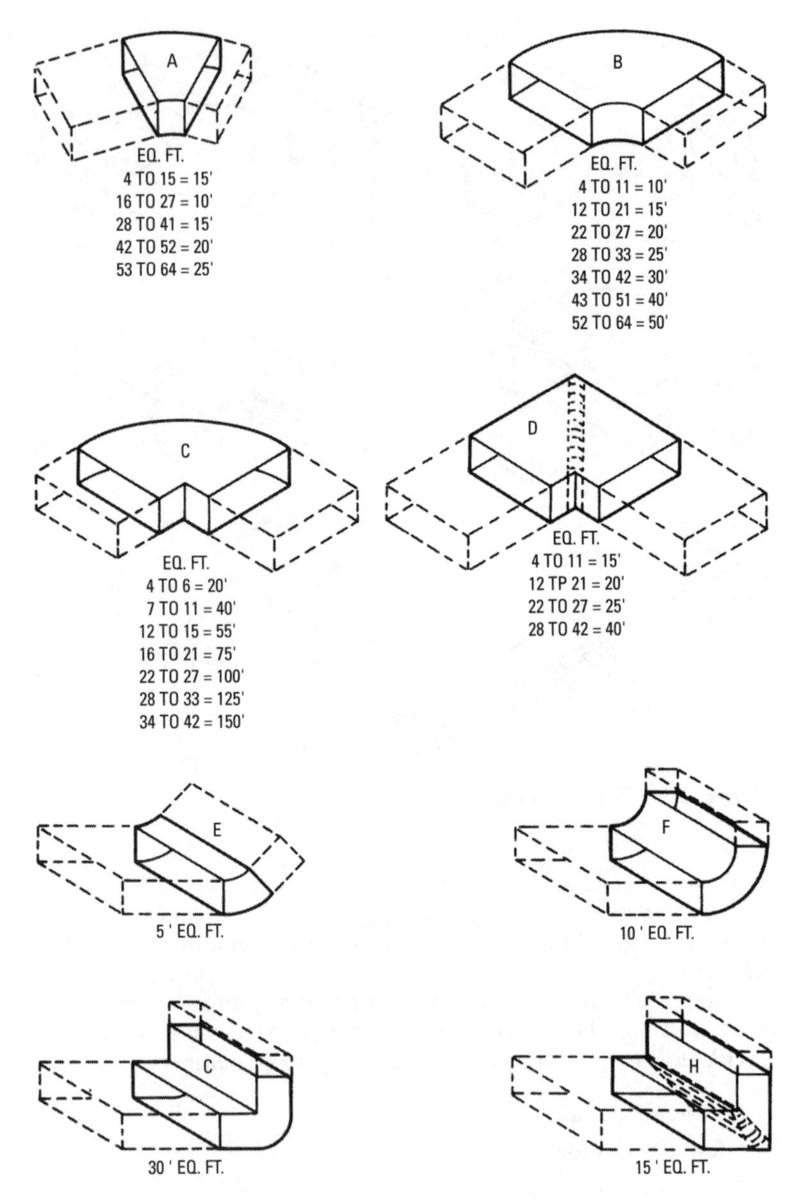

Figure 7-10 Angles and elbows for trunk ducts.

(Courtesy ASHRAE 1952 Guide)

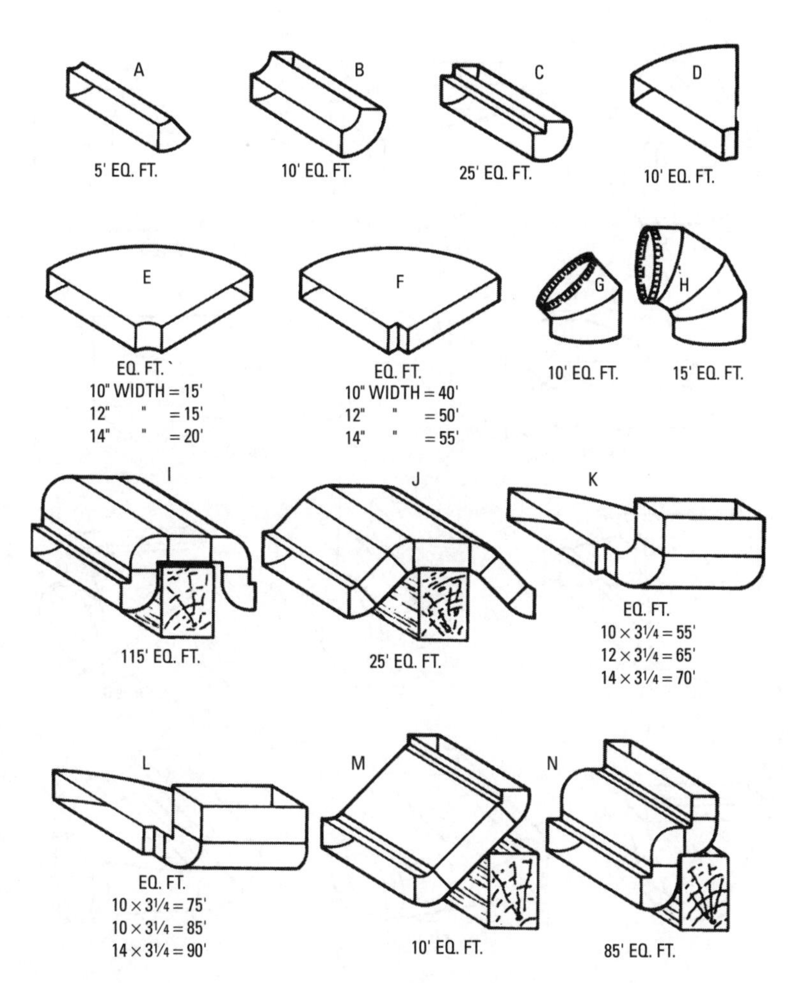

Figure 7-11 Stock angles and elbows. *(Courtesy ASHRAE 1952 Guide)*

of air in the duct. Adjustments of the volume of airflow and resistance can be made with a *squeeze damper.* A *splitter damper* is a directional device consisting of a blade hinged at one end and used at locations where a branch run leaves the main duct.

Three examples of volume dampers are (1) slide dampers, (2) hit-and-miss dampers, and (3) butterfly dampers. The *slide damper* is operated by sliding or pushing a metal plate across the duct. It operates at either full open or full closed position, there being no

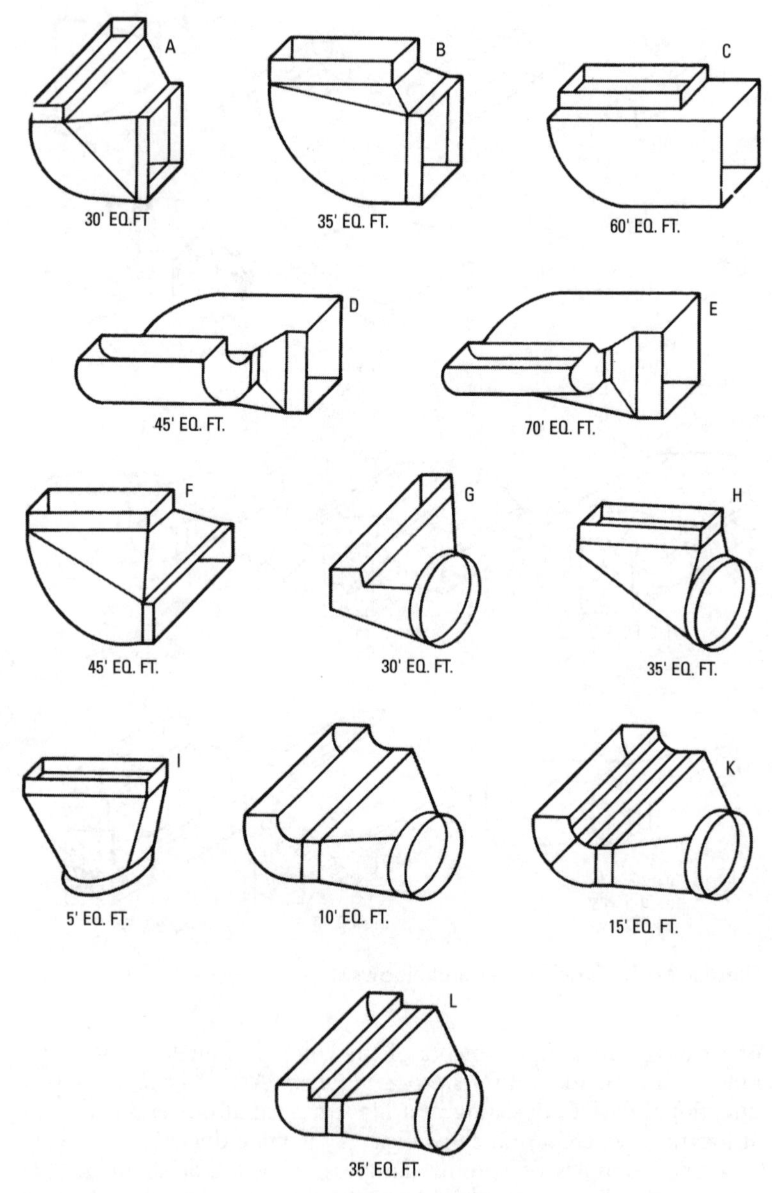

Figure 7-12 Boot fittings from branch to stack. *(Courtesy ASHRAE 1952 Guide)*

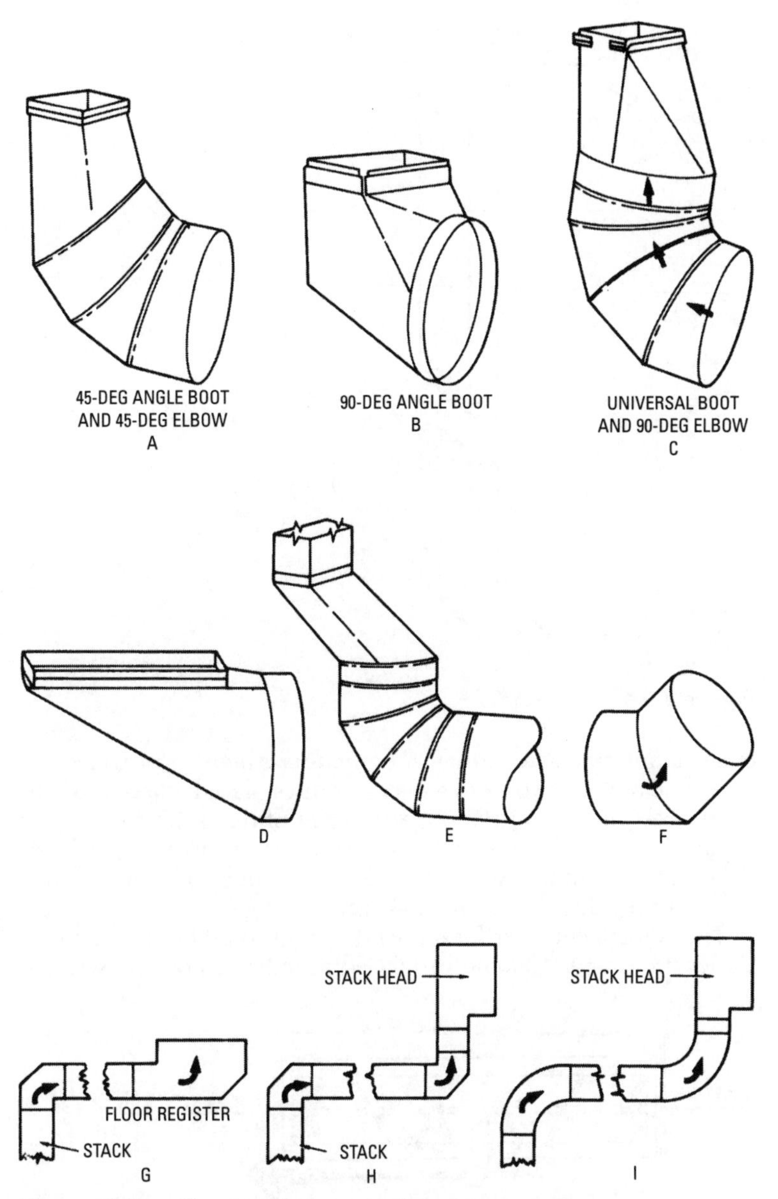

Figure 7-13 Combination warm-air boots. *(Courtesy ASHRAE 1952 Guide)*

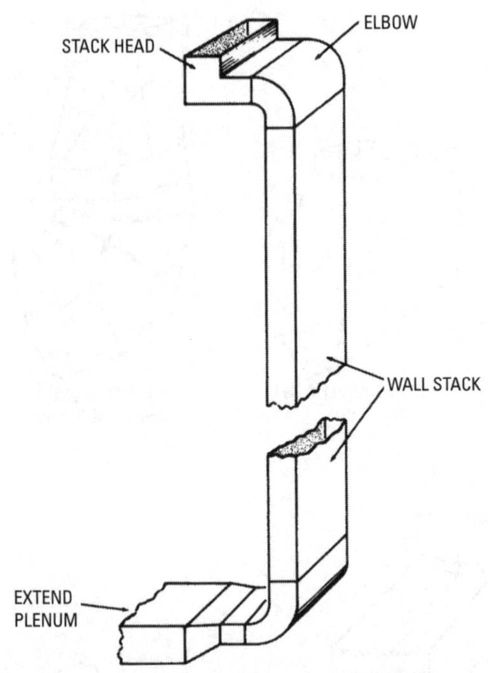

Figure 7-14 Wall stack.

practical intermediate setting. A *hit-and-miss damper* provides volume control with two slotted plates or discs placed adjacent to one another. At full open position, 50 percent of the duct cross-section remains blocked. A *butterfly damper* consists of a single blade hinged in the middle. At full open position, almost the entire cross-section of the duct is free of blockage.

The adjustment lever for a duct damper should be in an accessible location, and it should be labeled to indicate position settings,

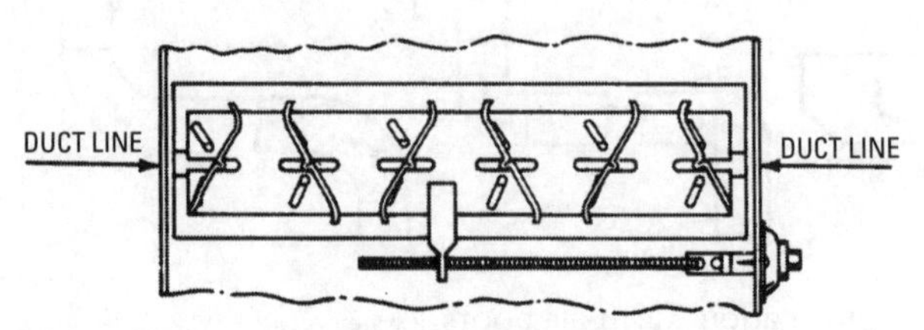

Figure 7-15 Volume-control damper used in branch ducts to control the volume of air.

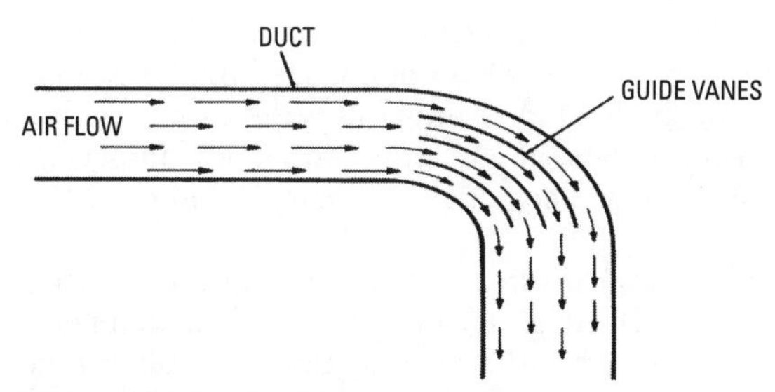

Figure 7-16 Guide vanes in an air duct.

function, and area served. The damper should also be equipped with a positive locking device.

Dampers properly placed in the supply and return ducts of an air distribution system will control the supply of air to a room; however, care should be taken in the placement of these dampers. A damper in a supply duct placed too close to the supply air outlet may disturb the airflow. Excess noise is often the result of improperly placed dampers.

Splitters and *guide vanes* (see Figure 7-16) are nonmovable sheet-metal partitions placed in stack heads or elbows to reduce air turbulence and to guide the airflow. *Air directors* are similar devices but are used to direct a portion of the airflow to a branch duct.

A *fire damper* is a damper designed to close off a duct in order to prevent the spread of flames and smoke. Fire dampers should be placed in branch duct connections that pass through fire walls, fire-rated partitions, or floors. *Never* install a fire damper in main exhaust ducts and risers. Installation of fire dampers should comply with local ordinances and National Fire Protection Association Standards.

Damper Motors and Actuators

Damper motors or actuators are devices used in heating, ventilating, and air-conditioning systems for the following applications:

- Diverting airflow
- Decreasing airflow
- Controlling ventilation
- Zoning

Motorized dampers are available in a number of different designs, but they can be grouped into two general classes: (1) single-blade dampers and (2) opposed-blade dampers. There are two types of single-blade dampers. One type is opened and closed by the damper motor. It remains closed when not engaged by the damper motor. Another type of single-blade damper uses the damper motor to open it and a spring to close it. An opposed-blade damper has narrow panels covering slots. The panels are opened and closed by the damper motor.

A damper motor or actuator can be used to position a diverting damper when a parallel airflow pattern is required. The damper is used to direct airflow through either the heating or cooling unit. The damper motor is connected to the heat-cool switch on a semi-automatic changeover thermostat or in parallel with the cooling contact of an automatic changeover thermostat.

In modern forced-air heating systems, zoning is commonly accomplished by a thermostat for each zone. When there is a call for heat, the thermostat sends a signal to a control panel. The control panel then opens the appropriate zone damper and turns on the furnace or air handler. Some of the older systems do not have a control panel. Instead, the furnace or air handler is turned on by damper drip switches. A control panel is used in modern heating and air-conditioning systems to control the changeover from heating to cooling or from cooling to heating.

A damper motor or actuator can also be used to operate a bypass damper for the following purposes:

- To bypass a cooling coil for dehumidification
- To bypass the heat exchanger on cooling
- To bypass the heat exchanger on heating

It is sometimes necessary to decrease the airflow during the heating cycle in some systems. This can be accomplished by using a damper motor or actuator to position a resistance damper.

Systems that require the introduction of outdoor air during the cooling season, but not during the heating season, can use damper motors or actuators to control the ventilation.

Damper motors or actuators are also frequently used in zoning a duct system by opening or closing dampers to the various zones.

The small, compact electric motors used to provide proportional control of dampers can also be used to drive valves or step controllers in heating, cooling, and ventilating systems. For this reason, additional information about these motors is included in Chapter 9 of Volume 2 ("Valves and Valve Installation").

An example of a motor used to drive a damper is found in Figure 7-17. This is a cutaway of a Honeywell modutrol motor,

Figure 7-17 Cutaway view of a modutrol motor.
(Courtesy Honeywell Tradeline Controls)

which functions as the drive unit in a *modulating control circuit.* The following are the basic components of this motor:

- Reversible motor
- Balancing relay
- Feedback potentiometer
- Gear train

The balancing relay controls the motor, which turns the motor drive shaft through the gear train. The motor is equipped with switches that limit its rotation to 90° or 160°. The gear train and other moving parts are immersed in oil to eliminate the need for periodic lubrication.

The motor is started, stopped, and reversed by the single-pole, double-throw contacts of the balancing relay. The balancing relay consists of two solenoid coils with parallel axes, into which are inserted the legs on the U-shaped armature. The armature is pivoted at the center so that it can be tilted by the changing magnetic flux of the two coils to energize the relay. A contact arm is fastened to the armature so that it may touch either of the two stationary contacts as the armature moves back and forth on its pivot. When the relay is balanced, the contact arm floats between the two contacts, touching neither of them.

A feedback potentiometer consisting of a coil of wire and a sliding contact is included in the Honeywell modulating motor. The sliding contact is moved by the motor shaft so that it travels along the coil and establishes contact wherever it touches, according to the position of the motor.

Figure 7-18 shows a typical wiring diagram for a Honeywell modutrol motor. Note that there are two separate circuits in the modulating motor powered from T_1 and T_2. The *motor circuit* consists of the reversible motor, the rotational limit switches, and the contacts of the balancing relay. The *control circuit* includes the feedback potentiometer, the coils of the balancing relay, and the controller potentiometer.

The control circuit offers two paths for current flow—done through each side of the balancing relay. Increasing resistance in the *B* leg of the motor control circuit by changing the setting of the controller potentiometer will run the motor toward the closed position. Adding resistance to the *W* leg runs the motor open.

As the motor shaft turns, it moves a wiper over the feedback potentiometer. This makes the resistance in each side of the circuit the same. When the resistances are equal, the current flow through both sides of the balancing relay is equal. The balancing relay contacts open, stopping the motor. The circuit is said to be balanced.

The motor of a damper actuator should be completely sealed and immersed in oil. Such a motor can operate without maintenance or service for the life of the unit. An example of this type of motor is found on the ITT General Controls DHO series damper actuator shown in Figure 7-19. These DHO series motors are available in two-position type (energized and deenergized) or three-position type (deenergized, first stage open, and second stage open). The first-stage intermediate position can be adjusted.

It is important that the DHO series damper motors work within the manufacturer's specified load limit during all phases of operation in order to ensure delivery of rated forces under usual voltage variation. Maximum workload can be determined from the data given in Table 7-3. Load is the deadweight pull at the particular stroke position. The imposed load can be measured with a small spring scale.

The Honeywell M833A damper actuator shown in Figure 7-20 is used to regulate duct damper condition according to zone thermostat requirements. It attaches directly to a damper shaft ½ inch in diameter or a ⅜-inch shaft with adapter provided. It mounts in any position directly on a duct, or inside a standard wiring junction box where Class 1 wiring is required.

Installing Damper Motors

The location of a damper motor will be determined by the design of the ductwork. If possible, the motor should be easily accessible for

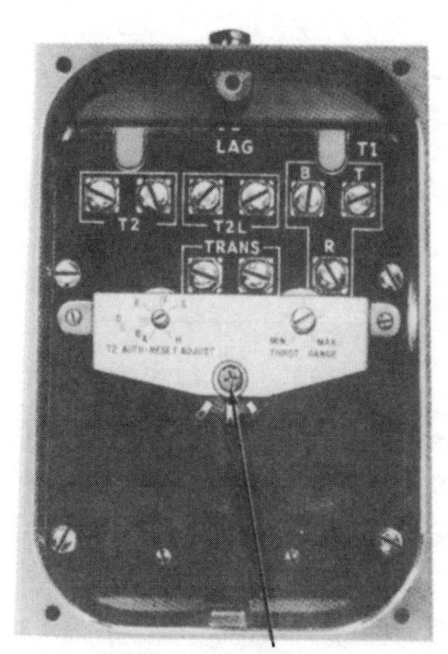

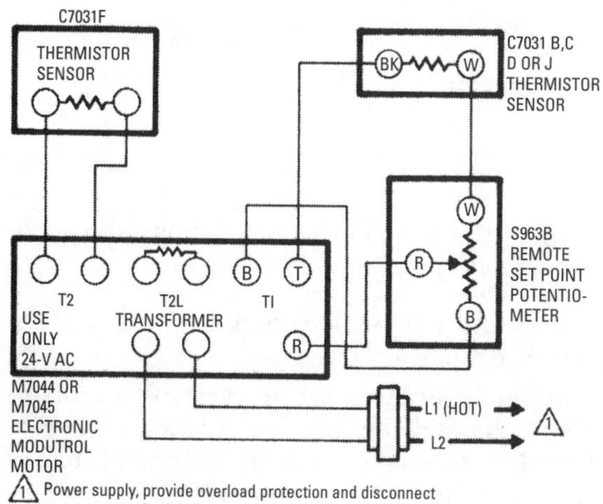

Figure 7-18 Wiring diagram for an electronic modutrol motor.

(Courtesy Honeywell Tradeline Controls)

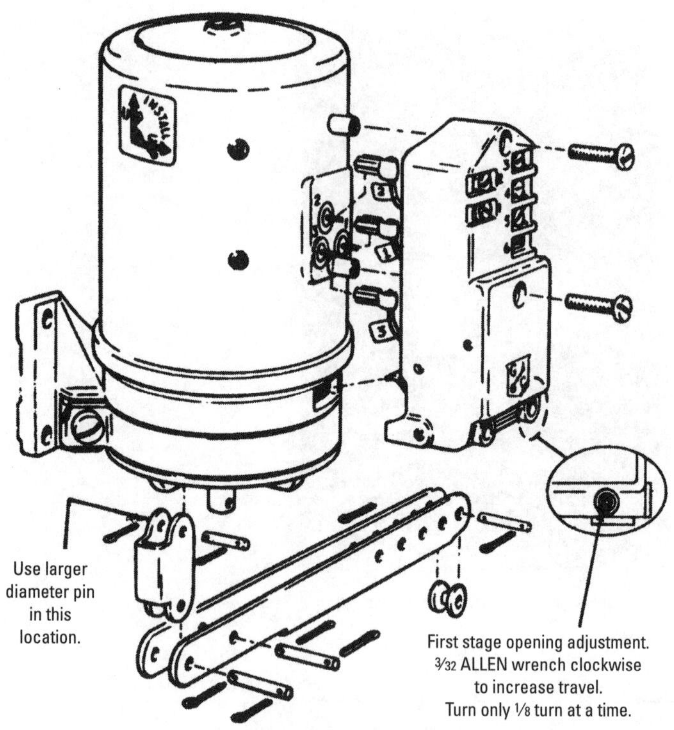

Figure 7-19 Damper actuator. *(Courtesy ITT General Controls)*

servicing and maintenance. The following suggestions should be considered when seeking a location for a duct motor:

- The ambient temperature must be within the ratings of the motor being installed.
- A motor should not be installed where corrosive vapors or sprays, or other contaminants, may present a problem to motor operation. It is especially necessary to prevent dust or dirt from getting on the feedback potentiometer.
- The linkage between the motor and the damper must be free to operate without binding on pipes or other obstructions.
- The motor must be installed so that it is free to run from one end of its stroke to the other.

Table 7-3 Maximum Workload Data

Connection Position	1		2		3		4		5	
Length of Travel (in)	3.25*		3.00*		2.75*		2.50*		2.25*	
Specifications	Max Load (lbs)	Return Force (lbs)	Max Load (lbs)	Return Force (lbs)	Max Load (lbs)	Return Force (lbs)	Max Load (lbs)	Return Force (lbs)	Max Load (lbs)	Return Force (lbs)
25 VA	15	8½	17	9½	19	10	20	11	22	13
40 VA	30	8½	34	9½	37	10	42	11	45	13
120 VA	30	8½	34	9½	37	10	42	11	45	13
Length of Travel (in)	2.625†		2.375†		2.125†		1.875†		1.625†	
25 VA	18	10	21	11	24	13	27	15	30	17
40 VA	36	10	42	11	48	13	54	15	60	17
120 VA	36	10	42	11	48	13	54	15	60	17

** When using dual damper arm, combined load must not exceed load maximums shown.*
† Table for reversed damper arm position.
(Courtesy ITT General Controls)

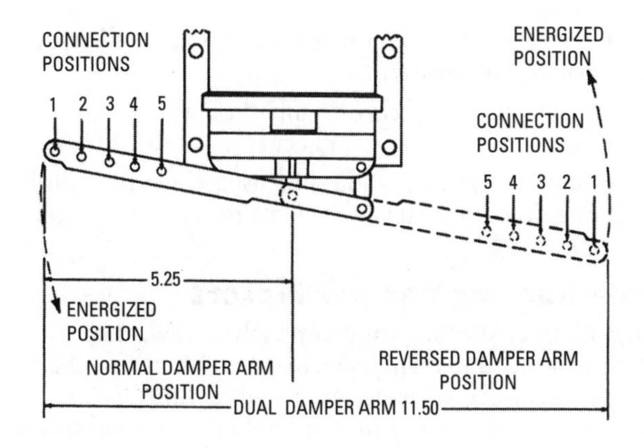

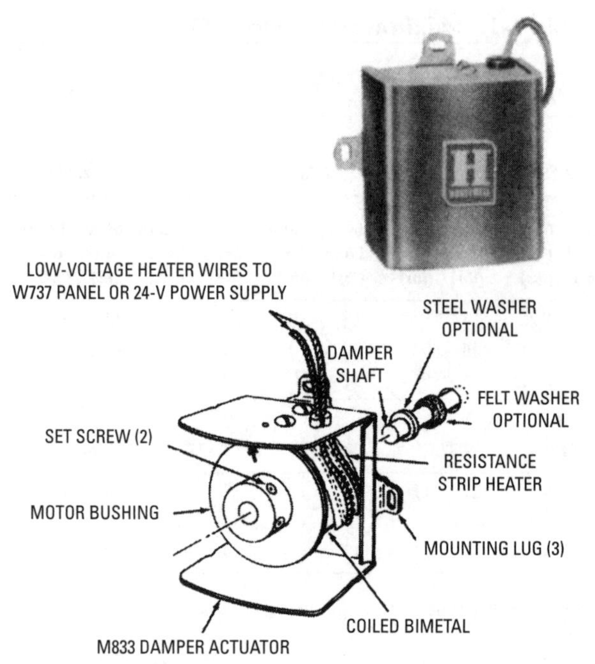

Figure 7-20 Damper actuator. *(Courtesy Honeywell Tradeline Controls)*

- Honeywell modutrol motors must be mounted with the shaft in a horizontal position.
- A *remote* balancing relay mounted on a firm support must be used if motor location is subject to severe vibration.
- Weatherproof motors should be used when the unit is located in a position exposed to the elements.

Troubleshooting Damper Motors

Although damper motors are completely sealed, they do malfunction from time to time. The principal malfunctions, their possible causes, and suggested remedies are listed in Table 7-4.

Some manufacturers produce special compact control units for mounting on damper motors. These control units (for example, the Honeywell W859 Economizer shown in Figure 7-21) give damper motors the capacity to regulate the outside and recirculated air dampers for the following applications:

Table 7-4 Troubleshooting Damper Motor

	Possible Cause	*Remedy*
Motor not operating.	Power off	Check switches and fuse.
	Loose wiring	Check connections.
	Motor damaged	Replace actuator.
Motor operates but output shaft does not move.	Linkage binding	Check for free movement.
	Actuator damaged	Replace actuator.
Motor stalls at maximum travel position.	Limit switch damage	Inspect and replace if necessary.

(Courtesy ITT General Controls)

- To bring in outside air (when it is colder than the inside air) for cooling or ventilating
- To close outdoor air dampers to minimum position during the heating cycle
- To prevent short cycling of the cooling compressor to avoid coil icing

Blowers (or Fans) for Duct Systems

The ductwork and blower (or fan) are sized for either heating or cooling (whichever requires the greater air volume); however, their correct sizing depends on an accurate determination of the following facts about a structure or space:

- The total heat loss
- The total cooling load
- Air delivery required (CFM)
- External static pressure for ductwork

Once these facts are known, it is simply a matter of referring to the performance data provided by the blower or fan manufacturer and selecting the most suitable equipment for the system.

Figure 7-21 Economizer control. *(Courtesy Honeywell Tradeline Controls)*

Designing a Duct System

The purpose of a duct system is to convey air from the blower or fan to the air supply outlets located in the various rooms and spaces of the structure, and then to return it to its point of origin.

The design of a duct system is determined by the cfm output of the furnace blower. The size (Btu/h output) of the furnace and its blower is determined by calculating the heat loss for the house or structure. The furnace must have an output capacity capable of replacing the heat loss. The specification sheet for the furnace will list the blower speeds (cfm) and outputs (Btu/h).

Accuracy in estimating the resistance to the flow of air through the duct system is important in the selection of a suitable blower or fan. Resistance should be kept as low as possible in the interest of economy; however, underestimating the resistance will result in the failure of the blower or fan to deliver the required volume of air.

Every precaution should be taken in the design of a duct system to ensure a smooth and efficient flow of air. Careful study should be made of the building drawings with consideration being given to the construction of duct locations and clearances.

The following recommendations may be of use to you in designing a duct system:

- Keep all duct runs as short as possible, bearing in mind that the airflow should be conducted as directly as possible between its source and delivery points, with the fewest possible changes in direction.
- Select location for duct outlets that will ensure proper air distribution. For example, locate the supply outlets in the floors along or near the exterior walls and the returns along or in the interior walls.
- Provide ducts with cross-sectional areas that will permit air to flow at suitable velocities. The furnace blower must have the capacity to overcome the friction between the moving air and the duct surface.
- Use moderate velocities in all ventilating work to avoid waste of power and to reduce noise.
- Use lower velocities in schools, churches, theaters, and so on, than in factories and other places where noise due to airflow is not objectionable.

Duct System Calculations

An *ideal* duct system would be one in which total pressure remained constant. In other words, there would be no pressure losses anywhere in the system. Under ideal operating conditions, any change in velocity pressure would be compensated for by an equal change in static pressure, thereby maintaining a constant total pressure (that is, the sum of the velocity and static pressures). A drop in velocity pressure would trigger a corresponding rise in static pressure, and vice versa. Unfortunately, this is not the way things work in actual practice because pressure losses very definitely do occur.

In the ductwork, pressure losses result from the resistance of the ducts to the passage of the air. This resistance occurs as a result of two effects: (1) friction loss and (2) dynamic loss. The former is caused primarily by the friction of the moving air against the surface of the duct. Dynamic loss results from sudden changes of direction (for example, in sharp elbows) in the air stream.

An important aspect of duct system calculations is determining the *total* external static pressure drop (that is, total resistance) of the duct system. In large part, a blower or fan is selected on the basis of its capacity to operate against the total resistance of the ductwork. This resistance of the ductwork to the flow of air is referred to as the *external* static pressure (or external static pressure *drop*), because it represents the pressure drop occurring *outside* the heating or cooling unit. In addition to external static pressure, a blower or fan must also overcome resistance due to *internal* static pressure drop caused by the passage of air through heaters, coils, filters, and washers. This data can be obtained from rating tables provided by manufacturers. Duct resistance (that is, external static pressure drop) must be calculated for each duct system.

The equal friction method is frequently used to calculate the external static pressure of a duct system (see *Equal Friction Method* in this chapter). Two other methods used for making these calculations are (1) the velocity-reduction method, and (2) the static-regain method. Detailed descriptions of both of these methods can be found in the *ASHRAE Guide.*

Duct Heat Loss and Gain

Duct heat loss or gain is another important factor to consider when designing a duct system. This aspect of heat transmission will depend on some or all of the following factors:

- Temperature of the air in the duct
- Ambient temperature
- Air velocity in the duct
- Duct insulation

If the temperature of the air inside the duct is different from the temperature surrounding it (that is, the ambient temperature), then either a loss or gain of heat will occur.

If the ducts are used to convey heat, excessive heat loss from the ducts will reduce the efficiency of the heating system. This will result in a total loss of heating effect if the heat loss occurs in ductwork passing through an unheated area, but it can be considerably reduced with proper duct insulation (see *Duct Insulation*). Such heat loss can also occur in heated spaces, but here it is a problem of poor air distribution.

The same conditions exist when the ducts carry cool air. When air passes through spaces subject to the cooling effect of air-conditioning, insulating the ducts will effectively reduce the amount of heat gain.

Proper air distribution will also minimize heat gain in spaces that are partially cooled.

Air velocity will also influence the amount of heat gain in the ducts. High air velocities are recommended when the ducts are carrying cool air because their effect is to reduce the amount of heat gain pickup in the ducts; however, they must be maintained consistent with the acoustic requirements of the installation.

Air Leakage

Air leakage through duct seams and holes will result in the loss of a portion of the air flowing through the duct and a proportionate reduction in the heating or cooling effectiveness of the system. Depending on the seriousness of the problem, the ductwork can lose up to ⅓ of the air supply in this manner. It is usually a matter of poor workmanship and can be corrected by sealing the seam cracks and holes by caulking or soldering.

Duct Insulation

Ducts are insulated to prevent excessive heat loss or gain. If the ducts are used to convey heat, excessive heat loss from the ducts will reduce the efficiency of the heating system. The reverse is true of ducts used in cooling systems. If the ducts are not insulated, they will absorb heat from the air around them, and system performance will be impaired.

To maintain the proper level of performance in a heating or cooling system, the following ducts should be insulated:

- Supply ducts running through spaces that are neither heated nor cooled (for example, attics, basements, garages, crawl spaces).
- Long supply ducts (particularly those over 45 feet in length).
- All ducts located on the outside of buildings.
- Cool-air return ducts passing through hot areas (for example, furnace and boiler rooms, kitchens).

Round ducts are insulated with flexible fiberglass insulation. Both flexible and slab (board) insulation are used for rectangular ducts. The latter is made of spun fibrous glass wool. These lightweight, semirigid panels are available with a variety of facings (for example, 0.0025 embossed aluminum foil) for appearance and functional use.

Flexible and slab insulation are also produced as duct liners for absorbing duct system noise. One or both sides are coated to reduce air friction loss and bind surface fibers. They also insulate thermally.

Flexible insulation is secured in place with light-gauge wire ties. Slab insulation is secured to the duct surface with adhesive or mechanical clips.

It is recommended that a vapor barrier be placed between the insulating material and the duct surface to prevent the formation of condensation. The vapor barrier should be used when the temperature of the air inside the duct is lower than the dew-point temperature of the air surrounding the duct.

Equal Friction Method

The *equal friction* method of sizing ducts is recommended because it does not require a great deal of experience in the selection of proper velocities in the various sections of the duct system. It is necessary to select the main duct velocity consistent with good practice from a standpoint of noise for a particular building or application. In this duct-sizing method, the duct design is based primarily on a consistent pressure loss for each foot of duct.

Proportioning for equal friction is more advantageous than reducing the velocity in a haphazard manner because the friction calculation is greatly simplified. In calculating the friction, it is necessary to know only the length of the longest run, the number and size of elbows, and the diameter and velocity of the largest duct. The friction loss is exactly the same as though the entire amount of air were carried the whole distance through the largest duct.

The air velocities listed in Table 7-5 have been found to accomplish satisfactory results in engineering practice. Where quiet operation is essential, the blower or fan should be selected on the basis of a low outlet velocity. This will also result in lower operating costs.

The following steps are involved in sizing ducts by the equal friction method:

1. Compute the total volume (in cubic feet) of the structure.
2. Compute the cubic-foot volume of each room in the structure to be supplied with heated or cooled air. The volume of each room should be expressed as a percentage of the total volume of the structure. For example, a room having a volume of 6000 cubic feet would represent 10 percent of a structure having a total volume of 60,000 cubic feet.

Table 7-5 Recommended and Maximum Air Velocities

	Recommended Velocities (fpm)		
Designation	***Residences***	***Schools, Theaters, Public Buildings***	***Industrial Buildings***
Outdoor air intakes*	500	500	500
Filters*	250	300	350
Heating coils*	450	500	600
Air washers	500	500	500
Fan outlets	1000–1600	1300–2000	1600–2400
Main ducts	700–900	1000–1300	1200–1800
Branch ducts	600	600–900	800–1000
Branch risers	500	600–700	800
	Maximum Velocities (fpm)		
Outdoor air intakes*	800	900	1200
Filters*	300	350	350
Heating coils*	500	600	700
Air washers	500	500	500
Fan outlets	1700	1500–2200	1700–2800
Main ducts	800–1200	1100–1600	1300–2200
Branch ducts	700–1000	800–1300	1000–1800
Branch risers	650–800	800–1200	1000–1600

These velocities are for total face area, not the net free area; other velocities in table are for net free area.
(Courtesy ASHRAE 1960 Guide)

3. Compute the total amount of air to be handled by the blower or fan. This will be the total CFM for the entire structure and can be computed by the air change method:

$$\text{CFM} = \frac{\text{Building Volume in Cubic Feet}}{\text{Minutes Air Change}}$$

4. Determine the portion of the total amount of air to be delivered to each room. This is computed by multiplying the total CFM for the structure by the room volume percentage (see step 2). For example, if the blower or fan handles a required 15,000 CFM, then the room described in step 2 would receive 1500 CFM (15,000 CFM × 10% = 1500 CFM).

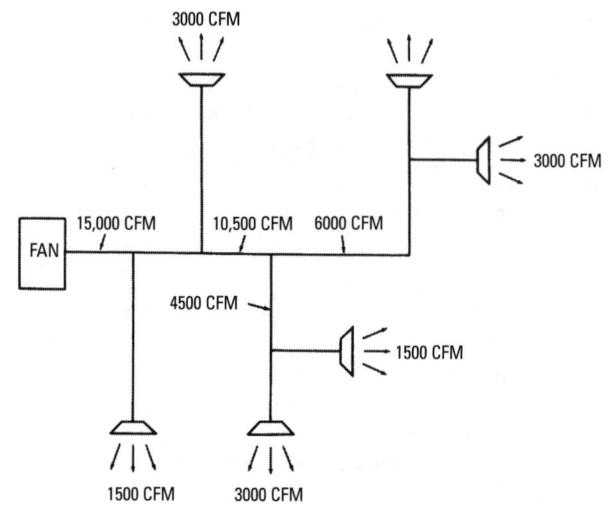

Figure 7-22 Air velocities of a duct system.

5. Determine the design and location of the duct runs (supply, branch, and return runs) (see Figure 7-22) and then locate the supply air outlets and return air openings for each room to give the most uniform distribution of air. Each outlet should be selected for suitable air velocity and throw (the manufacturer's catalog will provide the necessary data).
6. Determine the total CFM for the main supply duct *before* any branch ducts are reached. This will be equal to the total CFM for the entire structure.
7. Determine the allowable air velocity in the main supply duct (see Table 7-5). For a commercial building such as the one represented by this example (that is, 60,000 cubic feet), the air velocity in the main duct will be 1300 fpm.
8. Determine the static pressure drop from Figure 7-23. Since the supply duct must carry 15,000 CFM, locate 15,000 on the vertical scale on the left side of the friction chart in Figure 7-23. This will be found a little over halfway between 10,000 and 20,000 on the scale. Draw a line horizontally across the chart to where it intersects at a point halfway between the diagonals representing 1200 and 1400 fpm (that is, 1300 fpm). This gives a static pressure drop for the main duct of approximately 0.04

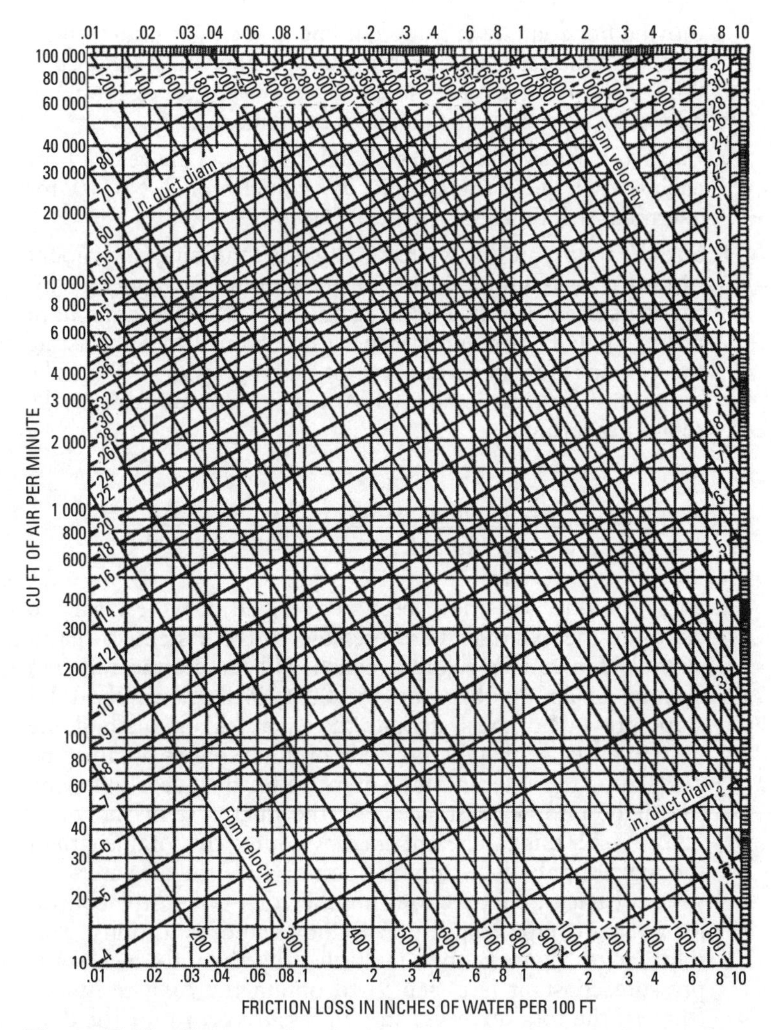

Figure 7-23 Friction of air in ducts.

inch of water per 100 feet of duct. Use the static pressure drop of 0.04 inch as a constant for the entire duct system.

9. Use Figure 7-23 to compute the round duct diameter for each branch duct. For example, branch No. 10 in Figure 7-22 will have a diameter of approximately 19 inches. This is determined by finding the 1500 CFM on the left-hand column of the chart in Figure 7-23 (halfway between 1000 and 2000 CFM) and moving horizontally in a straight line to the right until it intersects with the vertical line representing a static pressure drop of 0.04 inch. The two lines intersect at a point representing an approximate round duct diameter of 19 inches.

10. Using the same data in step 9, determine the air velocity for each branch duct. The air velocity of duct branch No. 10, for example, will be approximately 750 fpm.
11. Size the return air duct system by first determining the amount of return air required. In step 4, the amount of supply air for the structure was stated as being 15,000 CFM. The amount of return air (CFM) required can be determined by subtracting the amount of fresh air intake from the supply air (CFM). The amount of fresh air intake (CFM) can be determined from the following formula:

$$\text{CFM} = \frac{\text{Volume of Structure}}{\text{60 Min./Hr.}} \times \begin{array}{l}\text{Number of Air}\\ \text{Changes per Hour}\end{array}$$

12. After the duct sizes have been determined, it is necessary to compute the external static pressure of the system so that a suitable fan can be selected which will handle the required volume of air (in this case 15,000 CFM) against the total static pressure of the system. In the equal friction method, the total external static pressure drop of the system is obtained by calculating the external static pressure for the duct run having the highest total resistance. For the duct system shown in Figure 7-22, this can be obtained by adding the total length of the main supply duct from the point where the air enters the system (that is, duct sections numbered 1–6) *and* the equivalent straight pipe length of all elbows and transitions. This total length figure is then multiplied by the 0.04 inch static pressure constant (see step 9) to obtain the *total* resistance (that is, the total *external* static pressure drop) for the duct system. If, for example, the total length of duct was found to be 406 feet, the total resistance would be 0.1624 inch (406 ft ÷ 100 ft = 4.06 ft × 0.04 = 0.1624 in).
13. The total resistance for the return duct run is determined in the same manner as described for the supply air run. Because it represents less air volume than the supply air (return air = supply air – fresh air intake), it will always be a smaller figure than the static pressure determined for the supply duct run.

As a rule-of-thumb, the return-run static pressure will be about 25 percent of the supply-run static pressure.

14. The total static pressure against which the blower or fan must operate includes the following:
    a. Total resistance of the supply air duct system (that is, the total *external* static pressure drop).
    b. Total resistance of the return air duct system (that is, the total *external* static pressure drop).
    c. Total *internal* static pressure losses (that is, resistance through filters, cooling coils, and other forms of equipment).

Balancing an Air Distribution System

After the air distribution system has been installed, it should be tested to determine whether the air delivery and distribution correspond with the system design. Engineers and experienced field workers use a number of different instruments to make the various measurements required for balancing the system. Because these measuring devices are not generally available to the average person, the following simplified balancing procedure is suggested:

1. Open all duct and outlet dampers.
2. Check drafts, noise, and temperature differences from room to room while the fan or blower is operating.
3. Adjust the dampers to provide the greatest uniformity in operating characteristics.

This procedure is an abbreviated form of the one recommended for balancing a warm-air heating system. Read the appropriate sections of Chapter 6 of Volume 1 ("Warm-Air Heating Systems") for more information.

Duct Maintenance

The ducts used in heating, ventilating, and air-conditioning systems require very little maintenance other than periodic cleaning. Evidence that dust and dirt will accumulate in the ducts is often indicated by dirt streaks from ceiling and wall air supply outlets. Accumulations of dust and dirt in the ducts can be dangerous because they represent potential fire hazards. These are combustible materials that can be ignited when conditions are right. Ducts should be cleaned periodically to prevent these accumulations from forming. Doors should be placed in the ducts to provide access for cleaning.

Corrosion is very seldom a problem unless there is an accumulation of condensation over a long period of time. This can be prevented by keeping the ducts dry. A vapor barrier will accomplish this purpose (see *Duct Insulation*).

Check the ducts for cracks, holes, or other damage causing leaks. Leaks are usually found in the return ducts located outside the conditioned spaces (for example, in basements, crawl space, attic, wall, floor, and ceiling cavities). These cracks or holes in the return ducts draw in outside air, creating excess pressure inside the house and forcing conditioned air back out through the same cracks or holes. This can result in lowered comfort levels and higher energy costs. Most houses, especially older ones, will also have cracks or holes in the return and supply ducts. Making only partial repairs, such as patching the cracks and holes in the return ducts but not in the supply ducts, can create a pressure imbalance.

A properly designed duct system is one with the supply and return ducts in balance. In other words, the amount of air supplied to the ducts should be equal to the amount of air returned to the furnace air handler. If the supply and return ducts are in balance, the pressure inside the house will be neutral. A pressure imbalance can draw in combustion gases from the furnace, allergens from the outdoor air, dust mites, mold, and other contaminants from the basement and attic or attic craw space.

All ducts running through unconditioned spaces should be insulated to minimize conductive heat loss.

Roof Plenum Units

Air distribution systems for commercial and industrial buildings sometimes require variable amounts of outside air for ventilation purposes to be mixed with return air from conditioned spaces within the structure. A roof-mounted insulated plenum to which distribution supply and return air ducts may be connected through a roof opening is required for such an installation. Outside air dampers are designed to adjust from closed to full open and may be manually controlled by a multiposition remote potentiometer. When full open, the dampers will permit the introduction of a maximum of 80 percent outside air for ventilation purposes.

The Janitrol roof plenum unit shown in Figure 7-24 is located next to the control box to which it is connected by means of a flexible conduit. The construction of the roof base and curb is illustrated in Figure 7-25. Roof plenum units that do not provide for the addition of outside air to the system are also available.

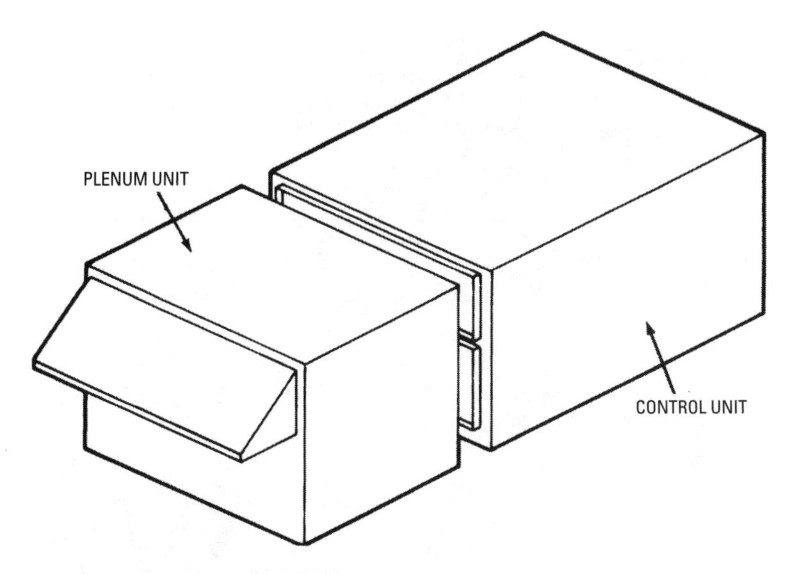

Figure 7-24 Roof plenum unit. *(Courtesy Janitrol)*

Mobile Home Duct Systems

Duct kits for adding central air-conditioning to forced warm-air heating systems are available from some manufacturers for installation in mobile homes, and particularly for installation in the double-wide combination mobile homes.

A typical air-conditioning duct kit installation is illustrated in Figure 7-26. The air conditioner is located outdoors and the cool air is supplied to the mobile home by a round (12-inch diameter) flexible duct that leads from the air-conditioning unit to the long metal heating ducts located in the center of the structure. A transition

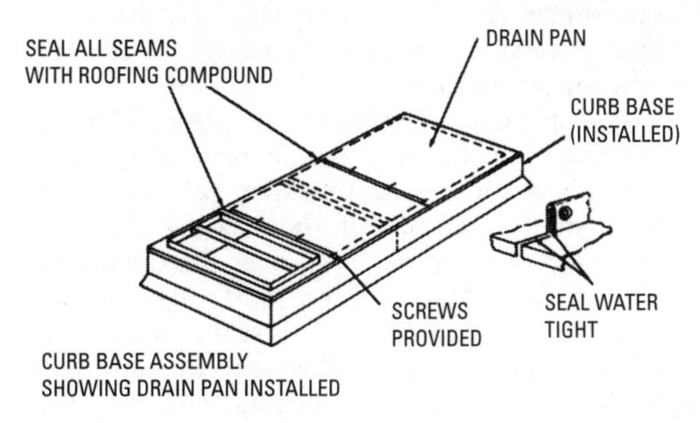

Figure 7-25 Roof base and curb assembly. *(Courtesy Janitrol)*

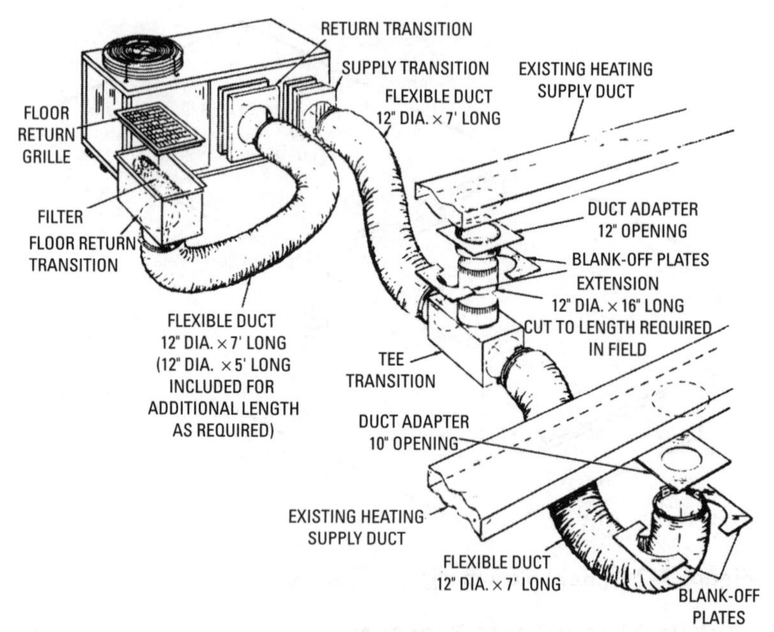

Figure 7-26 Duct kit for installation in the double-wide combination mobile home. *(Courtesy Dornback Industries, Inc.)*

tee with a short length of round, nonflexible metal duct connects the flexible duct leading from the air conditioner to the metal heating duct running the length of the mobile home through a *supply air opening* (not to be confused with the warm-air *registers* on the top of the heat supply duct and which are a part of the existing heating system).

A return air grille is located in the floor near the outside wall closest to the air conditioner. As shown in Figure 7-27, the return grille is located over an enclosure that contains a filter for cleaning the air returning to the air conditioner.

The flexible cool-air supply duct is connected by cutting a 16-inch by 16-inch square opening in the flooring directly beneath the main heat supply duct at the chosen location. A smaller 12½-inch-wide hole is then cut through the bottom of the heating duct. An adapter plate with a round opening is placed over the opening in the heating duct and secured in place with screws. Tape is then run around the outer edge of the adapter plate to seal it against air leaks (see Figure 7-28). Supply air outlets closest to the air conditioner use an adapter plate with a 12-inch opening. Those farthest away use an adapter plate with a 10-inch opening.

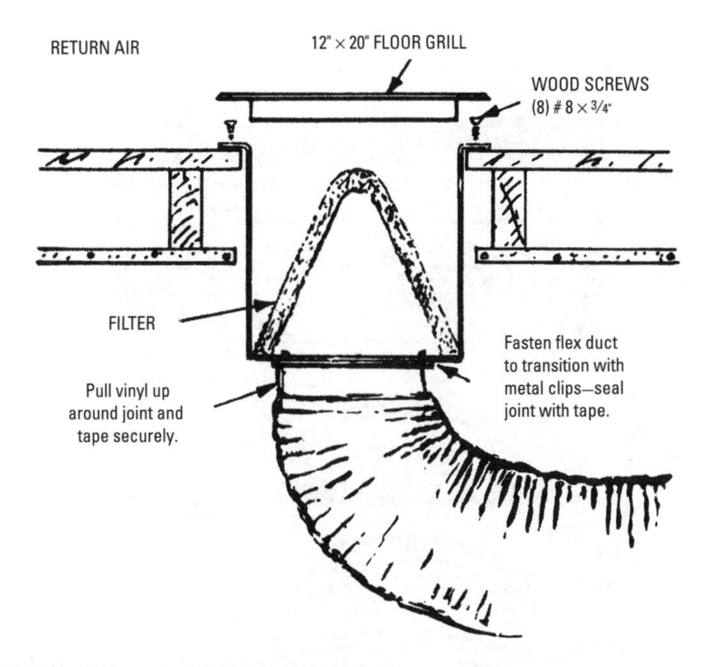

Cut 12 1/4" × 20 1/4" opening in both main and lower floors.
Lower return air duct enclosure through opening from inside of coach.
Screw duct to main floor with #8 × 3/4 wood screws.

Figure 7-27 Construction details of the return air grille and filter.
(Courtesy Dornback Industries, Inc.)

Before locating the best position for the air conditioner and routing the flexible duct, the positions of the supply air outlets and the return air grille should be determined. When doing this, the following recommendations are offered:

- *Never* locate the supply air opening closer than 4 feet from the furnace.
- *Never* locate the supply air opening directly under a supply register.
- Connect the *flexible* cool-air supply duct to the main heating duct as close to the center of the mobile home as possible.
- Locate the return air grille in a room or area open to the rest of the mobile home so that proper air circulation may be maintained.
- Provide adequate accessibility to the return air enclosure for cleaning or replacing the filter.

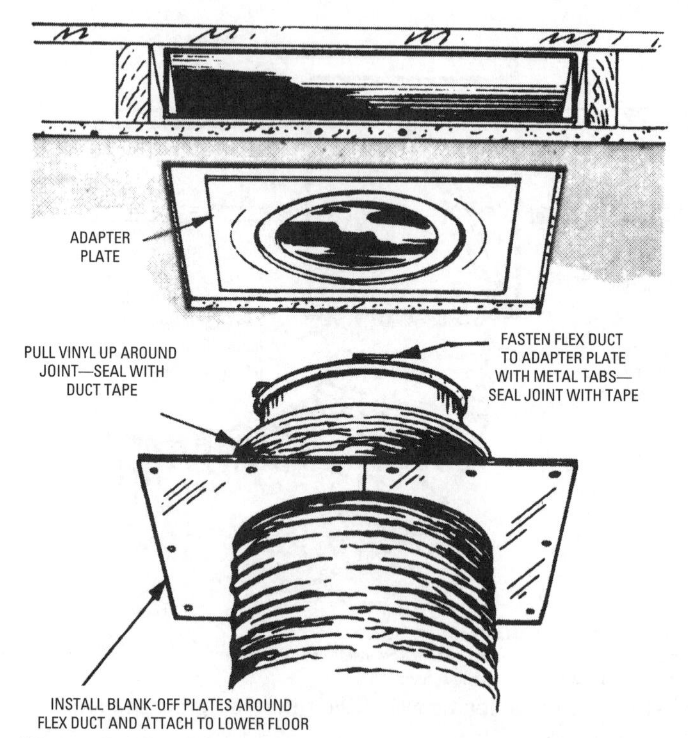

Figure 7-28 Construction details of a supply air grille.
(Courtesy Dornback Industries, Inc.)

Proprietary Air Distribution Systems

Proprietary air distribution systems are available from several manufacturers to provide supplementary heating or cooling in existing installations. They are also available as a complete heating and/or cooling system for installation in a structure where none previously existed or in which the existing system is clearly inadequate. They can also be used in new construction.

The Dunham-Bush Space-Pak air distribution system illustrated in Figure 7-29 is an example of a proprietary system that provides total comfort conditioning, including heating, cooling, air cleaning, and humidifying.

The Dunham-Bush air distribution system is generally an attic or overhead installation, but it may also be installed in basements or any other suitable area with no impairment of its operating performance.

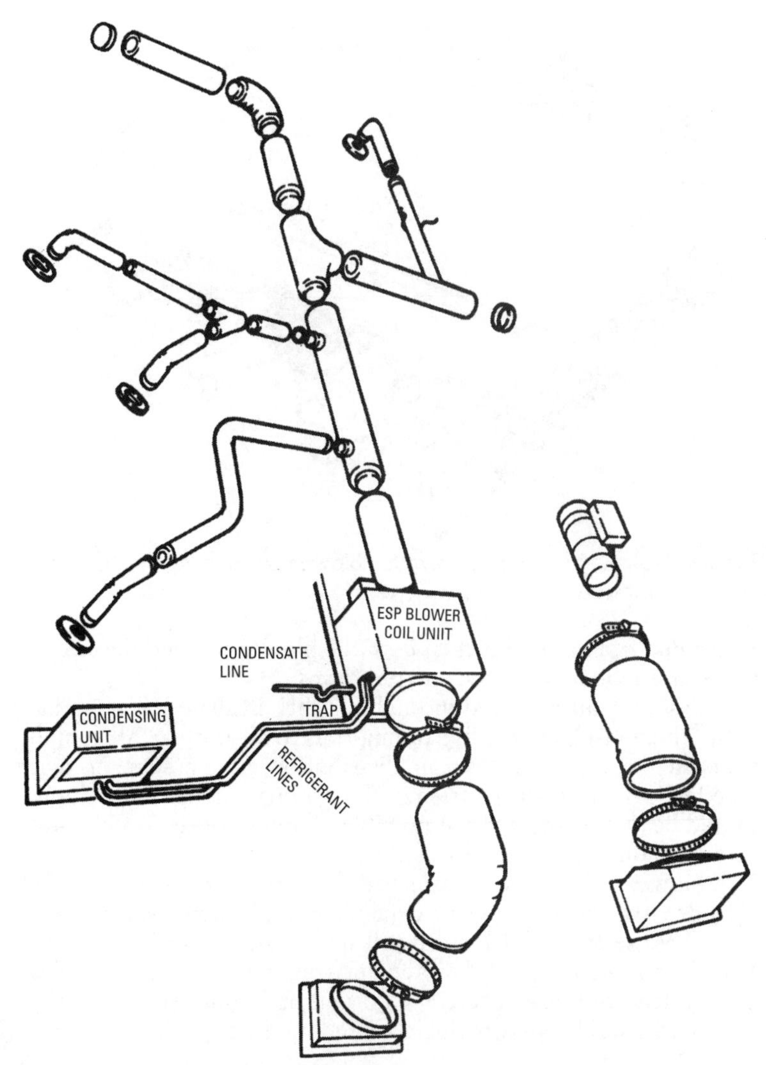

Figure 7-29 Space-Pak air distribution system. *(Courtesy Dunham-Bush, Inc.)*

Attic or overhead installation requires the construction of a mounting platform for the blower unit (see Figure 7-30). It is recommended that isolation pads or strips be placed between the blower coil unit and the mounting platform to prevent vibration transmission through walls or floors (see Figure 7-31). If the blower coil unit is

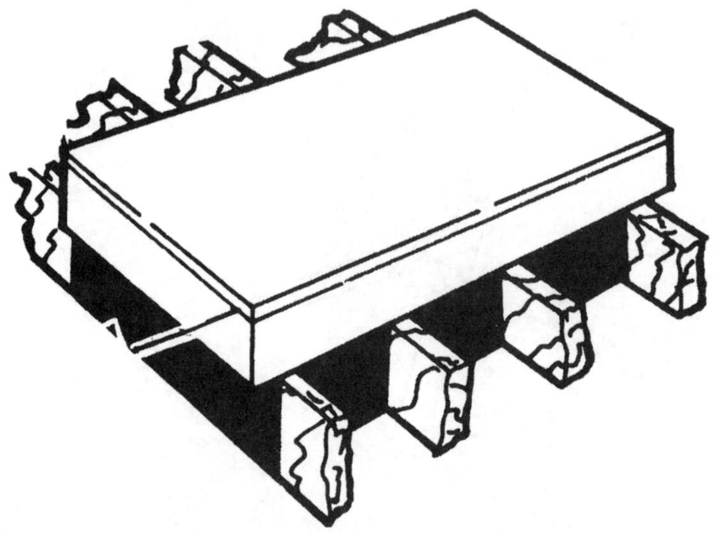

Figure 7-30 Mounting platform for blower coil in attic or overhead installation. *(Courtesy Dunham-Bush, Inc.)*

suspended from a ceiling, then both the blower coil and the electric heater are mounted on separate platforms.

The supply air is moved through a 7-inch insulated plenum duct with 2-inch insulated flexible tubing runs to air outlets. All components in the Dunham-Bush air distribution system snap or twist together. As in all duct systems, the number of tees and elbows should be limited to as few as possible in order to keep system pressure drop on larger layouts to a minimum.

A blower coil unit and an attached electric heater (see Figure 7-32) provide either heat or cool air depending on the control setting.

A cross-section of a blower coil unit is shown in Figure 7-33. The float switch is used to interrupt compressor operation when the condensation level exceeds a normal operating level. A further safeguard would be to provide a secondary drain pan.

Duct Furnaces

A *duct furnace* is essentially a unit heater designed for installation in a duct system. It is usually designed to operate on oil, natural gas, or propane gas, although electric duct heaters are also available and growing in popularity.

An example of a gas-fired (natural gas) furnace is the Janitrol 72 Series duct furnace illustrated in Figure 7-34, which is available in

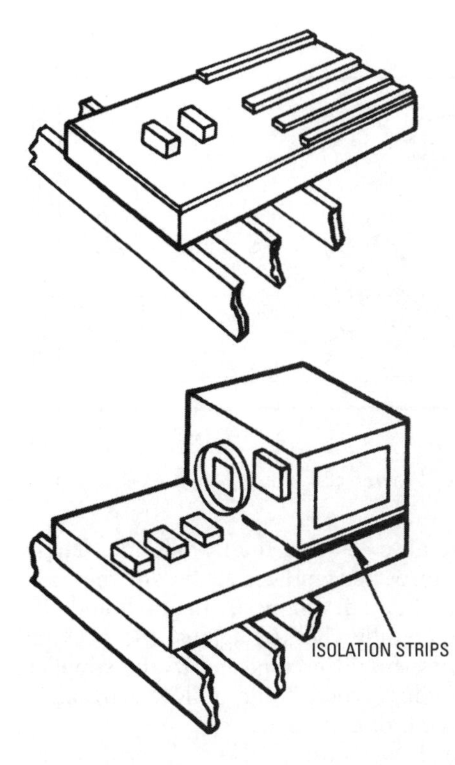

Figure 7-31 Using isolation strips for vibration control. *(Courtesy Dunham-Bush, Inc.)*

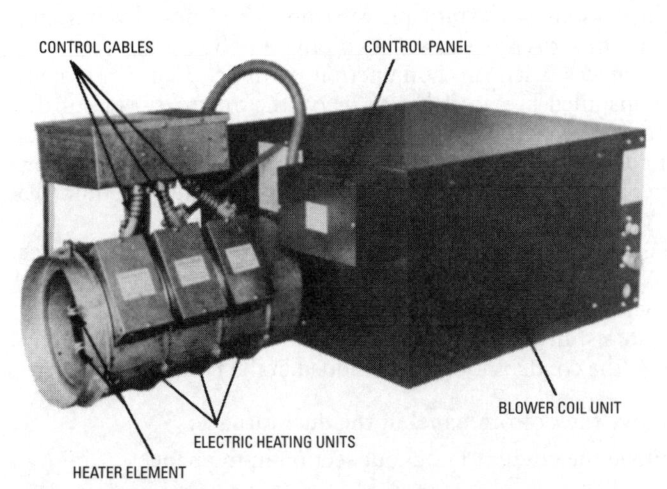

Figure 7-32 Blower-coil unit and electric heater. *(Courtesy Dunham-Bush, Inc.)*

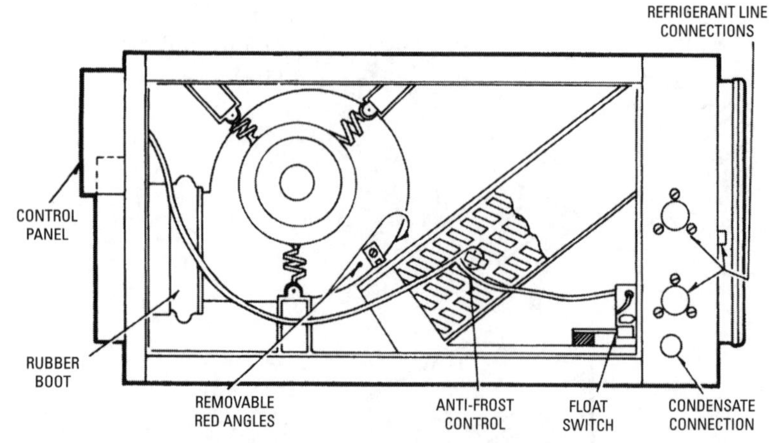

Figure 7-33 Cross-section of blower coil unit. *(Courtesy Dunham-Bush, Inc.)*

13 sizes depending on the requirements of the installation. These are factory-assembled units inspected and tested before they are shipped. Once they reach their destination, they should be unpacked and the contents carefully checked against the packing list. Missing or damaged parts should be reported to the supplier immediately. This is a procedure you should follow habitually whenever you receive a shipment of equipment.

The design of these Janitrol duct furnaces is AGA-certified for use in a duct system with static pressure up to 2 inches of water and with temperature rises as shown in column 4 of Table 7-6.

Duct furnaces used in conjunction with cooling equipment should be installed in parallel with or on the *upstream* side of the cooling coils. If a parallel flow arrangement is used, the dampers (or other means of controlling the airflow) should be made tight enough to prevent the circulation of cooled air through the unit (see ANSI Z21.30).

When equipped with a suitable condensation pan, a duct furnace may be AGA-certified for installation *downstream* from cooling coils, air washers, and evaporative coolers when operating as air-cooling systems. In the Janitrol 72 Series duct furnace illustrated in Figure 7-34, the condensation pan is added in the following manner:

1. Remove the bottom panel of the duct furnace.
2. Remove the circular knockout section in the panel.
3. Place the condensation drain connection of the condensation pan in the opening.

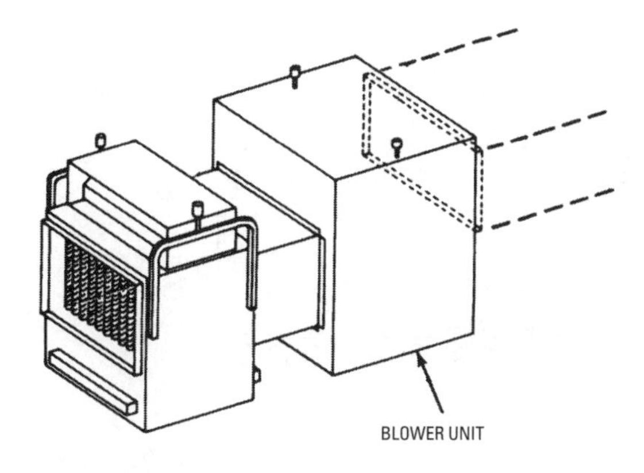

Figure 7-34 Series 72 gas-fired duct furnace. *(Courtesy Janitrol)*

4. Reinstall the bottom panel.
5. Connect the drain line.
6. Provide disconnect adjacent to bottom panel.

Table 7-6 Engineering Data for Series 72 Gas-Fired Duct Furnaces

	*AGA, Rating (Btu/h)				72-XXX-3		72-XXX-4	
Unit Size	Input†	Output	Temp. Rise	EDR‡ Steam	Net Wt. (lbs)	Approx. Ship Wt. (lbs)	Net Wt. (lbs)	Approx. Ship Wt. (lbs)
72–100	100,000	80,000	25–100	333	148	160	140	152
72–125	125,000	100,000	25–100	418	168	180	154	166
72–150	150,000	120,000	25–100	500	168	180	158	170
72–175	175,000	140,000	25–90	586	156	178	148	170
72–200	200,000	160,000	25–100	667	210	232	188	210
72–225	225,000	180,000	25–100	750	200	222	188	210
72–250	250,000	200,000	25–100	834	212	300	260	282
72–300	300,000	240,000	25–100	1000	270	298	256	284
72–350	350,000	280,000	25–90	1167	262	290	256	284
72–400	400,000	320,000	25–100	1333	399	440	383	424
72–500	500,000	400,000	25–100	1667	469	518	455	504
72–600	600,000	480,000	25–100	2000	565	622	525	582
72–700	700,000	560,000	25–90	2333	567	624	545	602

Tabled rating for elevations up to 2000 feet above sea level. From 2000 to 7000 feet, input must be reduced 4% per 1000 feet above sea level by manually adjusting manifold pressure. For elevation above 7000 feet, when ordering from factory, order must state elevation at which unit is to operate.

†*Input of unit (Btu/h) heat value of gas (Btu/ft^3) = gas consumption (ft^3/h)*

‡*Output ÷ 240 Btu*

(Courtesy Janitrol)

If space limitations prevent bottom-panel removal, access is possible through the front panel and burners (both of which must be removed *before* inserting the condensation panel).

Duct furnaces may be installed downstream of evaporative coolers or air washers considered as air-cooling systems (that is, operating with chilled water, which delivers air below the ambient air temperature at the duct furnace).

The minimum clearances between the duct furnace (and its draft hood) and the nearest adjacent walls, ceilings, and floors of combustible construction should be *at least* 6 inches (see ANSI Z21.30 and NEPA No. 31, *Installation of Oil Burning Equipment* 1972). Under certain circumstances the minimum clearance between the bottom of the duct furnace and the floor can be as close as 2 inches, provided the other clearances (that is, the 6-inch minimums) are maintained for the rest of the unit.

When planning duct furnace clearances, consideration must be given to accessibility for the following:

- Cleaning the heat exchanger
- Removal of burners
- Servicing the controls

The minimum clearance should be 18 inches for front removal of the burners and 10 inches for bottom removal.

The inlet and outlet ducts should be attached to the duct connection flange. An access panel (or removable duct section) should be provided at both the inlet and outlet ducts to provide for servicing the limit and fan control elements. Moreover, such access should be so constructed as to provide for visual inspection of the heat exchangers.

In gas-fired duct furnaces, *all* gas piping *must* be run in accordance with requirements outlined in the American Gas Association's publication *Installation of Gas Appliances and Gas Piping*.

It is recommended that a pipe joint compound certified for use with LPG be used on all pipelines. If possible, run a *new* gas supply line directly to the duct furnace from the meter. Support the gas line with hangers positioned close enough together to prevent strain on the unit. A trap consisting of a tee with a capped nipple should be provided in the gas line when the unit is installed. After installing the gas line, test it for leaks with a soap solution (*never* with a flame).

Oil- and gas-fired units *must* be properly vented. The draft head is built into these units, and the flue pipe must be the same size as the outlet of the flue collector on the duct furnace. *Never* reduce the size of the flue pipe or install a damper in it. Install the flue pipe to provide minimum clearances of 18 inches between it and combustible material. *Always* examine the chimney for proper construction and repair *before* connecting the flue pipe. Other details concerning venting practices are found in ANSI Z21.30 and NEPA No. 31, *Installation of Oil Burning Equipment 1972*. It is not necessary to vent electric-fired duct furnaces.

All electrical wiring for duct furnaces must be done in accordance with the *National Electrical Code*, ANSI CI-1971, and local code requirements. The unit must be grounded in accordance with these codes.

If a pilot flame is used for ignition, the flame should extend 1 inch beyond the pilot burner. Flame adjustment is made either with the pilot flame adjustment device located on the gas valve (on units without a pilot gas valve) or with the built-in adjusting screw (on units supplied with a pilot gas valve). Pilot gas pressure regulators are used on natural gas-fired units in areas where gas pressure variations are great. If a pilot regulator is supplied with the unit, pilot flame length is adjusted by adjusting the gas pressure regulator. These pressure regulators are *not* supplied with propane gas-fired units. In any event, the length of the pilot flame on propane gas-fired units is not adjustable. A normal flame on a natural gas-fired unit will be blue in color with an inner cone approximately 1 inch high. On a propane gas-fired unit, a normal flame will be green in color with a distinct inner cone approximately ⅛ to ¼ inch high. Figure 7-35 shows a typical pilot assembly for a Janitrol gas-fired duct furnace.

The thermocouple pilot and thermopilot relay on Janitrol gas-fired duct furnaces can be checked for proper functioning as follows:

1. Read the steps outlined on the *operating instruction plate*.
2. Start the unit and allow the pilot to heat for at least 3 minutes (do *not* turn on the main burners at this time).
3. Close the pilot valve, and wait for the pilot relay switch to open and cause the electric gas valve to close.
4. Under normal circumstances, the length of time between the closing of the pilot valve and the opening of the pilot relay switch should be *less* than 2 minutes.
5. If the length of time is *greater* than 2 minutes, then the pilot relay must be replaced.

The gas input for a gas-fired duct furnace must not be greater than specified on the rating plate of the unit. Duct furnaces are

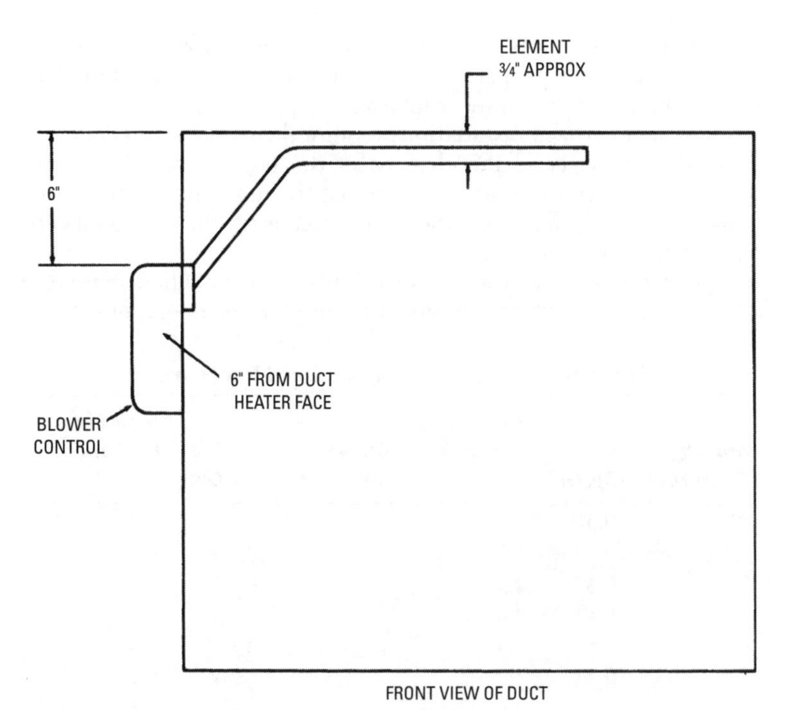

Figure 7-35 Location of fan control element. *(Courtesy Janitrol)*

shipped with spuds containing orifices sized for the particular gas (natural gas or propane) with which they are fired.

In natural gas-fired duct furnaces, the main burner gas-pressure regulator must be adjusted for the correct gas input. This may be done either by timing the test dial on the meter or by checking the manifold pressure (see Table 7-7). The following steps are recommended for adjusting the gas input:

1. Remove the cap on the top of the regulator.
2. Turn the adjusting screw in (or clockwise) to increase the gas input.
3. Back-out the adjusting screw (turn it counterclockwise) to reduce the gas input.

The gas pressure at the inlet to the regulator on natural gas-fired duct furnaces should *not* be allowed to exceed 12 inches of water.

Propane gas-fired units *must* maintain a manifold pressure of 11 inches of water for proper operation. These units are not supplied with appliance gas-pressure regulators.

Some duct furnace installations require the use of a fan control. When this is the case, the element of the fan control should be located as shown in Figure 7-35. Adjust the fan control for an off temperature as low as possible without causing the occupants to experience a feeling of cold air.

A proper maintenance schedule for duct furnaces will increase the life of the equipment and result in more efficient and economical

Table 7-7 Natural Gas Manifold Pressures

Btu per Cubic Foot	Sp. Gr.	Man. Press. (Inches of Water)	Btu per Cubic Foot	Sp. Gr.	Man. Press. (Inches of Water)
900	0.50	3.4	1000	0.55	3.0
	0.55	3.7		0.60	3.3
	0.60	4.1		0.65	3.6
	0.65	4.4		0.70	3.9
925	0.50	3.2	1025	0.55	2.9
	0.55	3.5		0.60	3.1
	0.60	3.9		0.65	3.4
	0.65	4.2		0.70	3.7
950	0.50	3.1	1050	0.55	2.7
	0.55	3.4		0.60	3.0
	0.60	3.7		0.65	3.2
	0.65	4.0		0.70	3.5
	0.70	4.3			
975	0.50	3.2	1075	0.55	2.6
	0.60	3.5		0.60	2.9
	0.65	3.8		0.65	3.2
	0.70	4.1		0.70	3.3
			1100	0.55	2.5
				0.60	2.7
				0.65	2.9
				0.70	3.2

Note: Manifold pressures on this table are based on orifice sizes as shown in orifice table in installation manual. Pressures given in this table apply to sizes of units. This table does not apply to units used in high-altitude areas. See supplement for high-altitude manifold pressure table.
(Courtesy Janitrol)

operating characteristics. Frequently check burners, pilot, and the interior and exterior of the heat exchanger for a buildup of residue. Clean the exterior of the heat exchanger with a brush, compressed air, or a heavy-duty vacuum. The following steps are recommended for the more complicated procedure of cleaning the interior of the heat exchanger:

1. Disconnect the flue pipe from the unit.
2. Remove the pilot and burners.
3. Remove the flue collector and heat exchanger tube baffles.
4. Brush the interior of the heat exchanger with a flexible 1½-inch-diameter bristle brush.
5. Remove debris with a vacuum.
6. Replace parts in reverse order to which they were removed.

On a gas-fired duct furnace, leave the pilot on during the summer (*except* when the unit is part of an air-conditioning system). Inspect the flue pipe for deterioration at the beginning of the heating season. Make a similar inspection of the heat exchanger (and related components) for carbon deposit, rust, or corrosion. Clean or replace when required.

Never light the pilot on gas-fired duct furnaces until you have first read the lighting instructions on the unit plate. Select and maintain a thermostat setting that provides adequate comfort. Do *not* keep changing the thermostat setting. It will only result in higher heating costs.

Electric Duct Heaters

Electric duct heaters are designed and manufactured to function in the same manner as oil- or gas-fired duct furnaces. These are prewired factory-assembled units available in a wide range of sizes and heating capacities for a variety of different installations.

Typical duct heater construction is illustrated by the Vulcan unit in Figure 7-36. The frame for the electric resistors is designed for insertion into the duct as shown in Figures 7-37 through 7-44. These Vulcan duct heaters are available in five standard supply voltage ratings (120, 277 single-phase; 208, 240, and 480 single- or three-phase) or a specific voltage depending on the requirements of the installation.

The control housing is located on the outside of the duct and should be located so that the control panel is accessible for inspection and service. Various standard and optional built-in control components are available, including (1) staging contactors, (2) control

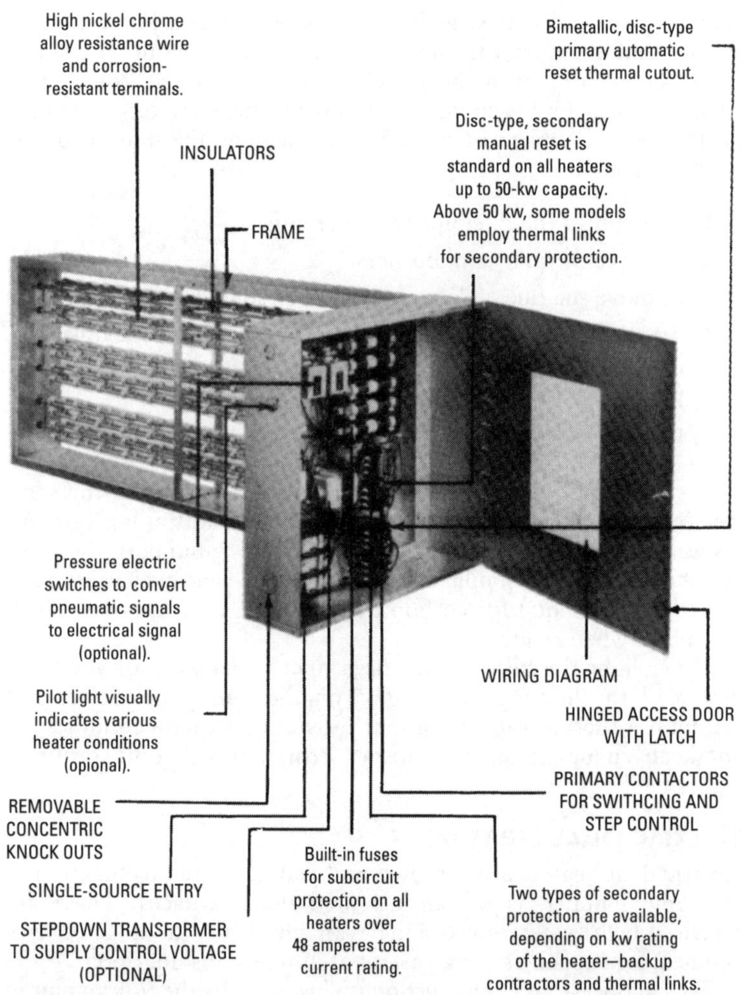

Figure 7-36 Construction of a Vulcan electric duct heater.
(Courtesy Vulcan Radiator Co.)

transformers, (3) sequencers, (4) fuses, and (5) primary and secondary protection.

There are three types of UL-listed thermal cutouts used as safety limit controls in Vulcan duct heaters. These are (1) primary automatic reset, (2) secondary manual reset, or (3) secondary

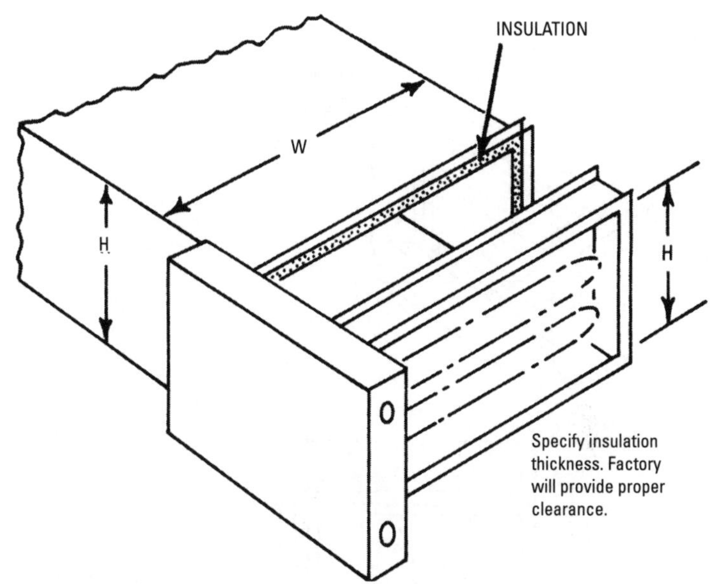

Figure 7-37 Flanged heater with internally insulated duct.
(Courtesy Vulcan Radiator Co.)

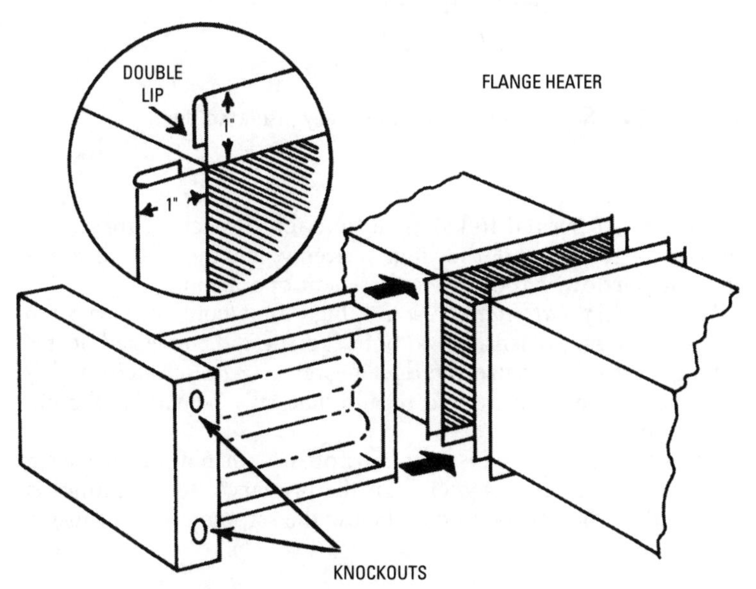

Figure 7-38 Flange heater. *(Courtesy Vulcan Radiator Co.)*

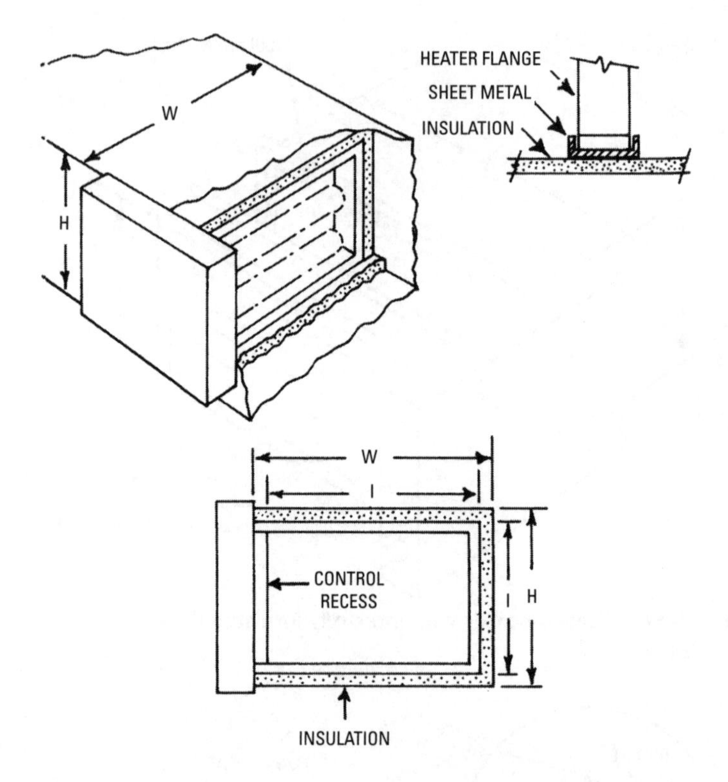

Figure 7-39 Slip-in heater with internally insulated duct.
(Courtesy Vulcan Radiator Co.)

heat limiters (thermal links). The function of each of these limit controls is to shut off the duct heater when there is no air or when the airflow is too low for efficient operation.

The *primary automatic reset*, a high-limit control, is a snap-action device that is sensitive to both radiant and convected heat. It is designed to deenergize the duct heater at a preselected, higher-than-normal temperature and to automatically reenergize the unit when it cools.

The *secondary manual reset* control is a warning device set to open at a temperature higher than the primary reset. It cannot be reset until the heater has cooled below the setpoint, and it must be reset manually.

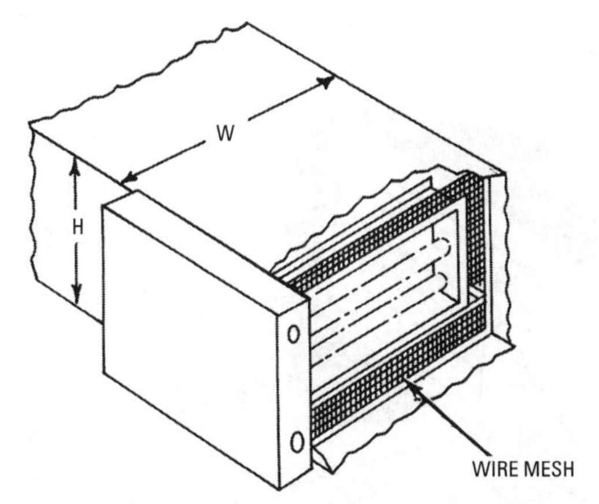

Figure 7-40 When a heater is smaller than the duct area, the opening between the heater frame and duct must be filled with wire mesh or expanded metal. *(Courtesy Vulcan Radiator Co.)*

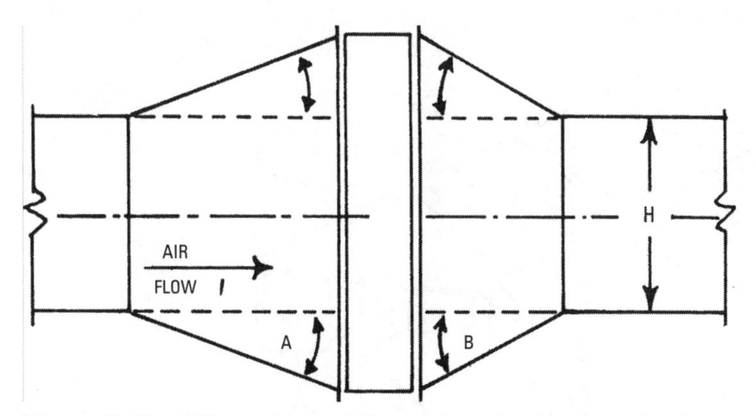

Figure 7-41 When a heater is larger than the duct area, the duct cross-section may be increased by a sheet-metal transition as shown in this drawing. *(Courtesy Vulcan Radiator Co.)*

The *heat limiters* (*thermal links*) belong to the secondary protection system. These are fusible, *one-time* protective devices that must be replaced in the event that they cut out.

Duct heater circuits may be subdivided into equal or unequal heating increments or stages in order to more closely match heater output with temperature variations.

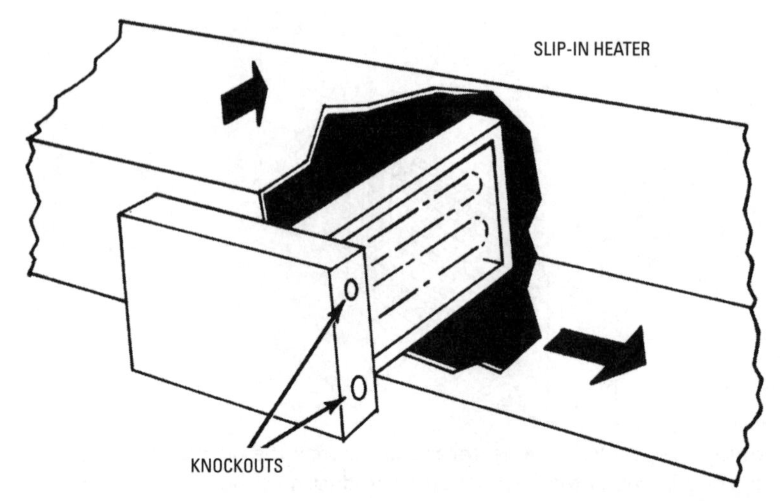

Figure 7-42 The installation of a slip-in heater.
(Courtesy Vulcan Radiator Co.)

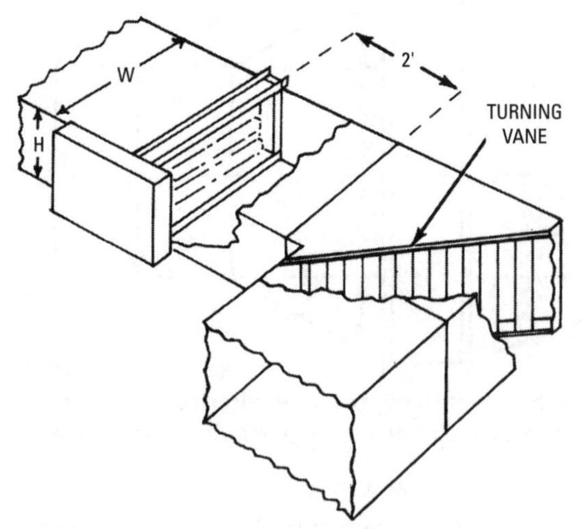

Figure 7-43 The installation of a turning vane.
(Courtesy Vulcan Radiator Co.)

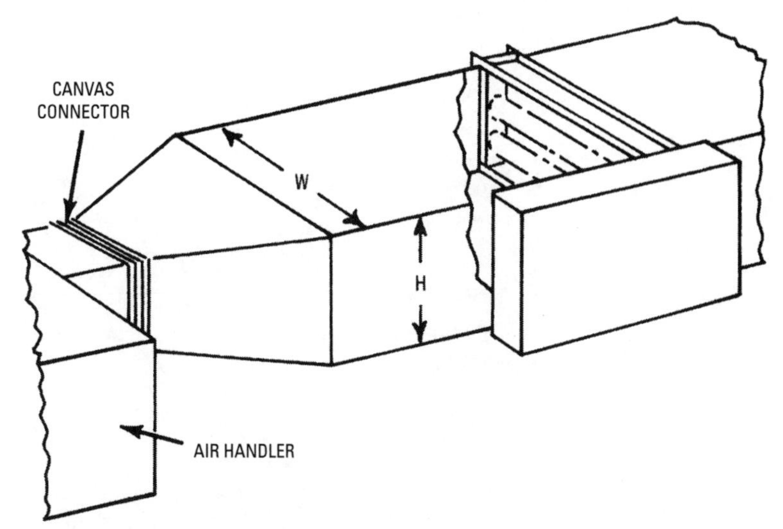

Figure 7-44 The installation of a canvas connector to minimize vibration. *(Courtesy Vulcan Radiator Co.)*

Chapter 8

Pipes, Pipe Fittings, and Piping Details

Pipes are available in a variety of materials, sizes, and weights. The type selected for a particular installation will depend on a number of factors, including government codes and standards, specifications, system requirements, availability of materials, and cost. Some of these factors, such as codes and standards, will have priorities over others, but all should be given equal consideration before deciding which pipe to use. The purpose of this chapter is to offer some guidance for making these decisions.

Types of Pipe Materials

The materials used in the manufacture of piping and tubing for steam and hydronic heating systems include the following:

- Wrought iron
- Wrought steel
- Galvanized steel
- Copper, brass, and bronze
- Plastic
- Synthetic rubber
- Composites

Steel piping leads in popularity and is available in the form of either wrought steel or galvanized steel. Wrought-iron and cast-iron pipe are also used, but less frequently because of their higher cost. Copper and brass are employed in the manufacture of both pipes and tubes and find their greatest application in radiant panel heating, air-conditioning, and refrigeration systems. Other metals (for example, aluminum, bronze, and alloy metals) have also been used in the manufacture of this pipe, but with limited application due to cost, lack of availability, and other factors. In addition to steel and copper tubing, modern hydronic heating systems also commonly use plastic tubing (cross-linked polyethylene or polybutylene), synthetic rubber hose, and composite tubing.

Note

> The terms pipe or piping and tube or tubing are sometimes used interchangeably, especially when referring to the piping/tubing systems of hydronic radiant panel heating systems. See *Copper and Brass Piping and Tubing* in this chapter.

This chapter concentrates on a description of iron, steel, copper, and brass pipes; the various types of nonmetal tubing; the pipe fittings used with them; and the methods employed in installing them.

Wrought-Iron Pipe

Wrought-iron pipe enjoyed widespread use in steam heating systems prior to World War II and was still being used, although on a much smaller scale, until 1968 when its production was discontinued in the United States.

Wrought-iron pipe has been shown to have an extremely long service life. Piping systems installed 50 to 80 years ago are still operating without any signs of deterioration. This is due not only to the internal grain structure of wrought-iron pipe, but also to its high resistance to corrosion.

Note

> Some ferrous metals, such as black iron (ordinary steel), are subject to corrosion if oxygen is allowed to enter the pipes. Certain types of cast-iron pipe fittings, as well as some valves and fittings, are also subject to corrosion if oxygen is present. The corrosion is irreversible.

Wrought-iron pipe differs very little in appearance from wrought-steel pipe. Indeed, were it not for special markings, the casual observer would not be able to distinguish between the two. Some wrought-iron pipe is stamped "genuine wrought iron" on each length. More frequently, the distinguishing mark is a spiral line marked into each length of pipe. The spiral identification line may be painted onto the surface in some bright color (usually red) or knurled into the metal.

The pipe used in heating installations is manufactured for different pressures and is available in various rated sizes (also referred to as the *nominal inside diameter*). The three *grades* of wrought-iron pipe are as follows:

- Standard
- Extra strong (or heavy)
- Double extra strong (or very heavy)

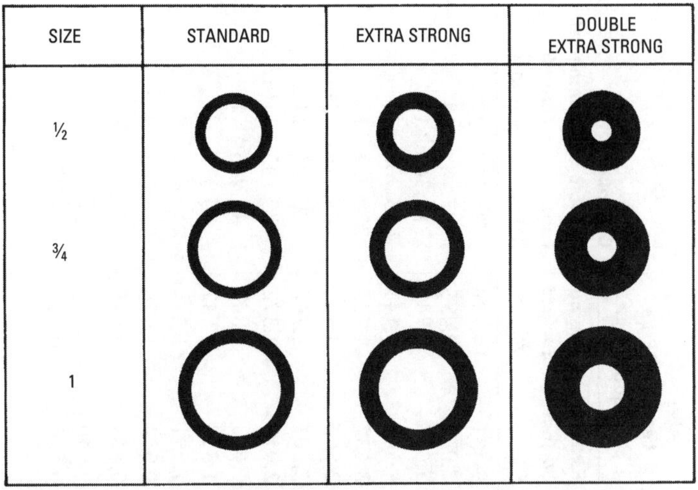

SIZE	STANDARD	EXTRA STRONG	DOUBLE EXTRA STRONG
½			
¾			
1			

Figure 8-1 Three sizes of standard, extra-strong, and double-extra-strong wrought pipe.

The pipe sections shown in Figure 8-1 are approximately half their actual size. Figure 8-2 illustrates their actual sizes, showing proportions of the three grades of wrought pipes.

The diameters given for pipes are far from the *actual* diameters, especially in the small sizes. Thus, a pipe known as ¼ inch (rated size) has an outside diameter of 0.54 inch and an inside diameter of 0.364 inch. Dimensions for standard, extra-strong, and double-extra-strong pipe are given in Tables 8-1, 8-2, and 8-3.

Wrought-iron pipes are adapted to higher pressures by making the walls thicker, but without changing the outside diameters. It is the inside diameter that is reduced.

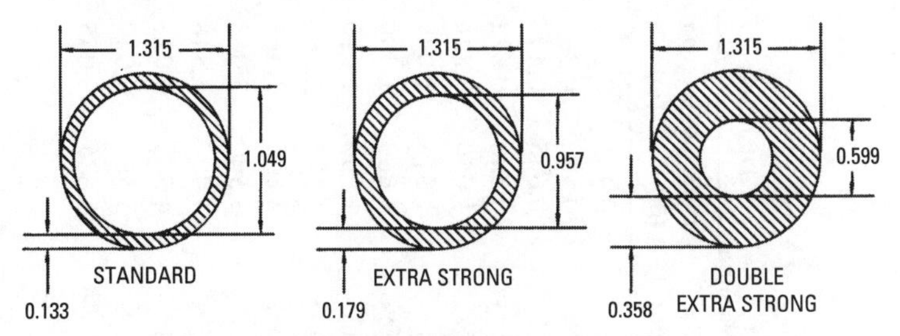

Figure 8-2 The three weights of wrought pipe shown actual size.

Table 8-1 Dimensions of Standard Pipe

Nominal Size (in)	Diameter External (in)	Diameter Internal (in)	Nominal Thickness (in)	Nominal Weight per Foot Threaded and Coupled (lbs)	Transverse Area External (in^2)	Transverse Area Internal (in^2)	Length of Pipe per Square Foot External Surface (ft)	Length of Pipe per Square Foot Internal Surface (ft)	Length of Pipe Containing 1 Cubic Foot (ft)	Number of Threads per Inch
1/8	0.405	0.269	0.068	0.245	0.129	0.057	9.431	14.199	2533.775	27
1/4	0.540	0.364	0.088	0.425	0.229	0.104	7.073	10.493	1383.789	18
3/8	0.675	0.493	0.091	0.568	0.358	0.191	5.658	7.748	754.360	18
1/2	0.840	0.622	0.109	0.852	0.554	0.304	4.547	6.141	473.906	14
3/4	1.050	0.824	0.113	1.134	0.866	0.533	3.637	4.635	270.034	14
1	1.315	1.049	0.133	1.684	1.358	0.864	2.904	3.641	166.618	11 1/2
1 1/4	1.660	1.380	0.140	2.281	2.164	1.495	2.301	2.768	96.275	11 1/2
1 1/2	1.900	1.610	0.145	2.731	2.835	2.036	2.010	2.372	70.733	11 1/2
2	2.375	2.067	0.154	3.678	4.430	3.355	1.608	1.847	42.913	11 1/2
2 1/2	2.875	2.469	0.203	5.819	6.492	4.788	1.328	1.547	30.077	8
3	3.500	3.068	0.216	7.616	9.621	7.393	1.091	1.245	19.479	8
3 1/2	4.000	3.548	0.226	9.202	12.566	9.886	0.954	1.076	14.565	8
4	4.500	4.026	0.237	10.889	15.904	12.730	0.848	0.948	11.312	8
5	5.563	5.047	0.258	14.810	24.306	20.006	0.686	0.756	7.198	8
6	6.625	6.065	0.280	19.185	34.472	28.891	0.576	0.629	4.984	8

8	8.625	8.071	0.277	25.000	58.426	51.161	0.442	0.473	2.815	8
8	8.625	7.981	0.322	28.809	58.426	50.027	0.442	0.478	2.878	8
10	10.750	10.192	0.279	32.000	90.763	81.585	0.355	0.374	1.765	8
10	10.750	10.136	0.307	35.000	90.763	80.691	0.355	0.376	1.785	8
10	10.750	10.020	0.365	41.132	90.763	78.855	0.355	0.381	1.826	8
12	12.750	12.090	0.330	45.000	127.676	114.800	0.299	0.315	1.254	8
12	12.750	12.000	0.375	50.706	127.676	113.097	0.299	0.318	1.273	8
14	15.000	14.250	0.375	60.375	176.715	159.485	0.254	0.268	0.903	8
17	17.000	16.214	0.393	72.602	226.980	206.476	0.224	0.235	0.697	8
18	18.000	17.182	0.409	80.482	254.469	231.866	0.212	0.222	0.621	8
20	20.000	19.182	0.409	89.617	314.159	288.986	0.190	0.199	0.498	8

Table 8-2 Dimensions of Extra-Strong Pipe

Nominal Size (in)	Diameter		Nominal Thickness (in)	Weight Nominal per Foot Plain Ends (lbs)	Transverse Area		Length of Pipe per Square Foot		Length of Pipe Containing 1 Cubic Foot (ft)
	External (in)	Internal (in)			External (in^2)	Internal (in^2)	External Surface (ft)	Internal Surface (ft)	
⅛	0.405	0.215	0.095	0.314	0.129	0.036	9.431	17.766	3966.393
¼	0.540	0.320	0.119	0.535	0.229	0.072	7.073	12.648	2010.290
⅜	0.675	0.423	0.126	0.738	0.358	0.141	5.658	9.030	1024.689
½	0.840	0.546	0.147	1.087	0.554	0.234	4.547	6.995	615.017
¾	1.050	0.742	0.154	1.473	0.866	0.433	3.637	5.147	333.016
1	1.315	0.957	0.179	2.171	1.358	0.719	2.904	3.991	200.198
1¼	1.660	1.278	0.191	2.996	2.164	1.283	2.301	2.988	112.256
1½	1.900	1.500	0.200	3.631	2.835	1.767	2.010	2.546	81.487
2	2.375	1.939	0.218	5.022	4.430	2.953	1.608	1.969	48.766
2½	2.875	2.323	0.276	7.661	6.492	4.238	1.328	1.644	33.976
3	3.500	2.900	0.300	10.252	9.621	6.605	1.091	1.317	21.801
3½	4.000	3.364	0.318	12.505	12.566	8.888	0.954	1.135	16.202

4	4.500	3.826	0.337	14.983	15.904	11.497	0.848	0.998	12.525
5	5.563	4.813	0.375	20.778	24.306	18.194	0.686	0.793	7.915
6	6.625	5.761	0.432	28.573	34.472	26.067	0.576	0.663	5.524
8	8.625	7.625	0.500	43.388	58.426	45.663	0.442	0.500	3.154
10	10.750	9.750	0.500	54.735	90.063	74.662	0.355	0.391	1.929
12	12.750	11.750	0.500	65.415	127.676	108.434	0.299	0.325	1.328

Table 8-3 Dimensions of Double-Extra-Strong Pipe

Nominal Size (in)	Diameter External (in)	Diameter Internal (in)	Nominal Thickness (in)	Nominal Weight per Foot Plain Ends (lbs)	Transverse Area External (in^2)	Transverse Area Internal (in^2)	Length of Pipe per Square Foot External Surface (ft)	Length of Pipe per Square Foot Internal Surface (ft)	Length of Pipe Containing 1 Cubic Foot (ft)
½	0.840	0.252	0.294	1.714	0.554	0.050	4.547	15.157	2887.165
¾	1.050	0.434	0.308	2.440	0.866	0.148	3.637	8.801	973.404
1	1.315	0.599	0.358	3.659	1.358	0.282	2.904	6.376	510.998
1¼	1.660	0.896	0.382	5.214	2.164	0.630	2.301	4.263	228.379
1½	1.900	1.100	0.400	6.408	2.835	0.950	2.010	3.472	151.526
2	2.375	1.503	0.436	9.029	4.430	1.774	1.608	2.541	81.162
2½	2.875	1.771	0.552	13.695	6.492	2.464	1.328	2.156	58.457
3	3.500	2.300	0.600	18.583	9.621	4.155	1.091	1.660	34.659
4	4.500	3.152	0.674	27.541	15.904	7.803	0.848	1.211	18.454
5	5.563	4.063	0.750	38.552	24.306	12.966	0.686	0.940	11.107
6	6.625	4.897	0.864	53.160	34.472	18.835	0.576	0.780	7.646
8	8.625	6.875	0.875	72.424	58.426	37.122	0.443	0.555	3.879

Wrought-Steel Pipe

Wrought-steel pipe is cheaper than wrought-iron pipe and consequently is used more widely in heating, ventilating, and air-conditioning than the latter. Depending on the method of manufacture, wrought-steel pipe is available as either *welded pipe* or *seamless pipe*. Seamless wrought-steel pipe finds frequent application in high-pressure work.

The wall thickness and weights of wrought-steel pipe are approximately the same as those for wrought-iron pipe. As with wrought-iron pipe, the two most commonly used weights are *standard* and *extra strong*. Theoretical bursting and working pressures for wrought-steel pipe are listed in Table 8-4.

In some systems, steel piping has lasted as long as 80 years without showing signs of deterioration. When installed properly, steel piping and tubing are not subject to leakage. Most problems with floor systems occur with steel pipe or tubing installed in a single-pour slab. The stresses that can develop in this type of construction can sometimes damage the pipes or tubing in the radiant floor panels. This does not seem to be the case if the pipes or tubing is installed between the two sections of a two-pour slab or above a concrete slab and beneath a wood floor.

Armco plastic-coated steel, cold-rolled or extruded steel, and stainless steel are among the types of steels currently used in the manufacture of tubing for hydronic heating systems.

Galvanized Pipe

Galvanized steel or iron pipe is covered with a protective coating to resist corrosion. This type of pipe is often used underground or in other areas subject to corrosion. The coating is not permanent, and care should be used when handling it to avoid nicks and scratches. If the surface coating is broken, corrosion will begin that much sooner. Galvanized pipe is cheaper than copper pipe but more expensive than either wrought-iron or wrought-steel pipe.

Caution

Galvanized pipe and galvanized fittings are not recommended for use in steam heating systems.

Copper and Brass Pipes and Tubing

Copper and brass are used in the manufacture of both pipes and tubes and find their greatest application in air-conditioning, refrigeration, and hydronic radiant floor heating systems. One major advantage of using copper or brass is that both metals are corrosion resistant.

Table 8-4 Wrought-Steel Pipe—Theoretical Bursting and Working Pressures (pounds per square inch)*

Size (in)	Size (mm)	*Standard* Bursting Pressure Barlow's Formula	*Standard* Working Pressure Factor 8	*Extra Strong* Bursting Pressure Barlow's Formula	*Extra Strong* Working Pressure Factor 8	*Double Extra strong* Bursting Pressure Barlow's Formula	*Double Extra strong* Working Pressure Factor 8	*Large O.D.* 3/8-Inch-Thick Bursting Pressure Barlow's Formula	*Large O.D.* 3/8-Inch-Thick Working Pressure Factor 8	*Large O.D.* 1/2-Inch-Thick Bursting Pressure Barlow's Formula	*Large O.D.* 1/2-Inch-Thick Working Pressure Factor 8
1/8	3	13,432	1679	18,760	2345						
1/4	6	13,032	1629	17,624	2204						
3/8	10	10,784	1348	14,928	1866						
1/2	13	10,384	1298	14,000	1750	28,000	3500				
3/4	19	8,608	1076	11,728	1466	23,464	2933				
1	25	8,088	1011	10,888	1361	21,776	2722				
1 1/4	32	6,744	843	9,200	1150	18,408	2301				
1 1/2	38	6,104	763	8,416	1052	16,840	2105				
2	50	5,184	648	7,336	917	14,680	1835				
2 1/2	64	5,648	706	7,680	960	15,360	1920				
3	76	4,936	617	6,856	857	13,714	1714				
3 1/2	90	5,610	701	7,950	994	15,900	1987				
4	100	5,266	658	7,480	935	14,970	1871				
4 1/2	113	4,940	618	7,100	887	14,200	1775				
5	125	4,630	579	6,740	842	13,480	1685				
6	150	4,220	528	6,520	815	13,040	1630				

7	175	3,940	493	6,550	819	11,470	1434				
8	200	3,730	466	5,780	722	10,140	1267				
9	225	3,550	444	5,190	649						
10	300	3,390	424	4,650	581						
12	300	2,940	368	3,920	490						
14	350							2,680	335	3,570	446
15	375							2,500	313	3,333	417
16	400							2,340	293	3,120	390
18	450							2,080	260	2,770	346
20	500							1,870	234	2,500	313
22	550							1,700	213	2,270	284
24	600							1,560	195	2,080	260

**Butt-welded pipe was figured on sizes 3 inches and smaller and lap-welded pipe on sizes 3½ inches and larger.*

Sometimes the terms copper tube and copper pipe are used interchangeably as if they were synonymous. Another semantic idiosyncrasy is to refer to the same product as tube or tubing when still on the inventory at the supply house, but as pipe or piping when installed. Both usages can be confusing because there are differences between the two. For example, copper pipe is often made thicker than tubing because it *can* be used with threaded fittings if soldering is not desired for making the joint connection. Another point to remember is that copper pipes have the same outside diameters as standard steel pipes. Copper tubing, on the other hand, is standardized on the basis of use into three standard wall-thickness schedules (see Table 8-5): (1) Type K, (2) Type L, and (3) Type M. Type L and Type M are used in heating, ventilating, and air-conditioning systems.

Copper tubing is available as hard-grained (drawn) copper tubes or soft (annealed) copper tubes. The former is subject to freezing, which can cause the tubes to twist in almost the same way as steel pipes. On the other hand, the stiffness of hard-grained copper tubes enables them to hold their shape better than the softer ones. They are, therefore, often used for exposed lines such as mains hung from the ceiling.

The soft-tempered Type L copper tubing is recommended for hydronic radiant heating panels. Because of the relative ease with which soft copper tubes can be bent and shaped, they are especially well adapted for making connections around furnaces, boilers, oil-burning equipment, and other obstructions. This high workability characteristic of copper tubing also results in reduced installation time and lower installation costs. Copper tubing is produced in diameters ranging from ⅛ inch to 10 inches and in a variety of different wall thicknesses. Both copper and brass fittings are available. Hydronic heating systems use small tube sizes joined by soldering.

The DIN Rating System

Oxygen from the outside air can permeate the tubing material and enter the hydronic system where it will corrode iron and steel components (boiler, fittings, valves, and so on). In the early 1990s, American tubing manufacturers decided to adopt the German DIN Standard 4726 as a uniform rating system. DIN stands for Deutsche (German) Industry Norm. The DIN Standard 4726 requires that hydronic systems not permit the entry of more than one-tenth of a milligram of oxygen per liter of water per day when the water is 104°F (40°C). All of the tubing used in hydronic heating systems must meet this standard.

Table 8-5 Sizes and Dimensions of Copper Water Tubes

Sizes		For Underground Services and General Plumbing Purposes, Used with Solder or Flared Fittings		For General Plumbing and House Heating Purposes, Used with Solder or Fittings		For General Plumbing and House Heating Purposes, with Normal Water Conditions. Used with Solder Fittings Only	
		Type K	Hard or Soft	Type L	Hard or Soft	Type M	Hard
Nominal Size (in)	Outside Diameter (in)	Wall Thickness	Pounds per Foot	Wall Thickness	Pounds per Foot	Wall Thickness	Pounds per Foot
⅛	0.250	0.032	0.085	0.025	0.068	0.025	0.068
¼	0.375	0.032	0.133	0.030	0.126	0.025	0.106
⅜	0.500	0.049	0.269	0.035	0.198	0.025	0.144
½	0.625	0.049	0.344	0.040	0.285	0.028	0.203
⅝	0.750	0.049	0.418	0.042	0.362	0.030	0.263
¾	0.875	0.065	0.641	0.045	0.455	0.032	0.328
1	1.125	0.065	0.839	0.050	0.655	0.035	0.464
1¼	1.375	0.065	1.04	0.055	0.884	0.042	0.681
1½	1.625	0.072	1.36	0.060	1.14	0.049	0.94
2	2.125	0.083	2.06	0.070	1.75	0.058	1.46
2½	2.625	0.095	2.92	0.080	2.48	0.065	2.03
3	3.125	0.109	4.00	0.090	3.33	0.072	2.68
3½	3.625	0.109	5.12	0.100	4.29	0.083	3.58
4	4.125	0.134	6.51	0.110	5.38	0.095	4.66
5	5.125	0.160	9.67	0.125	7.61	0.109	6.66
6	6.125	0.192	13.87	0.140	10.20	0.122	8.91
8	8.125	0.271	25.90	0.200	19.29	0.170	16.46
10	10.125	0.338	40.26	0.250	30.04	0.212	25.57
12	12.125	0.405	57.76	0.280	40.36	0.254	36.69

Plastic Tubing

Both cross-linked polyethylene tubing and polybutylene tubing are used in modern hydronic radiant panel heating systems. The former is by far the more popular of the two. The tubing is available in coils.

Some plastic tubing may become hardened and brittle after long use. If the tubing is then subjected to sudden unusual high pressures, such as those caused by boiler, valve, or other system component failures, cracks form and leakage occurs. Long cracks or fractures along a tubing circuit are not considered repairable to code.

Normal temperature differences in a hydronic system will cause the plastic tubing to expand and contract. This expansion and contraction over a long period of time eventually causes cracks or fractures to develop in the tubing, especially if it is already weakened by age or other factors. Expansion and contraction at the joints and connections between the tubing and the system boiler and manifolds may also cause leakage.

Cross-Linked Polyethylene Tubing

Cross-linked polyethylene (PEX) tubing is commonly used indoors in hydronic radiant heating panels or outdoors embedded beneath the surface of driveways, sidewalks, and patios to melt snow and ice. It is made of a high-density polyethylene plastic that has been subjected to a cross-linking process. It is flexible, durable, and easy to install. There are two types of PEX tubing:

- Oxygen barrier tubing
- Nonbarrier tubing

Oxygen barrier tubing (BPEX) is treated with an oxygen barrier coating (EVOH) to prevent oxygen from passing through the tubing wall. It is designed specifically to prevent corrosion to any ferrous fittings or valves in the piping system. BPEX tubing is recommended for use in hydronic radiant heating system.

Nonbarrier tubing should be used in a hydronic radiant heating system only if it can be isolated from the ferrous components by a corrosion-resistant heat exchanger, or if only corrosion-resistant system components (boiler, valves, and fittings) are used.

PEX tubing is easy to install. Its flexibility allows the installer to bend it around obstructions and into narrow spaces. A rigid plastic cutter tool, or a copper tubing cutter equipped with a plastic cutting wheel, should be used to cut and install PEX tubing. Both tools produce a square cut without burrs.

Caution

> PEX tubing is not resistant to ultraviolet (UV) rays. It should not be allowed to remain unprotected outdoors for long periods of time.

Polybutylene Tubing

Polybutylene tubing is offered in diameters and lengths comparable to PEX tubing but is more expensive, less durable, and not as easy to install.

Synthetic Rubber Hose

Synthetic rubber hose is very flexible and highly temperature resistant, but it is less durable than PEX tubing and has a low pressure rating. A number of different rubbers have been used to produce synthetic rubber hose for use in hydronic systems. Cross-linked EPDM is one of the most popular types.

Composite Tubing

Some manufacturers are producing composite tubing for hydronic heating systems. Composite construction commonly involves the combination of aluminum sandwiched between layers of plastic (PEX) or synthetic rubber.

PAX tubing is a typical three-layer composite tubing consisting of an inner layer of PEX tubing, a middle layer of aluminum, and an outer layer of PEX tubing. The aluminum middle layer, which provides an effective barrier to oxygen penetration through the tubing walls, is bonded to the inner PEX layer. PEX tubing is very flexible and, because of its aluminum middle layer, will retain its shape after bending much more easily than standard PEX tubing or BPEX tubing. PAX tubing is recommended for use in hydronic floor panel heating systems.

Brass tube and brass pipe should also be distinguished from one another. Brass tubing is generally manufactured from yellow brass (about 65 percent copper, 35 percent zinc). Brass piping is more frequently a red brass (about 85 percent copper, 15 percent zinc) and is much stronger than the tubing. As a result, brass pipe is sometimes used in heating, ventilating, and air-conditioning systems, particularly when it is necessary to use a pipe material that strongly resists corrosion.

Pipe Fittings

Pipe cannot be obtained in unlimited lengths. In most pipe installations, it is frequently necessary to join together two or more shorter lengths of pipe in order to create the longer one required by the blueprints. Furthermore, in practically all pipe installations there are numerous changes in directions and branches that require joining the pipes together in special arrangements. Pipe fittings have been devised for the necessary connections.

A *pipe fitting* may therefore be defined as any piece attached to pipes in order to lengthen a pipe, to alter its direction, to connect a branch to a main, to connect two pipes of different sizes, or to close an end.

Classification of Pipe Fittings

The pipe fittings used in heating, ventilating, and air-conditioning installations are most commonly either screwed or flanged types. Screwed pipe fittings use a male and female thread combination, which tightens together to form the joint. Screwed pipe fittings are designated as either male or female, depending on the location of the thread. A female thread is an internal thread, and a male thread is an external one.

A flanged pipe fitting has a lip or extension projecting at a right angle to its surface. This lip is bolted to the facing lip of the adjacent fitting for additional strength. As a result, flanged fittings are generally recommended for 4-inch pipe and above.

Screwed and flanged pipe fittings are used to make temporary joints. If soldering, brazing, or welding is used in joining the separate lengths of pipe, the joint is considered a permanent one. The advantage of a so-called temporary joint is that it can be easily disassembled for repair.

The great multiplicity of pipe fittings can be divided on the basis of their functions into the following six general classes:

- Extension or joining fittings
- Reducing or enlarging fittings
- Directional fittings
- Branching fittings
- Union or makeup fittings
- Shutoff or closing fittings

Extension or Joining Fittings

Nipples, locknuts, couplings, offsets, joints, and unions are all examples of *extension* or *joining fittings*. With the possible exception of an offset, these fittings are designed to join and extend (but not change the direction of) a length of pipe. An offset is used to reposition a length of piping so that it is parallel but not in alignment with another section of its length. The offset *itself* constitutes a change of direction; however, its function is to create a piping run parallel but not in alignment with the rest of its length.

Nipples

Nipples are classified as *close*, *short*, or *long* (see Figure 8-3). Standard lengths of nipples are listed in Table 8-6. The length of close nipples varies with the pipe size, it being determined by the length of thread necessary to make a satisfactory joint. There is one length of

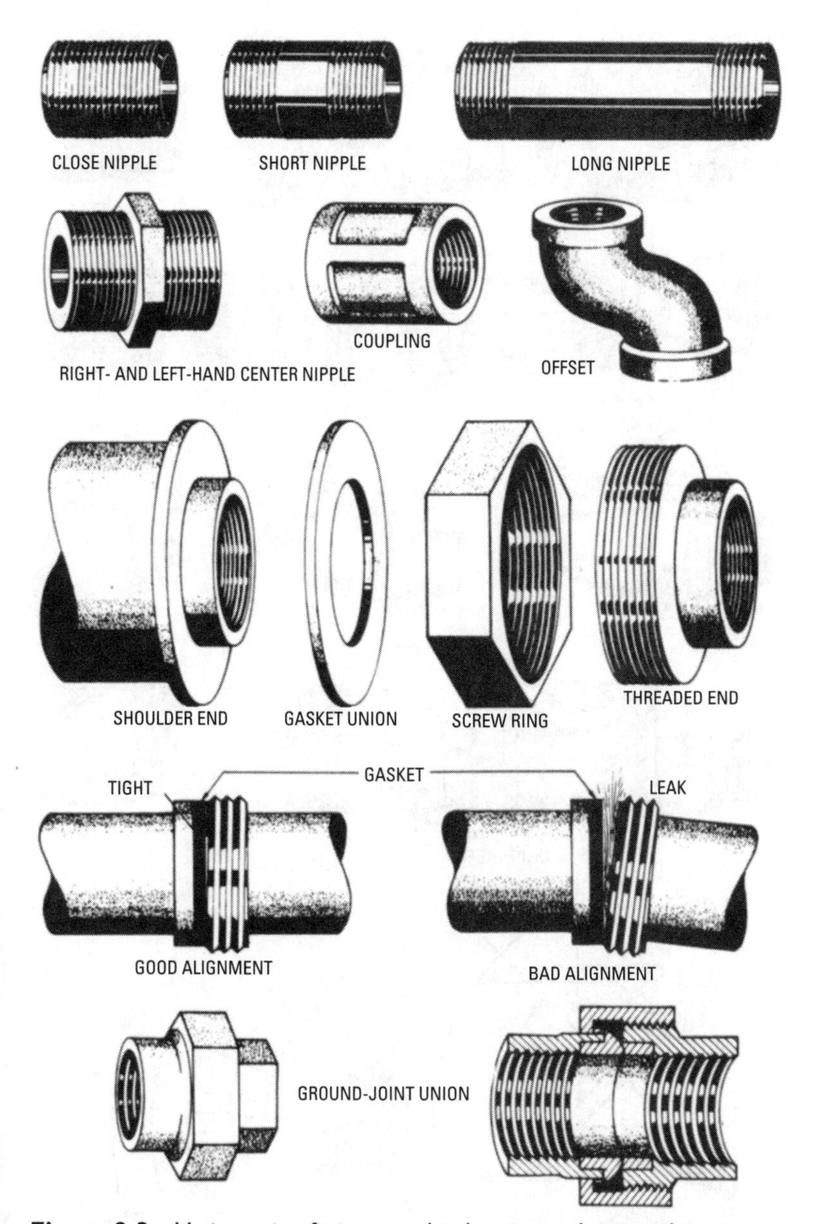

Figure 8-3 Various pipe fittings used in heating and air-conditioning installations.

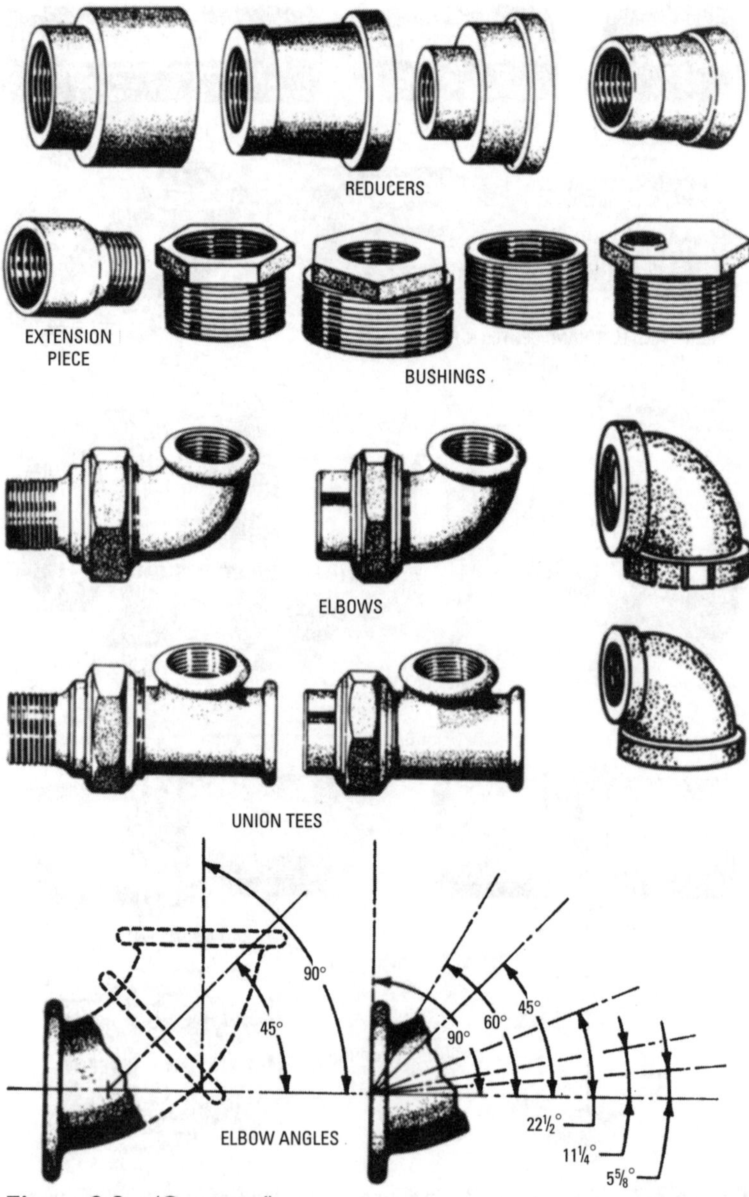

Figure 8-3 (Continued)

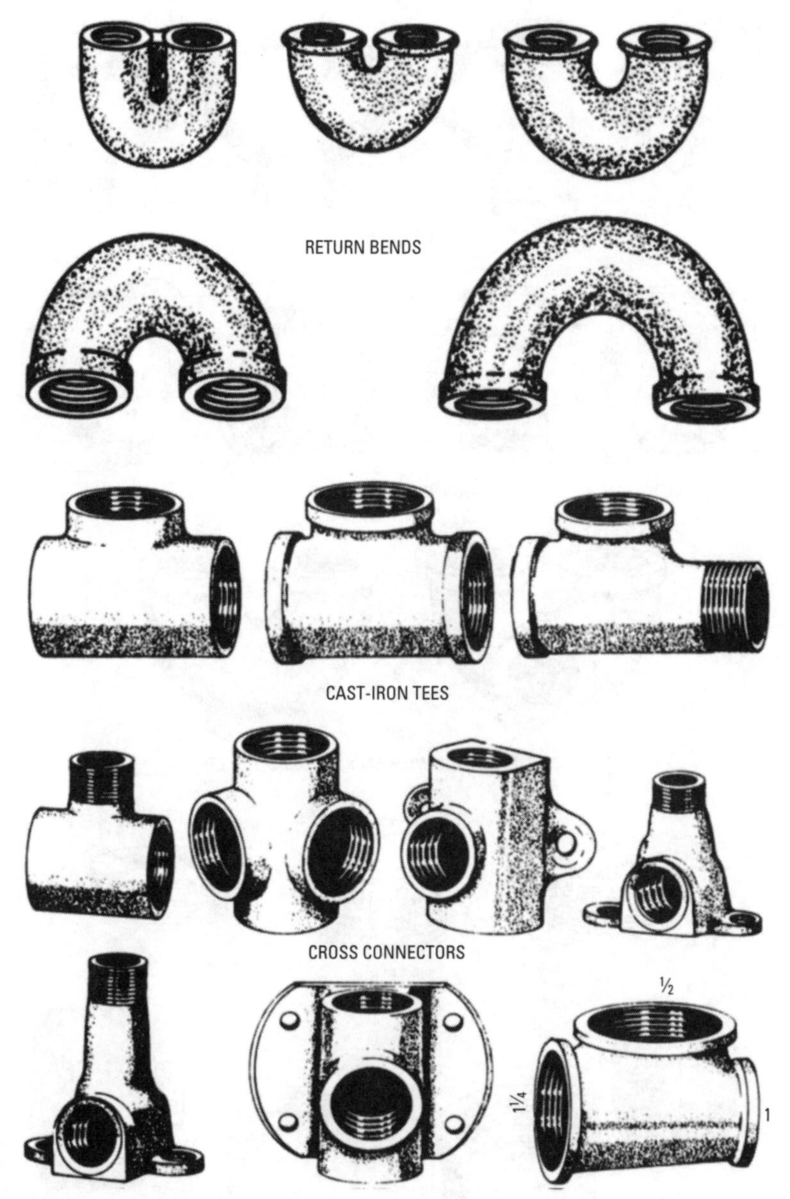

Figure 8-3 (Continued)

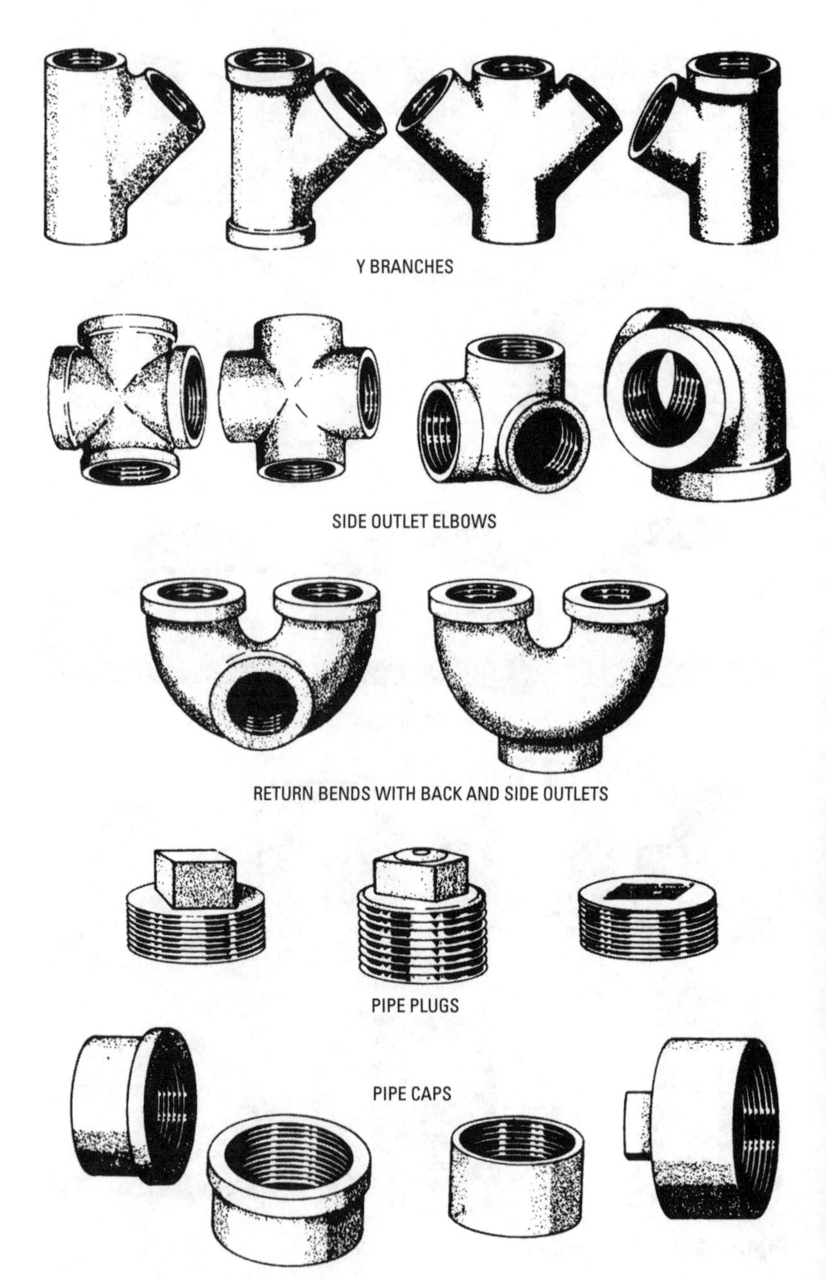

Figure 8-3 (Continued)

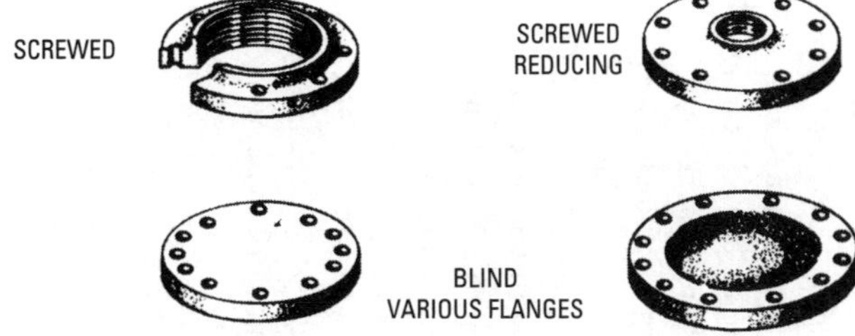

Figure 8-3 (Continued)

short nipple for each size of pipe. Long nipples are made in many lengths up to 12 inches. Anything over 12 inches is known as *cut pipe.*

Locknuts

Locknuts are commonly used on long screw nipples that have couplings (see Figure 8-3). A recessed or grooved end on the locknut fitting is used to hold packing when a particularly tight joint is required. If at all possible, a union is preferred to a locknut because a much tighter joint is obtained.

The standard length for all sizes of locknut (or tank nipples) is 6 inches. They are made from standard-weight pipe threaded (for locknut) 4 inches long on one end and with regular pipe thread on the other. Tank nipples longer than 6 inches are made-to-order only.

Couplings

A *coupling* is a pipe fitting used to couple or connect two lengths of pipe. They are available in numerous sizes and types. The *standard coupling* (see Figure 8-3) is threaded with right-hand threads. Others are available with both right-hand and left-hand threads. The *extension piece coupling* illustrated in Figure 8-3 has a male thread at one end. Couplings used as reducers (see Figure 8-3) are also common.

Offsets

An *offset fitting* (see Figure 8-3) is used when it is necessary to pass the pipeline around an obstruction that blocks its path. The new path will be parallel to the old one, but not aligned with it.

Joints

A *joint* (also referred to as an *expansion joint* or *bend*) is a pipe fitting designed to accommodate the linear expansion and contraction of the pipe metal caused by the temperature differences between the water or steam inside the pipe and the air on the outside of it. The

Table 8-6 Standard Lengths of Nipples Carried in Stock

Size (in)	Standard Black, Right Hand		
	Kind of Nipples		
1/8 to 1/2	Close, short, then by	1/2-inch lengths from 2	inches long to 6 inches long
	then by	1-inch lengths from 6	inches long to 12 inches long
3/4 and 1	Close, short, then by	1/2-inch lengths from 2 1/2	inches long to 6 inches long
	then by	1-inch lengths from 6	inches long to 12 inches long
1 1/4 to 2	Close, short, then by	1/2-inch lengths from 3	inches long to 6 inches long
	then by	1-inch lengths from 6	inches long to 12 inches long
2 1/2 and 3	Close, short, then by	1/2-inch lengths from 3 1/2	inches long to 6 inches long
	then by	1-inch lengths from 6	inches long to 12 inches long
3 1/2 and 4	Close, short, then by	1/2-inch lengths from 4 1/2	inches long to 6 inches long
	then by	1-inch lengths from 6	inches long to 12 inches long
5 and 6	Close, short, then by	1/2-inch lengths from 5	inches long to 6 inches long
	then by	1-inch lengths from 6	inches long to 12 inches long
8	Close, short, then by	1/2-inch lengths from 5 1/2	inches long to 6 inches long
	then by	1-inch lengths from 6	inches long to 12 inches long
10 and 12	Close, short, then by	1-inch lengths from 8	inches long to 12 inches long

Standard Galvanized, Right Hand
Up to and including 8-inch size, same lengths as black, right hand
Standard Black, Right and Left Hand
Up to and including 4-inch size, same lengths as black, right hand
Extra Strong Black, Right Hand
Up to and including 2-inch size, same lengths as standard, black, right hand

amount of expansion or contraction at different temperatures for a variety of metals used in steam pipes is listed in Table 8-7.

Unions

A *union* is another form of extension fitting used to join two pipes. The two most common types of unions are (1) the ground-joint union and (2) the plain or gasket union.

A *ground-joint union* (see Figure 8-3) consists of a composition ring pressing against iron or both contact surfaces of composition. A joint using a ground-joint union is characterized by spherical contact. Because no gasket is used, perfect alignment of the two pipes is not as important in making up the joint as it is when a plain or gasket union is used.

A disassembled *plain* or *gasket union* fitting is shown in Figure 8-3. It consists of three basic parts and a gasket. In assembling, the gasket (*A*) is placed over the projection on the shoulder so that it is in contact with its surface (*B*). The ring is slipped over the shoulder end and the threaded end placed in position so that the flat surface (*C*) of the threaded end presses against the gasket. The ring is then screwed firmly into the threaded end. Since the shoulder on the shoulder end cannot back off the ring, the two ends are pressed firmly together against the gasket by the ring, thus securing a tight joint.

The limitations of the plain or gasket union are also illustrated in Figure 8-3. The alignment *must* be good to secure a tight joint. In the illustration, the ring section is omitted for clearness. If both ends are in line and firmly pressed together against the gasket by the ring, the gasket will bear evenly over the entire contact surface and the joint will be tight. If the two ends are out of alignment when the ring is screwed tight, it will bring great pressure on the gasket at point *A*, and the surfaces will not come together at the opposite point *B*, thus causing a leak.

Reducing or Enlarging Fittings

Both bushings and reducers are examples of reducing or enlarging fittings. Their function in pipe installations is to connect pipes of different sizes.

Bushings and reducers may be distinguished by their construction. A reducer is a coupling device with female threads at *both* ends (see Figure 8-3). A bushing has both male and female threads.

A bushing (see Figure 8-3) is a pipe fitting designed in the form of a hollow plug, and it is used to connect the *male thread* of a pipe end to a fitting of larger size. Bushings are sold according to the pipe size of the male thread. Thus, a ¼-inch bushing (or, more specifically, a

Table 8-7 Expansion of Pipe (Increase in Inches per 100 Feet)

Temperature (°F)	Cast Iron	Wrought Iron	Steel	Brass and Copper	Temperature (°F)	Cast Iron	Wrought Iron	Steel	Brass and Copper
0	0.00	0.00	0.00	0.00	450	3.89	4.28	4.08	6.18
50	0.36	0.40	0.38	0.57	475	4.20	4.62	4.41	6.68
100	0.72	0.79	0.76	1.14	500	4.45	4.90	4.67	7.06
125	0.88	0.97	0.92	1.40	525	4.75	5.22	4.99	7.55
150	1.10	1.21	1.15	1.75	550	5.05	5.55	5.30	8.03
175	1.28	1.41	1.34	2.04	575	5.36	5.90	5.63	8.52
200	1.50	1.65	1.57	2.38	600	5.70	6.26	5.98	9.06
225	1.70	1.87	1.78	2.70	625	6.05	6.65	6.35	9.62
250	1.90	2.09	1.99	3.02	650	6.40	7.05	6.71	10.18
275	2.15	2.36	2.26	3.42	675	6.78	7.46	7.12	10.78
300	2.35	2.58	2.47	3.74	700	7.15	7.86	7.50	11.37
325	2.60	2.86	2.73	4.13	725	7.58	8.33	7.96	12.06
350	2.80	3.08	2.94	4.45	750	7.96	8.75	8.36	12.66
375	3.15	3.46	3.31	5.01	775	8.42	9.26	8.84	13.38
400	3.30	3.63	3.46	5.24	800	8.87	9.76	9.31	14.10
425	3.68	4.05	3.86	5.85					

Note: Expansion given is approximate but is correct to the best known information.
The linear expansion and contraction of pipe, due to difference of temperature of the fluid carried and the surrounding air, must be cared for by suitable expansion joints or bends.
In order to determine the amount of expansion or contraction in a pipeline, Table 8-8 shows the increase in length of a pipe 100 feet long at various temperatures.
The expansion for any length of pipe may be found by taking the difference in increased length at the minimum and maximum temperatures, dividing by 100, and multiplying by the length in feet of the line under consideration.

¼-inch×⅛-inch bushing) is one used to connect a ¼-inch fitting to a ⅛-inch pipe. This may be less confusing if you remember that a bushing has one male and one female thread and that the *female* thread of the bushing must be tightened into the *male* thread of the pipe end.

Ordinary bushings are sometimes used when a reducing fitting of proper size is unavailable. If the reduction is considerable, it may be necessary to use two bushings. A common application of eccentric bushings is to avoid water pockets on horizontal pipelines. This application is illustrated in *Eliminating Water Pockets* later in the chapter.

Directional Fittings

Directional fittings such as offsets, elbows, and return bends are used to change the direction of a pipe. Because of an overlap in function, offsets also may be considered to be a type of extension or joining fitting.

Elbows

An *elbow* (see Figure 8-3) is a pipe fitting used to change the direction of a gas, water, or steam pipeline. Elbows are available in standard angles of 45° and 90°, or special angles of 22½° and 60° (see Figure 8-3).

Return Bends

A *return bend* (see Figure 8-3) is a U-shaped pipe fitting commonly used for making up pipe coils for both water and steam heating boilers. They are commonly available in three patterns (close, medium, and open) with female threads at both ends. It is important to know the dimensions between centers when making up heating coils in order to avoid possible interference.

Branching Fittings

As the name implies, a branching fitting is used to join a branch pipe to the main pipeline. The principal branching fittings used for this purpose are as follows:

- Tees
- Crosses
- Y branches
- Elbows with side outlets
- Return bends with back or side outlets

Tees

Tees (see Figure 8-3) are made in a variety of sizes and patterns and represent the most widely used branch fitting. As the name suggests, a tee fitting is used for starting a branch pipe at a 90° angle to the main pipe.

A tee fitting is specified by first giving the run and then the branch. The *run* of a tee refers to its body with the outlets opposite each other (that is, at 180°).

Tees are available with all three outlets the same size, with the branch outlet a size different from the two outlets of the run, or with all three outlets different sizes. Whatever the size configuration of outlets, the run is always specified first. Tee specifications generally take the following form:

- 1″ (1-inch outlets on run and tee)
- 1″ × ½″ (1-inch outlets on run; ½-inch tee outlet)
- 1″ × ½″ (outlets of 1 inch and ½ inch on the run; ½-inch tee outlet)

Crosses

A *cross fitting* (see Figure 8-3) is simply a tee with two branch outlets instead of one. The branch outlets are located directly opposite one another on the main pipe so that the fitting forms the shape of a cross (hence its name). Both branch outlets are *always* the same size, regardless of the size of the outlets.

Y Branches

A *Y branch* (see Figure 8-3) is a pipe fitting with side outlets located at 45° or 60° angles to the main pipe. Y branches are available in a number of pipe sizes. They may be straight or reducing, and single- or double-branching.

Elbows with Side Outlets

An elbow with three outlets is classified as a branching rather than a directional fitting because the third outlet serves as the connection for a branch pipeline (see Figure 8-3). The branching outlet should be at a 90° angle to the plane of the elbow run.

Return Bends with Back or Side Outlets

An ordinary return bend is a U-shaped fitting with two outlets (see Figure 8-3). Some return bends are designed with three outlets, the third outlet being located on either the back or the side of the fitting and used for connecting to a branch pipeline. This type of return bend is more properly classified as a branching fitting.

Shutoff or Closing Fittings

Sometimes it is necessary to close the end of a fitting or pipe. This is accomplished with a shutoff or closing fitting, and the following two types are used for this purpose:

- Plugs
- Caps

Plugs

A *plug* (see Figure 8-3) is used to close the end of a pipe or fitting when it has a female thread. In other words, it is designed to be inserted *into* the end of the pipe or fitting. Plugs are made in a variety of sizes (⅛ inch to 12 inches), designs (hexagon, square head, or countersunk heads), and materials (for example, iron, brass).

Caps

A *cap* (see Figure 8-3) performs the same function as a plug except that it is used to close the end of a pipe or fitting that has a male thread. They are also available in a variety of sizes, designs, and materials.

Union or Makeup Fittings

Union or *makeup fittings* (see Figure 8-3) are represented by union elbows and union tees. This type of pipe fitting combines both a union and an elbow or tee in a single unit. They are available with female or both male and female threads.

Flanges

Flanges (see Figure 8-3) are pipe fittings used to close flanged pipelines or fittings. They are manufactured in the form of cast-iron discs and are available in many sizes, thicknesses, and types.

Pipe Expansion

The linear expansion and contraction of pipe due to the surrounding air must be provided for (especially in the case of long lines) by suitable expansion joints, bends, or equivalent provisions.

In order to determine the amount of expansion or contraction in a pipeline, Table 8-8 shows the increase in length of a pipe 100 feet long at various temperatures.

The expansion for any length of pipe may be found by taking the difference in increased length at the minimum and maximum temperatures, dividing by 100, and multiplying by the length in feet of the line under consideration. See Table 8-8.

Table 8-8 Expansion of Steam Pipes (Inches Increase per 100 Feet)

Temperature (°F)	*Steel*	*Wrought Iron*	*Cast Iron*	*Brass and Copper*
0	0	0	0	0
20	0.15	0.15	0.10	0.25
40	0.30	0.30	0.30	0.45
60	0.45	0.45	0.40	0.65
80	0.60	0.60	0.55	0.90
100	0.75	0.80	0.75	1.15
120	0.90	0.95	0.85	1.40
140	1.10	1.15	1.00	1.65
160	1.25	1.35	1.15	1.90
180	1.45	1.50	1.30	2.15
200	1.60	1.65	1.50	2.40
220	1.80	1.85	1.65	2.65
240	2.00	2.05	1.80	2.90
260	2.15	2.20	1.95	3.15
280	2.35	2.40	2.15	3.45
300	2.50	2.60	2.35	3.75
320	2.70	2.80	2.50	4.05
340	2.90	3.05	2.70	4.35
360	3.05	3.25	2.90	4.65
380	3.25	3.45	3.10	4.95
400	3.45	3.65	3.30	5.25
420	3.70	3.90	3.50	5.60
440	3.95	4.20	3.75	5.95
460	4.20	4.45	4.00	6.30
480	4.45	4.70	4.25	6.65
500	4.70	4.90	4.45	7.05
520	4.95	5.15	4.70	7.45
540	5.20	5.40	4.95	7.85
560	5.45	5.70	5.20	8.25
580	5.70	6.00	5.45	8.65
600	6.00	6.25	5.70	9.05
620	6.30	6.55	5.95	9.50
640	6.55	6.85	6.25	9.95
660	6.90	7.20	6.55	10.40

(continued)

Table 8-8 (continued)

Temperature (°F)	*Steel*	*Wrought Iron*	*Cast Iron*	*Brass and Copper*
680	7.20	7.50	6.85	10.95
700	7.50	7.85	7.15	11.40
720	7.80	8.20	7.45	11.90
740	8.20	8.55	7.80	12.40
760	8.55	8.90	8.15	12.95
780	8.95	9.30	8.50	13.50
800	9.30	9.75	8.90	14.10

Valves

Some authorities regard a valve as simply another type of pipe fitting and distinguish it from others by its capacity to control the flow of steam or hot water through the pipe. Be that as it may, the subject of valves is so extensive that it warrants a chapter of its own (see Chapter 9 of Volume 2, "Valves and Valve Installation").

Pipe Threads

The threads used on pipes are referred to as *pipe threads*. The distinguishing characteristic of pipe threads is that they are tapered. This results in a greater number of turns when screwing the pipe to another length of pipe or pipe fitting. When properly done, this will result in a tight, leak-free joint. Care must be taken, however, not to exceed the elastic limit or the joint will leak.

The total taper used on pipe threads in ¾ inch per foot. The total number of threads per inch will vary from 27 threads for ⅛-in pipe to 8 threads for 2½-in pipe and larger sizes. Standard pipe threads are listed in Table 8-9.

Pipe Sizing

Pipe sizing refers to the procedure of determining the projected capacities of a piping installation and selecting the pipe sizes most capable of meeting these capacities. Most methods used for determining pipe sizes are only approximate calculations, and they should be considered as such when you are using them.

Both the American Society of Heating, Refrigerating, and Air-Conditioning Engineers (ASHRAE) and the Institute of Boiler and Radiator Manufacturers (IBR) issue publications that contain considerable data for designing the piping arrangements of various steam and hot-water space heating systems. Manufacturers of proprietary heating systems, pipes, pipe fittings, and valves also provide data for sizing pipes and valves. Examples of these data are illustrated in Tables 8-10 and 8-11. Methods for pipe sizing are explained in detail with accompanying examples. These publications can be obtained by writing to these organizations. Their addresses are given elsewhere in this book. Sometimes copies are also available at a local library.

Because pipe sizing is specific to the piping arrangement and other variables within a system, no attempt is made in this chapter to cover the subject with the detail it requires. The basic principles of the methods used for sizing steam and hot-water heating pipes, along with recommendations for their application, are described in the sections that follow.

Sizing Steam Pipes

Many manufacturers of proprietary or patented steam heating systems provide their own pipe-sizing schedules. For nonproprietary systems, the projected capacities of the piping installation must be determined by a number of sizing calculations.

The principal factors used in determining pipe sizes for a given load of steam in a heating system are the following:

- Initial pressure
- Total pressure drop allowed between the boiler and the end of the return line
- Equivalent length of run from the boiler to the farthest radiator or convector
- Pressure drop per 100 feet of equivalent length

The total pressure drop should not exceed the initial gauge pressure of the system. As a general rule, it should not exceed 50 percent of the initial gauge pressure.

The equivalent length of run equals the actual measured length of pipe plus the equivalent straight pipe length of the fittings and valves. Table 8-12 lists equivalent lengths of the more common fittings and valves and is an example of the data provided by the ASHRAE for sizing pipes.

The pressure drop in pounds per square inch per 100 feet is determined by dividing 50 percent of the initial pressure by the equivalent length of the longest piping circuit.

For the sake of illustration, assume that you must calculate the pressure drop and determine the pipe size for a steam heating

Table 8-9 Standard Pipe Threads

	A	B	E	F	G	H		P	
Nominal Size of Pipe (in)	*Pitch Dia. at End of Pipe (in)*	*Pitch Dia. at Gauging Notch (in)*	*Length of Effective Thread (in)*	*Normal Engagement by Hand Between Male and Female Thread (in)*	*Outside Dia. of Pipe (in)*	*Actual Inside Dia. of Pipe (in)*	*Number of Threads per Inch*	*Pitch of Thread (in)*	*Depth of Thread (in)*
⅛	0.36351	0.37476	0.2638	0.180	0.405	0.269	27	0.0370	0.02963
¼	0.47739	0.48989	0.4018	0.200	0.540	0.364	18	0.0556	0.04444
⅜	0.61201	0.62701	0.4078	0.240	0.675	0.493	18	0.0556	0.04444
½	0.75843	0.77843	0.5337	0.320	0.840	0.622	14	0.0714	0.05714
¾	0.06768	0.98886	0.5457	0.339	1.050	0.824	14	0.0714	0.05714
1	1.21363	1.23863	0.6828	0.400	1.315	1.049	11½	0.0870	0.06954
1¼	1.55713	1.58338	0.7068	0.420	1.660	1.380	11½	0.0870	0.06954
1½	1.79609	1.82234	0.7235	0.420	1.900	1.610	11½	0.0870	0.06956
2	2.26902	2.29627	0.7565	0.436	2.375	2.067	11½	0.0870	0.06956
2½	2.71953	2.76216	1.1375	0.681	2.875	2.469	8	0.1250	0.10000

3	3.34063	3.38850	1.2000	0.766	3.500	3.068	8	0.1250	0.10000
3½	3.83750	3.88881	1.2500	0.821	4.000	3.548	8	0.1250	0.10000
4	4.33438	4.38713	1.3000	0.844	4.500	4.026	8	0.1250	0.10000
4½	4.83125	4.88594	1.3500	0.875	5.000	4.506	8	0.1250	0.10000
5	5.39073	5.44929	1.4063	0.937	5.563	5.047	8	0.1250	0.10000
6	6.44609	6.50597	1.5125	0.958	6.625	6.055	8	0.1250	0.10000
7	7.43984	7.50234	1.6125	1.000	7.625	7.023	8	0.1250	0.10000
8	8.43359	8.50003	1.7125	1.063	8.625	7.981	8	0.1250	0.10000
9	9.42734	9.49797	1.8125	1.130	9.625	8.941	8	0.1250	0.10000
10	10.54531	10.62094	1.9250	1.210	10.750	10.020	8	0.1250	0.10000
12	12.53281	12.61781	2.1250	1.360	12.750	12.000	8	0.1250	0.10000
14 O.D.	13.77500	13.87262	2.250	1.562	14.000	—	8	0.1250	0.10000
15 O.D.	14.76875	14.87419	2.350	1.687	15.000	—	8	0.1250	0.10000
16 O.D.	15.76250	15.87575	2.450	1.812	16.000	—	8	0.1250	0.10000
18 O.D.	17.75000	17.87500	2.650	2.000	18.000	—	8	0.1250	0.10000
20 O.D.	19.73750	19.87031	2.850	2.125	20.000	—	8	0.1250	0.10000
22 O.D.	21.72500	21.86562	3.050	2.250	22.000	—	8	0.1250	0.10000
24 O.D.	23.71250	23.86094	3.250	2.375	24.000	—	8	0.1250	0.10000

Data abstracted from the American Standard for Pipe Threads A.S.A.-B2—1919.

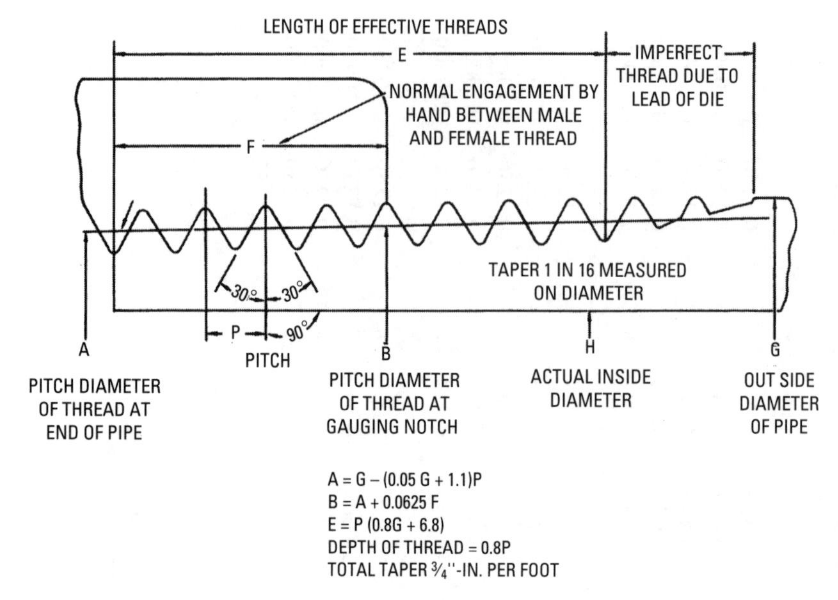

Illustration for Table 8-9.

system in which the initial pressure is 2 psig. In order to do this, the following steps are necessary:

1. Measure the length of the longest run of pipe. For our problem, use the figure 500 feet.
2. Determine the *assumed* equivalent length. This will not exceed twice the measured length (that is, 1000 feet).
3. Determine the pressure drop for the equivalent length of pipe (50 percent of the initial pressure, or 1 psig).
4. Determine the pressure drop per 100 feet of the equivalent length of pipe:

$$\text{Pressure drop per 100 ft. of equivalent length} = \frac{1 \text{ psig}}{1000 \text{ ft.}} = 0.1 \text{ psig per 100 ft.}$$

5. Size pipes for a desired capacity based on a pressure drop of 0.1 psig per 100 feet. These can be determined from ASHRAE charts and tables.

Table 8-10 Relative Discharging Capacities of Standard Pipe

Pipe Size (in)	Internal Diameter D (in)	$D^{5/2}$	Pipe Size in Inches 1/8	1/4	3/8	1/2	3/4	1	1 1/4	1 1/2	2	2 1/2	3	3 1/2	4	5	6	8	10	12
1/8	0.269	0.037530	1.0	—	—	—	—	—	—	—	—	—	—	—	—	—	—	—	—	—
1/4	0.364	0.079938	2.1	1.0	—	—	—	—	—	—	—	—	—	—	—	—	—	—	—	—
3/8	0.493	0.17065	4.5	2.1	1.0	—	—	—	—	—	—	—	—	—	—	—	—	—	—	—
1/2	0.622	0.30512	8.1	3.8	1.8	1.0	—	—	—	—	—	—	—	—	—	—	—	—	—	—
3/4	0.824	0.61634	16	7.7	3.6	2.0	1.0	—	—	—	—	—	—	—	—	—	—	—	—	—
1	1.049	1.1270	30	14	6.6	3.7	3.7	1.0	—	—	—	—	—	—	—	—	—	—	—	—
1 1/4	1.380	2.2372	60	28	13	7.3	3.6	2.0	1.0	—	—	—	—	—	—	—	—	—	—	—
1 1/2	1.610	3.2890	88	41	19	11	5.3	2.9	1.5	1.0	—	—	—	—	—	—	—	—	—	—
2	2.067	6.1426	164	77	36	20	10	5.5	2.7	1.9	1.0	—	—	—	—	—	—	—	—	—
2 1/2	2.469	9.5786	255	120	56	31	16	8.5	4.3	2.9	1.6	1.0	—	—	—	—	—	—	—	—
3	3.068	16.487	439	206	97	54	27	15	7.4	5.0	2.7	1.7	1.0	—	—	—	—	—	—	—
3 1/2	3.548	23.711	632	297	139	78	38	21	11	7.2	3.9	2.5	1.4	1.0	—	—	—	—	—	—
4	4.026	32.523	867	407	191	107	53	29	15	9.9	5.3	3.4	2.0	1.4	1.0	—	—	—	—	—
5	5.047	57.225	1526	716	335	188	93	51	26	17	9.3	6.0	3.5	2.4	1.8	1.0	—	—	—	—
6	6.065	90.589	2414	1163	531	297	147	80	40	28	15	9.5	5.5	3.8	2.8	1.6	1.0	—	—	—
8	7.981	179.95	4795	2251	1054	590	292	160	80	55	29	19	11	7.6	5.5	3.1	2.0	1.0	—	—
10	10.020	317.81	8468	3976	1862	1042	516	282	142	97	52	33	19	13	9.8	5.6	3.5	1.8	1.0	—
12	12.000	498.83	13292	6240	2923	1635	809	443	223	152	81	52	30	21	15	8.7	5.5	2.8	1.6	1.0

The figure that lies at the intersection of any two sizes is the number of smaller-size pipes required to equal one of the larger.
Example: How many 2-inch standard pipes will it take to equal the discharge of one 8-inch standard pipe?
Solution: Twenty-nine 2-inch pipes; the figure in the table that lies at the intersection of these two sizes is 29.

Table 8-11 Diagram Showing Resistance of Valves and Fittings of the Flow of Liquids

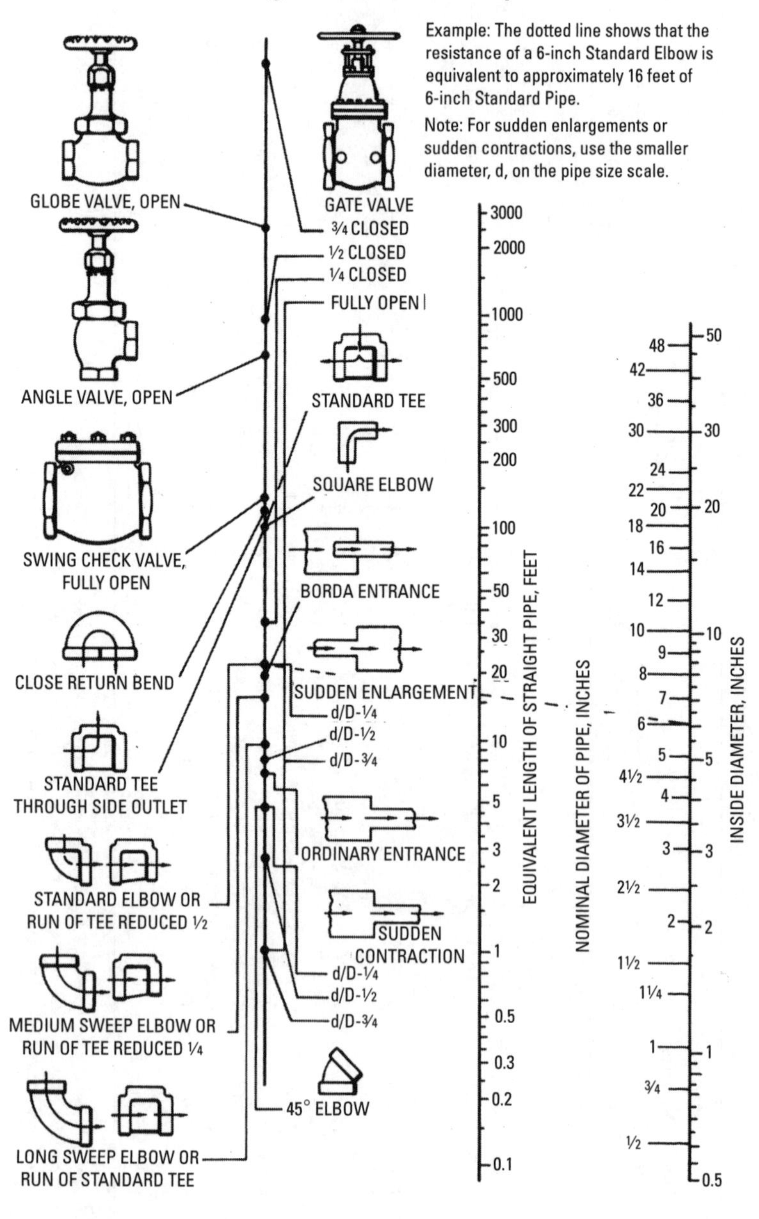

Table 8-12 Length of Pipe in Feet for Fittings to Be Added to Actual Length of Run in Order to Obtain Equivalent Length

Size of Pipe (in)	Length in Feet to Be Added to Run				
	Standard Elbow	Side Outlet Tee	Gate Valve*	Globe Valve*	Valve*
½	1.3	3	0.3	14	7
¾	1.8	4	0.4	18	10
1	2.2	5	0.5	23	12
1¼	3.0	6	0.6	29	15
1½	3.5	7	0.8	34	18
2	4.3	8	1.0	46	22
2½	5.0	11	1.1	54	27
3	6.5	13	1.4	66	34
3½	8	15	1.6	80	40
4	9	18	1.9	92	45
5	11	22	2.2	112	56
6	13	27	2.8	136	67
8	17	35	3.7	180	92
10	21	45	4.6	230	112
12	27	53	5.5	270	132
14	30	63	6.4	310	152

Valve in full open position.

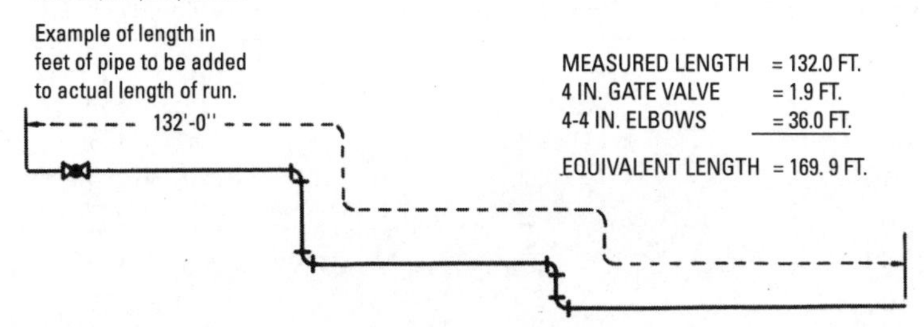

6. Check the pressure drop by calculating the equivalent length of run of the longest circuit from the pipe sizes determined. The pipe size determined in step 5 will be correct if the calculated pressure drop is less than the assumed pressure drop.

A steam supply main should not pitch less than ¼ inch per 10 feet of run, and its diameter should not be smaller than 2 inches. In gravity one-pipe systems, the diameter of the supply main at the farthest point should not be smaller than 50 percent of its largest diameter.

A rule-of-thumb method for determining the size of steam mains is to take the total amount of direct radiation and add to it 25 percent of the total for the piping allowance. Next, find the square root of this total and divide by 10. The result will be the size steam main to use for a one-pipe system. For a two-pipe system, one size smaller is generally sufficient for the supply main, and the return can be one or two sizes smaller than the supply main. A steam main should not decrease in size according to the area of its branches but much more gradually.

The aforementioned method for sizing steam mains can be illustrated by using a structure with an assumed direct radiation of 500 square feet. Adding 25 percent for piping allowance gives 625. The square root of 625 is 25, which divided by 10 gives 2½, or the size of the steam main (2½ inches). For reference and practical use, refer to Table 8-13 when making calculations.

The pitch of runouts to risers and radiators should be *at least* ½ inch per foot toward the main. Runouts over 8 feet in length but with *less* than ½-inch pitch per foot should be one size larger than specified in the pipe-sizing tables.

Table 8-13 Size of Steam Mains

Radiation (ft²)	*One-Pipe Work (in)*	*Two-Pipe Work (in)*
125	1½	1¼ × 1
250	2½	1½ × 1¼
400	3	2 × 1½
650	3½	2½ × 2
900	4	3 × 2½
1250	4½	3½ × 3
1600	5	4 × 3½
2050	5½	4½ × 4
2500	6	5 × 4½
3600	7	6 × 5
5000	8	7 × 6
6500	9	8 × 6
8100	10	9 × 6

Sizing Hot-Water (Hydronic) Pipes/Tubing

Simplified pipe-sizing tables for hot-water heating systems are also available from organizations such as the ASHRAE.

Pipe sizing hot-water lines is similar in some respects to the calculations used in duct sizing because the pipe sizes are selected on the basis of the quantity and rate of water flow (expressed in gallons per minute, or gpm) and the constant friction loss. This friction loss (or drop) is expressed in thousandths-of-an-inch per foot of pipe length.

The velocity of water in the smaller residential pipes should not exceed 4 fps (feet per second), or there will be a noise problem.

In *forced* hot-water heating systems, the problem of friction drop in the pipes can be overcome by the pump or circulator. In this respect, a forced hot-water heating system is much easier to size than a steam heating system.

The rule-of-thumb method used to size steam mains (see previous section) can also be used to determine the approximate sizes of hot-water mains; however, certain important differences should be noted.

When sizing hot-water mains, the mains may be reduced in size in proportion to the branches taken off. They should, however, have as large an area as the sum of all branches beyond this point. It is advisable that the horizontal branches be one size larger than the risers. Returns should be the same size as the supply mains. Table 8-14 lists sizes of hot-water mains and the equivalent radiation ranges in square feet. Sizes for mains and branches are given in Table 8-15.

Table 8-14 Sizes of Hot-Water Mains

Radiation (ft^2)	*Pipe (in)*
75 to 125	1¼
125 to 175	1½
175 to 300	2
300 to 475	2½
475 to 700	3
700 to 950	3½
950 to 1200	4
1200 to 1575	4½
1575 to 1975	5
1975 to 2375	5½
2375 to 2850	6

Table 8-15 Table of Mains and Branches

Main	*Branch*
1 inch will supply	2, ¾ inch
1¼ inch will supply	2, 1 inch
1½ inch will supply	2, 1¼ inches
2 inch will supply	2, 1½ inches
2½ inch will supply	2, 1½ inches and 1, 1¼ inches, or 1, 2 inches and 1, 1¼ inches
3 inch will supply	1, 2½ inches and 1, 2 inches, or 2, 2 inches and 1, 1½ inches
3½ inch will supply	2, 2½ inches or 1, 3 inches, and 1, 2 inches or 3, 2 inches
4 inch will supply	1, 3½ inches and 1, 2½ inches, or 2, 3 inches and 4, 2 inches
4½ inch will supply	1, 3½ inches and 1, 3 inches, or 1, 4 inches and 1, 2½ inches
5 inch will supply	1, 4 inches and 1, 3 inches, or 1, 4½ inches and 1, 2½ inches
6 inch will supply	2, 4 inches and 1, 3 inches, or 4, 3 inches or 10, 2 inches
7 inch will supply	1, 6 inches and 1, 4 inches, or 3, 4 inches and 1, 2 inches
8 inch will supply	2, 6 inches and 1, 5 inches, or 5, 4 inches and 2, 2 inches

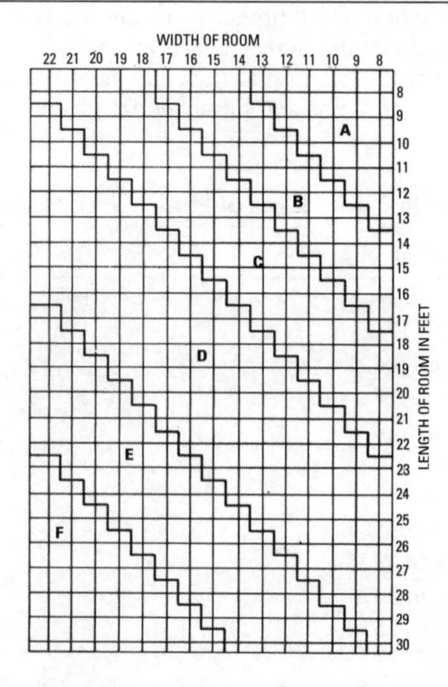

Table 8-16 General Dimensions of Straight Sizes

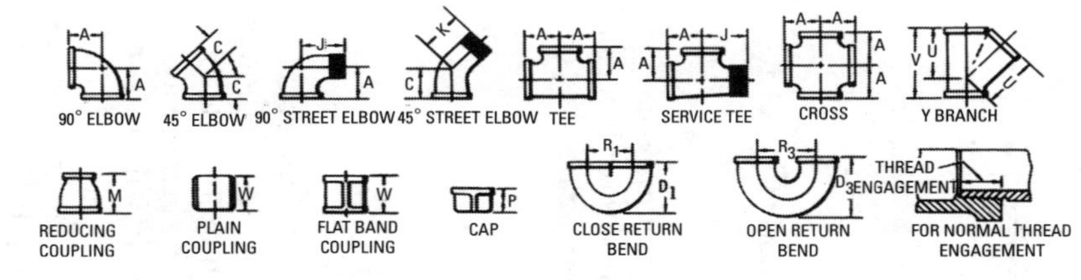

General dimentions of Crane standard malleable iron screwed fittings. The reference letters refer to the accompanying table.

Size (in)	*Dimensions (in)*												
1/8	11/16	3/4	—	—	1	7/8	—	11/16	—	—	—	—	31/32
1/4	13/16	3/4	—	—	1 3/16	15/16	1 1/4	11/16	—	—	—	—	1 1/16
3/8	15/16	13/16	—	—	1 7/16	1 1/32	1 1/4	11/16	—	—	1 23/32	2 15/32	1 5/32
1/2	1 1/8	7/8	1 3/4	1 7/8	1 5/8	1 5/32	1 3/8	25/32	1 1/8	1 1/2	1 23/32	2 15/32	1 11/32
3/4	1 5/16	1	2 1/32	2 5/16	1 7/8	1 5/16	1 7/16	1 1/32	1 3/8	2	2 1/16	2 7/8	1 1/2
1	1 1/2	1 1/8	2 5/16	2 25/32	2 1/8	1 15/32	1 11/16	1 7/32	1 3/4	2 1/2	2 7/16	3 3/8	1 11/16
1 1/4	1 3/4	1 5/16	2 13/16	3 5/16	2 7/16	1 23/32	2 1/16	1 9/32	2 1/8	3	2 15/16	4 3/32	1 15/16
1 1/2	1 15/16	1 7/16	3 3/16	3 11/16	2 11/16	1 7/8	2 5/16	1 11/32	2 1/2	3 1/2	3 9/32	4 17/32	2 5/32
2	2 1/4	1 11/16	3 7/8	4 1/4	3 1/4	7 7/32	2 5/8	1 7/16	2 3/4	4	4 1/32	5 17/32	2 17/32
2 1/2	2 11/16	1 15/16	—	5 5/32	4	—	3 1/16	2	—	4 1/2	4 3/4	6 1/2	2 7/8
3	3 1/16	2 3/16	—	5 13/16	4 11/16	—	3 9/16	2 1/8	—	5	5 11/16	7 3/4	3 1/2
3 1/2	3 27/32	2 19/32	—	—	—	—	4	2 3/16	—	—	—	—	—
4	3 13/16	2 5/8	—	6 15/16	5 11/16	—	4 3/8	2 5/16	—	6	6 15/16	9	4 5/16
5	5 1/2	3 1/16	—	6 15/16	—	—	5 1/8	2 17/32	—	—	—	—	—
6	5 1/8	3 15/32	—	—	—	—	5 7/8	2 11/16	—	—	—	—	—

General dimensions of Crane standard malleable iron screwed fittings.

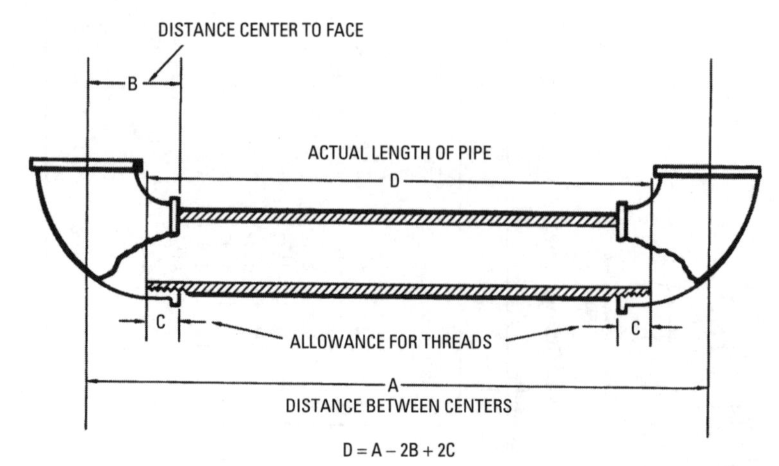

Figure 8-4 Diagram showing how to obtain the actual length of a pipe connecting two fittings.

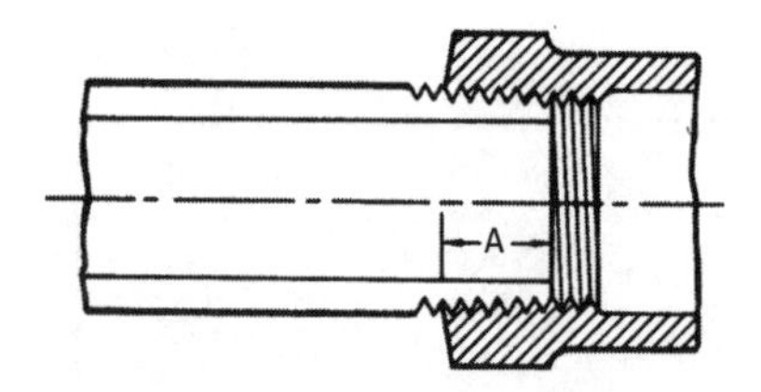

Figure 8-5 Joint made up showing length of thread on pipe (A) screwed into fitting.

Pipe Fitting Measurements

Pipe fitting may be done either by making close visual judgments or entirely by measurements scaled on a drawing. The first method is a hit-or-miss process and requires an experienced fitter to do a good job; the second method is one of precision and is the better way. In actual practice, a combination of the two methods will, in some cases, save time and give satisfactory results.

Working from a drawing with all the necessary dimensions has certain advantages. Especially in the case of a big job, all the pipe may be cut and threaded in the shop so that at the place of installation the only work to be done is assembling.

In making a drawing, the measurements are based on the distances between the centers of fittings. The data necessary to locate these centers are given in a more general dimension drawing with an accompanying table such as the one illustrated in Table 8-16. These dimension drawings and tables should be the ones corresponding to the make of fittings used; otherwise, there might be the possibility of slight variation. In general, however, the different makes are pretty well standardized.

Figure 8-4 illustrates how the *actual* length of pipe connecting the two fittings is obtained. The actual length of pipe is equal to the distance between centers (of fittings) *minus* twice the distance from the center of the face of the fittings *plus* twice the allowance for threads. This is expressed by the following equation:

$$D = A - 2B + 2C$$

The allowance for the length of thread that is screwed into the fitting (dimension C in Figure 8-4) is obtained from a table furnished by the manufacturer. This allowance (called *A* in Figure 8-5) corresponds to the values given in the accompanying table (see Table 8-17). Note that the dimension *A* in Figure 8-5 is the same as dimension *C* in Figure 8-4.

A working drawing showing centerlines and distances between centers of an installation is shown in Figure 8-6. This gives all the information *except* the actual length of the pipes.

Problem

Find the length of the pipe connecting the 2-inch elbows in Figure 8-7. Using Figure 8-4 as a guide, the following dimensions are provided: $A = 12$ ft, $B = 2\frac{1}{4}$ in, and $C = \frac{11}{16}$ in. The problem is solved as follows:

1. $D = A–2B + 2C$
2. $D = 12$ ft$–2 \times 2\frac{1}{4}$ in $+ 2 \times \frac{11}{16}$ in
3. $D = 12$ ft$–4\frac{1}{2}$ in $+ \frac{22}{32}$ (or $1\frac{3}{8}$) in
4. $D = 11$ ft $7\frac{1}{2}$ in $+ 1\frac{3}{8}$ in
5. $D = 11$ ft $8\frac{7}{8}$ in

Calculating Offsets

In pipe fitting, an *offset* is a change of direction (other than 90°) in a pipe bringing one part out of (but parallel with) the line of another.

An example of an offset is illustrated in Figure 8-8. As shown here, the problem is an obstruction (*E*), such as a wall, blocking the path of a pipeline (*L*). It is necessary to change the position of pipeline *L* at point *A* to some parallel position such as line *F* in order to move around the obstruction. When two lines such as *L* and *F* are to be piped with elbows other than 90° elbows, the pipe fitter is confronted with the following two problems: (1) finding the length of pipe *H* and (2) determining the distance *BC*. By determining the distance *BC*, the pipe fitter will be able to fix point *A* so that the two elbows *A* and *C* will be in alignment.

Table 8-17 Length of Thread on Pipe

Size (in)	Dimension A (in)	Size (in)	Dimension A (in)	Size (in)	Dimension A (in)	Size (in)	Dimension A (in)
1/8	1/4	1	9/16	3	1	6	1 1/4
1/4	3/8	1 1/4	5/8	3 1/2	1 1/16	7	1 1/4
3/8	3/8	1 1/2	5/8	4	1 1/16	8	1 5/16
—	—	—	—	—	—	9	1 3/8
1/2	1/2	2	11/16	4 1/2	1 1/8	10	1 1/2
3/4	1/2	2 1/2	15/16	5	1 3/16	12	1 5/8

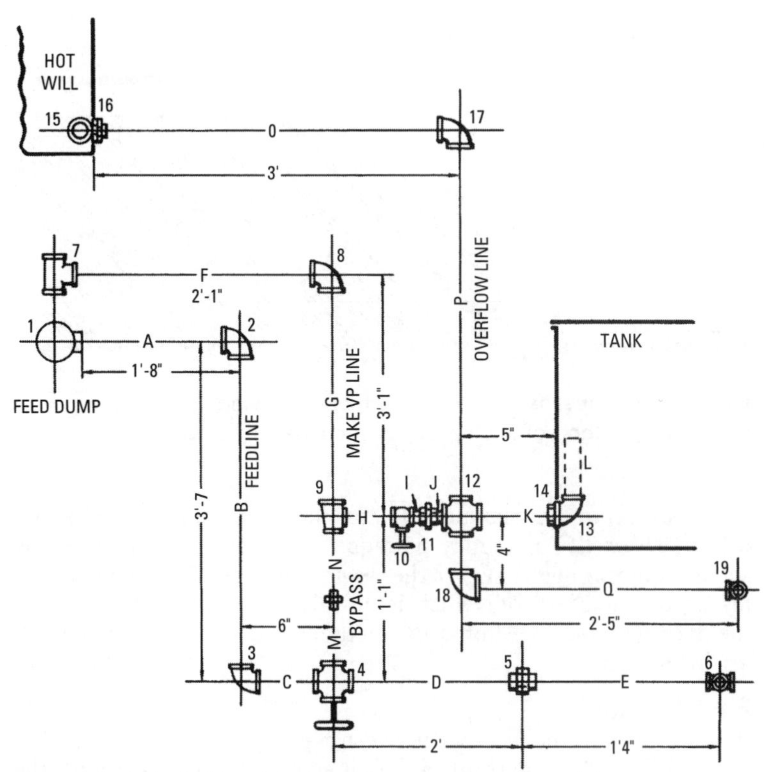

Figure 8-6 Working drawing showing centerline and distances between centers of an installation.

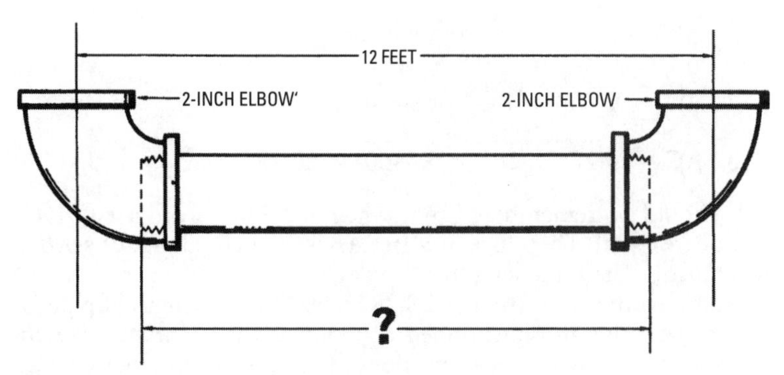

Figure 8-7 Pipe connecting two elbows.

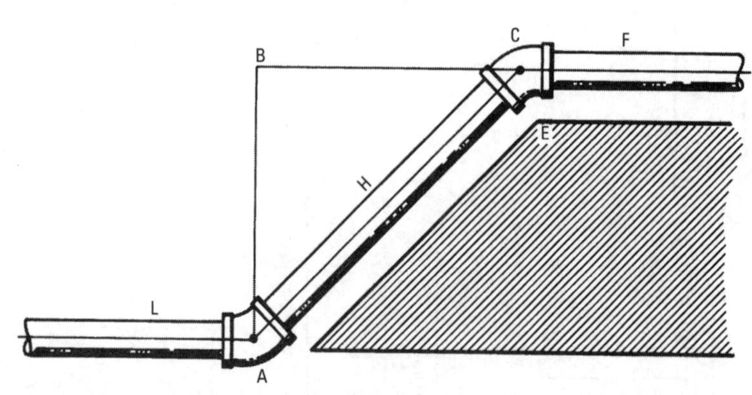

Figure 8-8 Pipeline connected with two 45° elbows illustrating offsets and method of finding length of connecting pipe H.

Of course, in the triangle *ABC*, the length of pipe *AC* and either offset (*AB* or *BC*) that may be required are quickly calculated by solving the triangle *ABC* for the desired member, but this involves taking the square root, which is not always easily understood by the average worker. Alternative methods are suggested in the following sections.

First Method

If, in Figure 8-8, the distance between pipelines *L* and *F* is 20 inches (offset *AB*), what length of pipe *H* is required to connect with the 45° elbows *A* and *C*?

Based on the triangle *ABC*, the following equation is offered for solving this problem:

1. $\overline{AC}^2 = \overline{AB}^2 + \overline{BC}^2$ from which:
2. $AC = \sqrt{\overline{AB}^2 = \overline{BC}^2}$ or substituting:
3. $AC = \sqrt{20^2 + 20^2} = \sqrt{800} = 28.28$ inches

It should be remembered that when 45° elbows are used, *both* offsets are equal. Therefore, if offset *AB* is 20 inches long, offset *BC* also must be the same length.

Note that the value (that is, 28.28 inches) for the length of pipe *H* obtained by the aforementioned equation is the *calculated length* and does not allow for the projections of the elbows. In other words, pipe *H* (as calculated by this equation) is too long and must be shortened so that the elbows will fit.

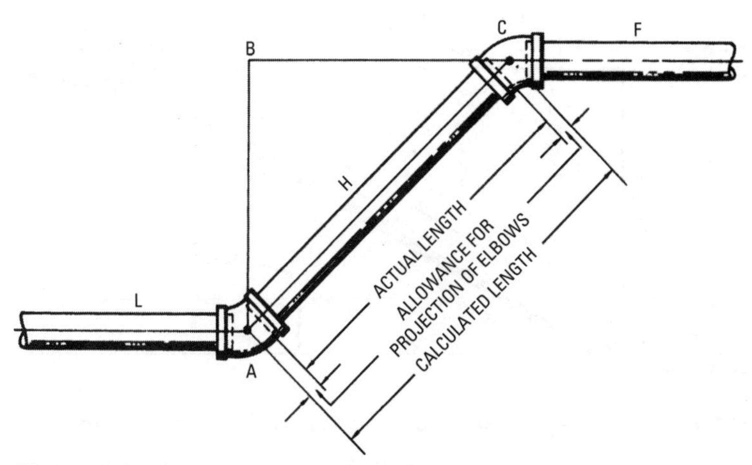

Figure 8-9 Second method of calculating pipe length.

Figure 8-8 illustrates the difference between the calculated length (that is, the measurement from point *A* to point *C*) and the actual length of connecting pipe *H* when used with elbows other than 90°. *Actual length* is obtained by deducting the *allowance for projection of the elbows* from the *calculated length*.

Second Method

Another method of calculating the connecting pipe length (that is, the length of pipe *H* in Figure 8-9) is by multiplying the offset by 53⁄128 inch and adding the product to the original offset figure. Thus, if offset *AB* is 20 inches, the following calculations are possible:

1. $20 \times \frac{53}{128} = \frac{1060}{128} = 8\frac{9}{32}$
2. $20 + 8\frac{9}{32} = 28\frac{9}{32}$ in.

Third Method

The pipe fitter will often encounter elbows of angles other than 45°. For these, the distance between elbow centers (points *A* and *C*) can easily be calculated with the following procedure:

1. Determine the angle of the elbow.
2. Determine the elbow constant equivalent to its angle.
3. Multiply the elbow constant by the known offset.

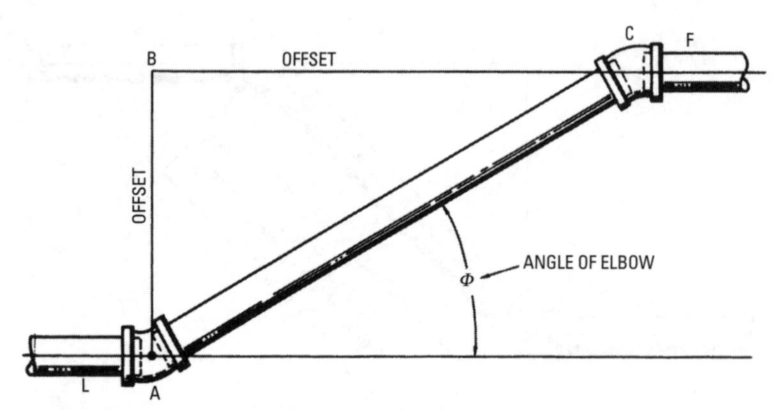

Figure 8-10 Diagram for elbow constant as given in Table 8-18.

In Figure 8-10, even though only offset *AB* is known, it is possible to determine the length of the other offset (*BC*) and the distance between elbow centers (*AC*). In order to do this, the following equations must be used:

$$AC = \text{offset } AB \times \text{constant for } AC$$
$$BC = \text{offset } AB \times \text{constant for } AB$$

Assume that the distance between pipelines *L* and *F* (offset *AB*) in Figure 8-10 is 20 inches and the angle of the elbow is 22½°. In Table 8-18 you will find that for a 22½° elbow, the elbow constant for *AB* is 2.41. Substituting values in the second equation, you have the following:

BC = 20 × 2.61 = 48.2 inches

The constant for the elbow centers of 22½° elbows is 2.61. Substituting values in the first equation gives the following results:

AC = 20 × 2.61 = 52.2 inches

Table 8-18 Elbow Constants

Angle of Elbow	*Elbow Centers AC*	*Offset AB*
60°	1.15	0.58
45°	1.41	1.00
30°	2.00	1.73
22½°	2.61	2.41
11¼°	5.12	5.02
5⅜°	10.20	10.15

Fourth Method

Offsets may also be calculated by using basic trigonometry. Using the example given in Figure 8-10, determine the length of the offset *AB* if *AC* is 8 feet and the angle $\phi = 60°$. From Table 8-19, sine 60° = 0.866:

$$\text{Length of offset AB} = 0.866 \times 8 = 6.93$$

Pipe Supports

If piping is to be run along the wall or ceiling, it should be attached to the surface with pipe supports (for example, hangers, straps, clamps). The type of pipe supports used and their spacing will be regulated in accordance with approved local standards.

Pipe straps (perforated metal straps) are used to support small size pipes (see Figure 8-11). Larger pipes require various types of hangers (for example, rod, spring), chains, or other devices capable of supporting the heavier weight.

Vertical pipe is best supported with a shoulder clamp attached to the flooring at the point through which the pipe passes, or by clamps attached to adjacent walls or columns.

Pipe hangers and anchors are also used for supporting suspended piping or securing it (as in the case of anchors) to adjacent surfaces. Hangers are similar in appearance and function to pipe straps (see previous).

Joint Compound

Joint compound (also referred to as *pipe dope*) is a substance applied to the male thread when making up screwed joints. The purpose of applying a joint compound is to lubricate the threads so that tightening is made easier. By lubricating the threads, the friction and heat produced by the tightening operation are greatly

Figure 8-11 Pipe strap.

Table 8-19 Natural Trigonometrical Functions

Deg	Sin	Cos	Tan	Sec	Deg	Sin	Cos	Tan	Sec
0	0.00000	1.0000	0.00000	1.0000	46	0.7193	0.6947	1.0355	1.4395
1	0.01745	0.9998	0.01745	1.0001	47	0.7314	0.6820	1.0724	1.4663
2	0.03490	0.9994	0.03492	1.0006	48	0.7431	0.6691	1.1106	1.4945
3	0.05234	0.9986	0.05241	1.0014	49	0.7547	0.6561	1.1504	1.5242
4	0.06976	0.9976	0.06993	1.0024	50	0.7660	0.6428	1.1918	1.5557
5	0.08716	0.9962	0.08749	1.0038	51	0.7771	0.6293	1.2349	1.5890
6	0.10453	0.9945	0.10510	1.0055	52	0.7880	0.6157	1.2799	1.6243
7	0.12187	0.9925	0.12278	1.0075	53	0.7986	0.6018	1.3270	1.6618
8	0.1392	0.9903	0.1405	1.0098	54	0.8090	0.5878	1.3764	1.7013
9	0.1564	0.9877	0.1584	1.0125	55	0.8192	0.5736	1.4281	1.7434
10	0.1736	0.9848	0.1763	1.0154	56	0.8290	0.5592	1.4826	1.7883
11	0.1908	0.9816	0.1944	1.0187	57	0.8387	0.5446	1.5399	1.8361
12	0.2079	0.9781	0.2126	1.0223	58	0.8480	0.5299	1.6003	1.8871
13	0.2250	0.9744	0.2309	1.0263	59	0.8572	0.5150	1.6643	1.9416
14	0.2419	0.9703	0.2493	1.0300	60	0.8660	0.5000	1.7321	2.0000
15	0.2588	0.9659	0.2679	1.0353	61	0.8746	0.4848	1.8040	2.0627
16	0.2756	0.9613	0.2867	1.0403	62	0.8820	0.4695	1.8807	2.1300
17	0.2924	0.9563	0.3057	1.0457	63	0.8910	0.4540	1.9626	2.2027
18	0.3090	0.9511	0.3249	1.0515	64	0.8988	0.4384	2.0503	2.2812
19	0.3256	0.0455	0.3443	1.0576	65	0.9063	0.4226	2.1445	2.3662
20	0.3420	0.9397	0.3640	1.0642	66	0.9135	0.4067	2.2460	2.4586
21	0.3584	0.9336	0.3839	1.0711	67	0.9205	0.3907	2.3559	2.5593

22	0.3746	0.0272	0.4040	1.0785	68	0.9272	0.3746	2.4751	2.6695
23	0.3907	0.9205	0.4245	1.0664	69	0.9330	0.3586	2.0051	2.7904
24	0.4087	0.9135	0.4452	1.0846	70	0.9397	0.3420	2.7475	2.9238
25	0.4220	0.9063	0.4663	1.1034	71	0.9455	0.3256	2.9042	3.0715
26	0.4386	0.8938	0.4877	1.1126	72	0.9511	0.3090	3.0777	3.2361
27	0.4540	0.8910	0.5095	1.1223	73	0.9563	0.2024	3.2709	3.4203
28	0.4695	0.8829	0.5317	1.1326	74	0.9613	0.2756	3.4874	3.6279
29	0.4848	0.8746	0.5543	1.1433	75	0.9650	0.2588	3.7321	3.8637
30	0.5000	0.8660	0.5774	1.1547	76	0.9703	0.2419	4.0108	4.1336
31	0.5150	0.8572	0.6009	1.1666	77	0.9744	0.2250	4.3315	4.4454
32	0.5200	0.8480	0.6249	1.1792	78	0.9781	0.2079	4.7046	4.9007
33	0.5446	0.8387	0.6494	1.1924	79	0.9816	0.1908	5.1446	5.2406
34	0.5592	0.8290	0.6745	1.2062	80	0.9848	0.1736	5.6713	5.7588
35	0.5736	0.8192	0.7002	1.2208	81	0.9877	0.1564	6.3128	6.3924
36	0.5878	0.8090	0.7265	1.2361	82	0.9903	0.1392	7.1154	7.1853
37	0.6018	0.7936	0.7536	1.2521	83	0.9925	0.12187	8.1443	8.2055
38	0.6157	0.7880	0.7613	1.2690	84	0.9945	0.10453	9.5668	—
39	0.6293	0.7771	0.8098	1.2867	85	0.9962	0.08716	11.4301	11.474
40	0.6428	0.7660	0.8391	1.3054	86	0.9976	0.06976	14.3007	14.335
41	0.6561	0.7547	0.8693	1.3230	87	0.9986	0.05384	10.0811	19.107
42	0.6691	0.7431	0.9004	1.3456	88	0.9954	0.03490	28.6363	28.654
43	0.6820	0.7314	0.9325	1.3673	89	0.9903	0.01745	57.2900	57.299
44	0.6947	0.7193	0.9657	1.3902	90	1.0000	Inf.	Inf.	Inf.
45	0.7071	0.7071	1.0000	1.4142		—	—	—	—

reduced. Moreover, the joint compound forms a seal inside the screwed joint, which prevents leakage and ensures a tight joint.

Joint compounds are commercially available, or they may be made on the job from a variety of different materials. Red lead, white lead, or graphite have frequently been used as a joint compound. Red lead produces a very tight joint, but it hardens to such an extent that it is difficult to unscrew the joint for repairs.

A tape material has also been developed for use in making up screwed joints. It functions in the same manner as a joint compound. The tape is made of Teflon and is so thin that it will sink into the threads when wrapped around them.

An old toothbrush is an excellent tool for applying joint compound to the thread. It is important to remember that the joint compound *must* be applied to the *male* thread only. If it is applied to the female thread, some of it will be forced into the pipe where it will lodge as a contaminating substance.

Pipe Fitting Wrenches

The numerous wrenches used in pipe fitting may be listed as follows:

- Monkey wrench
- Pipe wrench
  - Stillson wrench
  - Chain wrench
  - Strap wrench
- Open wrench

The important point to remember in pipe fitting is to select a suitable wrench for the job at hand. Each of the aforementioned wrenches is designed for one or more specific tasks. No wrench is suitable for every task encountered in pipe fitting.

A *monkey wrench* has smooth parallel jaws that are especially adapted for hexagonal valves and fittings (see Figure 8-12). Not only does it fit better on the part to be turned, it also does not have the crushing effect of a pipe wrench.

The operating principle of a *pipe wrench* is simple. The harder you pull, the tighter it squeezes the pipe. The pipe wrench was designed for use on pipe and screw fittings only. On parallel-sided objects, its efficiency is not up to that of a monkey wrench, and its squeezing action can do a great deal of damage.

Many inexperienced fitters have learned from experience that using a pipe wrench too large for the job can cause the fitting to

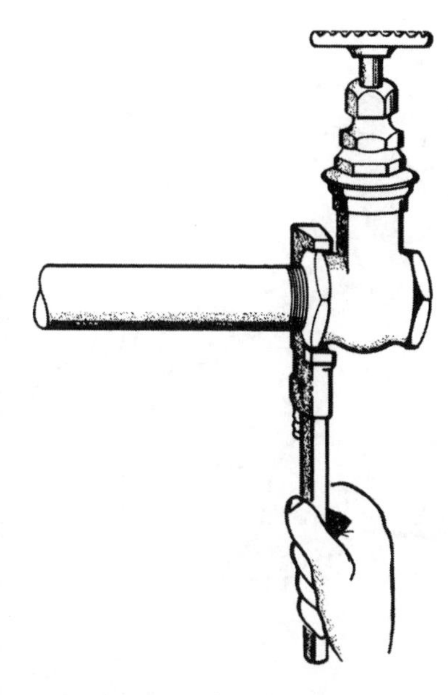

Figure 8-12 Using a monkey wrench.

stretch or crack. The result is a leaking joint that will require a new fitting to remedy the damage.

A Stillson wrench has serrated teeth jaws that enable it to grip a pipe or round surface in order to turn it against considerable resistance. The correct method for using a Stillson wrench is illustrated in Figure 8-13. Adjust the wrench so that the jaws will take hold of the pipe at about the middle part of the jaws. To support the wrench and prevent unnecessary lost motion when the wrench engages the pipe, hold the jaw at *A*, with the left hand pressing it against the pipe. At the beginning of the turning stroke *B*, with the jaw held firmly against the pipe with the left hand, the wrench will at once bite or take hold of the pipe with only the lost motion necessary to bring jaw C in contact with the pipe.

Figure 8-14 shows a *chain wrench* (or *pipe tongs*) and the method in which it is used. Although they are made for small sizes up, they are generally used for 6-inch pipe and larger.

A *strap wrench* (see Figure 8-15) is used when working with plated or polished-finish piping in order not to mar the surface. It also comes in handy in tight places where you cannot insert a Stillson wrench.

Open-end wrenches (see Figure 8-16) are used for making up flange couplings. The right size should be used in order to prevent wearing of the bolt heads or slippage that can cause bruised knuckles.

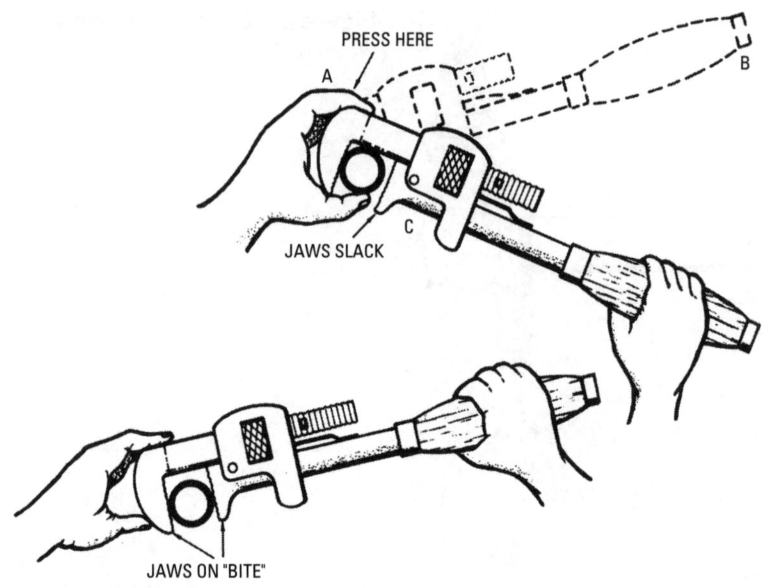

Figure 8-13 Method of using a Stillson wrench.

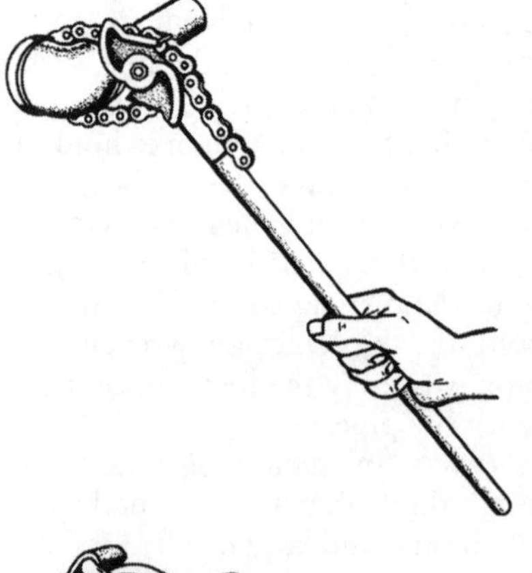

Figure 8-14 Method of using a chain wrench.

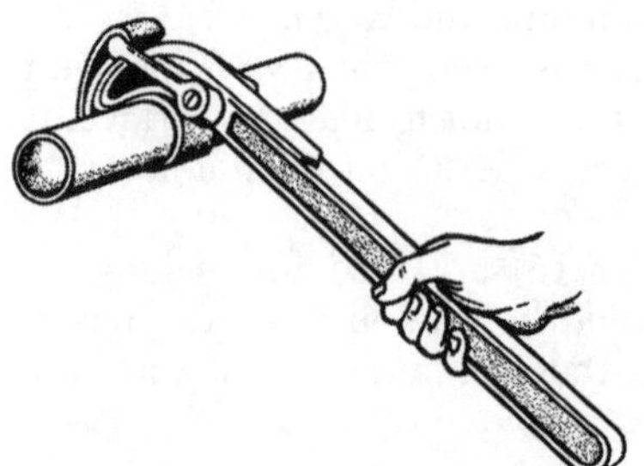

Figure 8-15 Method of using a strap wrench.

Figure 8-16 Use of open-end wrenches on a flange.

Pipe Vise

Either a pipe vise or a machinist vise (see Figure 8-17) can be used in pipe fitting. The pipe vise is used for pipe only. The machinist vise, on the other hand, has square jaws or a combination of square and gripper jaws, making it suitable for pipe as well as other work.

Several precautions should be taken when using a pipe vise. It exerts a powerful force at the jaws, which in some cases can do damage to the pipe. That is why experienced fitters never put a valve or fitting into a vise when making up a joint at the bench. There is too much danger of distorting the part by oversqueezing it or of putting the working parts of a valve out of line. The correct and incorrect methods of connecting a valve to a pipe are illustrated in Figure 8-18. Always hold a valve between lead- or copper-covered machinist vise jaws while unscrewing the bonnet (see Figure 8-18).

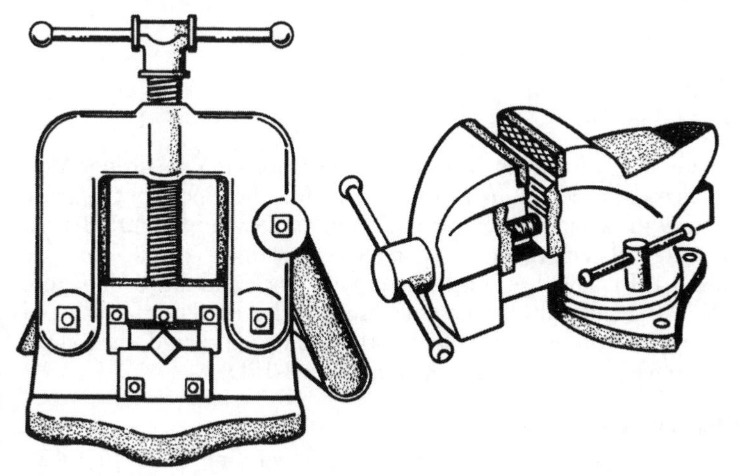

Figure 8-17 A pipe and machinist vise.

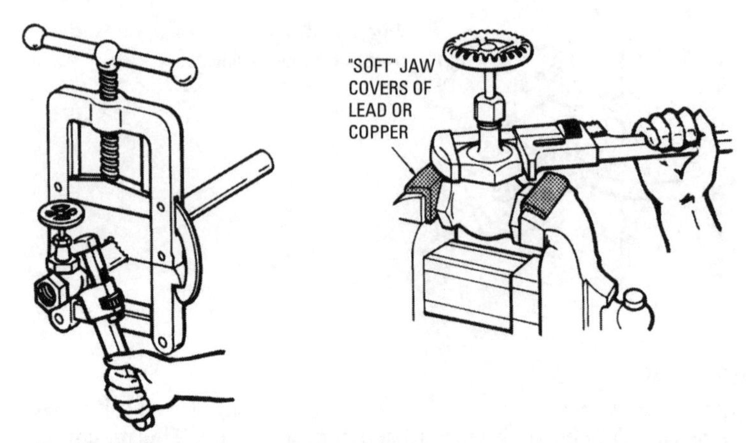

Figure 8-18 Correct method of using the pipe and machinist vise.

Installation Methods

Pipe fitting may be defined as the operations that must be performed in installing a pipe system made up of pipe and fittings. These pipe fitting operations can be listed as follows:

1. Cutting.
2. Threading.
3. Reaming.
4. Cleaning.
5. Tapping.
6. Bending.
7. Assembling.
8. Making up.

Figure 8-19 illustrates the principal operations in pipe fitting. After being marked to length by nicking with a file, the pipe is put in a vise and cut with a pipe cutter (or hacksaw) as shown in Figure 8-19A. Any external enlargement is removed with a metal file (see Figure 8-19B). The thread is next cut with stock and dies as in Figure 8-19C. After carefully cleaning the thread with a hard toothbrush and applying red lead or pipe cement to the freshly cut thread, the joint is made up with a Stillson wrench (see Figure 8-19D).

Pipe Cutting

Pipe is manufactured in different lengths varying from 12 to 22 feet. Accordingly, it must be cut to required length for the pipe

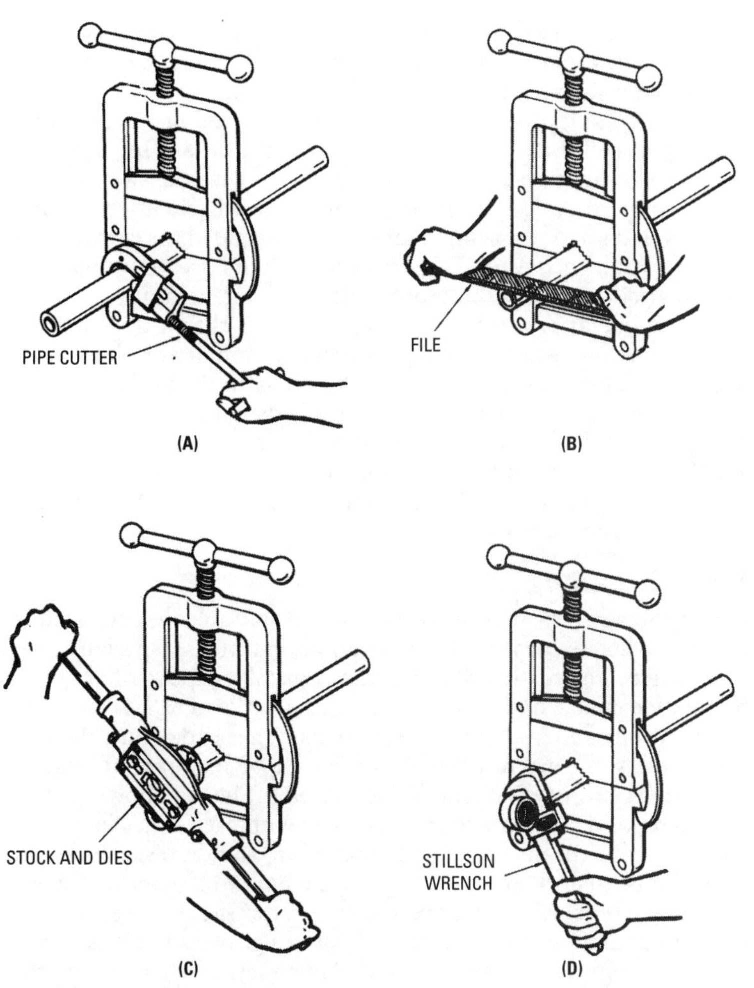

Figure 8-19 Pipe cutting.

installation. This may be done with either a hacksaw or a pipe cutter. The latter method is generally quicker and more convenient.

When a length of pipe is being cut or threaded, it is held firmly in a pipe vise. The pipe vise should be adjusted just tight enough to prevent the pipe from slipping, but not so tight as to cause the jaw teeth to unduly dig into the pipe.

A *pipe cutter* is a tool usually consisting of a hoop-shaped frame on whose stem a slide can be moved by a screw. On the side and frame, several cutting dies on wheels are mounted. In cutting, the pipe cutter is placed around the pipe so that the wheels contact the pipe. The tool is rotated around the pipe, tightening up with the screw stem each revolution until the pipe is cut.

The operating principles of a three-wheel cutter and a combined wheel and roller cutter are illustrated in Figure 8-20A. The cuts show the comparative movements necessary with the two types of cutters to perform their functions. The three-wheel cutter requires only a small arc of movement and is recommended for cutting pipe in inaccessible locations. The wheel cutter has a greater range than the roller cutter and is therefore preferred for general use.

The major disadvantage of a pipe cutter is that it does not cut but *crushes* the metal of the pipe, leaving a shoulder on the outside and a burr on the inside. This does not apply to the knife-type pipe cutter designed to actually cut (not crush) the pipe.

Figure 8-20B shows the appearance of pipe when cut by a hacksaw or knife cutter and when cut by a wheel pipe cutter. When the latter is used, the external enlargement must be removed by a file and the internal burr by a pipe reamer.

Pipe Threading

A pipe thread is cut with stock and dies. Adjustable dies are used in pipe threading because of slight variations in fittings, especially cast-iron fittings. Figure 8-21 illustrates an adjustable pipe stock and dies for double-ended dies. As shown, each pair of dies has one size thread at one end and another size at the other end. Thus, the two dies in the stock are in position for cutting ½-inch thread, and by reversing them they will cut ¾-inch thread. The cut shows plainly the reference marks, which must register with each other in adjusting the dies by means of the end setscrews to standard size.

A vise is used in conjunction with the pipe stock and dies when threading a pipe. After securing the pipe in the vise, use plenty of oil in starting and cutting the thread. In starting, press the dies firmly against the pipe until they take hold. After a few turns, blow out the chips and apply more oil. This should be done several times before completing the cut. When complete, blow out the chips as cleanly as possible and back off the dies. When drawing in your breath preliminary to blowing out the chips, turn your head away from the die to avoid drawing the chips into your lungs.

A nipple is short piece of pipe 12 inches in length or less and threaded at both ends. Nipples are properly cut by using a nipple holder designed for use with hand stock and dies. The holder is double ended and holds two sizes of nipples, one being for ½-inch nipples and the other for ¾-inch nipples. In construction, there is a

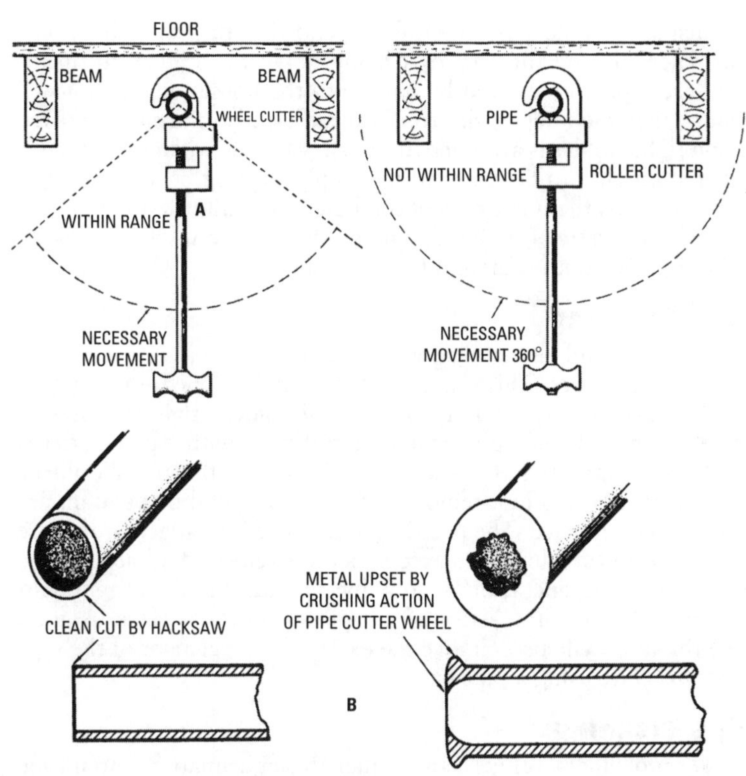

Figure 8-20 The appearance of a pipe cut by a hacksaw and a pipe cut by a cutter wheel.

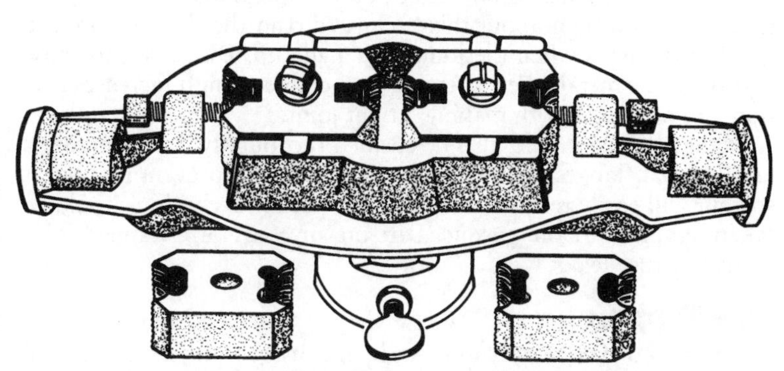

Figure 8-21 Adjustable pipe stock and dies for double-ended dies.

pin inside the holder having a fluted end that digs into the nipple end when pressed forward by driving down the wedge. In operation, the nipple is screwed by hand into the holder as far as it will go, and then the wedge is driven down sufficiently to firmly secure the nipple. The holder is arranged in such a way that when the thread is cut, the nipple can be removed by simply starting back the wedge, which loosens the inner part of the holder and allows the nipple to be easily unscrewed by hand. The holder can be used for making either right or both right and left nipples.

Pipe Reaming

The burrs should be removed with a reamer to avoid future trouble with clogged pipes. This is a job that should be done thoroughly.

The correct way of removing the shoulder (that is, external enlargement) left on a pipe end after cutting it with a pipe cutter is by using a flat file. Obviously at each stroke, the file should be given a turning motion, removing the excess metal through an arc of the circumference. The position of the pipe is changed in the vise from time to time until the excess metal is removed all around the pipe. When the operation is done by moving the file in a straight line, it will result in a series of flat places on the surface. A good pipe threader will also remove the external enlargement of the pipe end caused by using a pipe cutter.

Pipe Cleaning

Dirt, sand, metal chips, and other foreign matter should be removed from pipelines to prevent future problems. When a new pipeline is installed, flush it out completely with water to remove any loose scale or foreign matter.

Check the threads for any dirt or other foreign matter. It is important to be thorough about this because dirt in the threads can also get into the lines when the joints are made up. Dirt can also cause tearing of the metal when screwing up a connection. It increases friction and interferes with making a tight joint.

Flanged faces should also be cleaned thoroughly. Manufacturers usually coat flanges with a heavy oil or grease to prevent rusting. A solvent will easily remove this coating. Special precaution should be taken against dirt on gaskets. Dirt on any part of a flanged joint tends to cause leaks.

Pipe Tapping

An internal or female thread is cut by means of a pipe tap, a conical screw made of hardened steel and grooved longitudinally. A pipe tap and pipe reamer are illustrated in Figure 8-22.

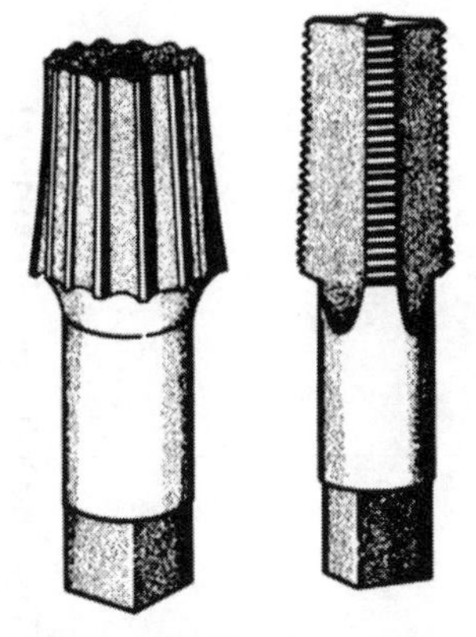

Figure 8-22 A typical pipe tap and pipe reamer.

Table 8-20 gives drill sizes that permit direct tapping without reaming the holes beforehand. Table 8-21 gives drill sizes for both the Briggs, or American Standard, and the Whitworth, or British Standard.

Pipe Bending

With the proper tools, pipes may be bent within certain limits without difficulty. An example of a pipe-bending tool is illustrated in Figure 8-23.

Pipes can also be bent by hand without the use of special tools. One method involves the complete filling of the pipe with sand and capping both ends so that none of the sand will be lost. Heat the part to be bent and then clamp the pipe in a vise as close to the part to be bent as possible. Now cool the outside with water so that the inside, being hot and plastic, is compressed as the bend is made.

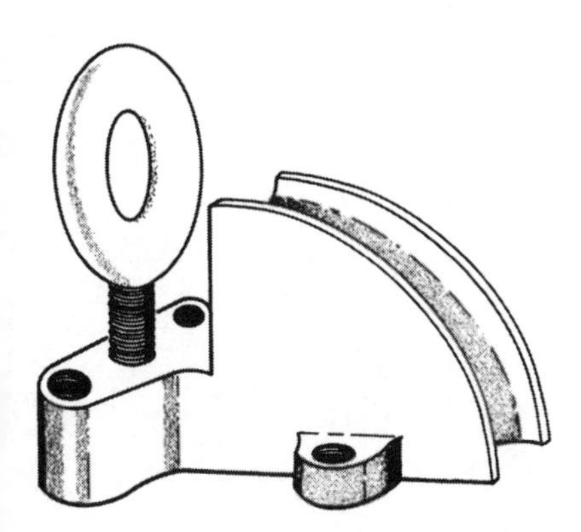

Figure 8-23 Pipe-bending tool.

Assembling and Make-Up

Assembling is the operation of putting together the various lengths of pipe and fittings used in an installation.

If no mistakes have been made in cutting the pipe to the right length or in following the dimensions on the blueprint, the pipe and fittings may be installed without difficulty. In other words, the last joint (either a right or left union, or long screw joint) will come together smoothly or, as they say in the trade, make up.

Table 8-20 Drill Sizes for Briggs Standard Pipe Taps (For Direct Tapping Without Reaming)

Size of pipe	1/8	1/4	3/8	1/2	3/4	1	1 1/4	1 1/2	2	2 1/2	3	3 1/2	4
Size of drill	21/64	7/16	9/16	45/64	29/32	1 9/64	1 31/64	1 47/64	2 13/64	2 5/8	3 1/4	3 47/64	4 15/64

Table 8-21 Drill Sizes for Pipe Taps

Size Taps (in)	*Briggs Standard*		*British (Whitworth) Standard*	
	Thread	*Drill*	*Thread*	*Drill*
1/8	27	21/64	28	5/16
1/4	18	27/64	19	7/16
3/8	18	9/16	19	9/16
1/2	14	11/16	14	23/32
5/8	—	—	14	25/32
3/4	14	29/32	14	29/32
7/8	—	—	14	1 1/16
1	11 1/2	1 1/8	11	1 5/32
1 1/4	11 1/2	1 15/32	11	1 1/2
1 1/2	11 1/2	1 23/32	11	1 23/32
1 3/4	—	—	11	1 31/32
2	11 1/2	2 3/16	11	2 3/16
2 1/4	—	—	11	2 13/32
2 1/2	8	2 9/16	11	2 25/32
2 3/4	—	—	11	3 1/32
3	8	3 3/16	11	3 9/32
3 1/4	—	—	11	3 1/2
3 1/2	8	3 11/16	11	3 3/4
3 3/4	—	—	11	4
4	8	4 3/16	11	4 1/4
4 1/2	8	4 11/16	11	4 3/4
5	8	5 1/4	11	5 1/4
5 1/2	—	—	11	5 3/4
6	8	6 5/16	11	6 1/4
7	8	7 5/16	11	7 5/16
8	8	8 5/16	11	8 5/16
9	8	9 5/16	11	9 5/16
10	8	10 7/16	11	10 5/16

Screwed joints are put together with red or white lead pigment mixed with graphite and linseed oil, or with some standard commercial joint compound.

It is unnecessary to put much material on the threads because it will be simply pushed out and wasted when the joint is screwed up.

It should be put on evenly and cover all the threads, with care being taken not to let any touch the reamed end of the pipe where it may get inside. The red lead is preferably obtained in the powder form and mixed with oil and a little dryer at the time the pipe is to be made up. Get a clean piece of glass on which to prepare the lead. The toothbrush should be laid on the glass after applying the lead in order to avoid getting grit on the brush and paint on the table. When grit becomes mixed with the lead, it prevents close contact of the filling and pipe, thus making the joint less efficient.

The following steps should be taken *before* making up a screwed joint:

1. Ream the pipe ends.
2. Remove burrs from both the inside and outside of the pipe.
3. Thoroughly clean the inside of the pipe.
4. Thoroughly clean the threads.

Clean threads, a suitable joint compound, and proper tightening (neither too much nor too little) are all necessary for a satisfactory screwed joint.

There are four requirements for satisfactorily making up a flanged joint. In the proper order of sequence, these four requirements are as follows:

- Thorough cleaning
- Accurate alignment
- Using the proper gasket
- Tightening bolts in proper order

Thoroughly clean the flange face with a solvent to remove any grease, and wipe it dry. The alignment must be accurate for satisfactory makeup. This is particularly important where valve flanges are involved. If the alignment is poor, a severe stress on the valve flanges may distort the valve seats and prevent tight closure (see Figure 8-24).

Selecting a suitable gasket for the service is also very important. Using the wrong gasket may result in a leak.

Figure 8-25 illustrates the recommended method of tightening the bolts with a wrench. This must be done not in rotation but in sequence as indicated by the numbers. This is the crossover method. Do not fully tighten on the first round but go over them two or preferably three times to fully tighten.

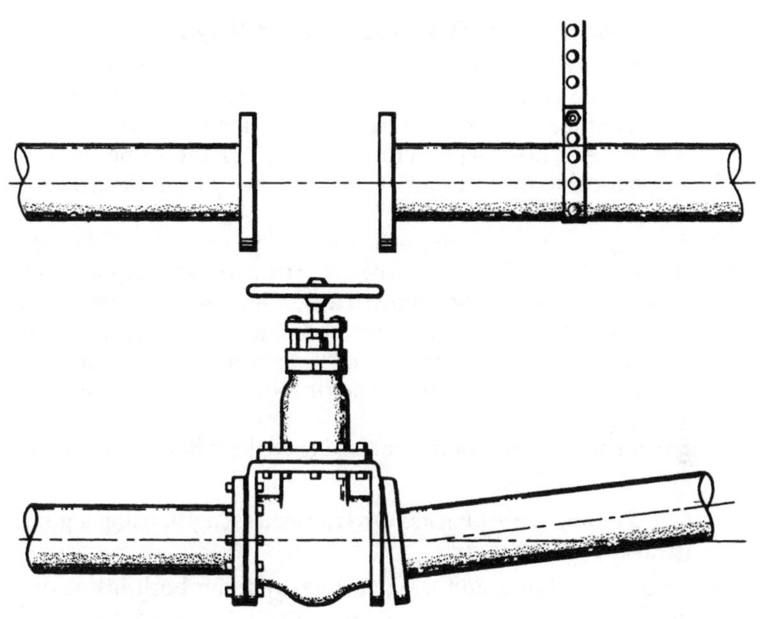

Figure 8-24 Flange and pipe alignment.

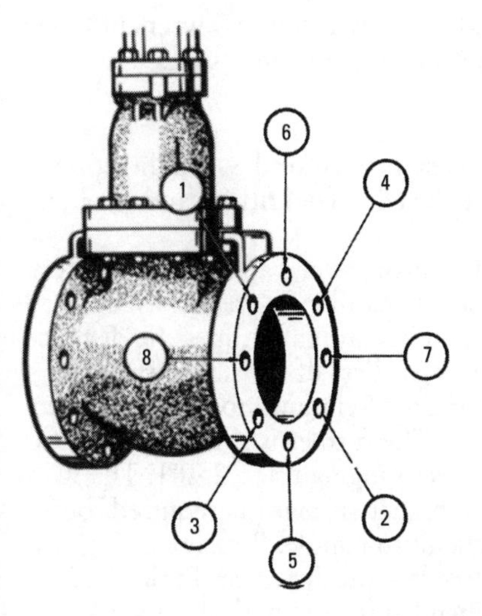

Figure 8-25 Recommended method of tightening bolts.

Nonferrous Pipes, Tubing, and Fittings

Both brass and copper are used as materials for this type of pipe and tubing. The construction may be either cast or wrought, and the methods of joining include screwed, flared, and soldered.

The technique used with screwed fittings is the same as with ordinary malleable-iron fittings. This has already been described in the first part of this chapter.

The fittings used for flared soft tube end joints are cast fittings. These are usually used on oil burner construction and supply lines. Flared joint fittings include elbows, tees, couplings, unions, and a full range of reducing and adapter combinations in all standard sizes and combinations of sizes from ⅛ inch to 2 inches inclusive (see Figure 8-26). A double-seal type of flared joint fitting is illustrated in Figure 8-27.

The sequence of operations required to make a flared joint are as follows:

1. Cut the tube with a hacksaw to the exact length, using a guide to ensure a square cut.
2. Remove all burrs and irregularities by filing both inside and outside.
3. Slip the coupling nut over the end of the tube and insert the flanging tool.
4. Drive the flanging tool into the tube with a few hammer blows, expanding the tube to its proper flare.
5. Assemble the fitting and tighten it by using two wrenches, one on the nut and the other on the body of the fitting.

Soldering Pipe

Solder fittings usually come in cast bronze and wrought copper. The two kinds of fittings used are the edge-feed fitting and the hole-feed fitting (see Figure 8-28).

The basic principle of solder fittings is capillary attraction. Because of capillary attraction, solder can be fed vertically upward between two closely fitted tubes to a height many times the distance required to made a soldered joint, regardless of the size of the fitting.

Either a 50-50 tin-lead solder or a 95-5 tin-antimony solder is recommended for joining copper tube. The former is generally used for moderate pressures with temperatures ranging up to 250°F. The 95-5 tin-antimony solder is used where higher strength is required, but it has the disadvantage of being difficult to handle. Pressure ratings for soldered joints using these two solders are listed in Table 8-22. A suitable paste-type flux is recommended for use with these solders.

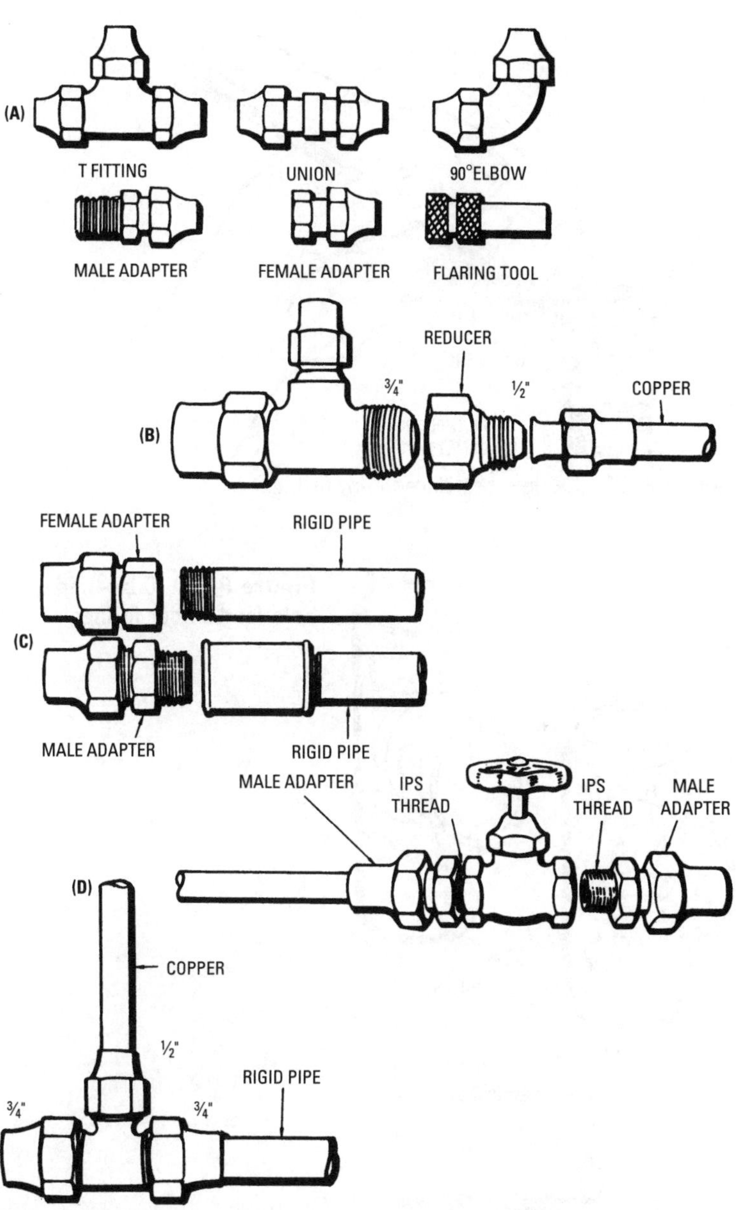

Figure 8-26 Various flare-type copper tube fittings.

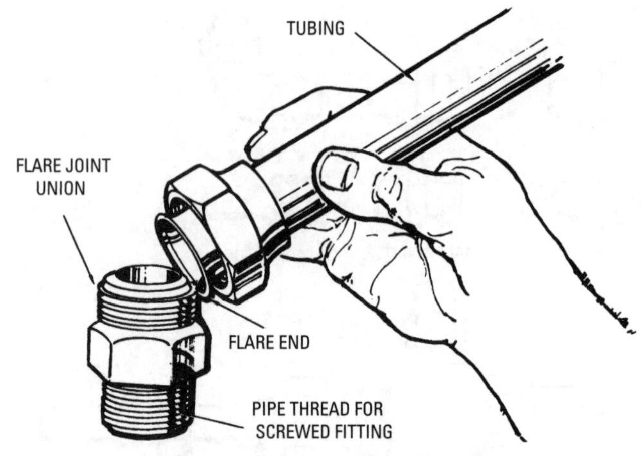

Figure 8-27 Double-seal flared joint fitting.

Figure 8-28 Edge- and hole-feed solder fittings.

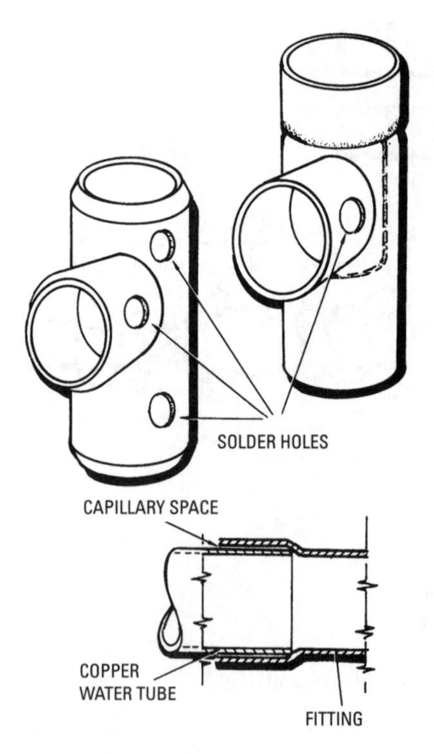

Table 8-22 Safe Strength of Soldered Joints—Pressure Ratings, Maximum Service Pressure (psi)

		Water‡		
Solder Used in Joints	***Service Temperatures (°F)***	***¼ to 1 Inch Incl.****	***1¼ to 2 Inches Incl.****	***2½ to 4 Inches Incl.****
50-50 tin-lead†	100	200	175	150
	150	150	125	100
	200	100	90	75
	250	85	75	50
95-5 tin-antimony	100	500	400	300
	150	400	350	275
	200	300	250	200
	250	200	175	150
Brazing filler metal melting at or above 1000°F§	250	300	210	170
	350	270	190	155

**Standard copper water tube sizes*
†ASTM B32, alloy grade 50A
‡Including refrigerants and other noncorrosive liquids and gases
§ASTM B260, brazing filler metal
(Courtesy Copper & Brass Research Assoc.)

The series of operations necessary in making a solder fitting joint are as follows:

1. Measure the tube to proper length so that it will run the full length of the socket of the fitting.
2. Cut the tube end squarely.
3. Clean tube end and socket of fitting.
4. Apply soldering flux to the cleaned areas of tube and fitting socket.
5. Assemble the joint.
6. Revolve the fitting if you can, to spread the flux evenly.
7. Apply heat and solder.
8. Remove residual solder and flux.
9. Allow joint to cool.

With a hole-feed fitting that has a feed hole for the solder and a groove inside, the procedure is just the same as for an edge-feed fitting except that solder is fed into the feed hole until it appears as a ring at the edge of the fitting. Be sure the hole is kept full of solder, as it shrinks on cooling and solidifying.

Do not select fittings that are oversize because they will result in a loose fit. The capillary action is dependent on a fairly tight fit, although a certain amount of looseness can be tolerated. A loose fit causes the greatest difficulty when working with large-size copper tubes.

A *thorough* cleaning of the tube surface and the fitting socket is absolutely essential for a strong, tight, and durable joint. *This cannot be emphasized too strongly.*

Brazing Pipes

Brazing is rapidly taking the place of many operations formerly performed by soldering because it is simpler and quicker and results in a stronger joint. Like soldering, the brazing alloy is applied at temperatures *below* the melting point of the metal being brazed. In this respect, both brazing and soldering differ from welding, which forms a joint by melting (fusing) the metal at temperatures *above* its melting point.

During the brazing process, the brazing alloy is heated until it adheres to the pipe surface and enters into the porous structure of the metal. The brazed joint is almost always as strong as the brazed metal surrounding it.

Brazing can be used to join nonferrous metals, such as copper, brass, or aluminum, or ferrous metals, such as cast iron, malleable iron, or steel. The brazing alloy used in making the joint depends on the type of metal being brazed. For example, aluminum requires the use of a special aluminum brazing alloy, whereas a low-temperature brazing alloy or a silver alloy is recommended for brazing copper and copper alloys. A local welding supply dealer should be able to provide answers to your questions about which alloy to use.

The procedure for brazing consists essentially of the following operations:

1. Clean both surfaces.
2. Apply a suitable flux.
3. Align and clamp the parts to be joined.
4. Preheat the surface until the flux becomes fluid.
5. Apply a suitable brazing alloy.

6. Allow the surface time to cool.
7. Clean the surface.

The pipe surfaces must be thoroughly cleaned, or the result will be a weak bond or no bond at all. All dirt, grease, oil, and other surface contaminants must be removed, or the capillary attraction so important to the brazing process will not function properly.

Select a flux suited to the requirements of the brazing operation. Fluxes differ in their chemical compositions and the brazing temperature ranges within which they are designed to operate. Apply the flux to the joint surface with a brush.

Align the parts to be joined, securing them in position until after the brazing alloy has solidified. Preheat the metal to the required brazing temperature (indicated by the flux reaching the fluid stage).

After the flux has become fluid, add the brazing alloy. If conditions are right, the brazing alloy will spread over the metal surface and into the joint by capillary attraction. Do *not* overheat the surface. Remove the heat as soon as the entire surface has been covered by the brazing alloy. Allow time for the joint to cool and then clean the surface.

Braze Welding Pipe

Braze welding is another bonding process that does not melt the base metal. In this respect, it resembles both soldering and brazing.

Brazing and braze welding operate under essentially the same basic principles. For example, both use nonferrous filler metals that melt above 880°F but below the melting point of the base metal. They differ primarily in application and procedure.

The braze-welding process follows most of the steps previously described for brazing but will differ principally as follows:

1. Edge preparation is necessary in braze welding and is essentially similar to that employed in gas welding. The edges must be prepared before the surfaces are cleaned.
2. A suitable filler metal is used instead of a brazing alloy.

The filler metal will flow throughout the joint by capillary attraction. Flanges that are to be brazed to copper pipes must be of copper or what is known as brazing metal (98 percent copper and 2 percent tin), as gunmetal flange would melt before the brazing alloy ran.

Welding Pipe

Oxyacetylene welding (gas welding) is probably the most common welding process used for joining pipe and fittings, particularly on smaller installations. Arc welding is also very popular.

Pipe welding should be done only by a skilled and experienced worker. The equipment and the procedure used are much more complicated than those used with soldering, brazing, or braze welding.

Welding forms a joint by melting (fusing) the metal at temperatures above its melting point. The filler metal, electrodes, or welding rods used must be suitable for use with the base metal to be welded, and the procedure used should be such as to ensure complete penetration and thorough fusion of the deposited metal with the base metal. The welding process is employed in many piping installations, but especially those in which large-diameter pipes are used.

Manufactured steel welding fittings are available for almost every conceivable type of pipe connection. These steel welding fittings can be divided into the following two principal categories:

- Butt-welding fittings
- Socket-welding fittings

Butt-welding fittings (see Table 8-23) have ends that are cut square or beveled 37½° for wall thicknesses under 3⁄16 inch. Wall thicknesses ranging from 3⁄16 inch to ¾ inch are beveled 37½°. For walls ranging from ¾ inch to 1¾ inches thick, the bevel is U-shaped.

Socket-welding fittings (see Table 8-24) have a machined recess or socket for inserting the pipe. A fillet weld is made between the pipe wall and the socket end of the fitting. The fillet weld is approximately triangular in cross-section, the throat lying in a plane of approximately 45° with respect to the surfaces of the part joined. As shown in Figure 8-29, the minimum thickness of the socket wall (L) is 1.25 times the nominal pipe thickness (T) for the designated schedule number of the pipe. Socket-welding fittings are generally limited in use to nominal pipe sizes 3 inches and smaller.

In addition to butt- and socket-welding fittings, flange fittings are also available for welding in sizes ranging from ½ inch to 24 inches.

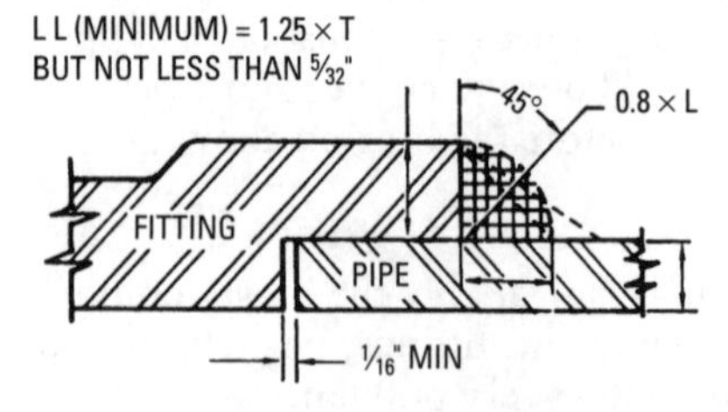

Figure 8-29 Fillet weld dimensions. *(Courtesy Dow Chemical Co.)*

Table 8-23 Dimensions of Steel Butt-Welding Fittings

Nominal Pipe Size	Long Radius Elbows: Outside Diameter at Bevel	Long Radius Elbows: Center to End 90° Elbows A	Long Radius Elbows: Center to End 45° Elbows B	180° Returns: Outside Diameter at Bevel	180° Returns: Center to Center O	180° Returns: Back to Face K	Straight Tees: Outside Diameter at Bevel	Straight Tees: Center to End Run C	Straight Tees: Center to End Outlet M
1	1.315	1½	⅞	1.315	3	2 3/16	1.315	1½	1½
1¼	1.660	1⅞	1	1.660	3¾	2¾	1.660	1⅞	1⅞
1½	1.900	2¼	1⅛	1.900	4½	3¼	1.900	2¼	2¼
2	2.375	3	1⅜	2.375	6	4 3/16	2.375	2½	2½
2½	2.875	3¾	1¾	2.875	7½	5 3/16	2.875	3	3
3	3.500	4½	2	3.500	9	6¼	3.500	3⅜	3⅜
3½	4.000	5¼	2¼	4.000	10½	7¼	4.000	3¾	3¾
4	4.500	6	2½	4.500	12	8¼	4.500	4⅛	4⅛
5	5.563	7½	3⅛	5.563	15	10 5/16	5.563	4⅞	4⅞
6	6.625	9	3¾	6.625	18	12 5/16	6.625	5⅝	5⅝
8	8.625	12	5	8.625	24	16 5/16	8.625	7	7
10	10.750	15	6¼	10.750	30	20⅜	10.750	8½	8½
12	12.750	18	7½	12.750	36	24⅜	12.750	10	10
14	14.000	21	8¾	14.000	42	28	14.000	11	
16	16.000	24	10	16.000	48	32	16.000	12	Not standard
18	18.000	27	11¼	18.000	54	36	18.000	13½	
20	20.000	30	12½	20.000	60	40	20.000	15	
24	24.000	36	15	24.000	72	48	24.000	17	

From American Standard for Butt-Welding Fittings, ASA B16.9-1958. All dimensions are in inches. Dimension A is equal to ½ of dimension O.
(Courtesy 1960 ASHRAE Guide)

Table 8-24 Dimensions of Socket-Welding Fittings

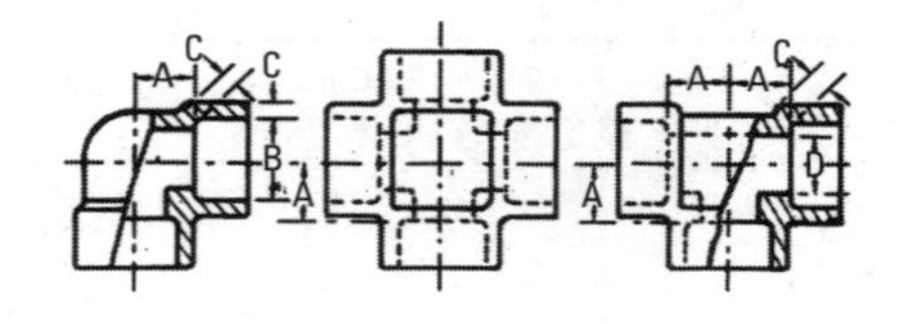

Nominal Pipe Size	Depth of Socket, Min	Center to Bottom of Socket A: Sched 40 and 80	Center to Bottom of Socket A: Sched 160	Bore Diameter of Socket, Min B	Socket Wall Thickness, Min C: Sched 40	Socket Wall Thickness, Min C: Sched 80	Socket Wall Thickness, Min C: Sched 160	Bore Diameter of Fitting D: Sched 40	Bore Diameter of Fitting D: Sched 80	Bore Diameter of Fitting D: Sched 160
1/8	3/8	7/16	—	0.420	0.125	0.125	—	0.269	0.215	—
1/4	3/8	7/16	—	0.555	0.125	0.149	—	0.364	0.302	—
3/8	3/8	17/32	—	0.690	0.125	0.158	—	0.493	0.423	—
1/2	3/8	5/8	3/4	0.855	0.136	0.184	0.234	0.622	0.546	0.466
3/4	1/2	3/4	7/8	1.065	0.141	0.193	0.273	0.824	0.742	0.614
1/2	1/2	7/8	1 1/16	1.330	0.166	0.224	0.313	1.049	0.957	0.815
1 1/4	1/2	1 1/16	1 1/2	1.675	0.175	0.239	0.313	1.380	1.278	1.160
1 1/2	1/2	1 1/4	1 1/2	1.915	0.181	0.250	0.351	1.610	1.500	1.338
2	5/8	1 1/2	1 5/8	2.406	0.193	0.273	0.429	2.067	1.939	1.689
2 1/2	5/8		2 1/4	2.906	0.254	0.345	0.469	2.469	2.323	2.125
3	5/8	2 1/4	2 1/2	3.535	0.270	0.375	0.546	3.068	2.900	2.626

(Courtesy 1960 ASHRAE Guide)

Gas Piping

The gas pipe installations in which gas-fired furnaces, boilers, heaters, or other gas appliances are used deserve special attention because of the volatile and highly flammable nature of the fuel. Special attention should be given to the installation of gas piping to ensure against leakage.

The installation and replacement of gas piping should be done only by qualified workers who have the necessary skills and experience.

The following recommendations are offered as a guide for installing and replacing gas piping:

- All work should be done in accordance with the building, heating, and plumbing codes and standards of the authorities having local jurisdiction. These take precedence over national codes and standards.
- In an existing installation, the gas supply to the premises and all burners in the system must be shut off before work begins.
- When installing a system, size the pipes according to the amount of gas to be delivered to each outlet and at the proper pressure. The length of pipe runs and number of outlets are the main determining factors.
- Use a piping material recommended by the local authorities having jurisdiction. *Never* bend gas piping because it may cause the pipe walls to crack and leak gas. Use fittings for making turns in gas piping. Take all branch connections from the top or side of horizontal pipes (never from the bottom).
- Locate the gas meter as close as possible to the point at which the gas service enters the structure.
- Make certain all pipes are adequately supported so that no unnecessary stress is placed on them.
- Offsets should be 45° elbows rather than 90° fittings in order to reduce the friction to the flow of gas.
- Check for gas leaks by applying a soap-and-water solution to the suspected area. *Never* use matches, candles, or any other flame to locate the leak.

Insulating Pipes

Insulating the *supply* pipes in a low-pressure steam or hot-water space heating system will reduce unwanted heat loss and improve the heating efficiency of the system. The return pipes in a hot-water heating system should also be insulated so that the water reaches the boiler with a minimum of heat loss. Do *not* insulate the return pipes in a steam heating system. Uninsulated pipes will aid in the condensation of any steam that has succeeded in bypassing the thermostatic traps and entering the returns.

The pipe insulation material must be noncombustible, durable, and resistant to moisture. Furthermore, it should be able to retain its original physical shape and insulating properties *after* becoming wet and drying out.

Fiberglass is used to insulate steam or hot-water heating pipes. It can be easily applied to the pipe, requiring little more than ordinary cutting shears or a sharp knife.

Another effective pipe-insulating material is expanded polystyrene (see Figure 8-30).

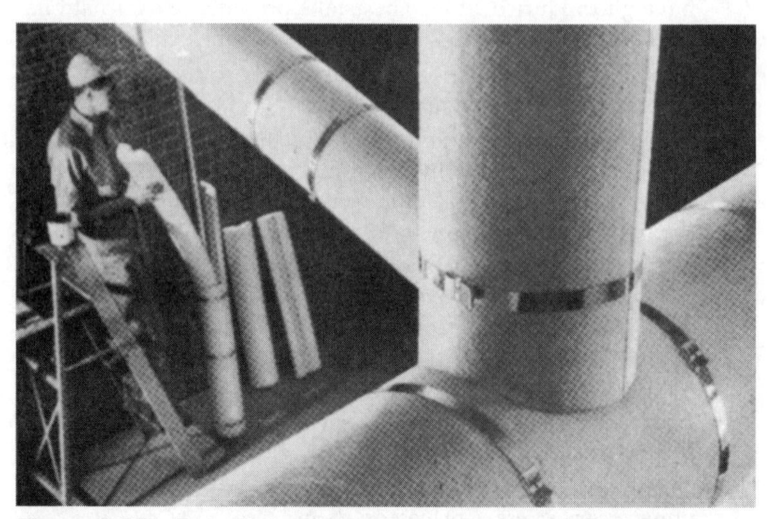

Figure 8-30 Expanded polystyrene as a pipe-insulating material.
(Courtesy Dow Chemical Co.)

Your local building supply dealer should be able to answer any questions you may have about pipe-insulting materials. The manufacturers of these materials also generally provide detailed instruction about how to apply them. You should not have any difficulty if you carefully read and follow these instructions.

Piping Details

Each piping installation will have design or layout problems that the fitter must solve. Many of these problems, such as installing dirt pockets or siphons, are quite simple for the pipe fitter to handle.

More-complicated layout problems are solved by calculating and installing lift fittings, swivels and offsets, and drips. These and other piping details are described in the sections that follow.

Connecting Risers to Mains

Many steam fitters connect risers directly to mains with a tee; although this method saves on extra labor and expense, it results in a more inefficient operation of the installation. When a tee is used, the condensation falls directly across the path of the steam flowing in the main and will be carried along and finally arrive at the radiator or convector with excess moisture. This problem is avoided with a 45° connection.

Using a 45° connection very effectively drains the condensation from the main, the path of the condensation being along the metal of the pipe and fittings instead of dripping directly into the steam.

The proper method of connecting a riser to a main where the riser has a direct connected drip pipe is by using a 45° connection downward. If the riser has no drip, the riser should be connected to the main with the leadoff being 45° upward connecting with a 45° elbow. The runout should pitch ½ inch per foot. Runouts over 8 feet in length, but with less than ½ inch per foot pitch, should be one size larger than specified in the pipe sizing tables.

If the condensation flows in the opposite direction to the steam, the runouts should be one size larger than the vertical pipe and pitched ¼ inch per foot toward the main. If the runouts are over 8 feet in length, use a pipe *two* sizes larger than specified.

Connections to Radiators or Convectors

The connections to radiators and convectors must have a proper pitch when installed and be arranged so that the pitch will be maintained under the strains of expansion and contraction. These connections are made by swing joints.

In two-pipe systems, radiators are connected either at the top and bottom opposite end or at the bottom and bottom opposite end. The top connection is not recommended for best performance. Short radiation may be top-supply and bottom-return connected on the *same* end.

Additional information about radiator and convector connections can be found in Chapter 2 of Volume 3 ("Radiators, Convectors, and Unit Heaters").

Lift Fittings

The lift fittings illustrated in Figure 8-31 are adapted for use on the main return lines of vacuum heating systems at points where it is desired to raise the condensation to a higher level. In operation, the

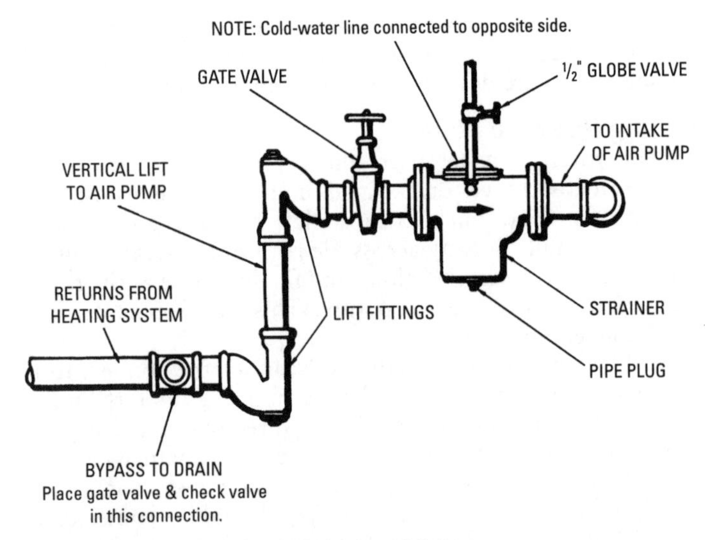

Figure 8-31 Typical installation of lift fittings.

momentum of the water is maintained and assists in making the lift with minimum loss of vacuum. The lift fitting is constructed with a pocket at the bottom of the lift into which the water drains. As soon as sufficient water accumulates to seal this pocket, it is drawn to the upper portion of the return by the vacuum produced by the pump. The shape of the fitting is such that dirt and scale are usually swept along by the current. Cleanout plugs are provided for use if necessary. A second fitting in a reversed position is recommended for use at the top of the lift to prevent water from running back while the pocket is filled.

Drips

A steam main in any steam heating system may be dropped to a lower level without dripping if the pitch is downward with the direction of the steam flow. By the same token, the steam main in any system may be elevated if properly dripped.

Various piping arrangements for dripping the main and riser are illustrated in Figures 8-32 through 8-39. Figure 8-39 shows a connection where the steam main is raised and the drain is to a wet return. If the elevation of the low point is above a dry return, it may be drained through a trap to the dry return in a two-pipe vapor, vacuum, or subatmospheric system.

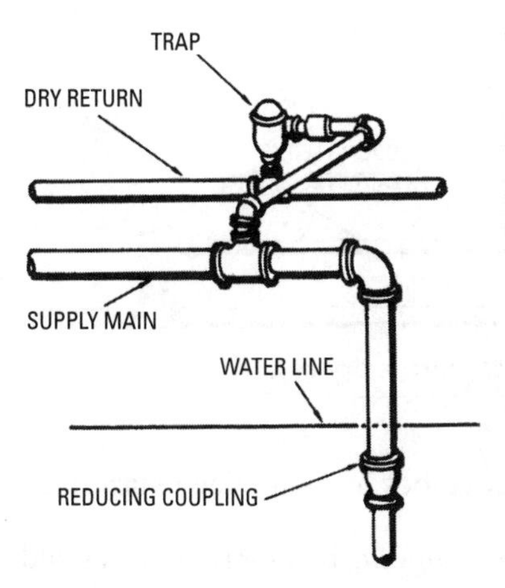

Figure 8-32 Dripping end of main into wet return.

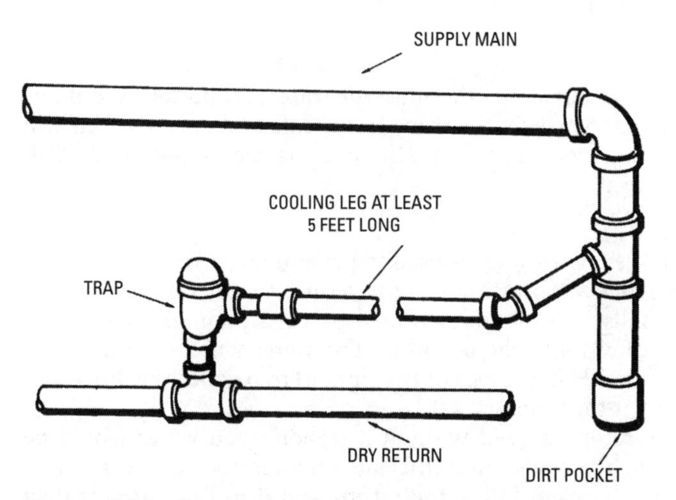

Figure 8-33 Dripping end of main into dry return.

A horizontal steam main can be run over an obstruction if a small pipe is carried below for the condensation with provisions for draining it.

In vacuum steam heating systems, drip traps for steam mains should be either thermostatic or combination float and thermostatic protected by dirt strainer or dirt pockets. Typical methods of making these connections are illustrated in Figures 8-40 through 8-43. The bases of supply risers are dripped through drip traps as

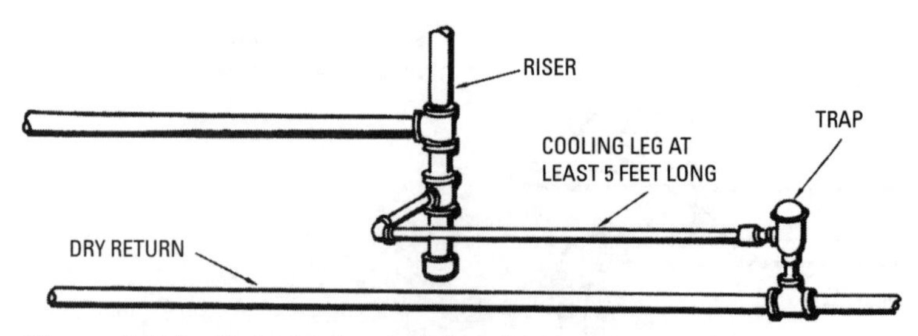

Figure 8-34 Dripping head of riser into a dry return.

shown in Figures 8-44 and 8-45. Methods of connecting return risers are also shown.

In vapor steam heating systems, runouts to supply risers should be dripped separately into a wet return.

Dirt Pockets

On all systems employing thermostatic traps, dirt pockets should be located so as to protect the traps from scale and muck, which will interfere with their operation. Dirt pockets are usually made 8 to 12 inches deep.

Siphons

A *siphon* (see Figure 8-46) is used to prevent water from leaving the boiler due to lower pressure in a dry return. Condensation from the drip pipe falls into the loop formed by the siphon and, after it is filled, overflows into the dry pipe. The water will rise to different heights (*G* and *H* in the legs of the siphon) to balance the difference in pressure at these points.

If a dry return is used without a siphon, then water would be drawn from the boiler in sufficient amounts to balance the low pressure in the riser, filling the return and drip to approximately point *M* (see Figure 8-46).

Hartford Connections

Another method employed to prevent water leaving the boiler is the Hartford connection (or loop) on the wet return (see Chapter 15 of Volume 1, "Boilers and Boiler Fittings").

Making Up Coils

In putting together lengths and return bends to form a coil heating unit, there is a right way and a wrong way to do the job. The essential requirement for the satisfactory operation of the coil is providing

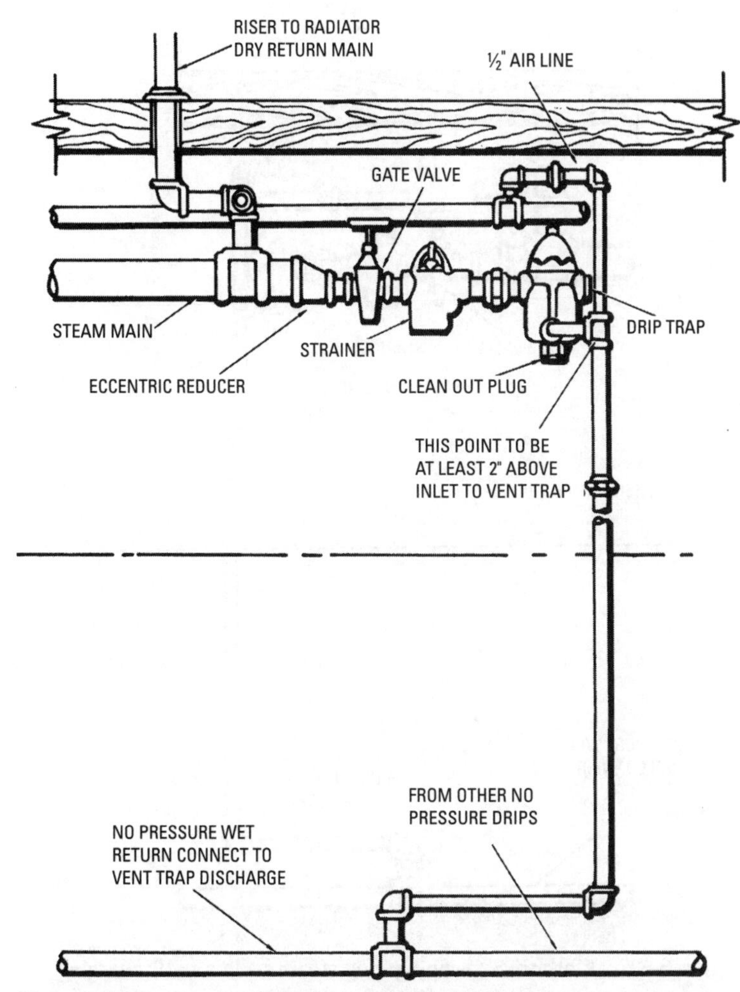

Figure 8-35 Method of dripping short steam main and discharging condensation into pressure wet return near floor.

for proper drainage. To obtain this, the pipes should not be parallel but should have a degree of pitch.

A pitch fitting should be used to obtain pitch in the coils rather than the so-called drunken thread method (see Figure 8-47). The drunken thread is obtained by removing the guide bushing from the stock and cutting the thread out of alignment. This gives a

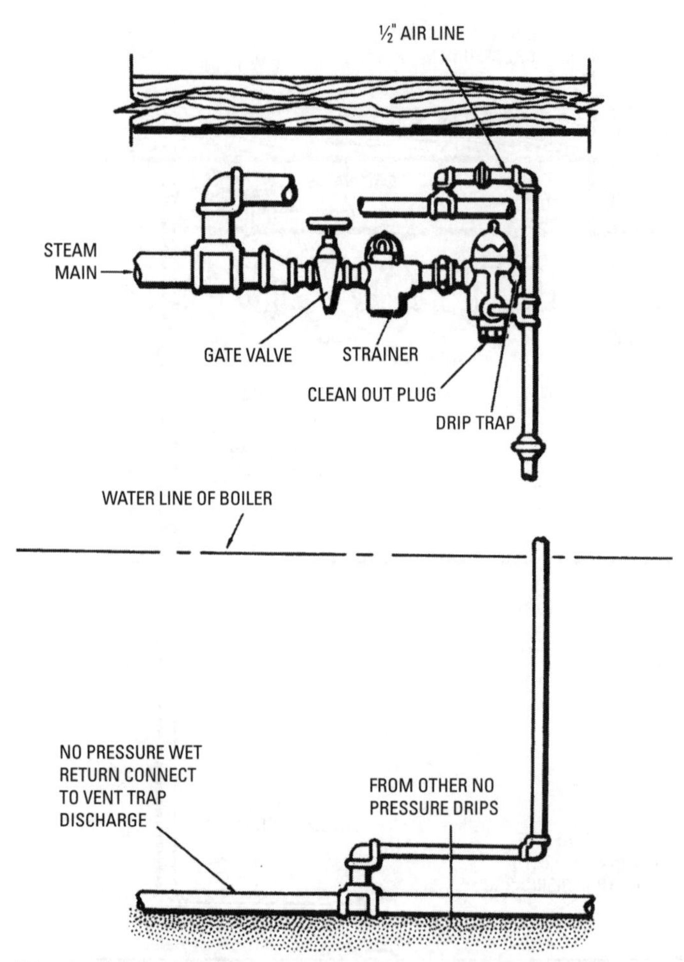

Figure 8-36 Method of dripping long steam main and discharging condensation into pressure wet return near floor.

poor joint—one that will eventually break because of corrosion. The corrosion is caused by the deep cut on one side of the pipe resulting from cutting the thread out of alignment.

Relieving Pipe Stress

Long runs of rigidly supported piping carrying steam or hot water, especially when they are at high pressures and temperatures, are

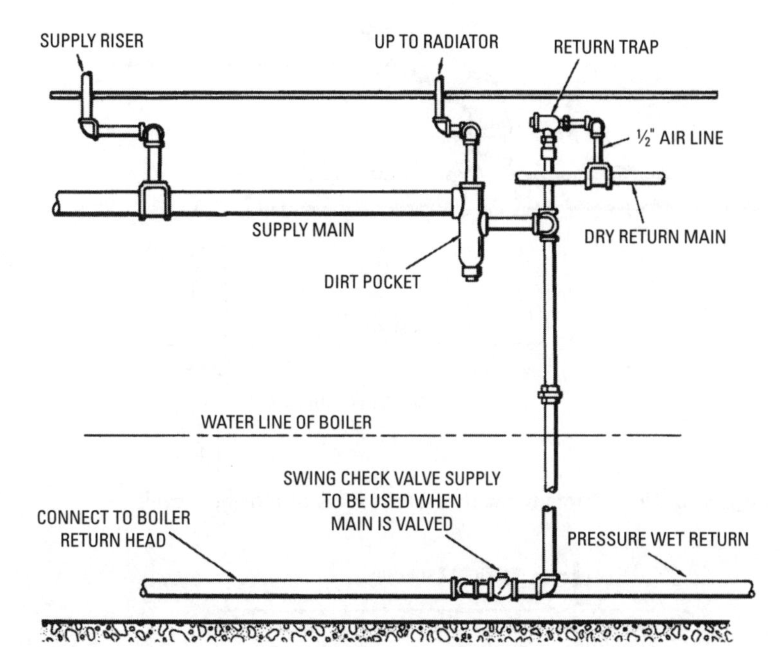

Figure 8-37 Alternative method of dripping end of supply main through dirt pocket to pressure wet return near floor and venting to overhead dry return.

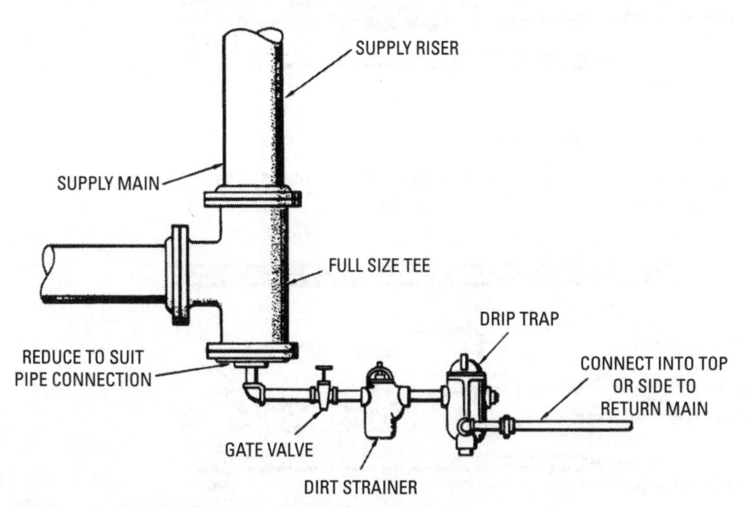

Figure 8-38 Method of dripping base of main supply riser of downfeed system through dirt strainer and drip trap.

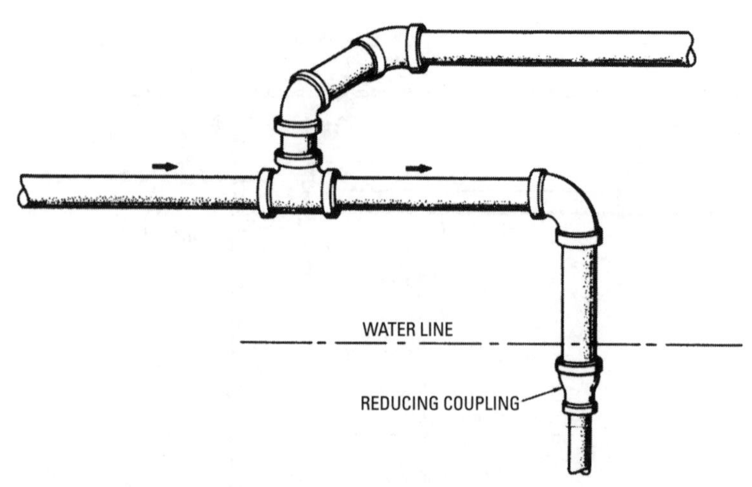

Figure 8-39 Dripping main where it rises to a higher level.

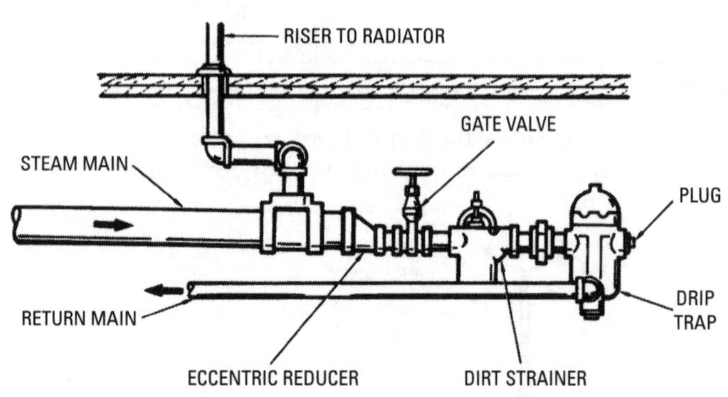

Figure 8-40 Method of dripping steam mains with drips on end of main.

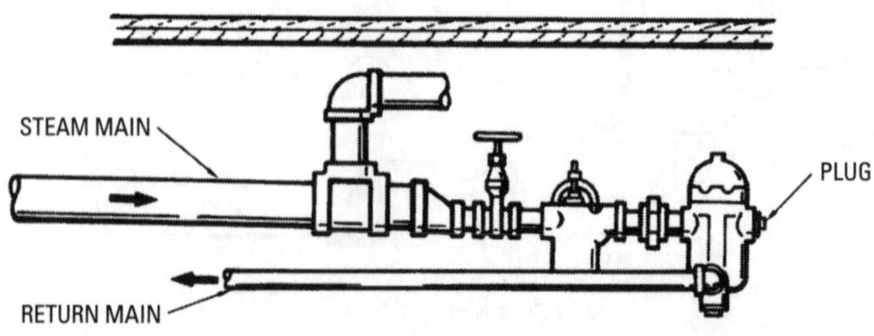

Figure 8-41 Method of dripping steam mains with relay drip in main.

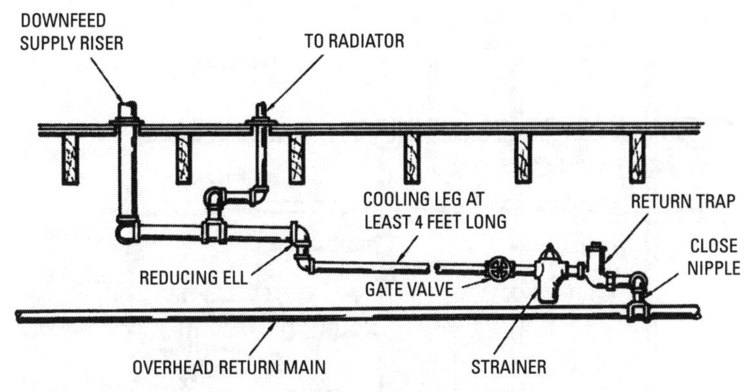

Figure 8-42 Method of installing drip connection at base of downfeed supply riser to overhead return main through cooling leg, dirt strainer, and return trap.

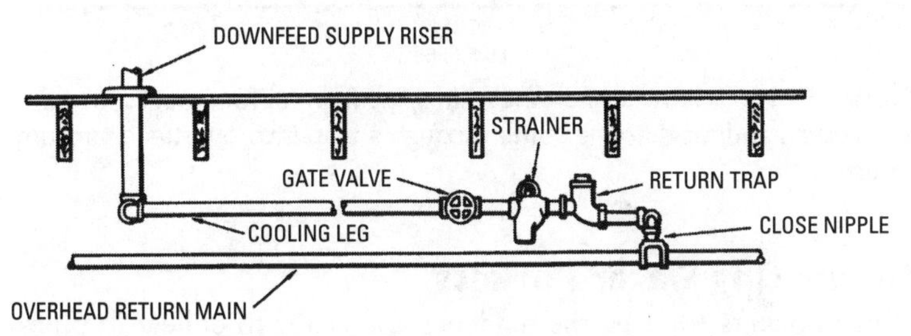

Figure 8-43 Method of installing drip connection at base of downfeed supply riser to overhead return main.

often subject to stresses caused by the expansion and contraction of the pipe. These pipe stresses can be relieved in a number of ways, including the installation of either a U-expansion bend or an expansion joint.

Swivels and Offsets

Another method of providing for pipe expansion in steam mains is through the use of swivels and offsets (see Figure 8-48). Allow at least 4 feet of offset for each inch of expansion to be taken up in the line. The offsets should be placed far enough apart to minimize any strain on the threads when expansion or contraction occurs.

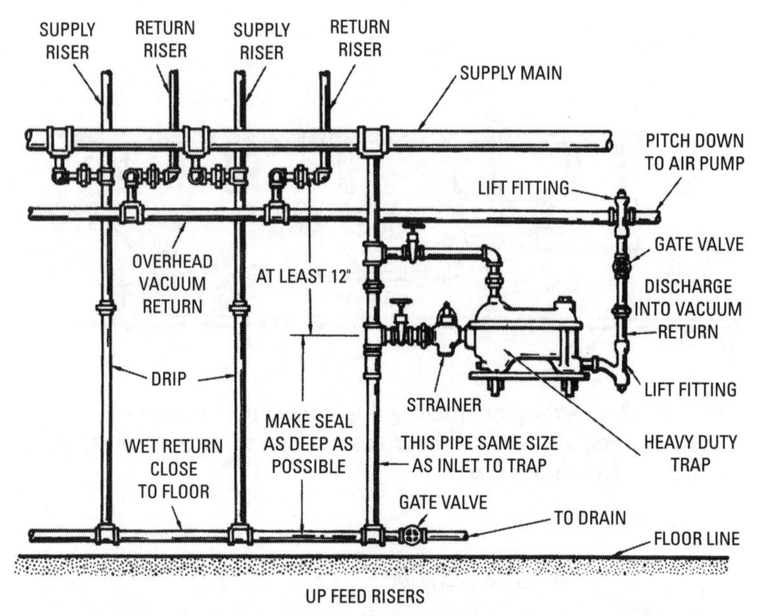

Figure 8-44 Method of connecting drips from upfeed risers into a wet return and discharging some through a trap into overhead vacuum return main.

Eliminating Water Pockets

Water pockets (that is, the tendency for water to collect in pipes) are caused by incorrect pipe installation methods (see Figures 8-49 and 8-50). Not only do these water pockets have the potential danger of freezing and damaging the pipes when the boiler is shut down, but they also cause that loud and disagreeable hammering in the pipes known as *water hammer.* Water hammer is caused by a sudden rush of steam picking up this undrained water when the radiator or convector is opened and forcing it against any turn in the direction of the main. Hammering in steam lines can be stopped by providing for the proper drainage of the condensation. This can be accomplished by installing eccentric fittings or by installing traps of suitable capacity. For example, the water pocket shown in Figure 8-50 can be avoided by using an eccentric reducing tee (see Figure 8-51).

Water trapped in a supply line can sometimes be the cause of radiators failing to heat properly. In this case, the trapped water is

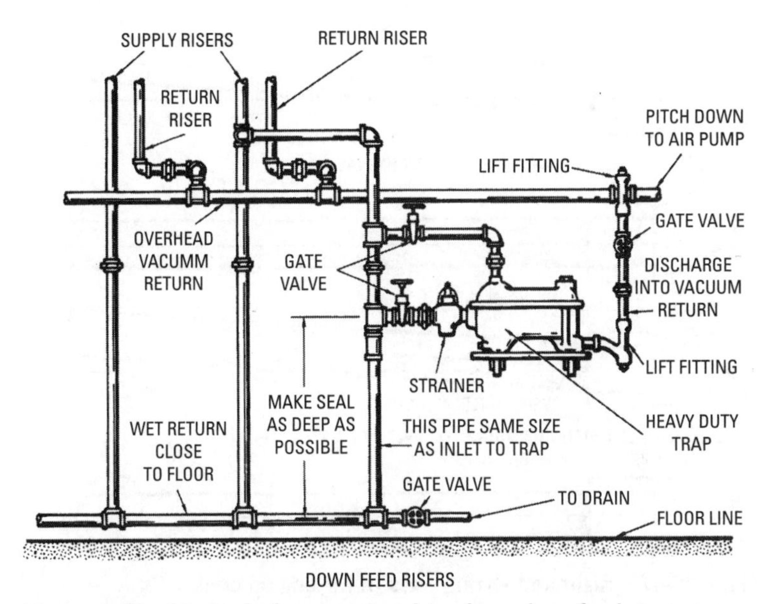

Figure 8-45 Method of connecting drips from downfeed risers into wet return and discharging some through a trap into overhead vacuum return main.

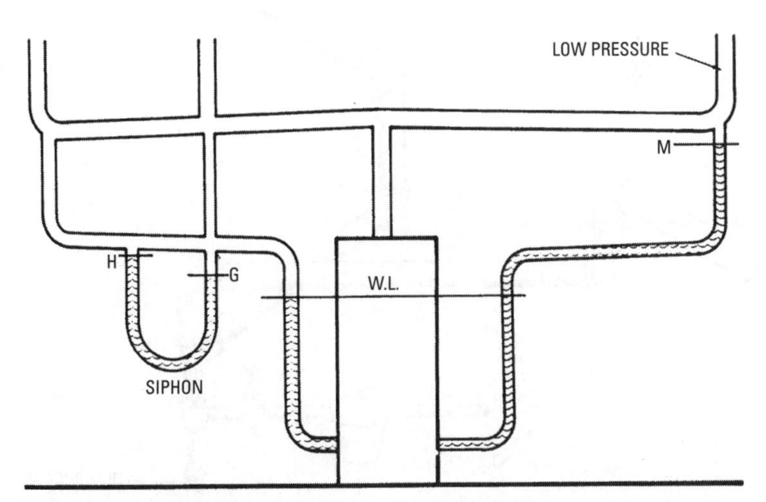

Figure 8-46 Siphon installed on a dry return.

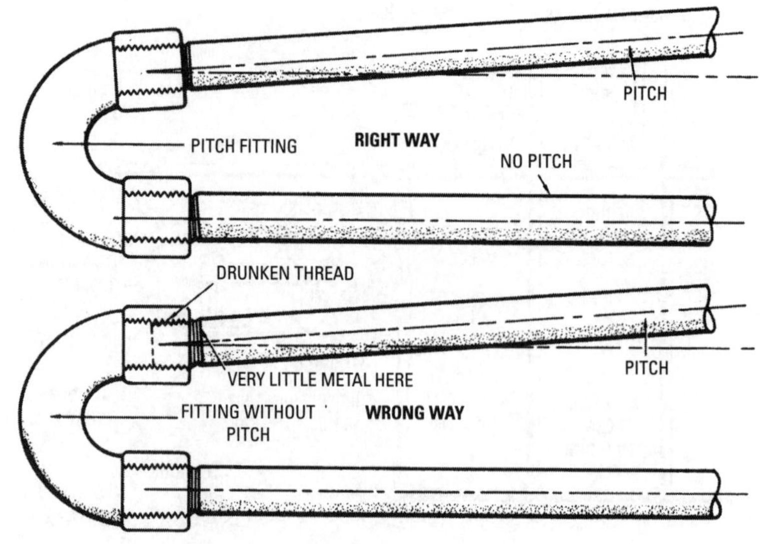

Figure 8-47 Right and wrong ways of making up coils.

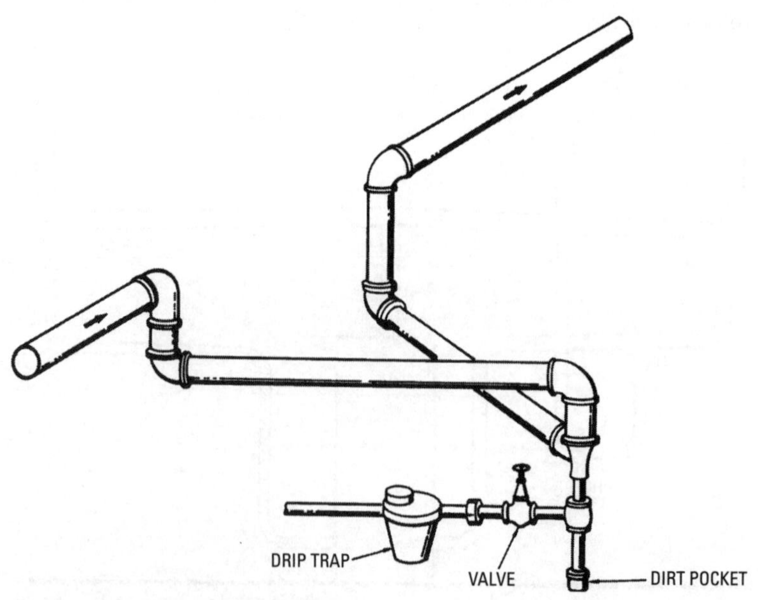

Figure 8-48 Swivels and offsets.

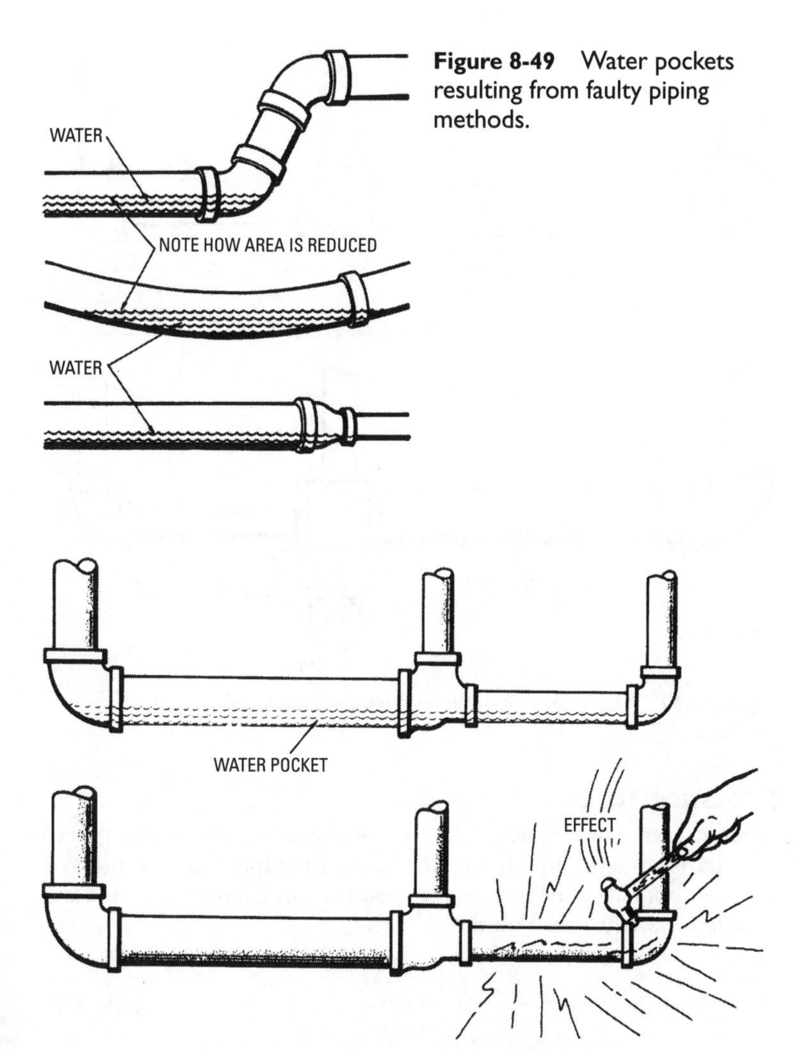

Figure 8-49 Water pockets resulting from faulty piping methods.

Figure 8-50 Water hammer caused by using ordinary fittings.

usually caused by an improperly pitched supply line, rather than fittings. You can create a slight pitch in the line by slipping wedges under the radiator. Space the wedges so that the radiator remains level (check it with a carpenter's level); otherwise, the radiator will not operate properly.

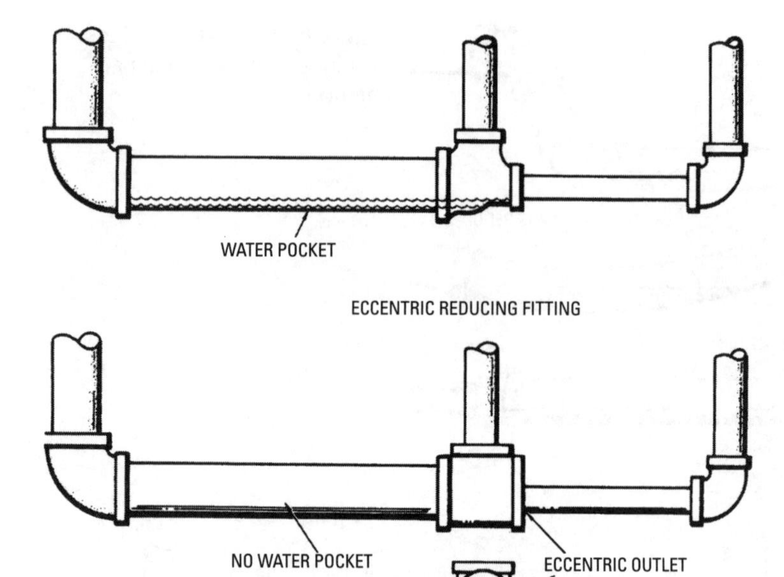

Figure 8-51 Using an eccentric reducing fitting to eliminate water pockets.

Pressure Tests

Pressure tests are performed on hydronic heating systems on a periodic basis to determine the condition of the piping/tubing and system components. These tests are described in Chapter 1 ("Radiant Heating Systems") in Volume 3.

Chapter 9

Valves and Valve Installation

A valve is a device used for controlling the flow or pressure of a fluid in the pipes of steam and hot-water heating systems. Valves are generally restricted in use to the control function for which they were designed. Specifically, a valve may be used to control a fluid in one of the following ways:

- Stopping its flow
- Checking its flow
- Throttling its flow
- Diverting its flow
- Reducing its temperature
- Relieving its pressure
- Reducing or regulating its pressure

Valve Components and Terminology

The operation of a valve may be controlled either internally or externally, depending on the type of valve. For example, the operation of a check valve depends on the fluid flow in the pipe. Because the operation of this type of valve is controlled internally and is therefore automatic in nature, no means are provided for external adjustment. Access to the working parts (disc, seat rings, ball, and so on) of a check valve is usually obtained through a cap secured to the valve bolt by bolts or a threaded connection. The operation of most other types of valves is controlled externally, either manually or automatically.

Generally speaking, an externally controlled valve consists of a body to which an extension (a bonnet, yoke, or yoke and bonnet) is attached. The bonnet contains the valve stem, packing nut, and stuffing box. The yoke consists of upright arms mounted on the bonnet or on a gasket placed on the neck of the valve body. When a yoke is used, the stem is threaded and guided through the upper yoke arm.

The valve stem is an adjustable screw or shaft inside the bonnet that opens or closes the valve by moving a disc holder and disc attached to the end of the stem up or down inside the valve body (see Figure 9-1). The valve closes when the disc makes full contact with the valve seat, a stationary surface in the valve body. The valve opens when the stem pulls the disc holder and disc up off the valve seat. Some valves use a wedge or plug to vary the flow of the medium through the valve by varying the size of the opening

One method of classifying these externally controlled valves is according to the design of the bonnet. The following are four common valve bonnet designs:

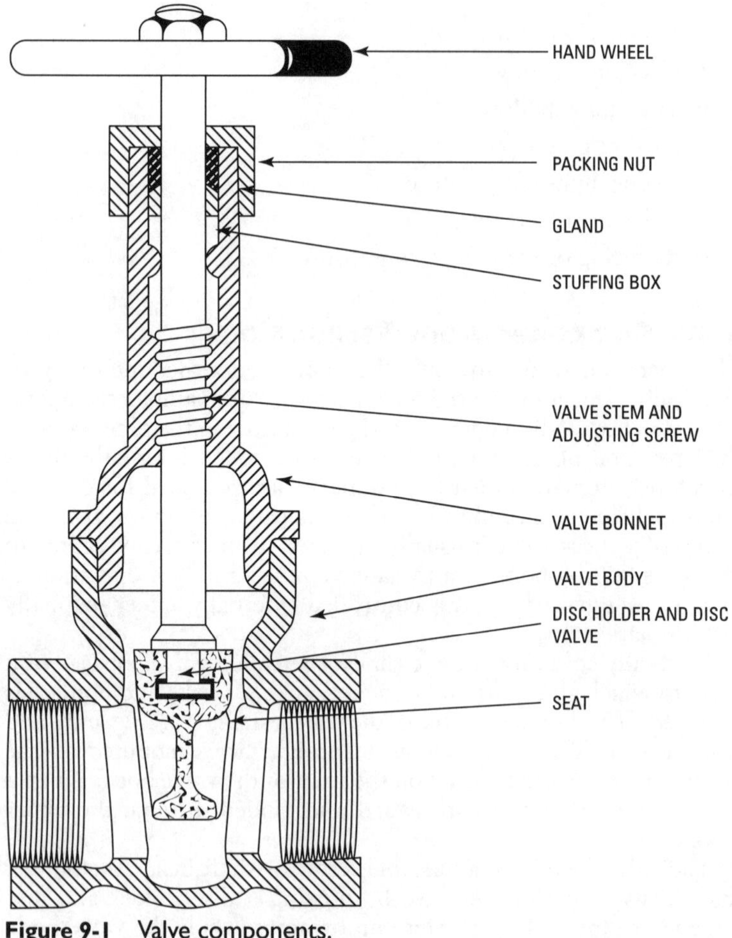

Figure 9-1 Valve components.

- Union bonnet
- Screwed bonnet
- Bolted flanged bonnet
- Bolted flanged yoke bonnet

A *union bonnet* (see Figure 9-2A) is secured to the valve body by a heavy ring nut. A bevel on the bottom of the bonnet engages with a corresponding bevel in the body neck. The ring nut fits down over the bonnet, covering the body neck and the bottom portion of the bonnet. A tight seal can be obtained by wrenching down the ring nut. The joint is additionally sealed by the pressure from within the valve body.

The advantage of a union bonnet is that its construction provides a quick and easy method of coupling and uncoupling the bonnet and valve body.

Figure 9-2A Union bonnet. *(Courtesy Wm. Powell Co.)*

A *screwed bonnet* (see Figure 9-2B) is secured to the valve body by a threaded connection. There are two basic types of screwed bonnets. One type has threads on the outside of the base of the bonnet and on the inside of the body neck and is sometimes referred to as a *screwed-in bonnet*. The other type has threads on the inside of the bonnet and on the outside of the body neck. It is referred to as a *screwed-on bonnet*.

A *bolted flanged bonnet* (see Figure 9-2C) is attached to the top of the valve body neck by bolts. The joint formed by the connection between these two surfaces may be one of the following three types:

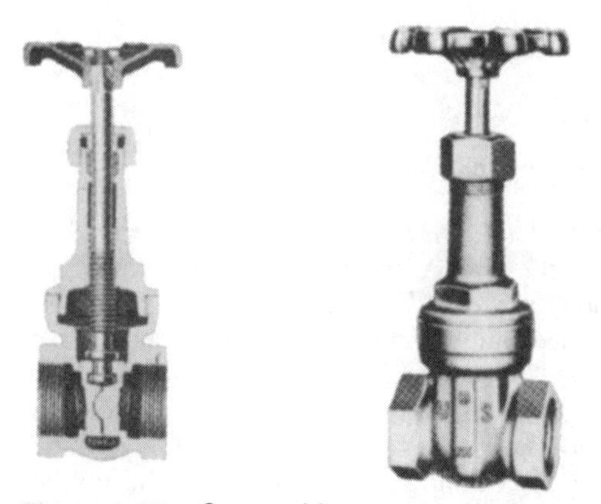

Figure 9-2B Screwed bonnet. *(Courtesy Wm. Powell Co.)*

Figure 9-2C Bolted flanged bonnet. *(Courtesy Wm. Powell Co.)*

(1) flat-faced joint, (2) male and female joint, or (3) tongue-and-groove joint. Each of these joints is illustrated in Figure 9-2D.

A flat-faced joint is generally used only in low-pressure service. The possibility of a gasket blowing out is eliminated when a male and female joint is used because it provides correct gasket compression and ensures alignment between the bonnet and the valve body. A tongue-and-groove joint is used in high-pressure and high-temperature installations. Its construction also ensures perfect alignment of the bonnet onto the valve body.

Some valves may also be classified according to the type of valve stem used with the bonnet valves, which sometimes includes screwed and union bonnet types. The stems used with these valves include the following:

Figure 9-2D Bolted flanged bonnet. *(Courtesy Wm. Powell Co.)*

- Outside screw, rising stem with bonnet.
- Inside screw, rising stem with bonnet.
- Inside screw, nonrising stem with bonnet.
- Outside screw, rising stem with yoke.

The *stem* is the shaft connecting the handwheel at the top of the valve to a wedge or disc in the valve body. The handwheel turns the stem, which changes the position of the wedge or disc and opens or closes the valve.

An *outside screw stem* (see Figure 9-3) is one that has the threaded portion of the stem shaft outside the valve. The advantage of this stem is that the threads do not come in contact with the

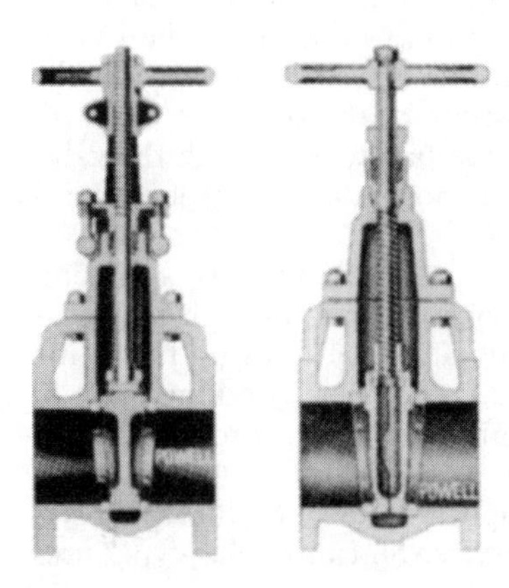

Figure 9-3 Rising and nonrising stem valves. *(Courtesy Wm. Powell Co.)*

media inside the valve and thereby avoid the possibility of erosion and corrosion. If the stem threads are located inside the valve, it is called an *inside screw stem.*

A stem may also be referred to as a rising or nonrising stem. A *rising stem* is a stem shaft that moves upward as the valve handwheel is turned in a counterclockwise direction (that is, to the *open* position). A *nonrising stem* does not rise but merely turns with the handwheel. The stem rotates in the bonnet or yoke and is threaded into a disc holder, which is threaded into the wedge or disc. Nonrising stem valves are especially suited where clearance is limited.

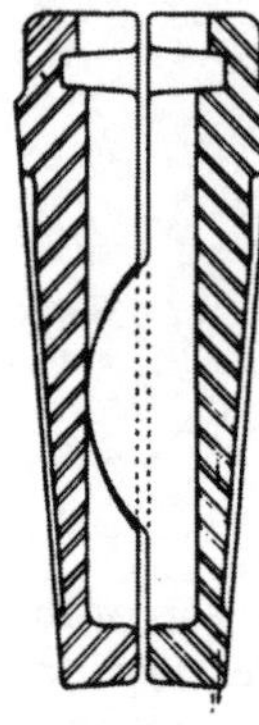

Figure 9-4 Sectional view of a split wedge.
(Courtesy Wm. Powell Co.)

A *wedge* (see Figure 9-4) is a device used to control the flow of steam or water through a gate valve. When the valve is wide open, the wedge is lifted entirely out of the main chamber of the valve body, providing a straightway flow area through the valve.

Wedges for gate valves are available in the following types: (1) split wedge, (2) solid wedge, (3) double wedge, and (4) flexible wedge. These wedges are illustrated in Figure 9-5.

The wedge is attached to the end of the valve stem. When the valve is closed, the wedge is in full contact with the seat.

In globe, angle Y globe and angle, and check valves, a disc is used to control the flow of the steam or water. A *disc* is a flat device attached to the end of the valve stem. The exception to the rule is the swing-check valve, which has no stem. In this valve, the position of the disc is controlled internally by the flow of the steam or water. In operation, the disc opens or closes the pathway through the valve depending on the position of the stem. Some different types of discs and seats are illustrated in Figure 9-6.

Both composition and metal discs are used in valves. The composition discs are suitable for service other than high temperature. While they do not last as long as metal discs on throttling service, they will enable the seat to last longer by taking most of the wear of wire drawing. They are also easier and cheaper to replace.

Metal discs last longer than composition discs on throttling service and on higher-temperature service. So-called plug-type discs are suitable for close throttling service.

Figure 9-5 Wedges for gate valves. *(Courtesy Wm. Powell Co.)*

The wedge or disc closes tightly against a seat to close off the valve. As illustrated in Figure 9-7, the valves and seats may be either flat or beveled. For equal discharge capacity, a beveled valve must be opened more than a flat valve.

Valve Materials

Valves are made of a number of different metals and metal alloys, and the choice of metal often will mean the difference between good or bad service. This is a *very* important point to remember, because it is not simply a matter of going down to the local supply house with only the valve design and function in mind. You must also consider the *service* for which the valve is intended. Regardless of the type of valve you select, it should be rated for its pressure and service temperature range. You would not want to select valves rated for 125 psi saturated steam and a steam service temperature range to 500°F maximum if your system is capable of producing pressures and temperatures in excess of these limits.

The standards and specifications prescribing the rules and regulations for valve construction and use will be found in the *latest* publications of the following associations:

- American Petroleum Institute (API Specifications)
- American National Standards Institute (ANSI Codes and Standards)

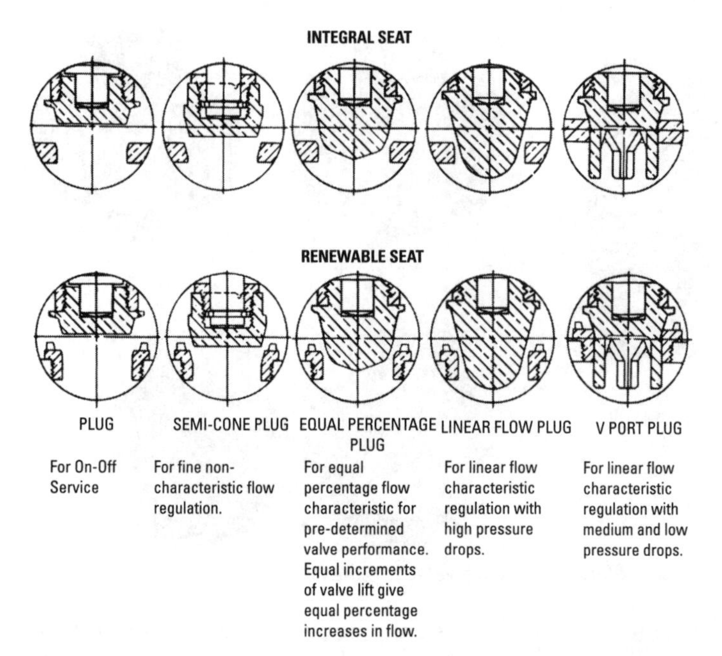

Figure 9-6 Examples of discs and seats provided for globe, angle, and Y valves.

- American Society of Mechanical Engineers (ASME Boiler Construction and Unfired Pressure Vessel Code)
- American Society for Testing Materials (ASTM Material Specifications)
- Manufacturers' Standardization Society of the Valve and Fittings Industry (MSS Standard Practices)

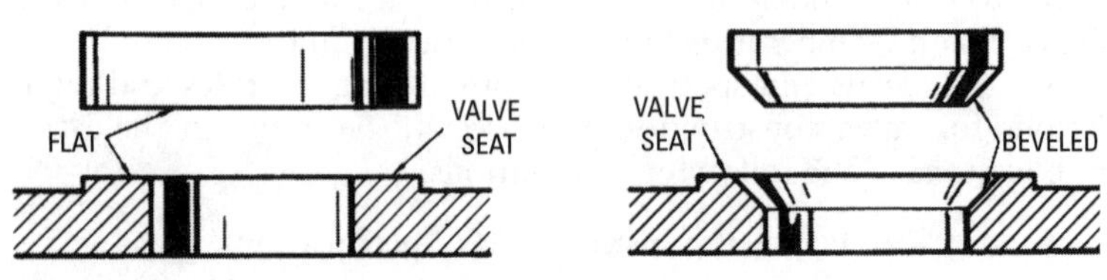

Figure 9-7 Valve seats.

Valves are available in the following metals and metal alloys:

- Bronzes
- Cast iron
- Semisteel
- Stainless steel
- Low-carbon and low-alloy steel
- Special alloys and pure metals

Bronze valves are suitable for steam pressures of 125 psi and 150 psi at 500°F and 200 psi, 300 psi, and 350 psi at 550°F. For steam pressures of 125 psi and 250 psi at 450°F, cast-iron valves are recommended. Steel valves should be used for higher pressures and temperatures. Valves made of pure metals and metal alloys are used in processing systems where resistance to the corrosive and erosive action of the fluids is of prime consideration.

Information About Valves

Valve manufacturers are especially willing to help you select the most suitable valve for your needs. They produce excellent informative literature to this end, including illustrations, specifications, and installation and operating procedures for their valves. This information is available by downloading it from their Web sites or by writing them directly.

- Wm. Powell Co.
  2503 Spring Grove Avenue
  Cincinnati, Ohio 45214
  513-852-2000
  www.powellvalves.com
- Spirax Sarco Inc.
  Northpoint Business Park
  1150 Northpoint Blvd.
  Blythewood, South Carolina 29016
  803-714-2000
  www.spiraxsarco-usa.com
- Honeywell, Inc.
  101 Columbia Road
  Morristown, New Jersey 07962
  973-455-2000
  www.honeywell.com

(continues)

- Hoffman Specialty
  3500 N. Spaulding Avenue
  Chicago, Illinois 60618
  773-267-1600
  www.hoffmanspecialty.com
- Bell & Gossett
  8200 N. Austin Avenue
  Morton Grove, Illinois 60053
  847-966-3700
  www.bellgossett.com
- Watts Industries Inc.
  815 Chestnut Street
  North Andover, Massachusetts 01845
  978-688-1811
  www.wattsind.com

Globe and Angle Valves

Globe and angle valves are used primarily as throttling devices to control the *rate* of flow. Flow characteristics of both globe and angle valves are illustrated in Figure 9-8.

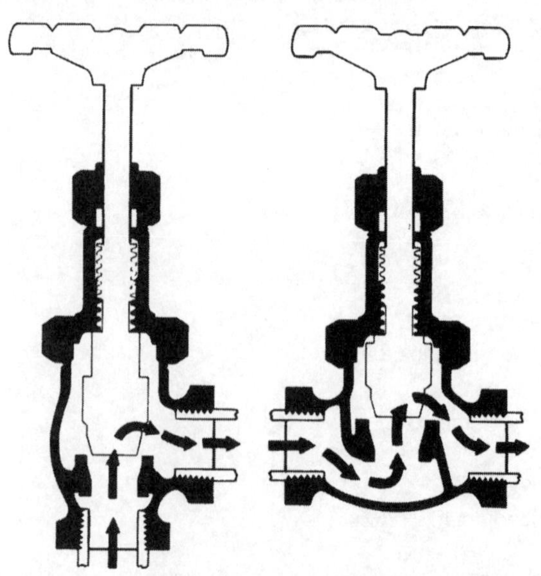

Figure 9-8 Sectional view of a globe and angle valve. *(Courtesy Wm. Powell Co.)*

A *globe valve* is a valve that has a round ball-like shell with a stuffing box extension through which passes the screw spindle, which operates the valve disc. The valve seat is parallel to the line of flow, and the inlet and outlet branches are opposite one another.

The design and construction of an *angle valve* is similar to the globe valve, except that the outlet is at right angles to the inlet branch, thus combining in itself a valve and an elbow.

Globe and angle valves are preferred to gate valves (see following section) if the valve is to be operated frequently or if it is to be used as a throttling device (that is, operated partially open), because either type can open with fewer turns. An additional feature of the angle valve is that it can control both a *change* and the direction of flow. It is designed to create a 90° change in direction. As a result, it saves an elbow and nipple and reduces the pressure loss.

Globe valves are commonly used in pipelines that extend through basement areas or areas of limited access. *Offset* globe valves are often recommended for radiators on second floors or higher. Angle valves are commonly used on first-floor radiators.

Ordinarily, globe and angle valves should be installed with the pressure *under* the disc so that the stuffing box may be repacked without escape of steam or water. Installing these valves in this manner not only promotes easier operation but also provides a certain degree of protection to the packing and reduces the erosive action on the disc and seat faces. High-temperature steam service proves to be the exception. If high-temperature steam is the medium to be controlled, the globe or angle valve should be installed so that the pressure is *above* the valve disc. If the usual installation method (that is, with the pressure under the disc) is used, then the high-temperature steam will be under the disc when the valve is closed, and the valve stem will be out of the fluid. As a result, the valve stem will cool and contract, causing the disc to lift slightly off the valve seat. The leaks that then occur result in wire drawing on seat and disc faces.

Figure 9-9 illustrates both the correct and incorrect way to place globe valves on horizontal lines. When the globe valve is placed in an upright position, considerable water will remain in the pipeline and be subject to freezing. When it is placed in a horizontal position, most of this water will drain through the seat opening with less danger of damage by freezing.

Figures 9-10, 9-11, and 9-12 illustrate a number of commonly used globe and angle valves. These illustrations should be referred to when reading the section on troubleshooting valves.

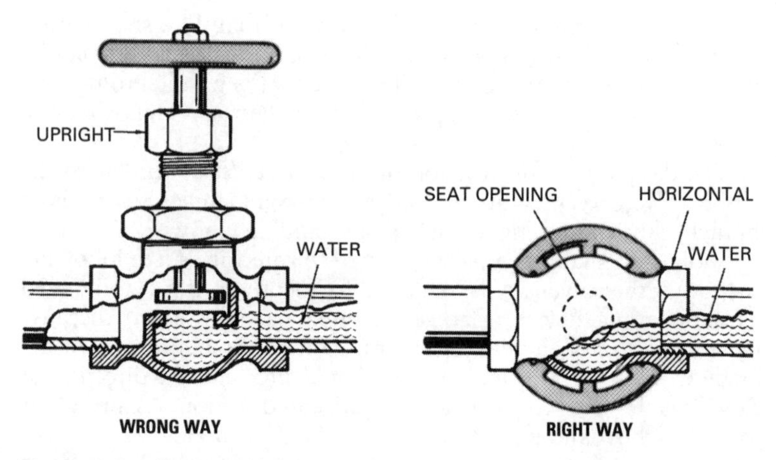

Figure 9-9 Correct and incorrect way of placing a globe valve on a horizontal line.

Gate Valves

A *gate valve* (see Figure 9-13) is one having two inclined seats between which the valve (consisting of a single or double disc) wedges down in closing. In opening, the valve is drawn up into a dome or recess, thus leaving a straight passage the full inside diameter of the pipe. This lifting of the wedges completely out of the waterway, thereby creating an unobstructed passage for the fluid, is a significant feature of the gate valve. As a result, turbulence is minimized, and there is very little pressure drop.

Gate valves are the most frequently used valves and are preferable to globe or angle valves for lines on which it is important to minimize resistance to flow and friction losses. They are also necessary where complete drainage of the pipe must be provided.

The gate valve is primarily suited for locations where valves are to be generally wide open or shut tight. Globe valves are preferable for throttling. When gate valves are used for throttling, the high velocity of flow tends to damage the surface of the seats.

Wedge gate valves are preferable where the valve is to be installed with the stem extending downward. Nonrising stem gate valves are usually tighter at the stuffing box than those with rising stems and require less headroom.

The double-disc gate valve enjoys certain advantages over the single-disc type (see Figure 9-14). If the fluid contains foreign matter, the flexibility of operation of the double disc enables one disc to

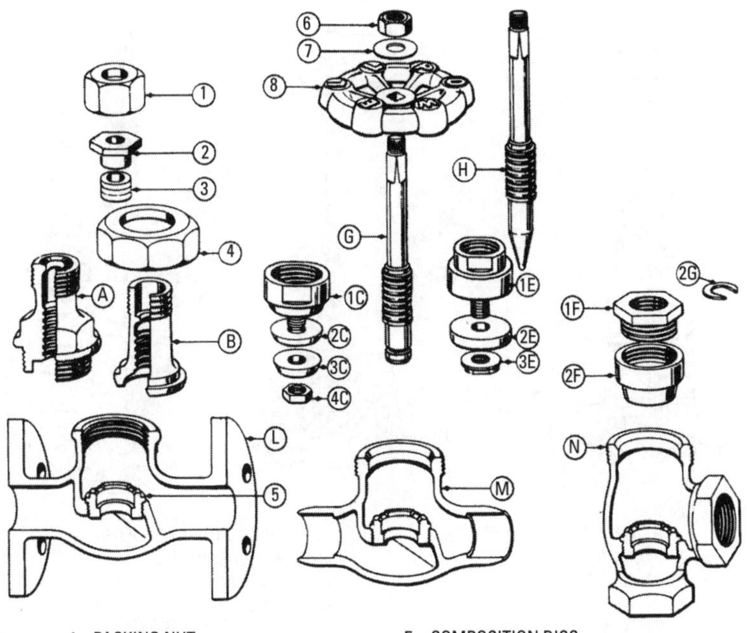

1. PACKING NUT
2. PACKING GLAND
3. PACKING
4. BONNET RING
5. SEAT RING
6. HANDWHEEL NUT
7. IDENTIFICATION PLATE
8. HANDWHEEL

A. SCREWED-IN BONNET
B. UNION BONNET
C. HI-LO DISC
   1C-DISC HOLDER
   2C-NON-METALLIC DISC
   3C-DISC PLATE
   4C-DISC NUT
E. COMPOSITION DISC
   1E-DISC HOLDER
   2E-NON-METALLIC DISC
   3E-DISC-LOCKNUT WASHER
F. DISC LOCKNUT
   1F-DISC NUT
   2F-DISC
G. STEM-DISC LOCKNUT (HORSE SHOE RING) TYPE
   2G-HORSE SHOE RING
H. STEM-NEEDLE DISC TYPE
L. BODY-GLOBE-FLANGED ENDS
M. BODY-GLOBE-SOLDER JOINT ENDS
N. BODY-ANGLE-THREADED ENDS

Figure 9-10 Screwed and union bonnet globe and angle valve.
(Courtesy Wm. Powell Co.)

seat tightly if dirt or some other type of foreign matter is present. Moreover, the discs and seats are more readily refaced than those of the single-disc type.

Outside screw and yoke valves should be used when it is necessary to know at a glance whether the valve is open or shut tight.

Some examples of gate valves used in heating and cooling installations are illustrated in Figures 9-15, 9-16, and 9-17. Refer to

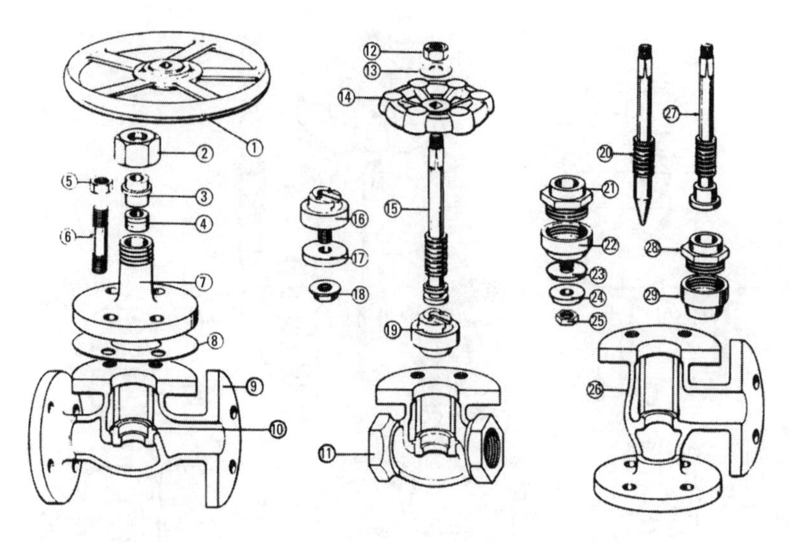

1. HANDWHEEL-ROUND
2. PACKING NUT
3. PACKING GLAND
4. PACKING
5. BODY NUT
6. BODY STUD
7. BONNER
8. GASKET
9. BODY-GLOBE-FLANGED ENDS
10. SEAT RING
11. BODY-GLOBE-THREADED ENDS
12. HANDWEEL NUT
13. IDENTIFICATION PLATE
14. HAND WHEEL-NON-HEATING
15. STEM-SLIP-ON DISC TYPE
16. COMPOSITION DISC HOLDER
17. NON-METALLIC DISC
18. DISC LOCKNUT
19. SLIP-ON DISC
20. STEM-NEEDLE DISC TYPE
21. HI-LO DISC LOCKNUT
22. DISC HOLDER
23. NON-METALLIC DISC
24. DISC PLATE
25. DISC NUT
26. BODY-ANGLE-FLANGED ENDS
27. STEM-DISC LOCKNUT TYPE
28. DISC LOCKNUT
29. DISC

Figure 9-11 Bolted bonnet inside screw globe and angle valve.
(Courtesy Wm. Powell Co.)

these illustrations when reading the section on troubleshooting valves in this chapter.

Gate valves are available with either a rising or nonrising stem, the former being more commonly used on smaller gate valves.

Check Valves

A *check valve* is one that automatically opens to permit the passage of liquid in one direction and automatically closes to prevent any flow in the opposite direction. In other words, it specifically operates to prevent the reversal of water flow in the line.

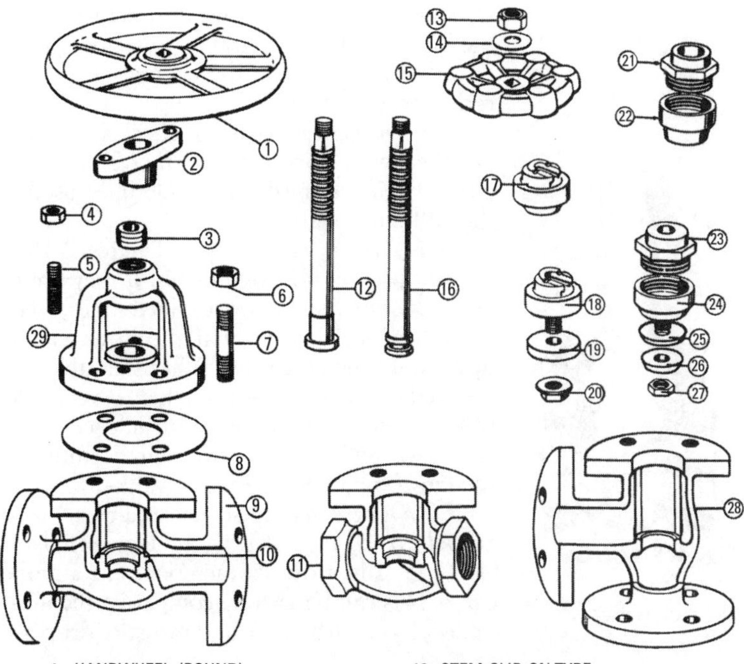

1. HANDWHEEL (ROUND)
2. PACKING GLAND
3. PACKING
4. GLAND STUD NUT
5. GLAND STUD
6. YOKE STUD NUT
7. YOKE STUD
8. GASKET
9. BODY-GLOBE-FLANGED ENDS
10. SEAT RING
11. BODY-GLOBE-THREADED ENDS
12. STEM-DISC LOCKNUT TYPE
13. HANDWHEEL NUT
14. IDENTIFICATION PLATE
15. HANDWHEEL
16. STEM-SLIP-ON TYPE
17. DISC-ONE-PIECE-SLIP-ON
18. COMPOSITION DISC HOLDER
19. COMPOSITION DISC
20. DISC NUT
21. DISC LOCKNUT
22. DISC
23. DISC LOCKNUT-HI-LO DISC
24. DISC HOLDER
25. NON-METALLIC DISC
26. DISC PLATE
27. DISC NUT
28. BODY-ANGLE-FLANGED ENDS
29. YOKE

Figure 9-12 Bolted bonnet outside screw and yoke globe and angle valve. *(Courtesy Wm. Powell Co.)*

A check valve is sometimes placed on the return connection to prevent the accidental loss of boiler water to the returns with consequent danger of boiler damage. When feasible, the Hartford connection is preferred over the check valve because the latter is apt to stick or not close tightly. Moreover, the check valve offers additional

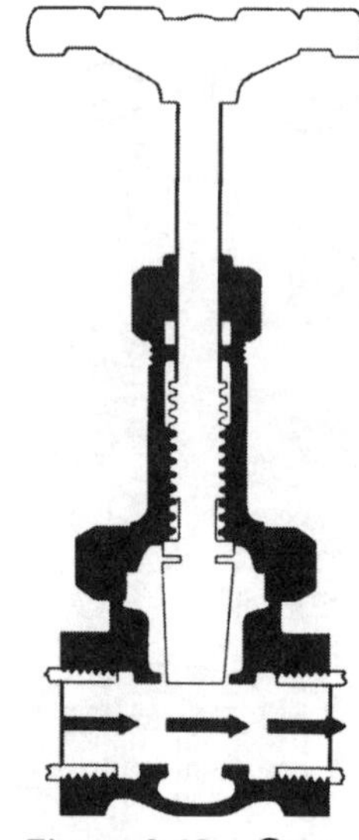

Figure 9-13 Gate valve flow characteristics.
(Courtesy Wm. Powell Co.)

resistance to the condensation returning to the boiler, which in gravity systems would raise the water line in the far end of the wet return several inches.

The two general types of check valves are (1) the swing-check valve and (2) the lift-check valve. A third type sometimes used is the ball-check valve.

Swing-check valves (see Figure 9-18A) operate best on horizontal lines. If they are to be used on vertical or inclined lines, they should be installed so that the flow will be upward through the valve. When the valve disc is raised in the open position, there is very little obstruction in the flow area. Turbulence is also very low. Swing-check valves are commonly used in combination with gate valves.

Lift-check valves (also referred to as *horizontal-lift check valves*) (see Figure 9-18B) have the same flow characteristics as a globe valve. This can be seen by comparing the illustrations. The turbulence and pressure drop are

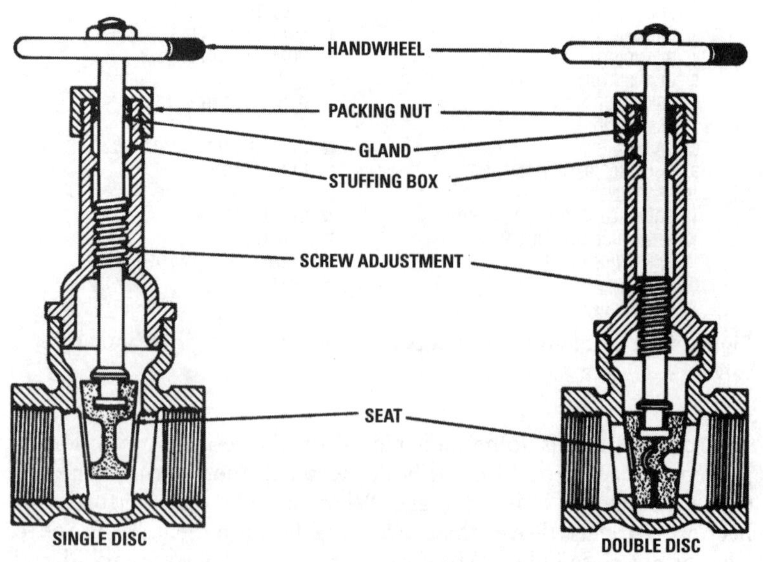

Figure 9-14 Comparison of single- and double-disc gate valves.

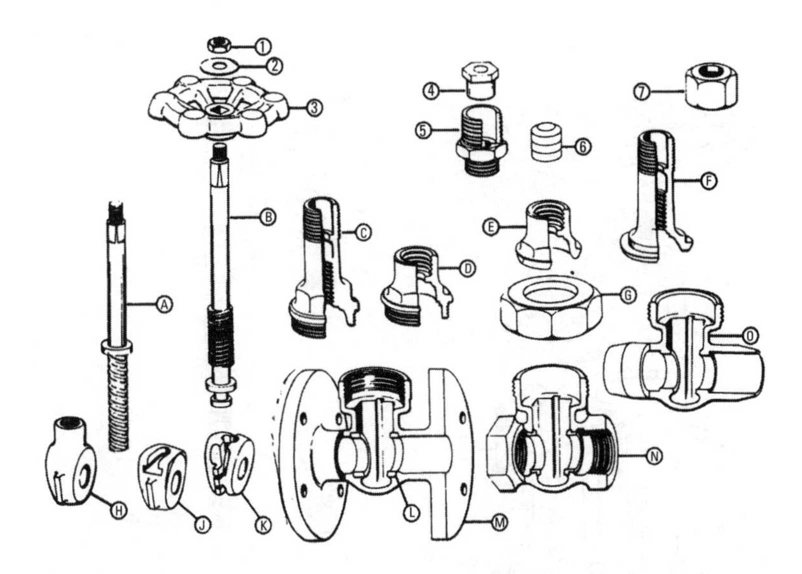

1. HANDWHEEL NUT
2. IDENTIFICATION PLATE
3. HANDWHEEL
4. PACKING GLAND
5. PACKING BOX SPUD (NON-RISING STEM VALVES ONLY)
6. PACKING
7. PACKING NUT

A. STEM-NON RISING STEM VALVES
B. STEM- RISING STEM VALVES
C. SCREWED-IN BONNET-RISING STEM VALVES
D. SCREWED-IN BONNET-NON RISING STEEM VALVES
E. UNION BONNET- NON-RISING STEM VALVES
F. UNION BONNET-RISING STEM VALVES
G. BONNET RING
H. SOLID WEDGE-NON-RISING STEM VALVES
J. SOLID WEDGE -RISING STEM VALES
K. DOUBLE WEDGE-RISING STEM VALVES
L. SEAT RING
M BODY-FLANGED ENDS
N. BODY-THREADED ENDS
O. BODY-SOLDER JOINT ENDS

Figure 9-15 Screwed and union bonnet rising and nonrising stem gate valve. *(Courtesy Wm. Powell Co.)*

therefore similar for these two valves. Lift-check valves are commonly used in combination with globe valves.

A *vertical check valve* should be used on vertical lines only and with the flow upward through the valve.

The *Watts Backflow Preventer* is a safety device containing two check valves and is designed to prevent backflow in boiler feed lines when the supply pressure falls below system pressure. The primary check valve in the backflow device utilizes a disc that seats against a rubber mating part to ensure tight closing. A secondary check valve utilizes a disc-to-metal seating. If the downstream check valve malfunctions (for example, fouls), leakage will be vented to

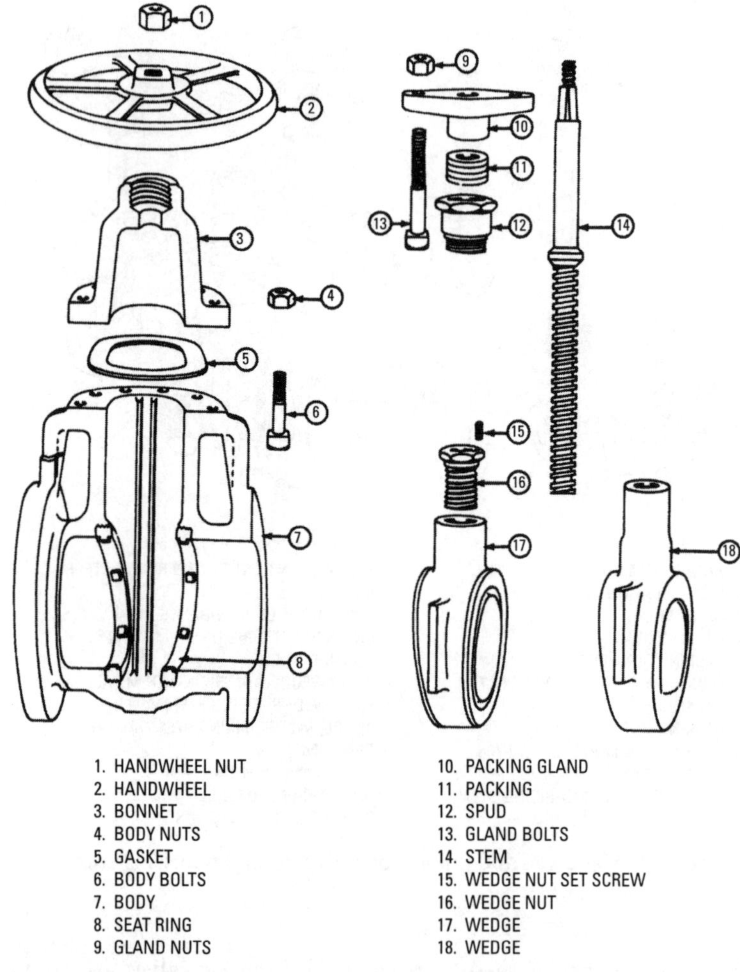

Figure 9-16 Bolted bonnet inside screw nonrising stem gate valve.
(Courtesy Wm. Powell Co.)

atmosphere through the vent port, thereby safeguarding the potable water from contamination.

Figures 9-19 through 9-23 illustrate a number of different check valves used in heating and cooling installations. Refer to these illustrations when reading the section on troubleshooting valves in this chapter.

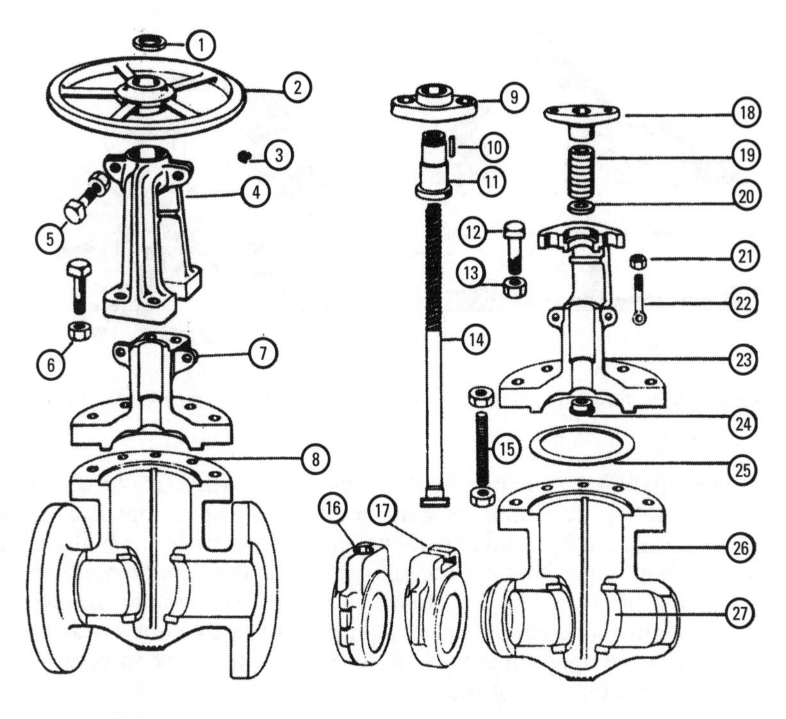

1. STEM BUSHING NUT
2. HANDWHEEL
3. LUBRICANT FITTING
4. YOKEARMS
5. YOKEARM EAR BOLT AND NUT
6. BONNET BOLT AND NUT
7. TWO-PIECE BONNET
8. BODY-FLANGED ENDS
9. BEARING CAP
10. HANDWHEEL KEY
11. STEM BUSHING
12. BEARING CAP BOLT
13. BEARING CAP NUT
14. STEM
15. BODY STUD AND NUTS
16. SPLIT WEDGE
17. SOLID WEDGE
18. PACKING GLAND
19. PACKING
20. PACKING WASHER
21. EYEBOLT NUT
22. EYEBOLT
23. ONE-PIECE BONNET
24. PACK-UNDER-PRESSURE BUSHING
25. GASKET
26. BODY-WELDED ENDS
27. SEAT RING

Figure 9-17 Bolted bonnet outside screw and yoke rising stem gate valve. *(Courtesy Wm. Powell Co.)*

Stop Valves

Globe, gate, and angle valves are used on steam and hot- or cold-water pipelines as *stop valves* (also referred to as *nonreturn valves, nonreturn stop valves,* or *boiler check valves*).

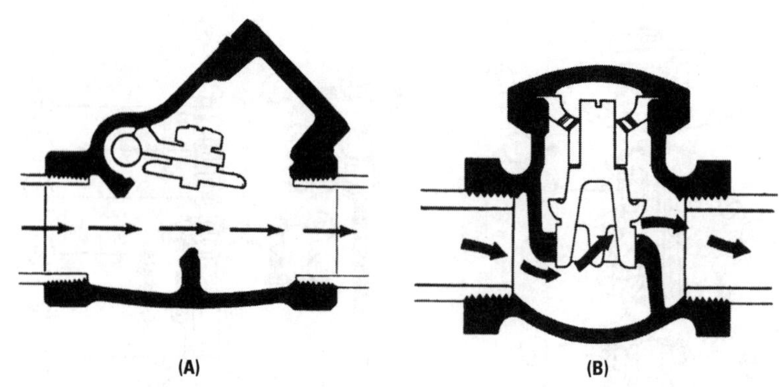

Figure 9-18 Flow characteristics of a swing- and lift-check valve. *(Courtesy Wm. Powell Co.)*

The operating principle of the typical stop valve is shown in Figure 9-24. It functions as a form of check valve that can be opened or closed by hand control when the pressure in the boiler is greater than in the line, but it cannot be opened when the pressure within the boiler is less than that in the line. The counterbalance spring slightly overbalances the weight of the valve and tends to hold the valve open, thus preventing movement of the valve with every fluctuation of pressure.

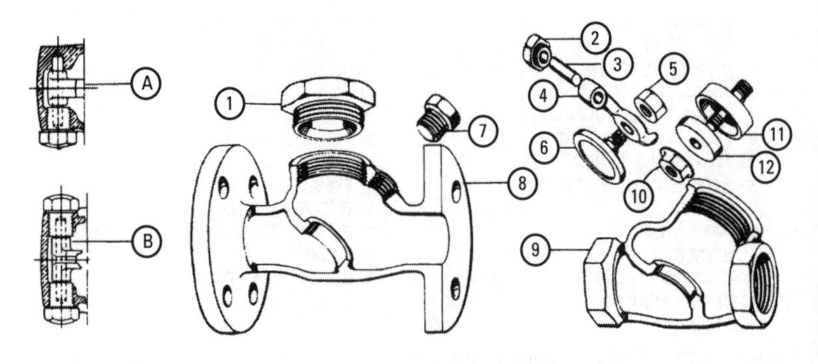

1. CAP
2. SLIDE PLUG
3. CARRIER PIN
4. CARRIER
5. DISC NUT
6. DISC
7. BUMPER PLUG
8. BODY-FLANGED END
9. BODY-THREADED END
10. DISC LOCKNUT-COMPOSITION DISC
11. DISC HOLDER
12. COMPOSITION DISC

A. DETAIL-ONE SIDE PLUG
B. DETAIL-TWO SIDE PLUGS

Figure 9-19 Screwed-cap swing-check valve. *(Courtesy Wm. Powell Co.)*

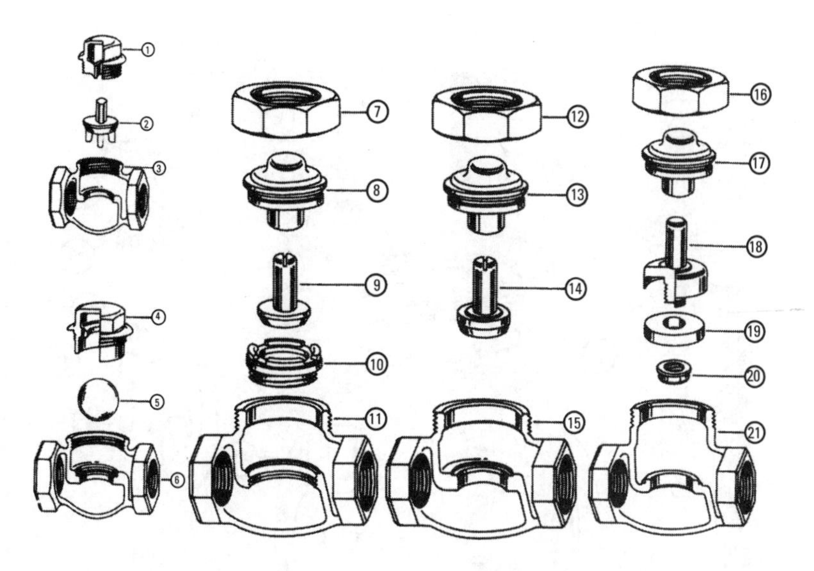

1. CAP
2. DISC
3. BODY-THREADED ENDS
4. CAP
5. BALL
6. BODY-THREADED ENDS
7. RING NUT
8. DISC GUIDE
9. DISC
10. SEAT RING
11. BODY-THREADED ENDS
12. RING NUT
13. DISC GUIDE
14. DISC
15. BODY-THREADED ENDS
16. RING NUT
17. DISC GUIDE
18. DISC HOLDER
19. DISC-COMPOSITION DISC
20. DISC NUT
21. BODY-THREADED ENDS

Figure 9-20 Screwed-cap horizontal-lift check valve. *(Courtesy Wm. Powell Co.)*

Stop valves are often used as a safety device in steam power plants where *more* than one boiler is connected to the same header. They are connected directly to the boiler nozzle outlet between the boiler and the header. They function as a safety device to prevent backflow of the steam into the boiler. This is particularly important if the boiler is cold.

The stop valve should be installed so that the valve stem is in a vertical or upright position. The pressure of the steam or water should be *under* the disc.

Butterfly Valves

A butterfly valve consists of a round (cylindrical) body, a shaft (stem), and a disc that rotates on the shaft (see Figure 9-25). The

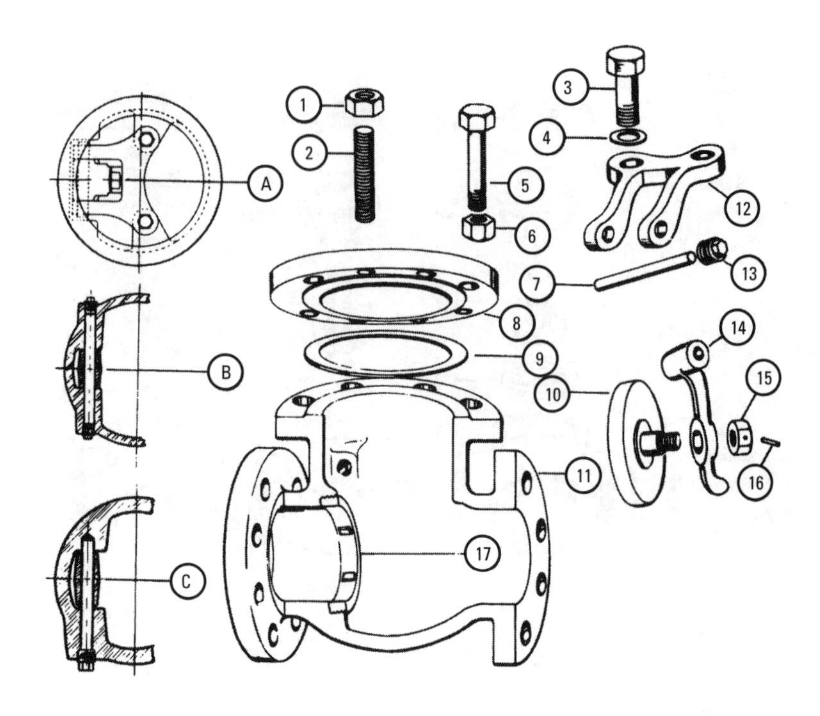

1. BODY NUT
2. BODY STUD
3. CAP SCREW
4. LOCKWASHER
5. BODY BOLT
6. BODY BOLT NUT
7. DISC HOLDER PIN
8. CAP
9. GASKET
10. DISC
11. BODY-FLANGED ENDS
12. DISC HOLDER HANGER
13. PIPE PLUG
14. DISC HOLDER'
15. DISC NUT
16. DISC NUT PIN
17. SEAT RING

A. DETAIL—HANGER TYPE DISC
B. DETAIL—PIN TYPE-TWO SIDE PLUGS
C. DETAIL—PIN TYPE-ONE SIDE PLUG

Figure 9-21 Bolted-cap swing-check valve. *(Courtesy Wm. Powell Co.)*

disc rotates 90 degrees from its open to closed positions. The valve disc provides tight shutoff, although a smaller disc size is available that does not provide tight shutoff.

A butterfly valve is used to control the flow of hot water or water from a condenser in two-position or proportional applications. These valves are available as two-way units. A three-way configuration can be achieved by connecting two two-way butterfly valves to a pipe tee.

Butterfly valves may be operated manually or by a unidirectional or reversing electrically operated actuator (see Figure 9-26). They

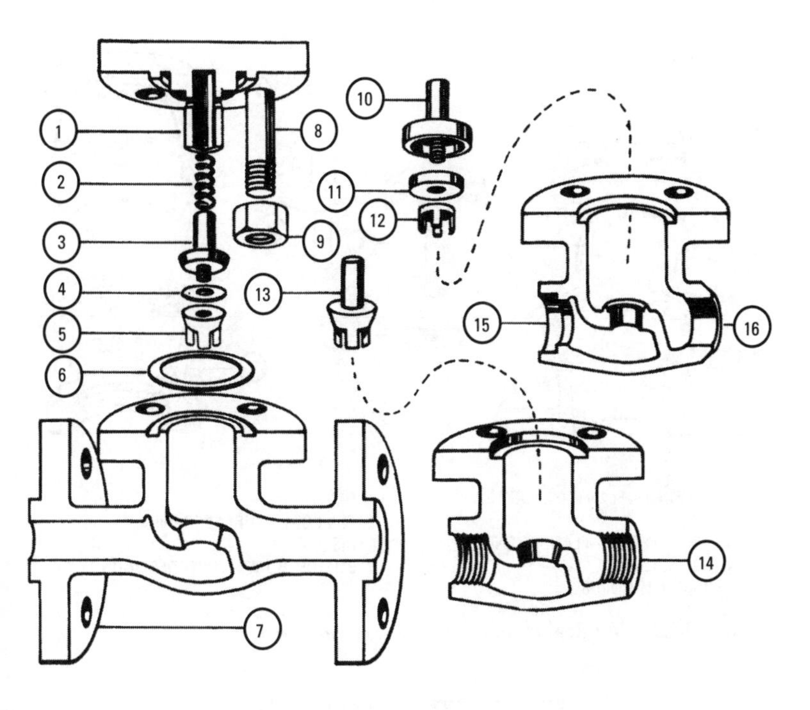

1. CAP—BOLTED	9. CAP BOLT NUT
2. SPRING (OPTIONAL)	10. COMPOSITION DISC HOLDER
3. DISC HOLDER	11. NON-METALLIC DISC
4. TEFLON DISC	12. DISC GUIDE
5. DISC GUIDE	13. METAL DISC
6. GASKET	14. BODY—THREADED ENDS
7. BODY-FLANGED ENDS	15. BODY—SILVER BRAZE ENDS
8. CAP BOLT	16. BODY—SOLDER JOINT ENDS

Figure 9-22 Bolted-cap horizontal-lift check valve. *(Courtesy Wm. Powell Co.)*

are available in sizes ranging from 2 to 12 inches and in either wafer or lug body designs.

Two-Way Valves

A two-way valve is a valve with only one inlet port and one outlet port. This category of valves is used to provide tight shutoff in straight and angled hydronic and steam piping systems. Two-way valves are designed to control the medium (water or steam) in either two-way or proportional applications. All two-way valves

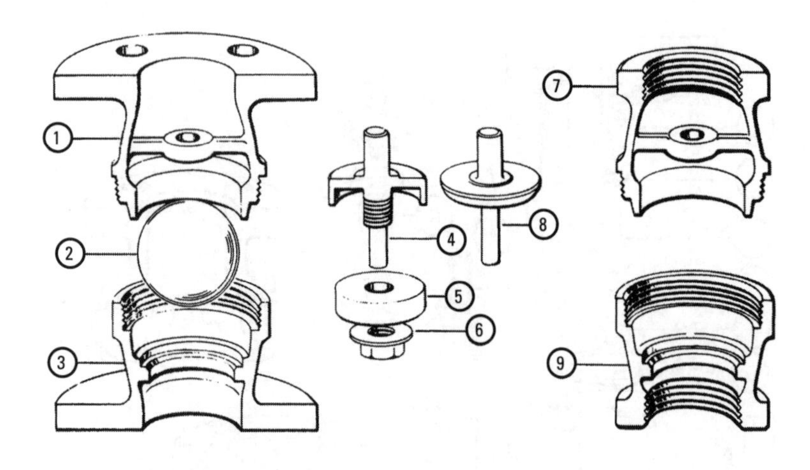

Figure 9-23 Vertical check valve. *(Courtesy Wm. Powell Co.)*

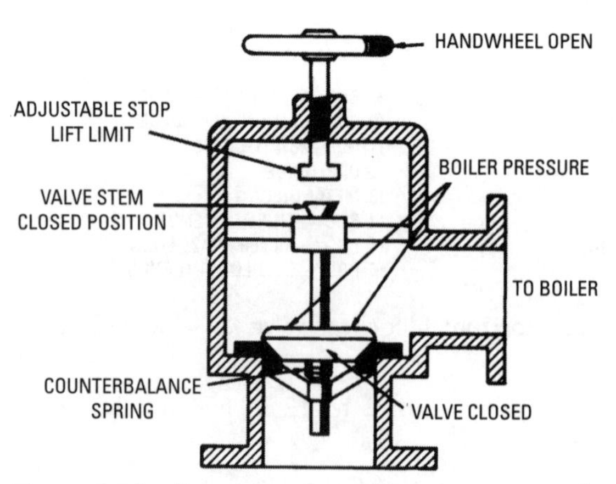

Figure 9-24 Functions of a stop or nonreturn valve.

provide tight shutoff in their closed position. Linear, quick-opening, and equal-percentage-flow two-way valves are available. The type selected will depend on the control requirements of the medium. For example, a linear valve is used for the proportional control of steam or chilled water, or for the control of a medium in applications that do not have wide load variations. A quick-opening two-way valve is used for the two-position control of steam. An equal-percentage two-way valve is used for the proportional control of hot water in hydronic systems.

Three-Way Valves

Three-way valves are globe valves with three ports used in either mixing or diverting applications. When used for mixing (for example, as a water-tempering valve), the valve uses two inlet ports and one outlet port (see Figure 9-27). If the three-way valve functions as a diverting valve, it has one inlet port and two outlet ports.

The *three-way* or *cross valve* shown in Figure 9-28 functions as a diverting valve. It is used to control the flow of the medium to or from a branch line at the junction of a branch and main line. It is essentially a globe valve with its seat located at the bottom of the body and the passage through the seat connected with a third opening at right angles to the other two openings.

Y Valves

A *Y valve* or *Y globe valve* (see Figure 9-29) is designed to combine the operating characteristics of globe and gate valves. In other words, it combines the throttling characteristics of the former with the straightway flow area of the latter. The valve disc can be easily reground and renewed.

Valve Selection

Valves are available in a wide range of designs, sizes, pressure capacity ranges, materials, operating characteristics, and other factors critical to specific applications. The valves must have adequate capacity to support the heating and cooling loads. At the same time, they must be able to efficiently control the flow of water or steam in the system.

The following factors must be taken into consideration when selecting valves for a heating/cooling system:

- Type of medium (water or steam) being controlled
- Pressure and temperature ranges of the medium
- Piping arrangement
- Piping size

The valve manufacturer publishes a specification sheet for each of its valves, which includes the following information:

- Valve size
- Operating pressure

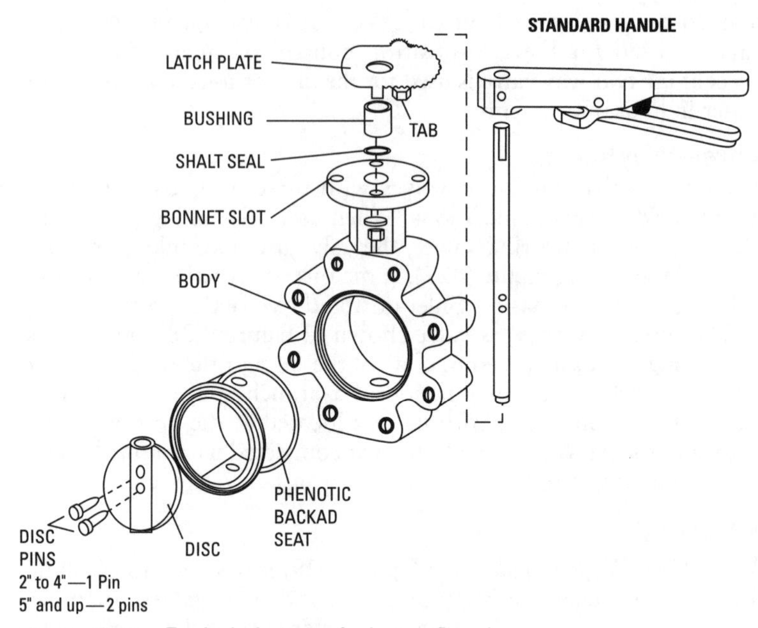

Figure 9-25 Exploded view of a butterfly valve. *(Courtesy Watts Industries, Inc.)*

- Maximum and minimum temperature range
- Reduced pressure range
- Valve material
- Valve capacity

The maximum and minimum temperature range of the valve must match the maximum and minimum temperature of the medium (water or steam) being controlled. The maximum pressure of the medium must never exceed the maximum pressure rating of the valve. If the medium is potentially corrosive (for example, chlorinated water), the valve components must be made of a corrosion-resistant material.

The full open and closed pressure drop across the valve is also an important consideration. The *pressure drop* is the measured difference in upstream and downstream pressures of the medium flowing through the valve. The flow of the fluid through the valve increases as the pressure drop increases until it reaches a critical point. This point is called the *critical pressure drop* and is the point at which increases of the pressure drop no longer increase the flow rate of

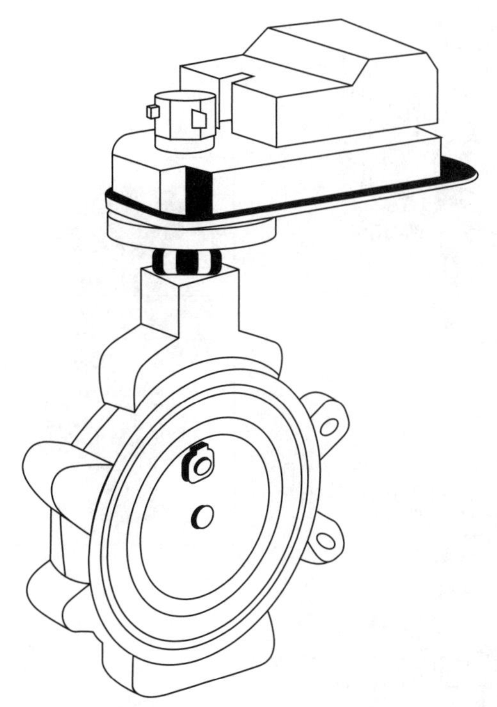

Figure 9-26 Butterfly valve with electrically operated actuator. *(Courtesy Honeywell, Inc.)*

the medium. Instead, it is dissipated as noise and vibration. The vibration can eventually destroy the valve and adjacent pipe fittings if allowed to continue.

The pressure drop of the valve in its full open position must not exceed the valve rating for quiet operation and normal service life. On the other hand, the full open pressure drop must be high enough to enable the valve to operate efficiently. Finally, the closed pressure drop of a valve must not exceed its close-off rating and, if used, that of its operator (actuator).

Valves are available with screwed (threaded) ends or flanged ends. The piping size (diameter) will determine the valve end selection. Larger-diameter pipes require bolted flange ends.

Still another consideration in valve selection is whether the valve is normally closed or open when at rest. This will depend on a number of factors, including the type of load being controlled, the medium (water or steam) being controlled, the system configuration, and

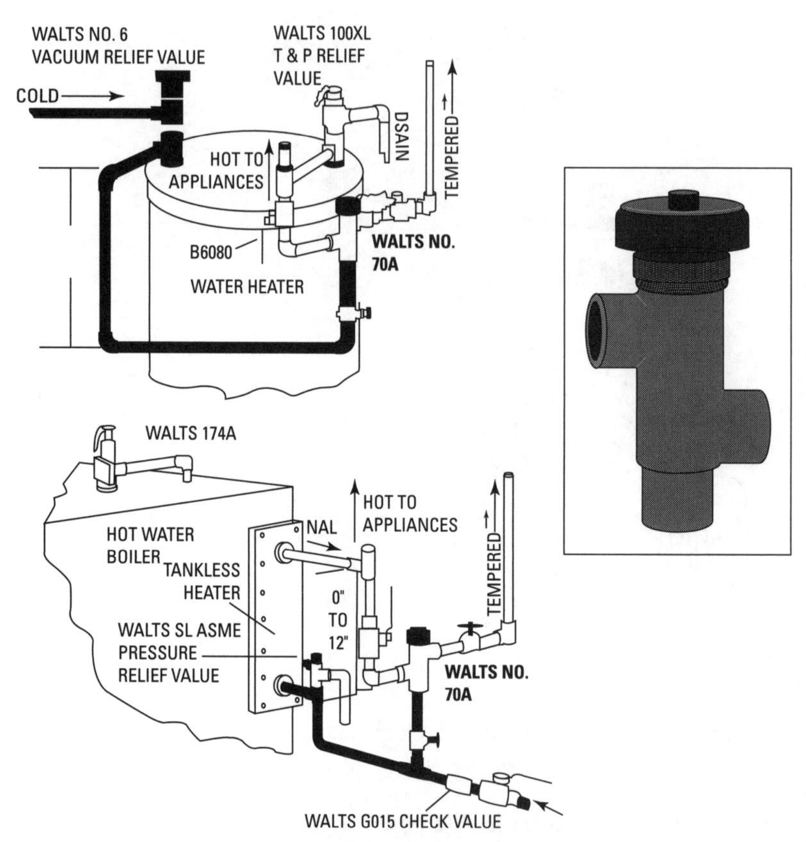

Figure 9-27 Three-way valve flow.

the valve application. For example, converter control valves should be normally closed, whereas outdoor preheat valves should be normally open.

Troubleshooting Valves

Valve manufacturers generally provide information for servicing and repairing their valves. When the information is not available, as is often the case on older heating and cooling systems, troubleshooting the valve problem generally depends on experience or educated guesswork. Experience has shown that most valve problems can be traced to one of the following three sources:

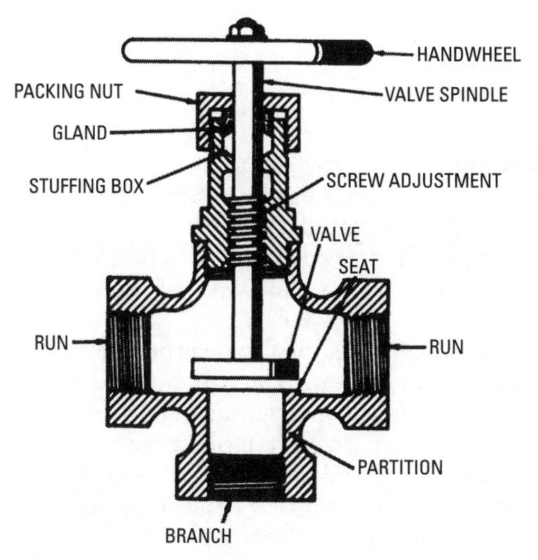

Figure 9-28 A three-way cross valve.

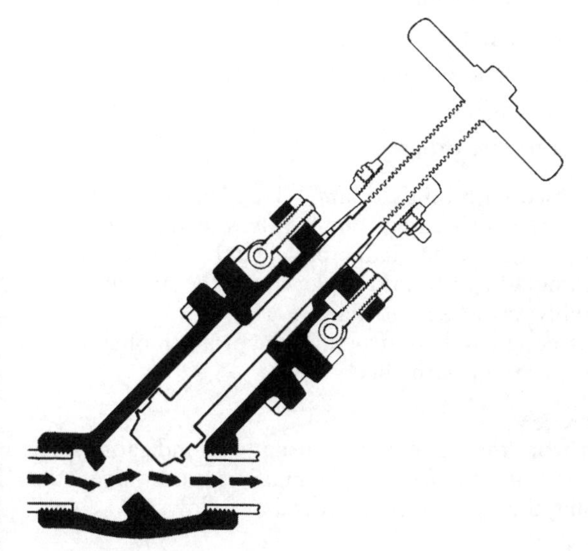

Figure 9-29 A Y valve. *(Courtesy Wm. Powell Co.)*

- Stuffing-box leakage
- Seat leakage
- Damaged stem

Valve Stuffing-Box Leakage

Leakage around the valve stuffing box is usually an indication that the stuffing must be adjusted or replaced. This leakage does not occur when the valve is completely opened or closed. Therefore, an absence of leakage is not necessarily an indication that the valve is functioning normally.

Once you have detected leakage, check first to determine whether or not adjusting the packing will stop it. If it is a bolted bonnet valve, turn the packing gland nuts (or gland stud nuts) clockwise alternately with no more than 1/4 turn on each until leakage stops. If you are dealing with a screwed and union bonnet valve, turn the packing nut clockwise until the leakage stops. If the leakage will not respond to adjustment, the packing must be replaced.

The procedure for replacing the packing in most valves may be summarized as follows:

1. Remove the handwheel nut and the handwheel.
2. Remove the packing nut.
3. Slip the packing gland off of the stem.
4. Replace the packing.
5. Reassemble in reverse order.

The procedure used with bolted bonnet outside screw and yoke valves is a little more complicated. On Y valves of this type, it is necessary to remove the gland flange and gland follower before replacing the packing. On globe and angle valves, the stud nuts and upper valve assembly must be removed.

Because of their design and construction, the problem of stuffing-box leakage does not occur with check valves.

Valve Seat Leakage

Leakage of water from the valve body is usually an indication that the wedge, disc, or seat ring needs replacing. For most valves, the procedure for doing this may be summarized as follows:

1. Open the valve.
2. Remove the bonnet and other components of the upper valve assembly.
3. Run the stem down by turning it in a clockwise direction.

4. Remove the wedge or disc from the stem and replace if necessary.
5. Remove the seat ring with a seat ring wrench and replace if necessary.
6. Reassemble in reverse order.

The disc, disc assembly, and ball are all possible sources of seat leakage in check valves. Access to these components is gained by removing the valve cap (counterclockwise), side plugs, and pins. Reassembly is in reverse order.

Damaged Valve Stems

Sometimes the threads on valve stems become worn or damaged, making the valves inoperable. When this occurs, the stems must be replaced. Before the stem can be replaced, however, all pressure must be removed. Then, with pressure removed, disconnect and remove the bonnet and upper valve assembly. The remainder of the procedure depends on the type of stem used (that is, rising or nonrising stem) and other design factors. Basically, the procedure may be summarized as follows:

1. Run the stem down by turning it in a clockwise direction.
2. Rotate the stem in a clockwise direction until the stem threads are completely out of the threaded portion of the upper bushing.
3. Pull the stem out of the stuffing box.
4. Remove the wedge or disc from the stem.
5. Replace the old stem with a new one.
6. Reassemble in reverse order with new packing and gasket (when applicable).

Automatic Valves and Valve Operators

An *automatic valve* is a controlled device designed to regulate the flow of steam, water, gas, or oil in response to impulses from a controller such as a thermostat. The controller measures changes in the surrounding variable conditions (for example, temperature, humidity, and pressure) and activates the valve in order to compensate for these changes. The controller, controlled device, and a feedback device constitute a closed-loop automatic control system. A more detailed description of the operating principles of an automatic control system is contained in Chapter 4 ("Thermostats and Humidistats").

A number of different automatic valves are used in heating and cooling systems. They are generally available in two-way or three-way models and can be used for either two-position or modulating operation or both. The type of automatic valves used must be carefully sized and selected for the specific application. Valve manufacturers generally provide information to aid you in making a suitable selection.

When selecting an automatic valve, do not confuse the valve body rating with the close-off rating. The former is the actual rating of the valve body, whereas the latter indicates the maximum allowable pressure drop to which the valve may be subjected while fully closed. The close-off rating is a function of the power available from the valve actuator for holding the valve closed against pressure drop (although structural parts such as the valve stem are also sometimes the limiting factor).

A *valve operator* (also called an *actuator*) is a device designed to automatically operate a valve. The operator (actuator) may be an integral part of the valve, or the valve and operator may be two separate devices. It receives the electrical, electronic, or pneumatic impulses from the controller and activates the valve stem. There are three types of valve operators:

- Solenoid operators
- Pneumatic operators
- Motorized operators

A *solenoid operator* is commonly used with a valve designed for two-position operation (that is, fully opened or fully closed). It consists essentially of a magnetic coil, which electrically operates a movable plunger. The plunger controls the opening and closing of the valve. In design, a solenoid-operated valve is less complicated than either the pneumatic or motorized types, but it is limited in size.

The operating principle of a *pneumatic operator* is based on changes in air pressure. Essentially, it consists of a spring-equipped flexible bellows or diaphragm attached to the valve stem. When the air pressure increases, the bellows or diaphragm moves the stem and compresses the spring. A reduction in air pressure results in the operation reversing itself with the spring returning the valve stem to its original position. Pneumatic-operated valves are designed for proportional control of a medium and are available as *normally open* or *normally closed* valves.

Springless pneumatic operators are available with opposing flexible diaphragms located on either side of a single diaphragm. This type of pneumatic operator is limited in use to special applications (for example, those involving high pressures or large valves).

A *motorized operator* consists of an electric motor that operates the valve stem through a gear train and linkage. Based on their operating principles, electric motors used in motorized operators can be classified as follows:

- Unidirectional motors
- Spring-return motors
- Reversible motors

The reversible motor illustrated in Figure 9-30 contains both a balancing relay and a feedback potentiometer. The balancing relay controls the motor, which turns the motor drive shaft connected to the gear train.

The balancing relay consists of two solenoid coils with parallel axes, into which are inserted the legs on the U-shaped armature. The armature is pivoted at the center so that it can be tilted by the changing magnetic flux of the two coils to energize the relay. A contact arm is fastened to the armature so that it may touch either of the two stationary contacts as the armature moves back and forth on its pivot. When the relay is balanced, the contact arm floats between the two contacts, touching neither of them. In operation, the motor is started, stopped, and reversed by the single-pole double-throw contacts of the balancing relay.

The feedback potentiometer contained in the motor consists of a coil of wire and a sliding contact. It is similar in construction to the internal view of the Honeywell Q181A auxiliary potentiometer illustrated in Figure 9-31. In operation, the motor shaft moves the sliding contact of the potentiometer along the coil, establishing contact wherever it touches according to the position of the motor.

Motorized valves are commonly used in residential zoning applications. Examples of valves used for this purpose are illustrated in Figures 9-32 and 9-33. Both of these valves contain replaceable O-rings and are equipped with a manual means of opening the valve in case of power failure.

The zone valve shown in Figure 9-32 is available in either line or low-voltage models. The latter requires a 24-volt power source. One advantage of this zone valve is that the powerhead can be replaced without the removal of the valve body from the pipeline.

Figure 9-30 A reversible motor with a balancing relay and a feedback potentiometer. *(Courtesy Honeywell, Inc.)*

Caution

Make sure the electrical power is shut off before attempting to install, remove, or repair an electrically operated valve.

Note that the manual opening lever on the powerhead (see Figure 9-32) has two settings: manual open (*man. open*) and

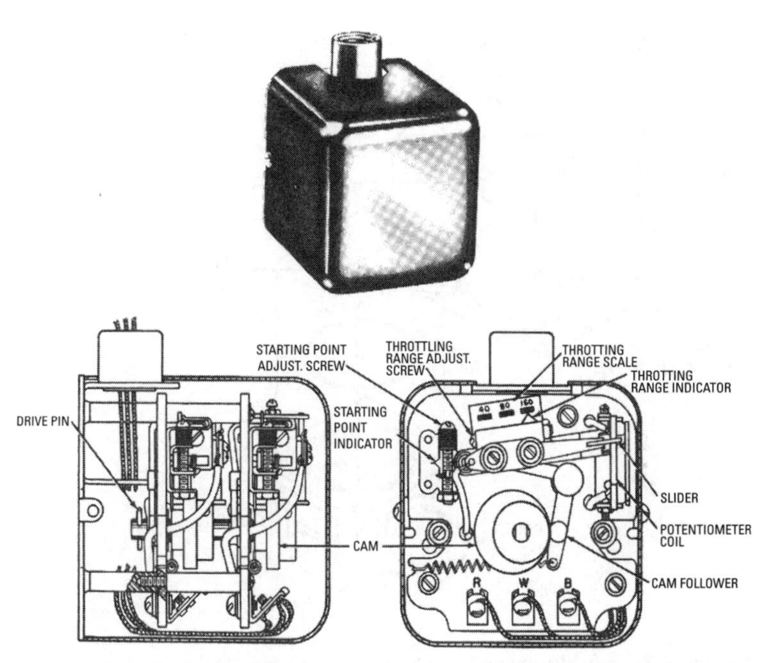

Figure 9-31 Internal view of a Honeywell auxiliary potentiometer.
(Courtesy Honeywell, Inc.)

automatic (*auto*). Always place the manual opening lever in the automatic position after the installation procedure has been completed and *before* running the valve. Always place the manual opening lever in the manual position before removing the valve from the pipeline or replacing a powerhead. *Never* attempt to cycle the valve electrically when the manual opening lever is in this position.

Before attempting to install a new powerhead, drain the water from the system, disconnect the valve from the electrical power source, and remove the conduit connections (if fitted). Then, place the manual opening lever on the *old* powerhead in the manual position and remove the cover. The remainder of the procedure may be summarized as follows:

1. Take out the four screws connecting the powerhead to the valve body and remove the old powerhead.
2. Replace the old O-ring in the valve body with a new one.
3. Place the manual opening lever on the new powerhead in the manual open position.

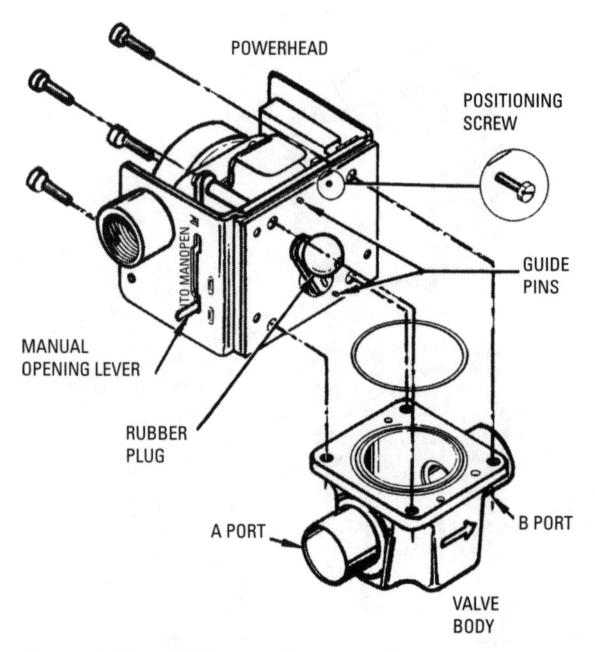

Figure 9-32 A Honeywell zone valve. *(Courtesy Honeywell, Inc.)*

4. Align the powerhead on the valve body in the same position as the old one, and screw it down.
5. Reconnect the conduit and electrical connections.
6. Adjust the thermostat or controller so that the valve runs smoothly through its cycle.

The motorized zoning valve shown in Figure 9-33 is energized by impulses from a standard two-wire thermostat. A wiring diagram for four such valves used in a residential installation is illustrated in Figure 9-34. Low-voltage DC power and circular-burner control is supplied by a unit containing a transformer, a rectifier, and two circuit breakers. When supplied for boilers with self-energizing controls, a built-in power failure manual switch is furnished that permits manual boiler operation.

The valve itself contains a water-damped piston that is lifted by a strong magnetic flux and opens a straight-through passage for unrestricted flow. The valve closes slowly under the action of gravity.

The butterfly valve is ideally suited for regulating the flow of water or steam in applications where tight close-off is not required.

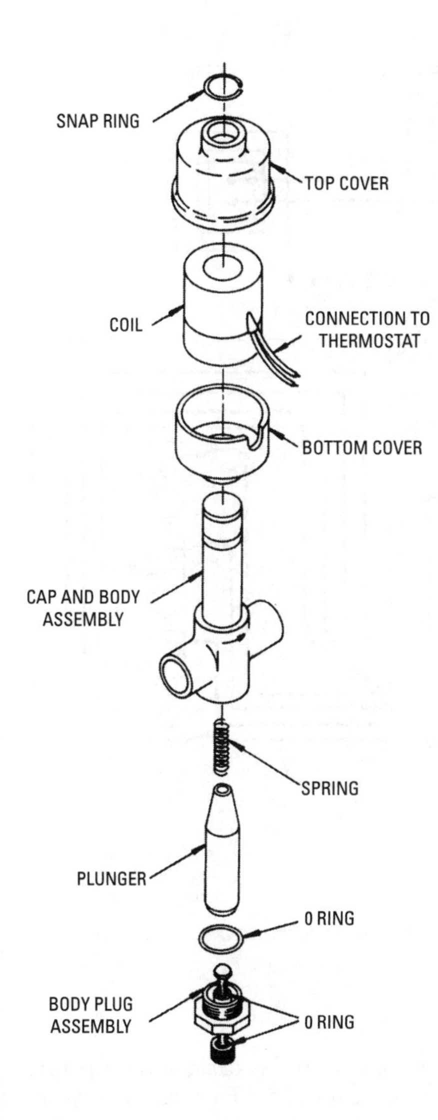

Figure 9-33 An electro-zone valve.
(Courtesy Hydrotherm, Inc.)

Typical applications include zone control of gravity hot-water or low-pressure steam heating systems. For tight close-off, a final shutoff valve must also be used.

The motorized valve illustrated in Figure 9-35 is a single-seated type used for two-position operational control of hot or chilled water, steam radiators, or zoned residential heating and cooling systems. It

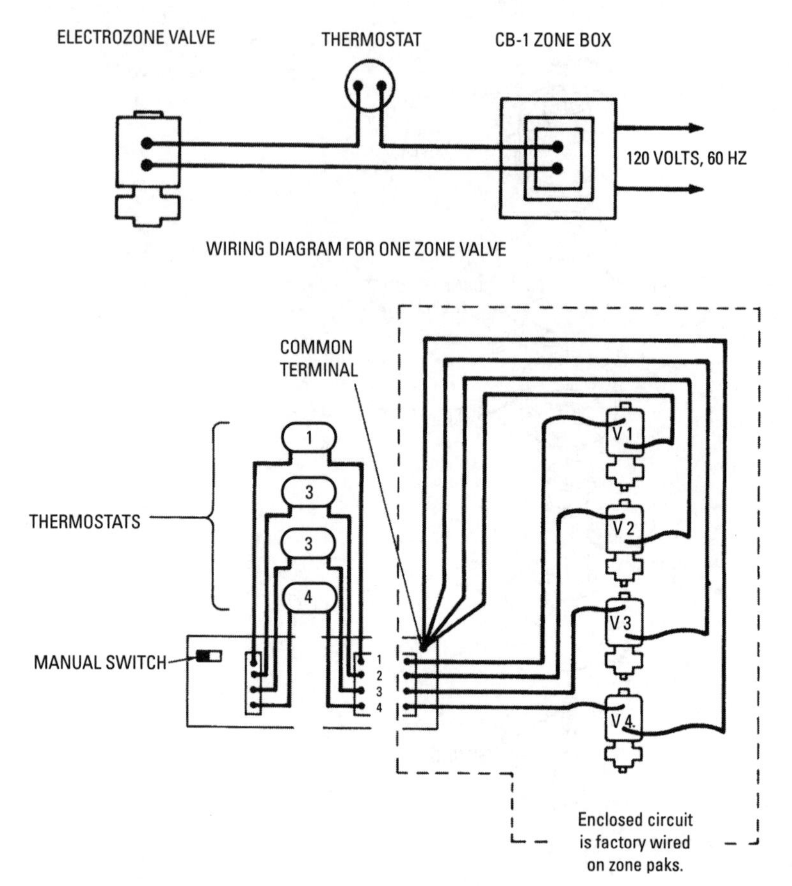

Figure 9-34 Wiring diagram for the zone valve. *(Courtesy Hydrotherm, Inc.)*

contains a replaceable composition disc and is designed to operate at fluid temperatures up to a maximum of 250°F or steam pressure up to 15 psi maximum.

Automatic room temperature control on hot-water or two-pipe steam heating systems can be accomplished by using self-actuating thermostatic radiator valves (see Figure 9-36). These valves can be used on freestanding radiators, convectors, and baseboard heating units. Electric power is not required for their operation.

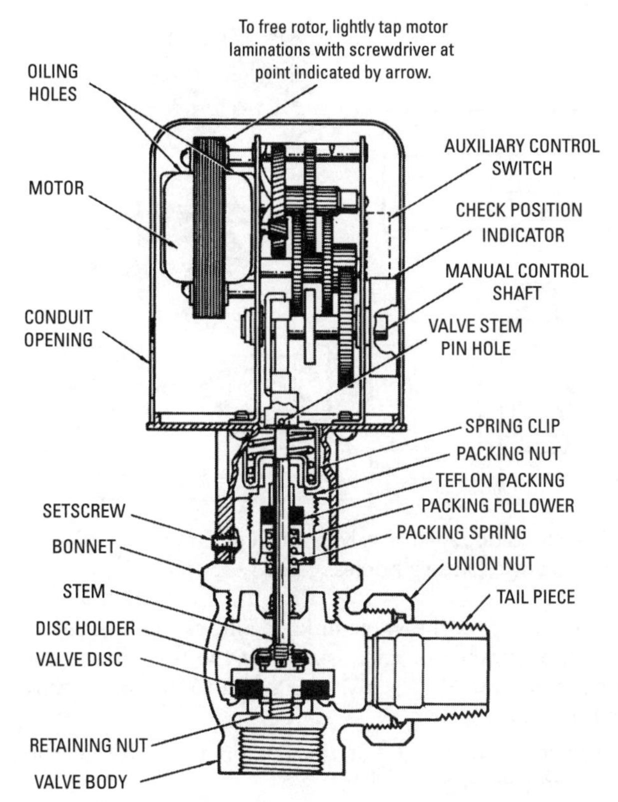

Figure 9-35 Honeywell V5045A valve body and V2045A operator.
(Courtesy Honeywell)

Valve linkage is used to connect an electric motor to an automatic valve. The *type* of valve linkage used will depend on the type of automatic valve used in the heating or cooling application. The valve manufacturer's recommendations should be followed to avoid problems.

The valve linkage in Figure 9-37 is an example of the type of linkage used with automatic steam or hot-water valves (see Figure 9-38). Installation and adjustment of this linkage for a two-position valve may be summarized as follows:

1. Follow the manufacturer's instructions and connect the motor and valve to the linkage. Be sure to align the key on the crank arm with the *keyway* on the motor shaft.

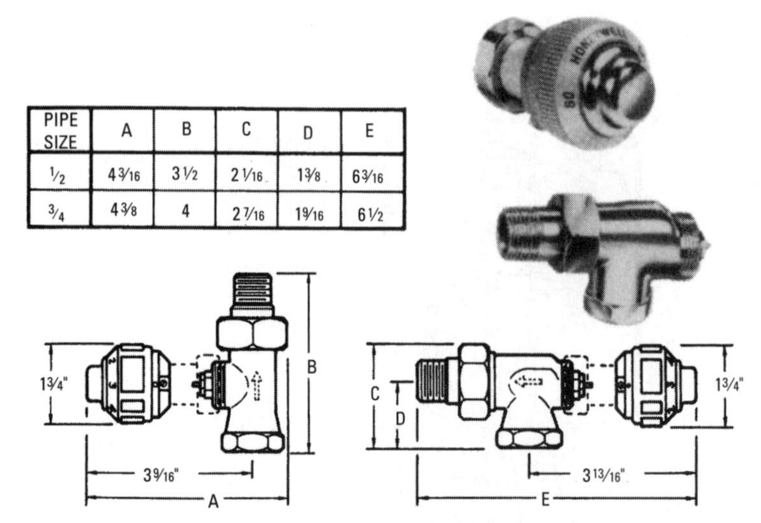

PIPE SIZE	A	B	C	D	E
1/2	4 3/16	3 1/2	2 1/16	1 3/8	6 3/16
3/4	4 3/8	4	2 7/16	1 9/16	6 1/2

Figure 9-36 Thermostatic radiator valves showing dimensions of straight-through and angle pattern bodies. *(Courtesy Honeywell)*

2. The lift adjustment on the valve linkage allows the linkage to be used with a number of valve sizes. Set the lift adjustment equal to the rated lift of the valve according to the three steps illustrated in Figure 9-39.
3. Use the adjusting screw to put tension on the valve in the closed position (see Figure 9-40).
4. Run the motor and valve all the way open. Check to see that the motor runs all the way to the end of the stroke without putting tension on the valve stem. The existence of tension is an indication that it may jam open.

Adjusting the same linkage for a three-way valve involves the same steps, particularly the alignment of the key on the crank arm with the keyway on the motor shaft. The lift adjustment, however, should be set 1/2 division above the rated valve lift (see Figure 9-39).

With the motor closed and the valve at the bottom of its stroke, loosen the locknut and turn the adjustment screw down until the top of the washer is even with the pointer (see Figure 9-40).

When the motor is run all the way open, the washer at the bottom of the strain-relief spring should move upward 1/16 inch (see Figure 9-41). If it moves more than 1/16 inch, reduce the lift adjustment. Increase the lift adjustment if it moves less than 1/16 inch. Any

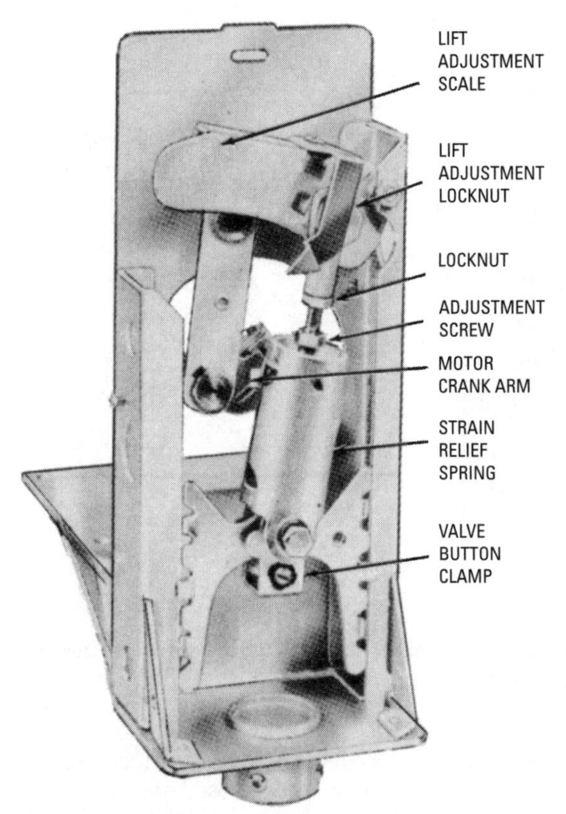

Figure 9-37 Honeywell model Q601 valve linkage. *(Courtesy Honeywell)*

change in the lift adjustment will require that the close-off tension at the bottom of the stroke be reset before continuing. In operation, the motor should be free to run through its entire stroke.

Another type of valve linkage that requires adjustment during its installation is the Honeywell Q455 valve linkage illustrated in Figure 9-42. This linkage is also used with automatic valves in many different types of heating and cooling applications. The method of connecting a Q455 valve linkage to a water or steam valve is illustrated in Figure 9-42. The adjustment procedure for a two-way automatic valve is as follows:

1. Set the lift adjustment on the crank arm to equal the rated lift of the valve (see Figure 9-43).

VALVE SIZE	DIMENSIONS IN INCHES		
	A	B	C
2½	9½	3½	4⅝
3	11	3 15/16	4⅞
4	13	4½	5⅝
5	15	5½	7⅜
6	16½	6⅛	5⅜

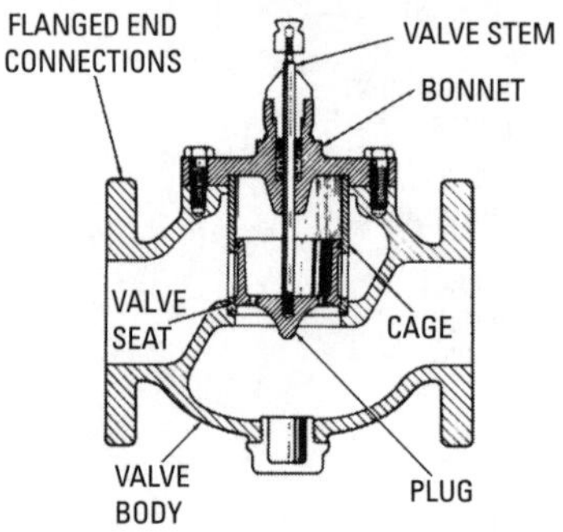

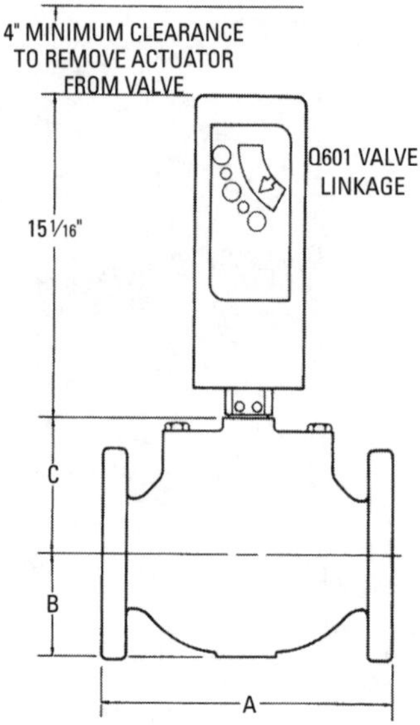

Figure 9-38 Cage-type single-seated valve used for control of steam, liquids, or noncombustible gases. *(Courtesy Honeywell)*

2. With the motor running closed, turn the adjustment screw down (see Figure 9-43) until the valve is closed and washer A (see Figure 9-44) has been pushed down 1/16 inch as indicated on the scale marks.
3. Run the motor all the way open. If the adjustment is correct, the valve will not reach the top of its stroke and jam. A tendency to jam will be indicated by tension in the linkage.

The procedure used for connecting a motor and an automatic three-way valve to a Honeywell Q455 valve linkage is identical to that used for the two-way valve. However, there are significant differences in the adjustment procedure.

The lift adjustment should be set at *one mark higher* than the rated lift of the automatic valve. Then, with the motor closed, turn the shaft adjustment screw down until the valve is closed and washer *A* has been pushed down 1/16 inch. Now, run the motor all the way open and check washer *B* for a 1/16-inch upward movement (see Figure 9-44).

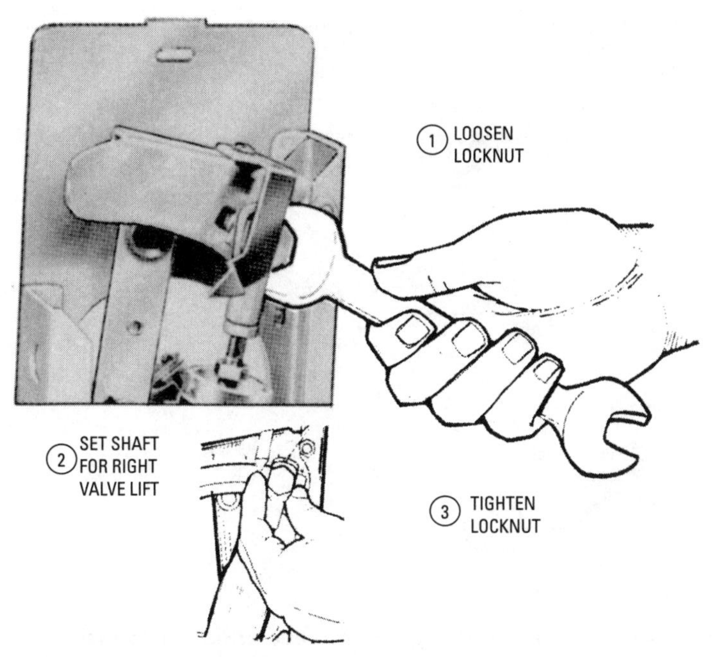

Figure 9-39 Setting the left adjustment. *(Courtesy Honeywell)*

Any readjustment of the lift requires a readjustment of the shaft adjustment screw for proper close-off force on the valve stem at the bottom of the stroke.

An example of a type of valve linkage that does *not* require adjustment during installation is the Honeywell Q618 valve linkage. This valve linkage can be used in most motorized valve applications.

Valve Pipe Connections

The pipe ends or connections of valves are commonly available in the following three types:

- Threaded ends
- Flanged ends
- Grooved ends

Threaded-end valves are tapped usually with female threads into which the pipe is threaded. These are the cheapest of the three types to use because less material and finishing are required. On

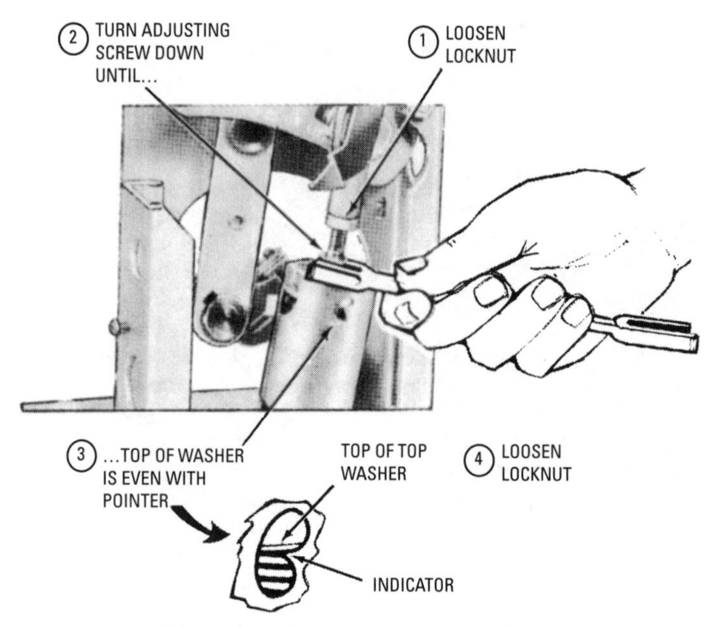

Figure 9-40 Using the adjusting screw to put tension on the valve. *(Courtesy Honeywell)*

the other hand, it is difficult to remove them without dismantling a considerable portion of the piping (unless extra fittings such as unions are used).

Flanged-end valves are used where heavy viscous media are to be controlled or where high-pressure steam service is the rule. The different types of flanged ends available for use are shown in Figure 9-45. Although flanged ends make a stronger, tighter, and more leakproof connection, their initial cost and the cost of installation is higher. This higher cost can be traced to a number of factors. For example, more metal is required in their manufacture. Moreover, flanged ends must be carefully and accurately machined before they are connected. Finally, additional parts such as gaskets, bolts, nuts, and companion flanges (to which the valve end flanges are bolted) must be provided to complete the connection.

Grooved ends are lighter in weight than flanged ends and require only an ordinary socket or speed wrench for installation or removal. The grooved end is connected to groove piping with a coupling patented by the Vitaulic Company of America.

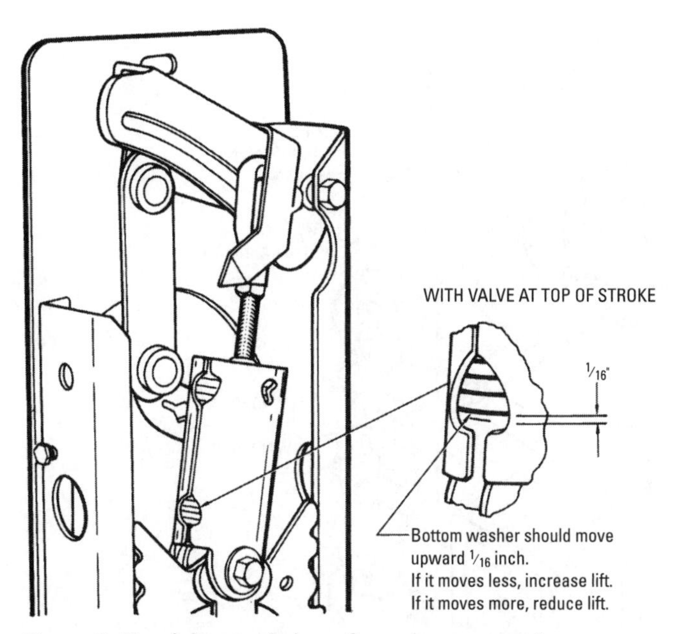

Figure 9-41 Adjusting linkage for a three-way valve. *(Courtesy Honeywell)*

Pipe threads in valve bodies are gauged to standard tolerances. Threaded-end valves should be installed with the proper-size wrenches. The wrench should be applied to the pipe side of the valve when installing. Doing so minimizes the possibility of distorting the valve body, particularly when the valve is made of a malleable material such as bronze. Closing the valve tightly before installation will also reduce the possibility of distortion.

Figure 9-46 illustrates the method of tightening flanged valves and fittings. Note the numbering sequence of the flange bolts in the illustration. The reason for tightening down bolts diametrically opposite one another is to create a gradual and uniform tightness. This results in uniform stress across the entire cross-section of the flange, which eliminates the possibility of a leaky gasket.

Valve Installing Pointers

Unless valves are installed properly, they will not operate efficiently and can cause problems in the system. There are certain precautions you should take when installing valves that will improve their

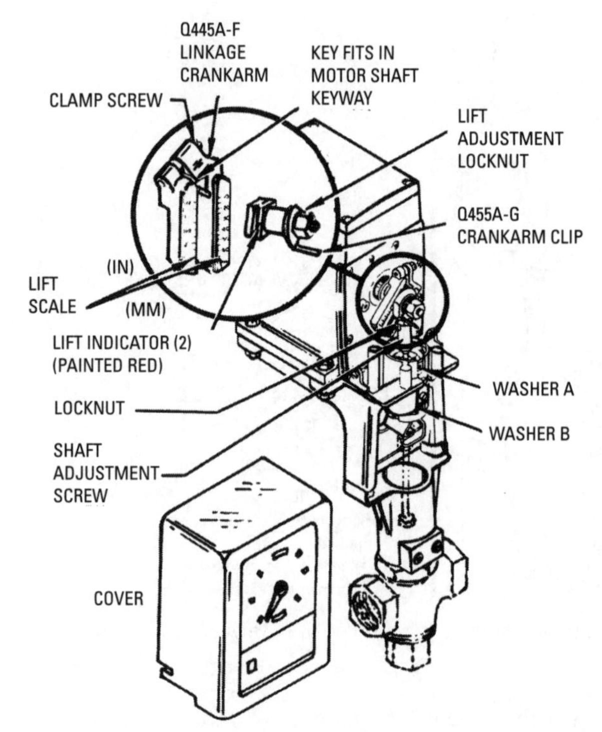

Figure 9-42 Honeywell model Q455 valve linkage connecting a motor to a water or steam valve. *(Courtesy Honeywell)*

performance and minimize the possibilities of a malfunction. These precautions may be summarized as follows:

1. Always clean out a valve before installation because dirt, metal chips, and other foreign matter can foul it. This can be done by flushing the valve with water or blowing it out with compressed air.
2. Clean the piping before installing the valve. If you cannot flush or blow out the foreign matter, the ends of the pipes should be swabbed with a damp cloth.
3. Only apply paint, grease, or joint sealing compound to the pipe threads (that is, the *male* threads). *Never* apply these substances to the valve body threads because you run the risk of their getting into the valve itself and interfering with its operation.

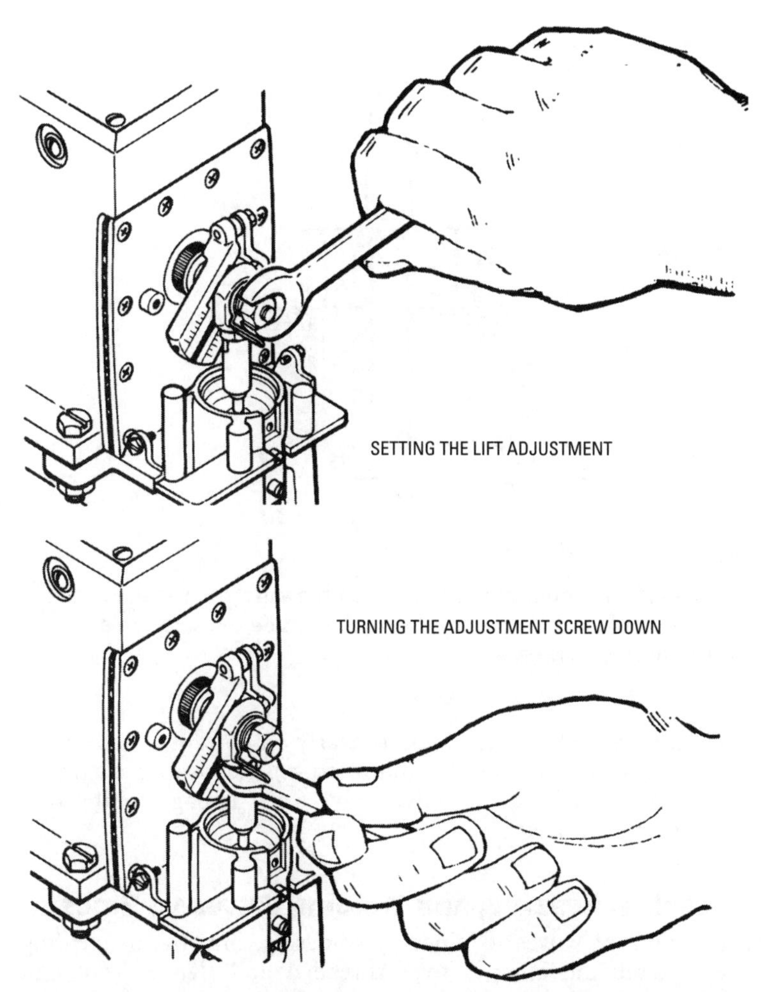

Figure 9-43 Setting the lift adjustment to equal the rated lift of the valve it is used with. *(Courtesy Honeywell)*

4. Install valves in a location that can be reached conveniently (see Figure 9-47). If the valve is placed so that it is awkward to reach, it sometimes may not be closed tightly enough. This can eventually cause leaks to develop in the valve.
5. When necessary, support the piping so that additional strain is not placed on the valve. Small or medium-size valves can be supported with hangers placed on either side. Large valves should always be independently suspended.

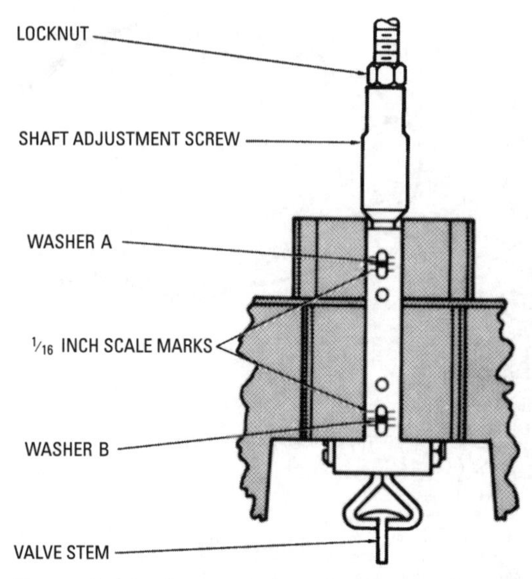

Figure 9-44 Amount of tension put on the valve stem at the ends of the valve stroke is indicated by the washer inside the strain release mechanism. *(Courtesy Honeywell)*

6. Sufficient clearance is particularly important when rising stem valves are used. Failure to ensure proper clearance before installing the valve may cause damage to the disc sealing surface.

Soldering, Brazing, and Welding Valves to Pipes

The joining of valves to pipes by soldering, brazing, or welding offers certain advantages over threaded and flanged joints in reduced maintenance and repairs. Furthermore, joints of this type enable you to use lighter pipe and require fewer flanges and gaskets. The major disadvantage is that an experienced and skilled worker is required to do the job. This can result in higher labor costs.

Fusion welding is the most commonly used method of welding a valve to a pipe. Essentially, it consists of fusing the metal of the valve and pipe connection by applying heat at a temperature above the melting point of the metal. The following two types of welds can be made: (1) a butt weld, and (2) a socket weld.

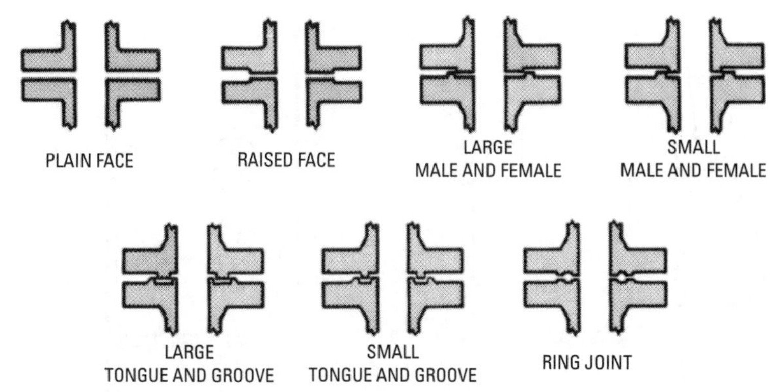

Figure 9-45 Various types of flanged ends used on valves.
(Courtesy Wm. Powell Co.)

Both soldering and silver brazing are procedures used to join two metals, but they differ from fusion welding in that temperatures *below* the melting point of the metal are used. Either a suitable soldering flux or a silver brazing alloy is heated to a temperature at which it will flow into the joint formed by the valve connection and pipe end. Silver brazing is replacing soldering in many applications because it is quicker and results in a stronger joint.

A more detailed description of the procedures used in soldering, brazing, and welding pipe connections is given in Chapter 8 ("Pipes, Pipe Fittings, and Piping Details"). These procedures are summarized in the following sections.

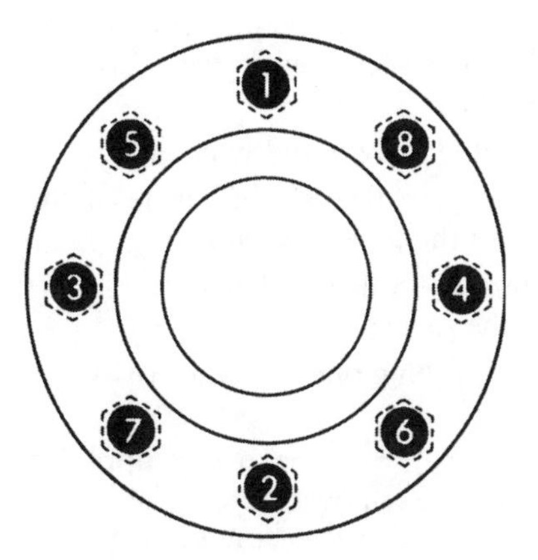

Figure 9-46 Recommended tightening sequence for flange bolts.

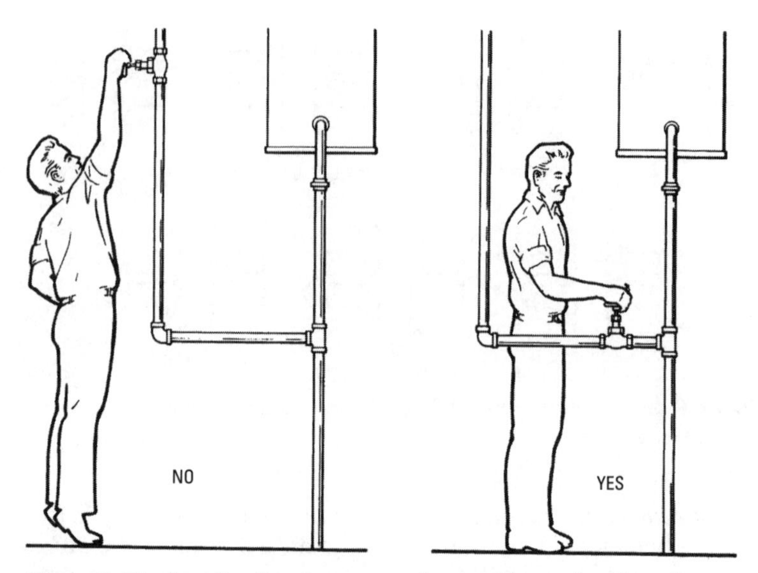

Figure 9-47 Install valves in a convenient, easily reached location.

Soldering or Silver-Brazing Procedure

The basic procedure for soldering or silver-brazing pipe or tubing to valves is as follows:

1. Cut the tube or pipe end square, and make sure the diameter is not undersize or out of round.
2. Remove all burrs with a metal file.
3. Clean the pipe or tubing end (at least to the depth of the socket) and the inside of the valve socket with steel wool and a cloth to wipe away the residue.
4. Clean all surfaces with a suitable solvent and wipe dry.
5. Apply solder flux or silver-brazing flux to the inside of the valve socket and the outside of the pipe or tubing.
6. Insert the pipe or tubing into the valve socket until it seats against the shoulder within the socket.
7. Turn the valve and the pipe or tubing once or twice to evenly distribute the flux.
8. Make certain the valve is in open position before applying heat. A nonmetallic disc should be removed before the heat is applied. After removing the disc, the valve bonnet or bonnet

ring should be replaced hand-tight to prevent distortion to the threaded sections when heating the valve.

9. Make certain the valve and pipe or tubing are properly supported during the soldering or silver-brazing process. Any strain on the joint while cooling will weaken it.
10. Apply flame evenly around piping or tubing adjacent to valve ends until solder or brazing alloy suitable for the service flows upon contact.
11. *Soldering:* Apply solder to the joint between the pipe or tubing and the end of the valve socket. Apply the flame toward the bottom of the valve socket until all the solder is absorbed. Control the direction of the flame away from the valve body to avoid excessive heating, which causes distortion and improper functioning of the valve.
12. *Silver brazing:* Apply brazing alloy to the joint between the pipe or tubing and the end of the valve socket. Wave the flame over the valve hexes to draw the metal alloy into the socket, leaving a solid fillet of brazing alloy at the joint. Control the direction of the flame away from the valve body to avoid excessive heating, which causes distortion and improper functioning of the valve.
13. Remove all excess and loose matter from the surface with a clean cloth or brush.

Butt-Welding Procedure

The procedure for making a butt weld is as follows:

1. Machine the pipe ends for the butt-welding joint.
2. Clean the pipe ends, valve joint, and the inside of the valve socket with a degreasing agent to remove oil, grease, or other foreign materials.
3. Align by means of fixtures, and tack weld in place.
4. Make certain the valve is in the open position before applying heat. The valve bonnet should be hand-tight to prevent distortion or damage to the threads. Nonmetallic discs should be removed before applying heat.
5. The valve and pipe should be supported during the welding process and must not be strained while cooling.
6. Preheat the welding area 400°F to 500°F.

7. Depending on the welding method used (gas, arc, and so on), a butt weld is normally completed in two to four passes. The first pass should have complete joint penetration and be flush with the internal bore of the pipe. Make sure the first pass is clean and free from cracks before proceeding with the second pass. The second pass should blend smoothly with the base metal and be flush with the external diameter. Avoid excessive heat because this can cause distortion and possible malfunctioning of the valve.
8. Use a wire brush and a clean cloth to remove discoloration.

Socket-Welding Procedure

The procedure for making a socket weld is as follows:

1. Cut the pipe end square, making sure the diameter is not undersize or out of round.
2. Remove all burrs with a metal file.
3. Clean the pipe end, valve joint, and the inside of the valve socket with a degreasing agent to remove any oil, grease, or other foreign material.
4. Insert the pipe end into the valve socket and space by backing off the pipe after it hits against the shoulder inside the spacing collar. Tack weld in place.
5. Make certain the valve is in the open position before applying heat. Valve bonnets should be hand-tight to prevent distortion or damage to the threads. Nonmetallic discs should be removed before applying heat.
6. The valve and pipe should be supported during the welding process and must not be strained while cooling.
7. Preheat the welding area 400°F to 500°F.
8. A socket weld can generally be completed in two to four passes, depending on the welding method used. Make sure the first pass is clean and free from cracks before proceeding with the second pass. Avoid excessive heat because it may cause distortion to the valve bonnet.
9. Use a wire brush and a clean cloth to remove discoloration.

Chapter 10

Steam and Hydronic Line Controls

The pipelines of steam and hot-water heating systems contain a number of different devices or controls designed to maintain and control the circulation of the steam or water through the steam and hot-water lines. These steam- and hot-water–line controls include such devices as pumps, manifolds, valves, flow switches, steam traps, and expansion tanks. Some of these devices, such as pumps, manifolds, and valves, are designed to directly control the rate and direction of flow of the steam or water. Others indirectly influence the flow rate.

Most of the line-control devices commonly used in steam and hot-water heating systems are covered in this chapter. However, with the exception of certain specialized applications, valves are described more fully in Chapter 9 ("Valves and Valve Installation").

Steam and Hydronic System Pumps

Heating and circulating pumps are used to maintain the desired flow rate of steam or hot water in a heating system.

The two types of pumps used in *steam* heating systems are (1) condensate pumps and (2) vacuum heating pumps. The condensate pump is used in a gravity steam heating system to return the condensation to low- or medium-pressure boilers. Vacuum heating pumps are used in either return-line or variable-vacuum heating systems to return condensation to the boiler *and* to produce a vacuum in the system by removing air and vapor along with the condensation.

Circulators are used in hot-water (hydronic) heating systems to maintain a continuous flow of the water in the system. They are smaller in size than either a condensate or a vacuum pump and are usually of the motor-driven centrifugal type.

Condensate Pumps

The design of a *conventional* gravity-flow steam heating system is such that the heat-emitting units cannot be placed at a level lower than the water level in the boiler. The movement of the condensation depends on gravity and, therefore, must move from a higher to a lower level until it reaches the boiler (at the lowest level). This is illustrated by the one-pipe steam heating system in Figure 10-1. Each heating unit has a single pipe through which it receives the hot steam and returns the condensate in the opposite direction. The dependence on gravity to return the condensate to the boiler places certain limitations on the design of the heating system unless mechanical means are used to compensate for the lack of gravity. A condensate pump serves this purpose.

Many different types of condensate pumps have been developed for use in steam heating systems. Screw, rotary, turbine, reciprocating, and centrifugal pumps are some of the types used for this purpose. One of the most common uses of condensate pumps in low-pressure steam heating systems is the motor-driven centrifugal pump equipped with receiver (tank) and float-control automatic switch.

In operation, condensation enters the receiver and fills the tank. A float connected to an automatic switch rises with the water until the tank is almost full. At that point, the float closes the switch and starts the pump motor. The water is pumped from the receiver and the float drops, causing the switch to open and shut off the pump motor.

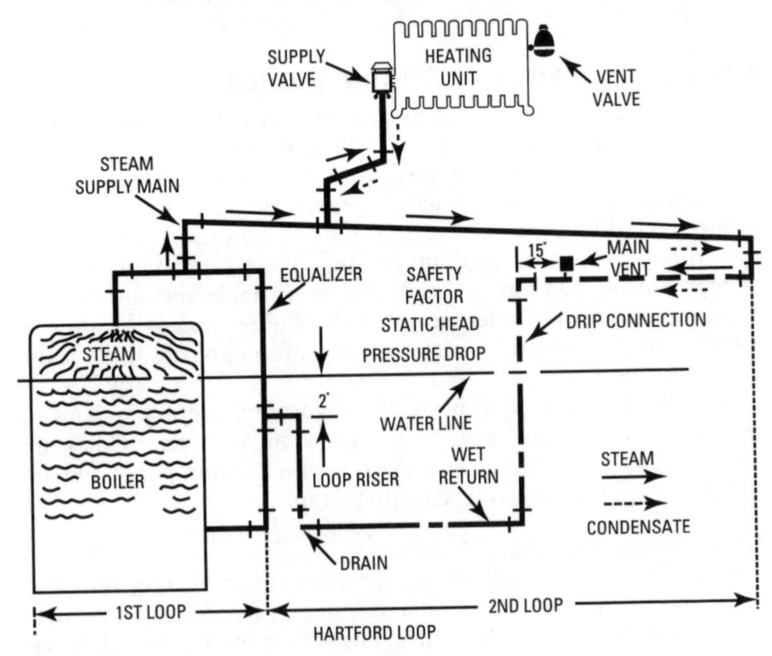

Figure 10-1 One-pipe steam heating system. *(Courtesy ITT Hoffman Specialty)*

The centrifugal pump shown in Figure 10-2 is used to pump condensation from a lower level return line to one at a higher level, or against a higher pressure. These units are also used to pump condensation from a flash tank to a boiler (see Figure 10-3) and for other special applications.

As shown in Figure 10-4, the basic components of this pump consist of an impeller (*A*), with an inlet at its center rotating on a shaft (*D*). The condensation enters the inlet orifice and flows radially through vanes to the outer periphery (*F*) of the impeller; it has approximately the same velocity as the periphery. The head of pressure developed by the pump is the result of the velocity imparted to the condensation by the rotating impeller.

When the condensation leaves the outer periphery of the impeller, it flows around the volute casing (*B*) and through the discharge orifice (*E*) of the pump. A wear ring (*C*) is provided to prevent bypassing of the condensation.

Condensate pumps are available in either single or duplex units. The latter are used in installations where it is necessary to have a pump available for use at all times. A duplex unit is actually two condensation pumps fitted with a mechanical alternator. Both pumps feed into the same receiver. If one pump malfunctions, the other starts automatically and continues to provide uninterrupted pumping service for the system.

Figure 10-2 Typical centrifugal condensate pump. *(Courtesy Spirax Sarco Co.)*

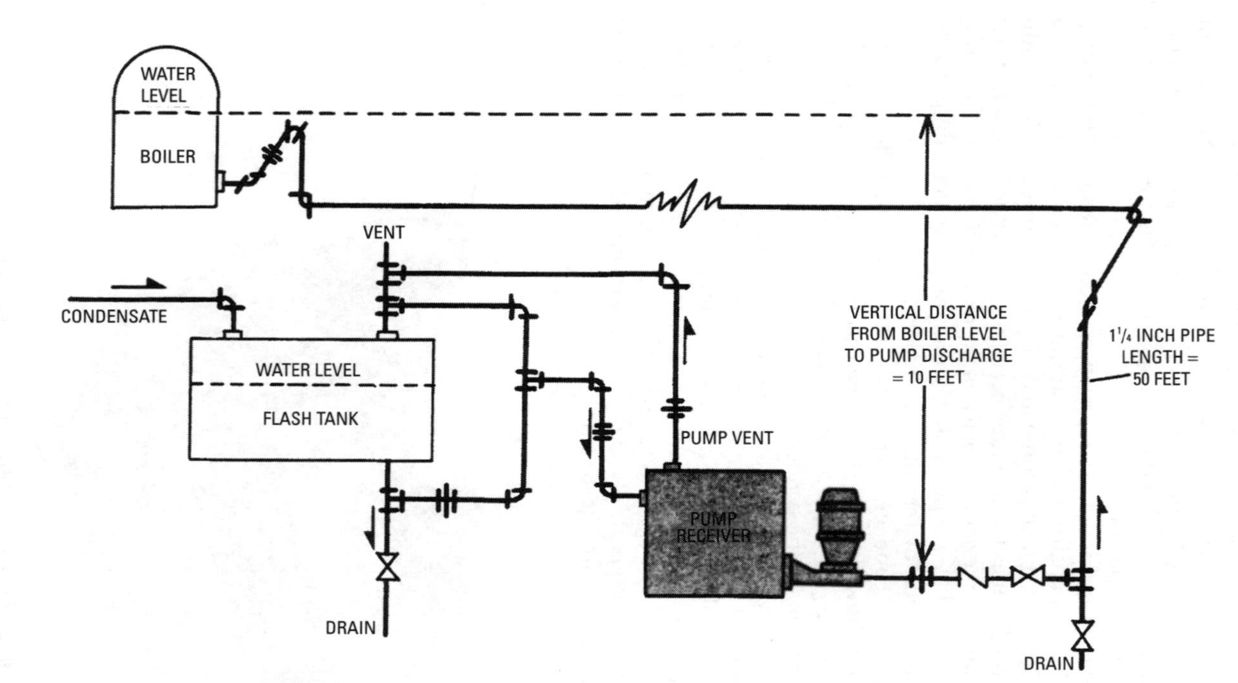

Figure 10-3 Pumping condensation from flash tank to boiler. *(Courtesy Spirax Sarco Co.)*

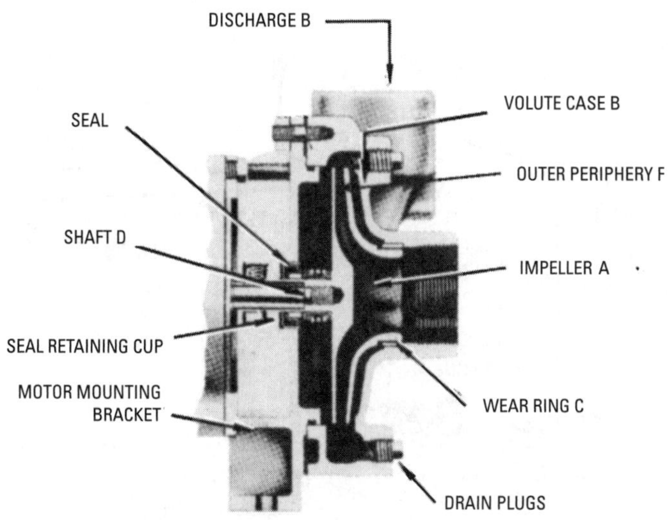

Figure 10-4 Basic components of a centrifugal condensate pump.
(Courtesy Spirax Sarco Co.)

Vertical condensation pumps are available for use in installations where the space for the pump is limited, where the returns run below the floor, or where it is undesirable to place a horizontal pump in a sunken area. Vertical condensation pumps are also available in both single and duplex units.

The use of a condensate pump in a two-pipe steam heating system is shown in Figure 10-5. Note the arrangement of gate and check valves on the discharge side of the pump. This type of system is commonly referred to as a *condensate-return steam heating system*. Using a condensate pump to return the condensation provides greater design flexibility for the system. The major disadvantage is that larger steam traps and piping must be used than in vacuum heating systems.

Condensate pumps can also be used as mechanical lifts in vacuum steam heating systems (see Figure 10-6). By connecting the vent outlet of the condensation pump to a return line above the level of the vacuum heating pump, the same vacuum return condition is maintained in the piping below the water level as in the rest of the system. In this arrangement, the only purpose of the condensate pump is to lift the condensation from the lower level to the higher one without reducing the capacity of the vacuum heating pump.

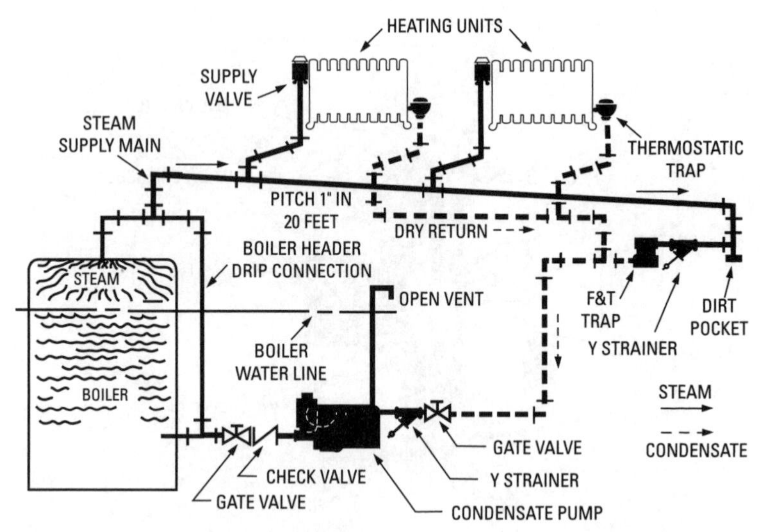

Figure 10-5 Condensate pump in a two-pipe steam heating system.
(Courtesy ITT Hoffman Specialty)

Vacuum Pumps

Vacuum heating pumps are used to maintain the vacuum in mechanical vacuum heating systems by removing air, vapor, and condensation from the lines. Many vacuum pumps are available in both single and duplex units, and are designed to automatically adjust themselves to the varying conditions of the system. The duplex units have an advantage over the single pumps, because they provide automatic standby service. If one of the pumps in a duplex unit should happen to malfunction, the other one cuts in and picks up the load.

A vacuum pump is operated either by steam or electricity. Steam-driven vacuum pumps are sometimes used in high-pressure steam systems, but these pumps have been generally replaced by automatic motor-driven return-line pumps. Examples of vacuum heating pumps are shown in Figures 10-7 and 10-8.

In operation, the pump is started before the steam enters the system. When the pump removes the air from the lines, steam quickly fills the radiators of the system. The radiators remain full of steam, because the air is automatically removed as fast as it accumulates. By quickly exhausting the air and condensation from the system, the vacuum pump causes the steam to circulate more rapidly, resulting in faster warm-up time and quieter operation.

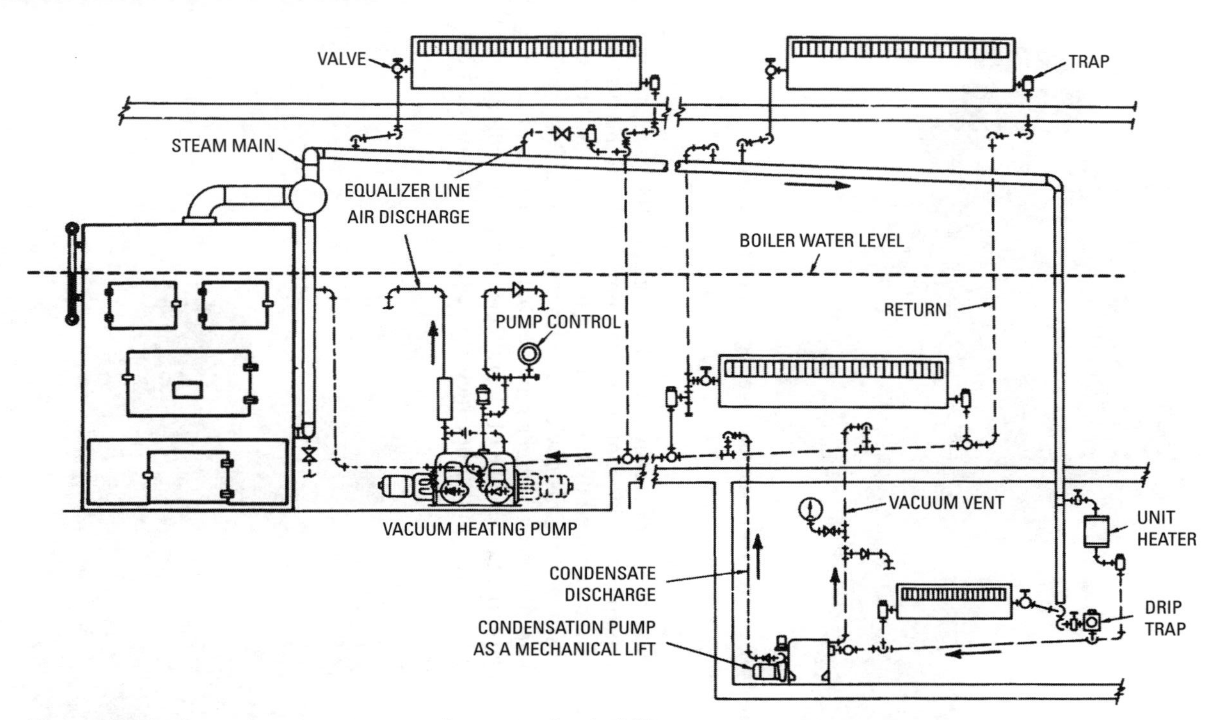

Figure 10-6 A condensate pump used as a mechanical lift. *(Courtesy Nash Engineering Co.)*

Figure 10-7 Return-line vacuum heating pump and receiver. *(Courtesy Nash Engineering Co.)*

The condensation of steam in the lines creates the vacuum, and the pump maintains it by continuing to pump air from the system. The vacuum maintained by the pump is only a *partial* one, because it is not possible with this device to extract all the air. Each stroke of the pump piston or plunger removes only a fraction of the air, depending on the percentage of clearance in the pump cylinder, the resistance of valves, and other factors; hence, an infinite number

Figure 10-8 Duplex return-line vacuum and boiler feed pump for a vacuum steam heating system. *(Courtesy Chicago Pump Co.)*

of strokes would theoretically be necessary to obtain a perfect vacuum, not considering line and pump resistance.

Vacuum pumps designed to remove only air from a system are referred to as *dry* pumps. Those that remove both air and condensation are called *wet* pumps. When a wet pump is used, the condensation is pumped back to the boiler. In operation, the air, being heavier than steam, passes off through thermostatic retainer valves to the pump. When the steam reaches the retainer valves, they close automatically to prevent the steam passing into the dry return line to the pump and breaking the vacuum. The air from the pump is passed into a receiver, where it is discharged through an air vent. The condensation is pumped back to the boiler generally by means of a centrifugal pump.

In most vacuum systems, the pump is controlled by a vacuum regulator and a float control. The vacuum regulator cuts in when the vacuum drops to a preset level and cuts out when the vacuum reaches its highest point. The float control operates independently from the vacuum regulator, starting the pump when condensation reaches a certain level in the receiver.

Two typical installations in which vacuum heating pumps are used are illustrated in Figures 10-9 and 10-10. In the vacuum air-line heating system, shown in Figure 10-10, thermostatic-type air-line valves are used instead of radiator air vents. The primary purpose of the vacuum heating pump is to expel air from the system.

The vacuum return-line system is very similar to a condensation-return steam heating system, except that a vacuum pump is used to provide a low vacuum in the pipes and to return the condensation to the boiler. Because of the vacuum condition, smaller steam traps and piping can be used.

An accumulator tank must be installed in a vacuum pump steam heating system if the returns are below the inlet connection of the vacuum pump receiver. As shown in Figure 10-10, the condensate flows by gravity from the baseboard heating units to the accumulator tank, where it is lifted to the vacuum pump receiver.

Circulators (Water-Circulating Pumps)

Hydronic heating systems use small compact pumps to provide the motive force to circulate the water in the pipes. They are usually referred to as *circulators* or *water-circulating pumps*. The circulator is used to move the water from the boiler to the heat-emitting units and back again. It is not used for lifting, as is the case with vacuum and condensate pumps, but simply for circulating the water through a closed loop.

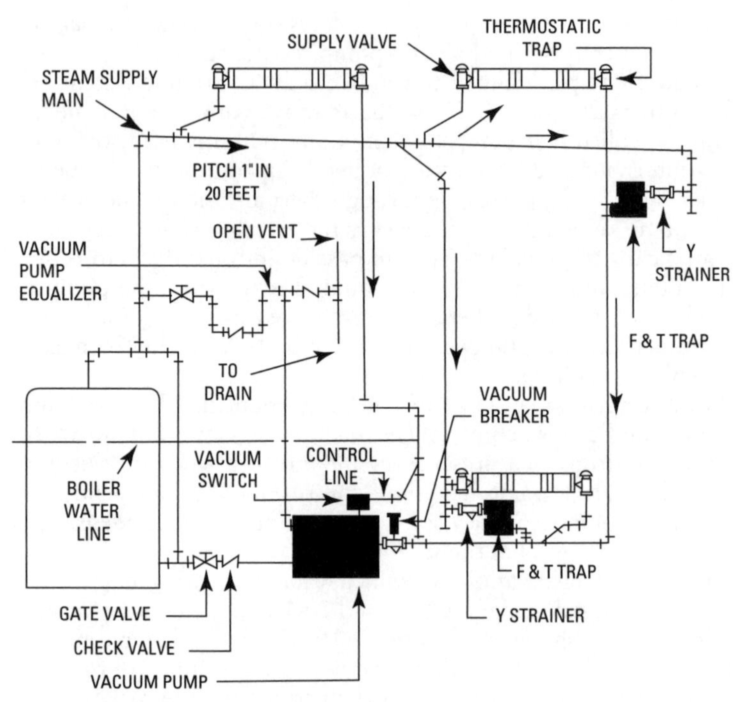

Figure 10-9 Vacuum pump in a two-pipe steam heating system.
(Courtesy ITT Hoffman Specialty)

Circulators were first introduced in the 1930s to augment water circulation in the traditional hot-water space-heating systems. Prior to their introduction, the hot-water systems relied on the density difference between cold and hot water to provide the motive force for water circulation. These systems were called *gravity* hot-water heating systems, and the pumps were added to boost circulation. These early pumps (sometimes called *three-piece circulators* or *booster pumps*) are still with us today, although as more technically advanced models. All modern hydronic heating systems are closed-loop installations that use one of several different types of pumps to circulate the water.

Three-Piece Booster Pumps

The three-piece booster pump illustrated in Figures 10-11 and 10-12 is an example of the circulators used in small-to medium-size residential and light commercial hydronic heating systems since the 1930s.

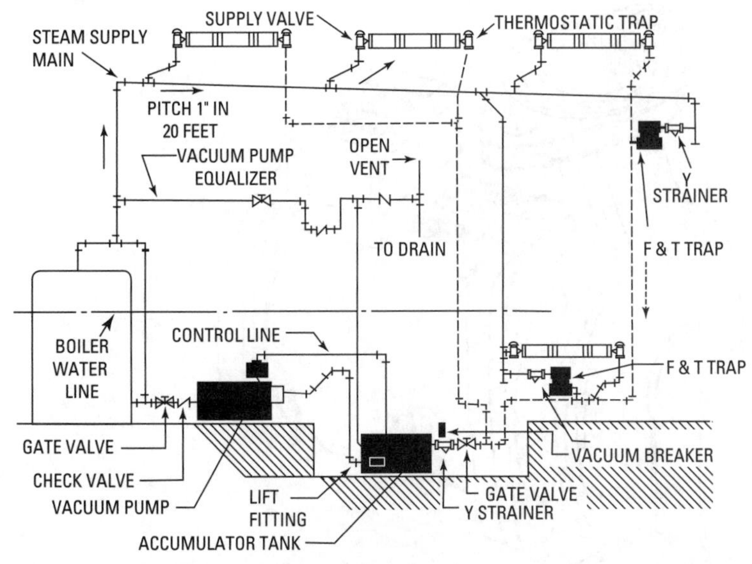

Figure 10-10 Vacuum pump with accumulator tank in two-pipe steam heating system. *(Courtesy ITT Hoffman Specialty)*

A typical three-piece booster pump consists of the following three sections: (1) the pump body (also called the *volute, body assembly,* or *waterway*), (2) the coupling assembly, and (3) the motor assembly (also called the *shaft-and-motor assembly*).

The three-piece booster pump contains hermetically sealed sleeve bearings, a carbon/ceramic seal, and a coupler that uses springs in tension to provide quiet operation. The motor of a three-piece booster pump can be serviced by removing it from the pump body. Consequently, there is no need to drain the system or disconnect the pump from the piping for servicing.

The three-piece booster pump has an inline volute, which means the inlet and discharge ports are located along the same centerline. It has a strong starting torque, which enables it to free a stuck impeller without any difficulty.

Note

The *volute* is just another term for the pump body. It contains the motor bracket, the impeller, the volute gasket, the inlet and discharge ports, and the pump mounting flanges. The shape of the volute will determine how the circulator is connected to the piping.

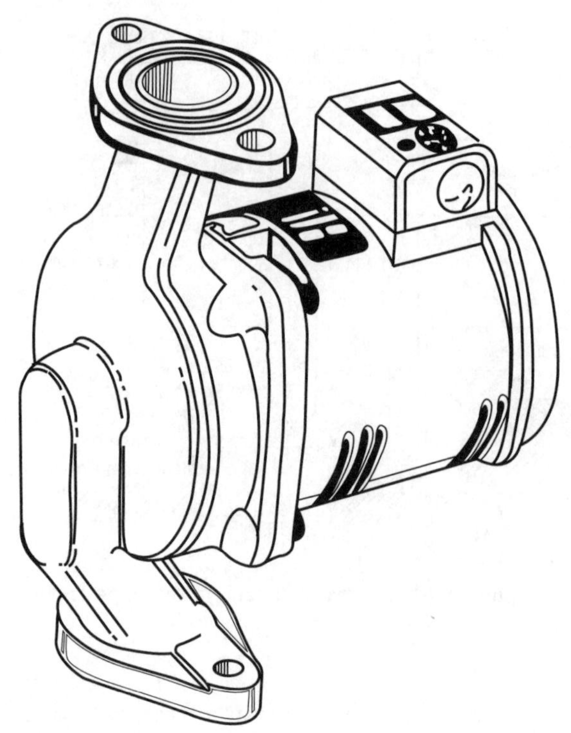

Figure 10-11 Bell & Gossett Series PL three-piece booster pump.
(Courtesy ITT Bell & Gossett)

A three-piece booster pump requires periodic inspection. The mechanical seal will sometimes need replacement. After removing the old seal, clean the shaft and sleeve before installing a new one. The pump manufacturer will provide step-by-step instructions for servicing the pump mode.

This pump also requires periodic lubrication. A wool wicking is used to draw the lubricating oil into the bearing assembly (see Figure 10-13). Check the pump manufacturer's operating and maintenance instructions for the recommended lubricating schedule. As shown in Figure 10-13, the three-piece booster pump must always be installed with the oil ports facing upward and with the motor, motor shaft, and bearing assembly in a horizontal position.

Caution

Never plug or cover the weep hole, or you will trap the excess oil in the pump body. Any dirt or sediments in the oil may damage the bearings and shorten their service life.

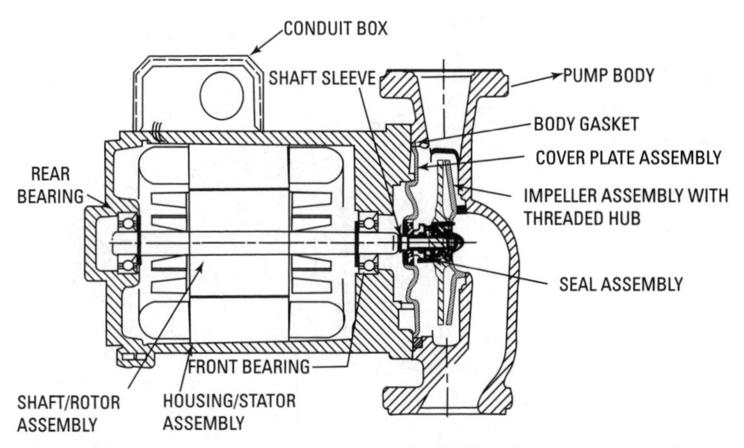

Figure 10-12 Cross sectional view of a Bell & Gossett Series PL three-piece booster pump. *(Courtesy ITT Bell & Gossett)*

Warning

Use only the lubricant specified by the pump manufacturer. An SAE 20 (nondetergent) or 10W-30–weight oil can be substituted if the pump manufacturer's recommended oil is unknown.

The three-piece booster pump shown in Figures 10-11 and 10-12 can be installed to discharge in any direction (e.g., up, down, horizontally, etc.), *but* the motor shaft must *always* be in the horizontal position, the arrow on the pump body must always point in the direction of flow, and the conduit box must be positioned on top of the motor housing.

Wet-Rotor Circulators

The wet-rotor circulator is also used in small- to medium-size residential and light-commercial hydronic systems (see Figure 10-14). This is a small, close-coupled pump with an integral 40–125-watt motor. It is a sealed pump that does not require lubrication. The wet-rotor pump combines the motor, the shaft, and the impeller in a single assembly housed in a chamber filled with system fluid. In other words, it is both cooled and lubricated by the system fluid, hence the name wet-rotor pump. These circulators do not have a mechanical seal, as is the case with booster pumps. Consequently, seal replacement is not a problem.

Wet-rotor circulators are small, require no maintenance, and provide a long service life before they have to be replaced. A principal disadvantage using the wet-rotor pump is that it cannot be serviced or repaired while connected to the piping; the system must be

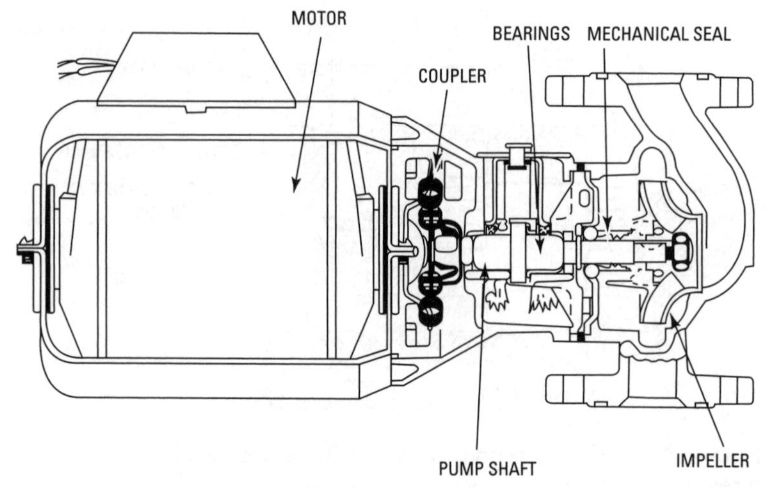

Figure 10-13 Bell & Gossett Series 100 booster pump.
(Courtesy ITT Bell & Gossett)

drained before it can be removed. They are sometimes called throw-away pumps, because it is less expensive to replace them than it is to repair them.

Note

The problem of having to drain the system in order to remove the pump can be avoided if shutoff (check) valves are installed in the return line on either side of the unit.

Inline Centrifugal Circulators

An inline centrifugal circulator is another type of pump commonly used in hydronic heating systems of small- to medium-size residential and light-commercial structures. The Bell & Gossett Series 90 inline centrifugal pump shown in Figures 10-15 and 10-16 is an example of this type of pump. It is a close-coupled, low-maintenance unit that can be mounted inline both vertically and horizontally. Because the motor can be disconnected from the pump body for servicing, there is no need to remove the pump body from the piping circuit for servicing.

End Suction Pumps

End suction pumps are used in the hydronic systems of hotels, stores, theaters, and other large structures. The Bell & Gossett Series VSC pump shown in Figure 10-17 is a double suction unit

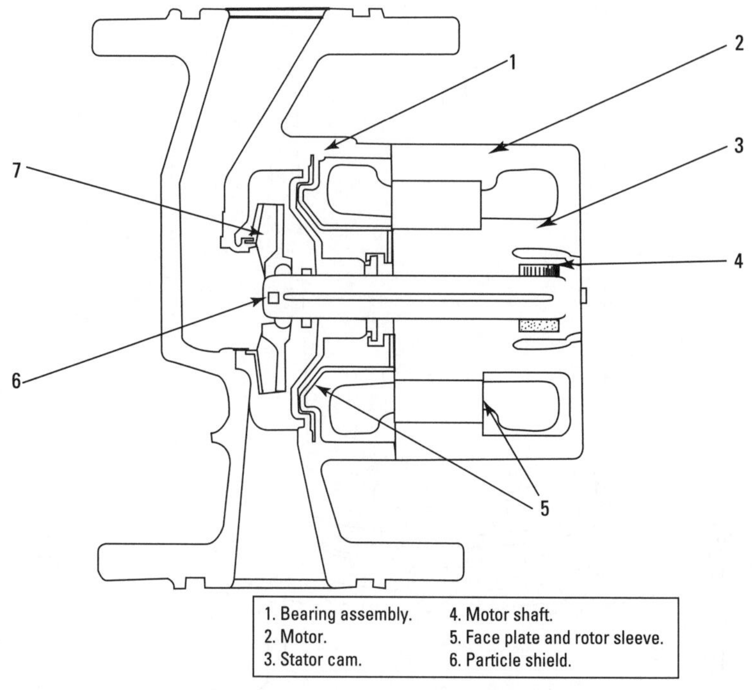

Figure 10-14 Bell & Gossett NRF wet-rotor circulator.
(Courtesy ITT Bell & Gossett)

characterized by a vertically split casing perpendicular to the pump shaft. It is used in hydronic heating and cooling systems, and for condenser water, cooling towers, refrigeration, and general service. These pumps have volutes that differ in design from the smaller circulators used in small to medium hydronic systems, and they are commonly installed at a 90° angle to the system piping with a lateral offset between the discharge and inlet ports.

Circulator Selection

A number of different factors must be considered when selecting circulators for both large and small heating systems. The selection factors include:

- Amount of water to be handled
- Temperature of the water to be handled
- Head against which the pump must operate

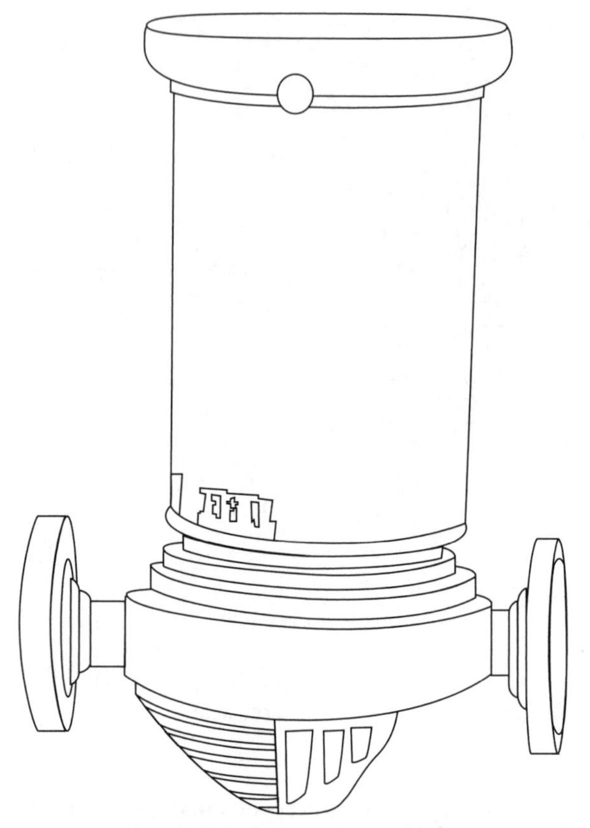

Figure 10-15 Bell & Gossett Series 90 inline centrifugal pump.
(Courtesy ITT Bell & Gossett)

- Working head of the system
- Pump suction head

Data necessary for selecting a suitable water-circulating pump are supplied by pump manufacturers. Always check the manufacturer's specifications for the maximum working pressure and the maximum operating temperature of the pump before deciding on which model to use. These operational limits can be found in the manufacturer's specification sheet, in the installation manual, or on the pump nameplate. The maximum working pressure and maximum operating temperature of the system *must not* exceed those of the pump. If they do, it can lead to possible property damage, serious injury, or even death.

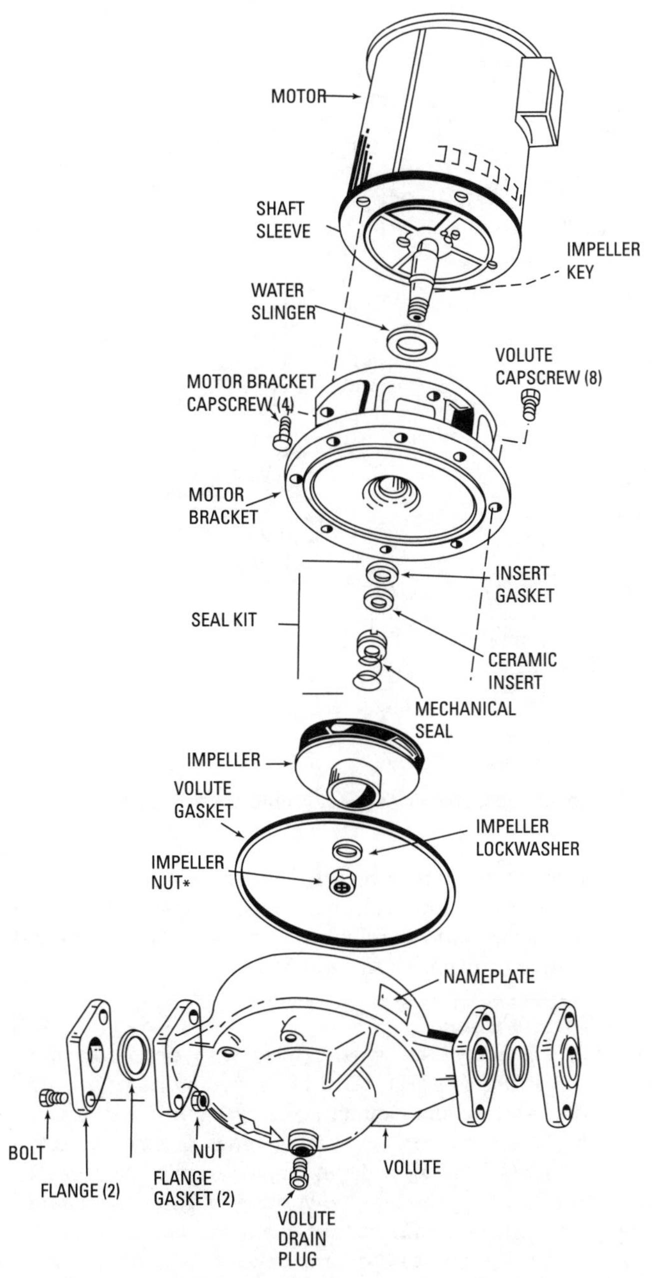

Figure 10-16 Exploded view of a Bell & Gossett inline centrifugal pump. *(Courtesy ITT Bell & Gossett)*

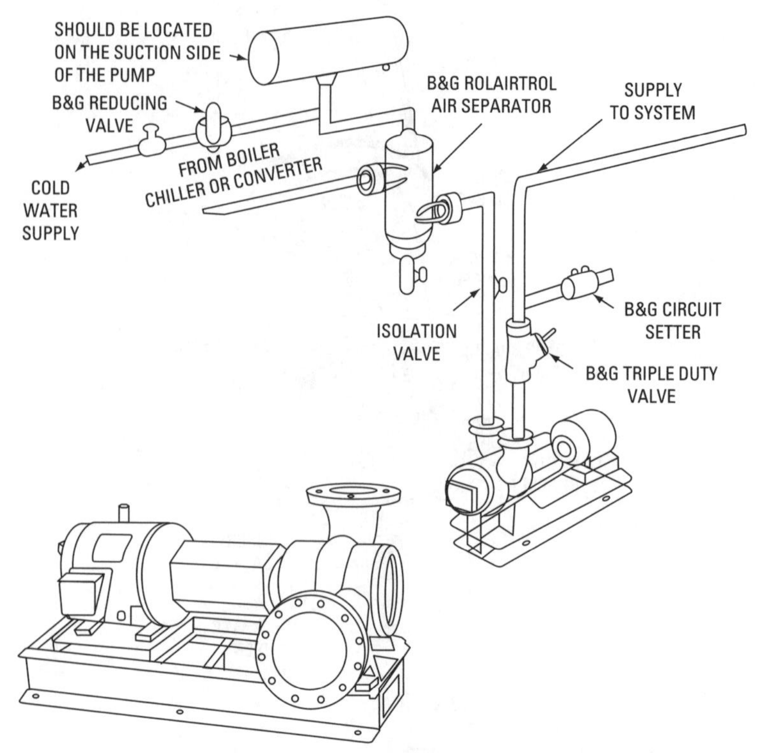

Figure 10-17 Bell & Gossett Series VSC double-suction pump.
(Courtesy ITT Bell & Gossett)

The temperature of the water handled by the pump will determine the type of pump packing selected. The working head of the system is the sum of the static head and the friction. The pump will have either a positive or negative suction head.

Pump Head and Pressure Drop

Two terms you will encounter when dealing with circulators are *pump head* and *pressure drop*. The term *pressure drop* refers to the friction created when the fluid (water) flows against the inside surface of the piping and passes through valves, fittings, or other system components. This friction will slow down the flow rate of the system fluid (water). The circulator must be powerful enough to overcome this friction and produce a steady, uniform flow rate. The term *pump head* (or *pressure head*) refers to the force developed by the circulator to overcome pressure drop.

The circulators used in small residential hot-water heating systems are installed directly in the return line. These are commonly single-suction pumps in which both the motor and the pump share a common shaft. The major objection to installing a pump directly in the return line is that the pipe connections may have to be broken to remove the pump for service and repair. However, some pumps such as the Bell & Gossett Series 90 model shown in Figures 10-15 and 10-16 can be serviced without removing the entire unit. If the circulator must be completely removed from the return line, a good idea is to install shutoff (check) valves above and below the unit so that the entire system does not have to be drained.

The operating head of circulators used in smaller hot-water heating systems is limited, and it is often general practice to size the pipelines of the system *after* selecting a pump capable of meeting the requirements of the system.

Pump manufacturers recommend that cast-iron pumps be selected for circulating water in a hydronic space heating system. Bronze pumps are recommended for pumping domestic (potable) hot water.

Circulator Installation

Always follow the pump manufacturer's installation instructions. These instructions will accompany the pump or can be obtained from the pump manufacturer (often by going online and downloading the manual from their web site).

Note

> Most circulator manufacturers recommend using their cast-iron models for circulating water in a hydronic space heating system and their bronze models for pumping domestic (potable) hot water.

Warning

> Never install a circulating pump with operational limits (maximum working pressure and maximum operating temperature) less than those of the hydronic system. Make certain the electrical rating of the circulator is appropriate for the installation.

In residential and light-commercial hydronic heating systems, the circulator should be mounted with its inlet port close to the point at which the expansion tank connects to the return line, which normally is the point of no pressure change in the system. Placing the inlet side of the circulator close to the expansion tank is important, because the latter controls the pressure of the water in the system. The circulator only circulates the water. It does not create pressure. The circulator should also be mounted in the return line as close as possible to the boiler. Install it with the flow arrow on its body facing the boiler. This will ensure that its discharge port also faces the boiler. Never mount a circulator in the highest part of the system.

The circulators used in residential and light-commercial hydronic systems should be installed with their motor shafts in a horizontal position. Installing the circulator with the motor shaft in a vertical position places unwanted weight on the bushings, rotor, and impeller.

When installing a circulator, it is also a good idea to install shut-off valves in the line above and below it. Doing so will allow removal of the circulator without having to first drain the system.

The header piping should be strong enough to hold the weight of the circulator. This can be a problem if it is installed in copper pipe or tubing, because copper has neither the strength nor the rigidity of steel. To prevent sagging from the weight of the circulator, support the header pipe or tubing with metal strap hangers or brackets, or use steel pipe or tubing for the header and connect the copper lines to either end of it.

Circulator Operation

A hydronic system circulator is always filled with system fluid (water). As soon as the pump moves water out of its discharge port, it is immediately replaced by an equal volume of water entering through its inlet port. Remember, these are closed systems that are *always* completely filled with water. If the system is not filled with water, either there is a leak or the operator has failed to check the water level before startup.

Warning

> A circulator pump should *never* be run without any water in it. Doing so will damage the pump. Always check to be sure it is filled with water *before* startup.

Water enters the inlet port of the circulator and flows directly to the *impeller* located in the volute (see Figure 10-18). The impeller is a rotating wheel that creates the centrifugal force required to move the water through the piping. The pump drive shaft enters the back of the impeller and exits the front through an opening called an *eye*. The centrifugal force of the rotating impeller accelerates the velocity of the water, forcing it away from the eye and around the inside of the pump body. The water is directed toward a discharge port that is much smaller than the pump inlet port. Squeezing the water through this smaller discharge port converts the water velocity to

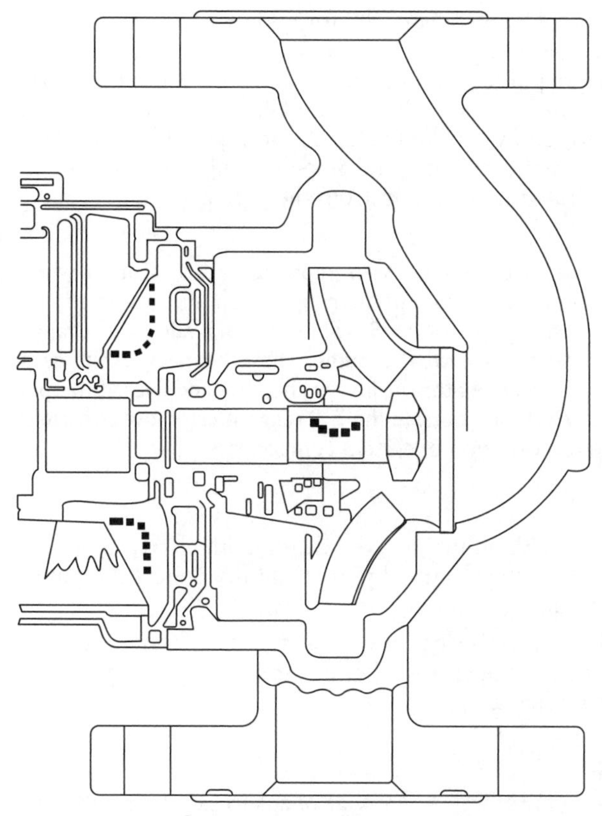

Figure 10-18. Circulator impeller and volute. *(Courtesy ITT Bell & Gossett)*

pressure. This is the *pressure head* (or *pump head*) developed by the pump to overcome the friction (*pressure drop*) created by the water flowing through the piping, valves, and other components of a hydronic heating system.

Circulator Troubleshooting, Service, and Maintenance

Hermetically sealed, self-lubricating pumps should never be oiled or lubricated. It is not only unnecessary, but could also damage the pump. Very little maintenance is required for these pumps.

Sometimes the failure of a hot-water (hydronic) heating system to produce heat can be traced to a malfunctioning circulator (pump). Before attempting to repair or replace the unit, check the electric power to the pump from the system controls. The problem may be due to an electrical failure, instead of the mechanical failure of the pump itself. Always first check for a blown fuse or a tripped circuit breaker.

If repairs are required, manufacturers commonly provide parts lists for their pump models so that the unit can be serviced on site. For the circulator shown in Figure 10-16, it is possible to replace the shaft sleeve, seals, gaskets, and impeller (all parts subject to wear) without removing the pump from the piping.

Note

The pH value of the water is an important consideration when operating circulators. For optimum operation, the water should have a pH ranging from 7 to 9. The water pH value can change during the service life of the pump. These changes occur as a result of a change in water quality or chemical additives. If the pH value of the water falls outside the 7–9 range, it can cause circulator seal failure or corrosion of system components.

Important Safety Tips

- To avoid possible injury or even death from electrical shock, always shut off the electrical power and disconnect the pump before attempting to service or repair it.
- To avoid scalding burns, always allow the system water to cool to room temperature before attempting to remove a pump for servicing.

Steam Traps

A steam trap is an automatic valve that opens to expel air and condensation from steam lines and closes to prevent the flow of steam. The functions of a steam trap are:

- Remove (vent) air from the system so that steam can enter. (Air in the pipes will block the flow of steam into the radiators.)
- Prevent steam from leaving the system until all of its latent heat is removed.
- Remove (drain) the condensate from the system after the latent heat has been removed. (Draining the condensate from the system prevents corrosion and water hammer.)

All steam traps operate on the fundamental principle that the pressure within the trap at the time of discharge will be slightly in excess of the pressure against which the trap must discharge. This includes the friction head, the velocity head, and the static head on the discharge side of the trap. The steam trap cannot operate unless the excess pressure of discharge is greater than the total back pressure.

Each steam trap used in steam heating is designed for a specific range of applications that its operating characteristics best suit. Although there is no universal steam trap per se, the many different types can be grouped into the following three classes on the basis of their operating characteristics:

- Separating traps
- Return traps
- Air traps

Separating traps are designed to release condensation but close against steam. They are float-operated, thermostatically operated, or float-and-thermostatically operated. Thermostatic traps are designed to release air and condensation but close against them.

Return traps may be operated to receive condensation under a vacuum and return it to atmosphere or a higher pressure. *Air traps* are generally operated by a float.

Steam Trap Information

The 80-page *Steam Traps & Repair Parts catalog* from State Supply Company, Inc. of St. Paul, Minnesota is filled with information necessary for troubleshooting and repairing steam traps. It contains repair part drawings, capacity tables, and product ordering guides from all the major steam-trap manufacturers (Dunham-Bush/MEPCO, Monash, Illinois, Marsh, Spirax-Sarco, Hoffman Specialty, Trane, Armstrong, Sterling, and Warren-Webster). A copy can be obtained by writing or calling State Supply:

The State Supply Company, Inc.
597 East Seventh Street
St. Paul, MN 55101-2477
800-772-2099
info@statesupply.com
www.statesupply.com

Sizing Steam Traps

Selecting the correct size steam trap for a system is an important factor in its operational efficiency. For example, an oversized trap will operate less efficiently than a correctly sized one, and will tend to create abnormal back pressure. Moreover, the installation cost will be higher and the operational life expectancy will be reduced.

Manufacturers of steam traps provide information in the form of capacity ratings and related data to make these selections easier. Data should be based on hot condensation under actual operating conditions rather than cold-water ratings. If possible, always try to determine the basis for a manufacturer's ratings. Table 10-1 and the sizing example were provided by Sarco Company, Inc., a manufacturer of steam traps.

Steam Trap Maintenance

Isolate the steam-trap assembly from the return and supply lines and allow the pressure to normalize (return to atmosphere) before starting any maintenance. The steam trap can be disassembled for maintenance as soon as it has cooled down.

Note

Many steam trap manufacturers produce repair kits for current models and even for traps no longer in production

Automatic Heat-Up

Select a steam trap with a pressure rating equal to or greater than the pressure in the steam supply main but with a capacity based on the estimated pressure at the trap inlet. The pressure at the inlet of the steam trap can be considerably less than the pressure in the steam supply main.

If the steam trap is connected into a common piping return system, it may have to operate against a certain amount of static pressure. This static (back) pressure can cause a reduction in the operating capacity of the steam trap. Table 10-2 illustrates the effect of back pressure on steam trap capacity.

The safety factor for a steam trap is the ratio between its maximum discharge capacity and the condensation load it is expected to handle. The actual safety factor to use for any particular application will depend upon the accuracy of the estimated condensation load, the accuracy of the estimated pressure conditions at trap inlet and outlet, and the operational characteristics of the trap.

The application for which a steam trap is to be used is also an important factor in its selection. For example, a float and thermostatic trap is recommended for use as a steam-line drip trap at pressures ranging from 16 psig to 125 psig. For the same application at pressures of 126 psig or above, an inverted bucket trap is suggested. Unusual operating conditions may also influence the choice of a steam trap for a particular application. A careful reading of the selection guide and related literature provided by the manufacturer will greatly reduce the possibility of error in choosing a suitable steam trap.

Table 10-1 Sizing Traps for Steam Mains—Condensation Load in Pounds per Hour per 1000 Feet of Insulated Steam Main*—Ambient Temperature 70°F—Insulation 80% Efficient

Steam Pressure (PSIG)	Main Size 2"	2½"	3"	4"	5"	6"	8"	10"	12"	14"	16"	18"	20"	24"	0°F† Correction Factor
10	6	7	9	11	13	16	20	24	29	32	36	39	44	53	1.58
30	8	9	11	14	17	20	26	32	38	42	48	51	57	68	1.50
60	10	12	14	18	24	27	33	41	49	54	62	67	74	89	1.45
100	12	15	18	22	28	33	41	51	61	67	77	83	93	111	1.41
125	13	16	20	24	30	36	45	56	66	73	84	90	101	121	1.39
175	16	19	23	26	33	38	53	66	78	86	98	107	119	142	1.38
250	18	22	27	34	42	50	62	77	92	101	116	126	140	168	1.36
300	20	25	30	37	46	54	68	85	101	111	126	138	154	184	1.35
400	23	28	34	43	53	63	80	99	118	130	148	162	180	216	1.33
500	27	33	39	49	61	73	91	114	135	148	170	185	206	246	1.32
600	30	37	44	55	68	82	103	128	152	167	191	208	232	277	1.31

** Chart loads represent losses due to radiation and convection for standard steam.*

† For outdoor temperature of 0°F, multiply load value in table for each main size by correction factor corresponding to steam pressure.

(Courtesy Spirax Sarco Co.)

Table 10-2 Effect of Back Pressure on Steam Trap Capacity (Percentage Reduction in Capacity)

%	*Inlet Pressure PSIG*			
Back Pressure	***5***	***25***	***100***	***200***
25	6	3	0	0
50	20	12	6	5
75	38	30	25	23

(Courtesy Spirax Sarco Co.)

Installing Steam Traps

The installation of steam traps requires the following modifications in the piping:

1. Install a long vertical drip and a strainer between the trap and the apparatus it drains. The vertical drip should be as long as the installation design will permit. *Exception:* Thermostatic traps in radiators, convectors, and pipe coils are attached directly to the unit without a strainer.
2. A gate valve should be installed on each side of a trap, along with a valved bypass around the traps if continuous service is required. This permits removal of the trap for servicing, repair, or replacement without interrupting service.
3. A check valve and gate valve should be installed on the discharge side of a trap used to discharge condensation against back pressure or to a main located above the trap (as for lift service).

Always carefully follow the steam trap manufacturer's instructions for installing a steam trap. If these instructions are not available, call the factory or an authorized representative for information before attempting to install the trap. The following are offered as guidelines for installing steam traps:

1. All work must be performed by qualified personnel trained in the correct installation of the trap.
2. Installation work must comply with all local codes and ordinances.
3. Allow the boiler to cool down to approximately 80°F and the pressure to drop to 0 before attempting to do any work on the trap.

4. Wear heat-resistant gloves to prevent serious burns when opening and shutting steam valves.
5. Cap off the gate valves if they are not connected to a drain or not in use for test or pressure-relief purposes to prevent property damage, serious injury, or death.
6. Connect a temporary pipe between the steam pipe opening and a drain to prevent injury from steam pipe blow-down. In lieu of installing a temporary pipe, stand at least 100 feet from the pipe opening.
7. Open supply valves *slowly* after installing the trap.

Note

Check the trap seat rating on the nameplate *before* installing it. The rating must be equal to or greater than the maximum pressure differential across the trap.

Float Traps

A *float trap* (see Figure 10-19) is operated by the rise and fall of a float connected to a discharge valve. The change of condensation level in the trap determines the level of the float. When the trap is empty, the float is at its lowest position and the discharge valve is closed. As the condensation level in the trap rises, the float also rises and gradually opens the valve. The pressure of the steam then pushes the condensation out of the valve. Because the opening of the valve is proportional to the flow of condensation through the trap, the discharge of condensation from the trap is generally continuous. On some float traps, a gauge glass is used to indicate the height of the condensation in the trap chamber.

One of the principal disadvantages of a float trap is the tendency of the valve to malfunction. Valve malfunctions can result from the

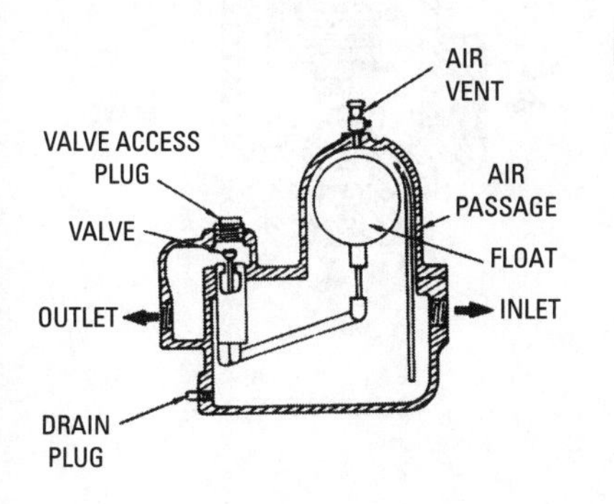

Figure 10-19 Float trap.

sticking of moving parts or excess steam leakage due to unequal expansion of the valve and seat.

Float traps are designed for steam pressures ranging from vacuum conditions to 200 psig and are used to drain condensation from heating systems, steam headers, steam separators, laundry equipment, and other steam process equipment. When used in heating systems, a float trap should be equipped with a thermostatic air vent (see "Float and Thermostatic Traps" in this chapter).

Thermostatic Traps

The operation of a *thermostatic trap* (see Figure 10-20) is based on the expansion or contraction of an element under the influence of heat or cold.

Thermostatic traps are of the following two types: (1) those in which the discharge valve is operated by the relative expansion of metals and (2) those in which the action of the liquid is utilized for this purpose. The latter is probably the most commonly used thermostatic trap found in modern steam heating systems. Thermostatic traps of large capacity for draining blast coils or very large radiators are called *blast traps*.

Modern thermostatic traps consist of thin, corrugated-metal bellows or discs enclosing a hollow chamber that is filled with a liquid or partially filled with a volatile liquid. When steam comes in contact with the expansive element, the liquid expands or becomes a gas and thereby creates a certain amount of pressure. The element expands as a result of this pressure and closes the valve against the escape of the steam.

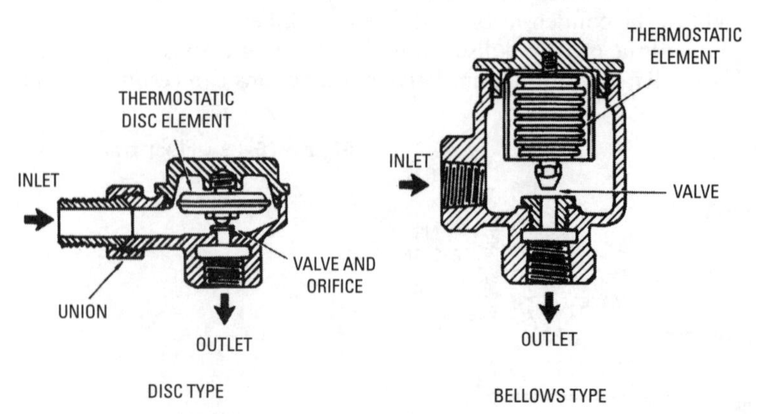

Figure 10-20 Thermostatic trap.

Balanced-Pressure Thermostatic Steam Traps

As shown in Figure 10-21, the principal parts of a *balanced-pressure thermostatic trap* consist of a flexible bellows, a valve head, and a valve seat. The bellows is partially filled with a volatile fluid and hermetically sealed. The fluid sealed in the bellows has a pressure–temperature relationship that closely parallels, but is approximately 10°F below, that of steam. When the condensation surrounding the bellows reaches approximately 10°F below saturated steam pressure, the fluid *inside* the bellows begins to build up pressure. When the temperature of the condensation approaches that of steam, the pressure inside the bellows exceeds the external pressure. This pressure imbalance causes the bellows to expand, driving the valve head to its seat and closing the trap (see Figure 10-22). When the condensation surrounding the bellows cools, the vaporized fluid condenses and reduces the internal pressure. The reduction of internal pressure causes the bellows to contract, opening the trap for discharge.

A balanced-pressure thermostatic steam trap is vulnerable to water hammer and corrosive elements in the condensation. The latter problem can be handled by fitting the trap with anticorrosive internal components. On the plus side, this type of trap has relatively large capacity and high air-venting capability. It is completely self-adjusting within its pressure range.

Additional information about balanced pressure steam traps can be found in Chapter 2 ("Radiators, Convectors, and Unit Heaters") in Volume 3.

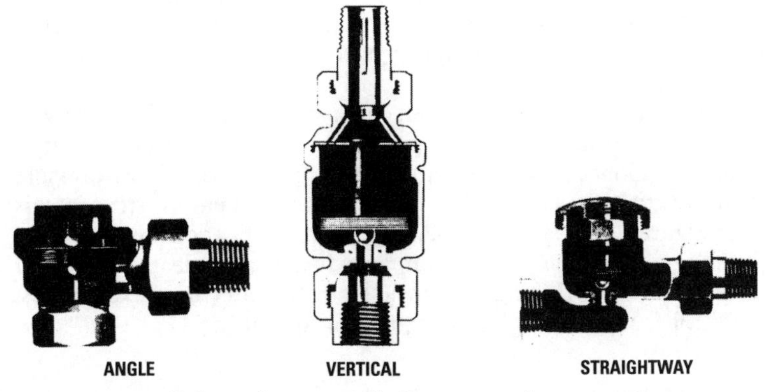

Figure 10-21 Balanced-pressure, bellows-type thermostatic steam traps. *(Courtesy Spirax Sarco Co.)*

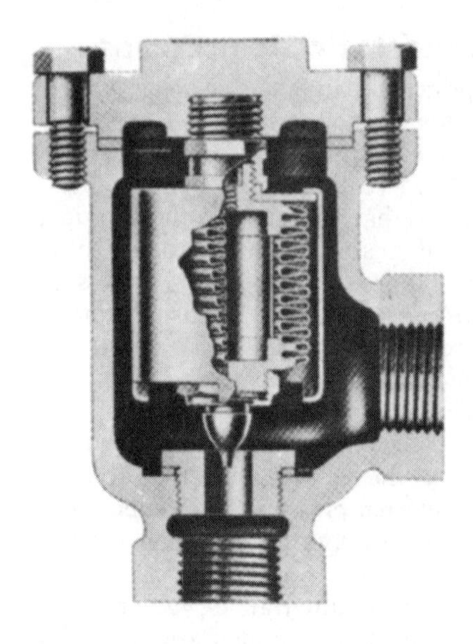

Figure 10-22 Balanced-pressure thermostatic steam trap. *(Courtesy Spirax Sarco Co.)*

Maintenance

These valves are factory sealed and no repair parts are available. If defective, they must be replaced. The valves must be completely isolated from both the supply and return lines before removal for replacement.

Float and Thermostatic Traps

A *float and thermostatic trap* (see Figure 10-23) has both a thermostatic element to release air and a float element to release the condensation. As such, it combines features of both the float trap and the thermostatic trap.

These traps are recommended for installations in which the volume of condensation is too large for an ordinary thermostatic trap to handle. Float and thermostatic traps are also used in low-pressure steam heating systems to drain the bottom and end of steam risers (see Figures 10-24 and 10-25). Other applications include the draining of condensation from unit heaters, preheat and reheat coils in air conditioning systems, steam-to-water heat exchangers, blast coils, and similar types of process equipment.

Note

Check the trap-seat pressure before installation to make sure its rating is equal to or greater than the steam supply of the boiler.

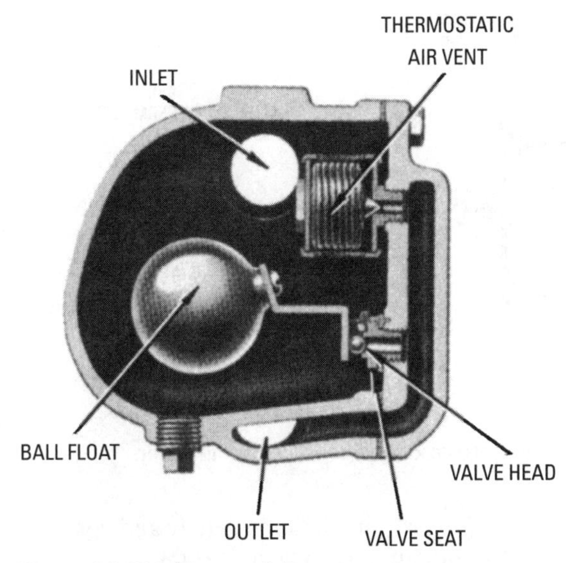

Figure 10-23 Float and thermostatic steam trap.
(Courtesy Spirax Sarco Co.)

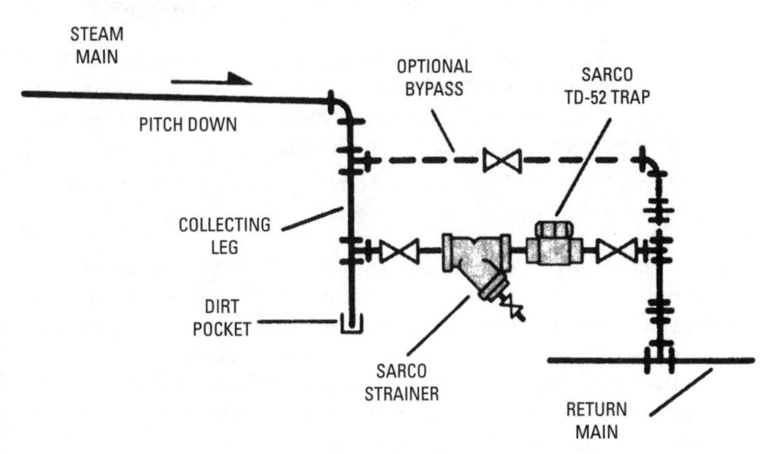

Figure 10-24 Draining end of low-pressure steam risers.
(Courtesy Spirax Sarco Co.)

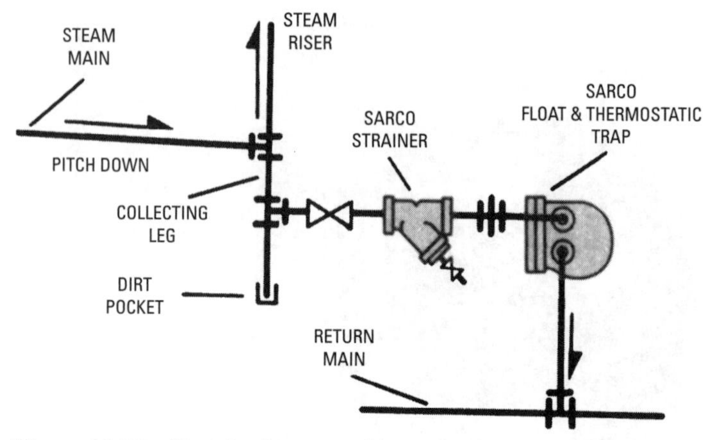

Figure 10-25 Draining bottom of low-pressure steam riser.
(Courtesy Spirax Sarco Co.)

If the trap is operating properly, it will immediately and continuously discharge condensation, air, and noncondensable gases from the system that enter the inlet orifice of the trap.

No Heat or Uneven Heat

Float and thermostatic traps are used in two-pipe steam heating systems to pass air and condensate into the return piping while simultaneously preventing the steam from moving past the radiators and ends of the steam mains. Failure of these traps results in the following problems:

- Trap fails in the open position causing the steam to pass into the return lines. As a result, the pressure in both the return and supply piping is the same and, without a pressure differential between the two, the steam cannot move to the radiators.
- Trap fails in the closed position, blocking the entry of air and steam into the radiators.

The condensation is handled by the ball float, which is connected by a level assembly to the main valve head. Condensation entering through the trap inlet causes the ball float to rise, moving the level assembly and opening the valve for discharge.

Air and noncondensable gases are discharged through the thermostatic air vent. The thermostatic element is also designed to prevent the flow of steam around the float valve.

Float and thermostatic traps operate under pressures ranging from vacuum to a maximum pressure of 200 psig; however, the great majority of them are designed for 40 psig or less.

These traps have limited resistance to water hammer. They are also vulnerable to corrosive elements in the condensation unless fitted with anticorrosive internal components.

Thermodynamic Steam Traps

A thermodynamic steam trap (see Figures 10-26 and 10-27) contains only one moving part, a hardened stainless-steel disc that functions as a valve. Because of its construction simplicity, this is an

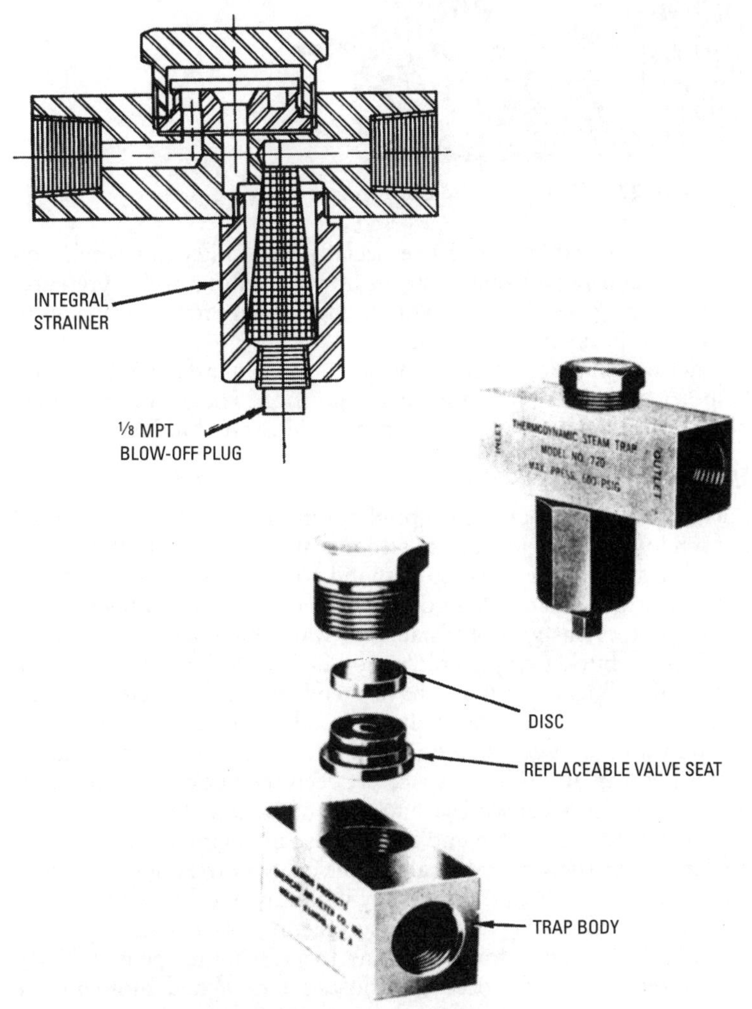

Figure 10-26 Thermodynamic steam trap. *(Courtesy American Air Filter)*

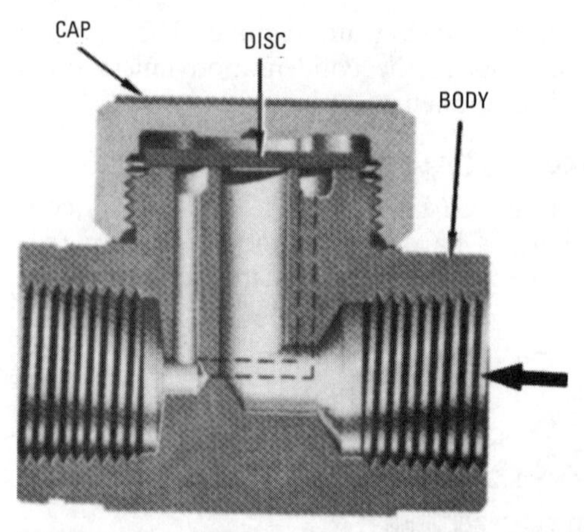

Figure 10-27 Thermodynamic steam trap. *(Courtesy Spirax Sarco Co.)*

extremely rugged trap and is especially well suited for service on medium- and high-pressure steam lines operating under pressures up to 600 psig. The minimum operating pressure for some makes of these traps is as low as 3.5 psig.

Thermodynamic steam traps are small, unaffected by water hammer, and can be mounted in any position. The operating principles of a thermodynamic steam trap are illustrated in Figure 10-28.

Bucket Traps

Bucket traps are either of the upright or inverted design and are used in both low- and high-pressure steam heating systems. Both types of bucket traps are designed to respond to the difference in density between steam and condensation. The construction of a bucket trap is such that it has good resistance to water hammer. On the other hand, most bucket traps, unless modified, have limited air-venting capabilities. Bucket traps also have a tendency to lose their waterseal and blow steam continuously during sudden pressure changes.

In an *upright bucket trap* (see Figure 10-29), the condensation enters the trap and fills the space between the bucket and the walls of the trap. This causes the bucket to float and forces the valve against its seat, the valve and its stem usually being fastened to the bucket. When the water rises above the edges of the bucket, it floats into it and causes it to sink, thereby withdrawing the valve from its seat. This permits the steam pressure acting on the surface of the water in the bucket to force the water to a discharge opening. When the bucket is emptied, it rises and closes the valve and another cycle begins. The discharge from this type of trap is intermittent.

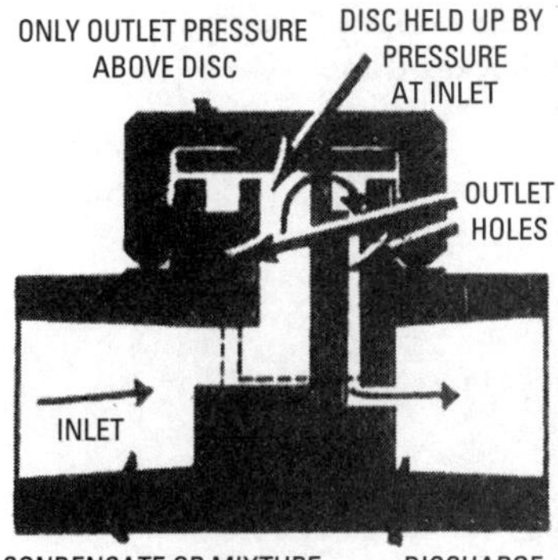

Pressure of condensate or air lifts the disc off its seats. Flow is across the underside of the disc to the three outlet holes. Discharge continues until the flashing condensate approaches steam temperature.

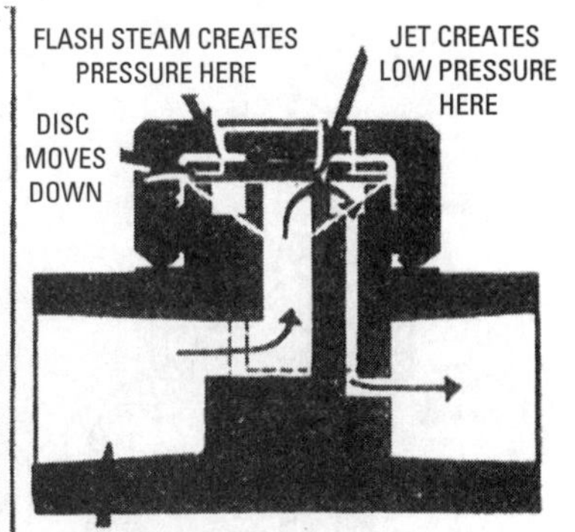

A high-velocity jet of flash steam reduces pressure under the disc and at the same time, by recompression, builds up pressure in the control chamber above the disc. This drives the disk to the seats ensuring tight closure without steam loss.

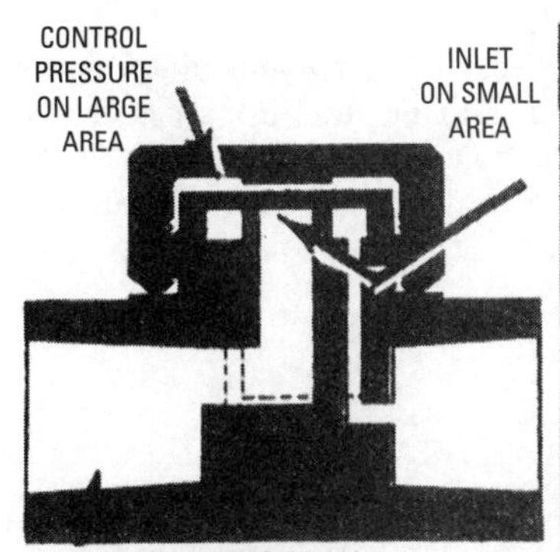

Steam pressure in the control chamber, acting over the total disc area, holds the disc closed against inlet pressure acting over the smaller inlet seat area.

STEAM CONDENSES HERE
DISC RISES
CONDENSATE ACCUMLATION REDUCES HEAT TRANSFER TO CONTROL CHAMBER

As soon as condensate collects, even at steam temperature, it reduces heat transferred to the control chamber Pressure in thechamber decreases as steam trapped there condenses. The disc is lifted by inlet pressure and condensate is discharged.

Figure 10-28 Operating principles of a thermodynamic steam trap.

(Courtesy Spirax Sarco Co.)

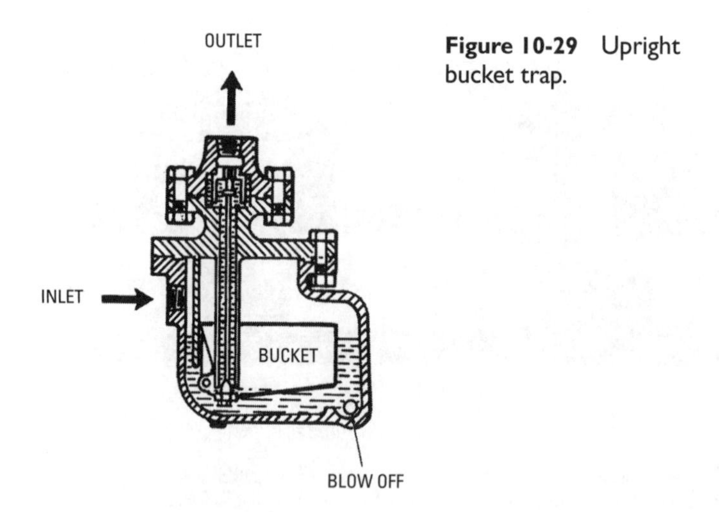

Figure 10-29 Upright bucket trap.

In the *inverted bucket trap* (see Figure 10-30), steam floats the inverted submerged bucket and closes the valve. Water entering the trap fills the bucket, which sinks, and through compound leverage opens the valve, and the trap discharges.

An inverted bucket trap with its seat open to vent air or drain condensate is shown in Figure 10-31. During start-up, air is vented into the return line through a bleed hole located at the top of the bucket. The condensate enters the trap, moves around the bucket, and drains from the open trap seat.

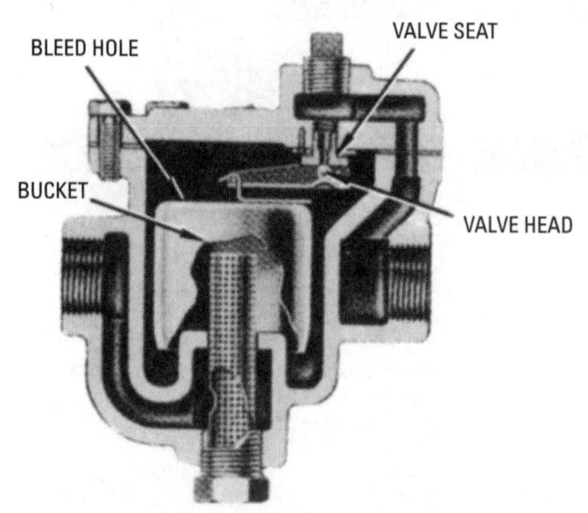

Figure 10-30 Inverted bucket trap. *(Courtesy Spirax Sarco Co.)*

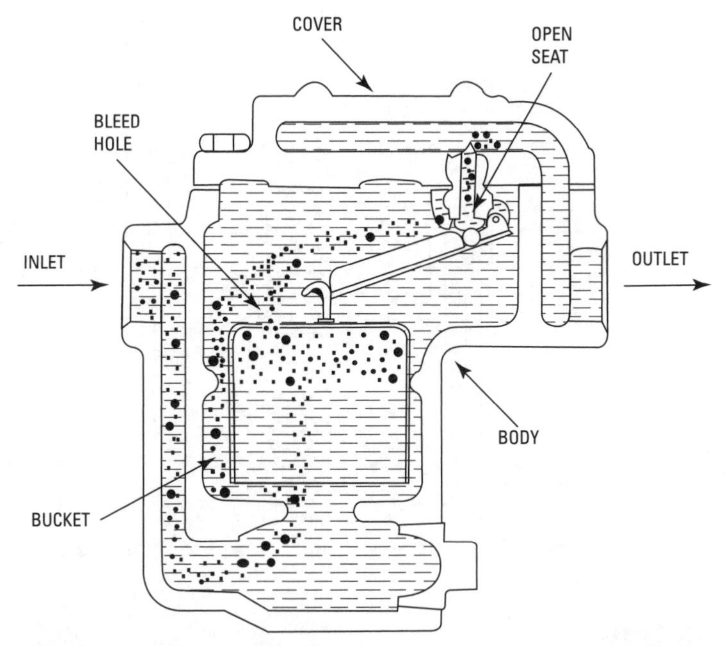

Figure 10-31 Inverted bucket trap seat open to vent air or drain condensate. *(Courtesy ITT Hoffman Specialty)*

As steam flows into the trap, it collects at the top of the bucket. When enough steam has collected there, its buoyancy causes the bucket to rise and close the trap seat (see Figure 10-32). The closed trap seat blocks the exit of the steam until more condensate enters the bucket trap. The cycle repeats itself as long as the boiler is producing heat. Some inverted bucket traps have a thermal vent at the top of the bucket, which produces faster venting during start-up.

Typical installations for bucket traps are shown in Figure 10-33. Note the following installation recommendations:

- Provide enough room around the trap to allow easy access for service and maintenance.
- Locate the trap as close as possible to and below the equipment being drained.
- Install the trap in a straight run of horizontal pipe that is slightly pitched to allow condensate to flow down into the trap inlet.

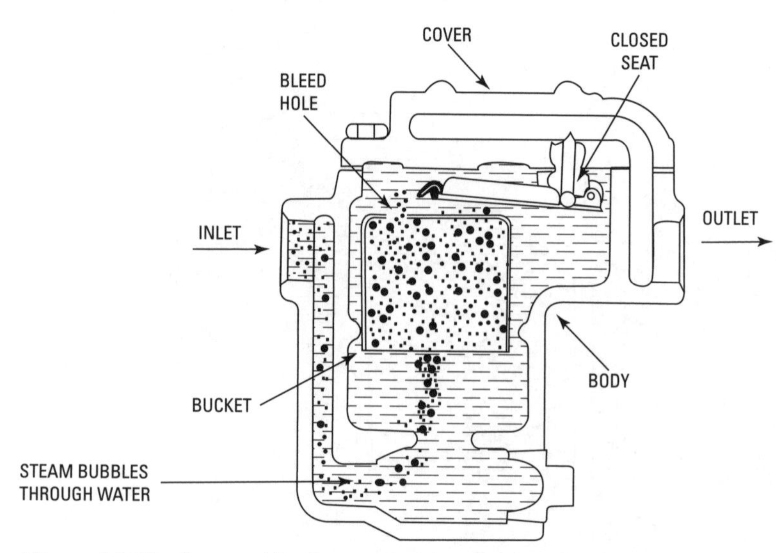

Figure 10-32 Inverted bucket trap with seat closed to retain steam.
(Courtesy ITT Hoffman Specialty)

Flash Traps

A *flash trap* (see Figure 10-34) is used to drain condensation from steam lines; steam, water, and oil heaters; unit heaters; and other equipment in which the pressure differential between the steam supply and condensation return is 5 psig or more.

The operation of a flash trap depends upon the property of condensation at a high pressure and temperature to flash into steam at a lower pressure. The condensation flows freely through the trap due to the pressure difference between the inlet and outlet orifices. The free flow of the condensation is interrupted by the introduction of steam into the inlet chamber, where it mixes with the remaining condensation. The steam heats the condensation and causes it to flash, thereby temporarily halting its flow through the orifice and allowing it to accumulate in the trap.

Except for an adjustable orifice used for adjusting the pressure differential, a flash trap contains no other moving parts. Flash traps operate intermittently. They are generally available for pressures ranging from vacuum to 450 psig.

Impulse Traps

An *impulse trap* (see Figure 10-35) operates with a moving valve actuated by a control cylinder. When the trap is handling condensation,

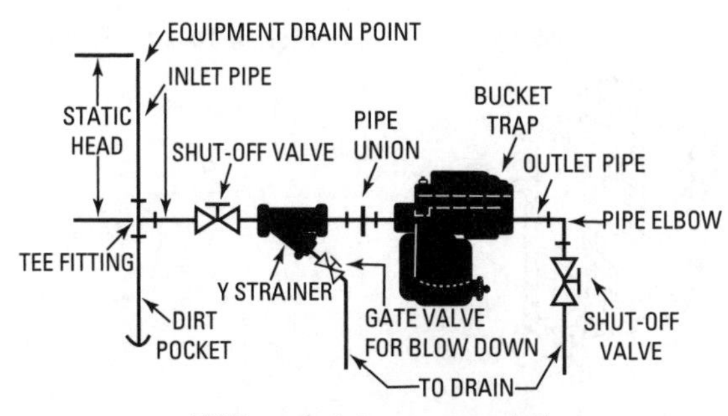

(A) Trap draining to open drain.

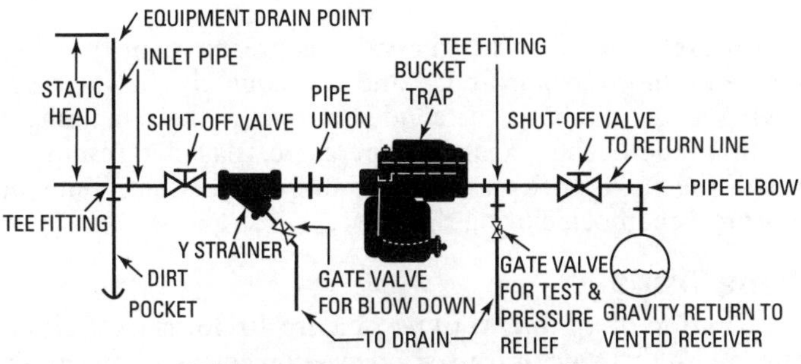

(B) Trap draining to gravity return line.

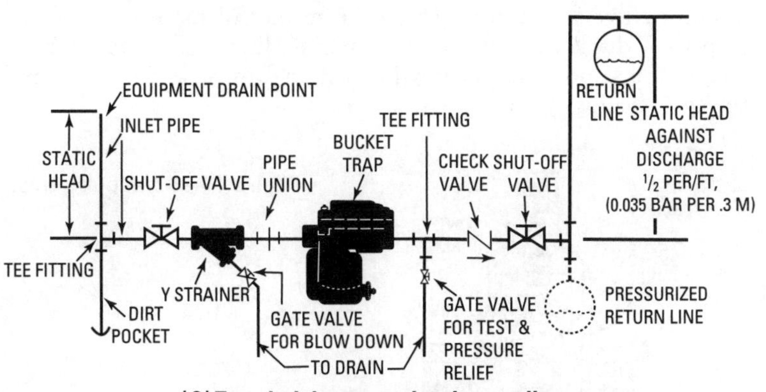

(C) Trap draining to overhead return line or pressurized return line.

Figure 10-33 Typical bucket-trap locations and piping diagrams.

(Courtesy ITT Hoffman Specialty)

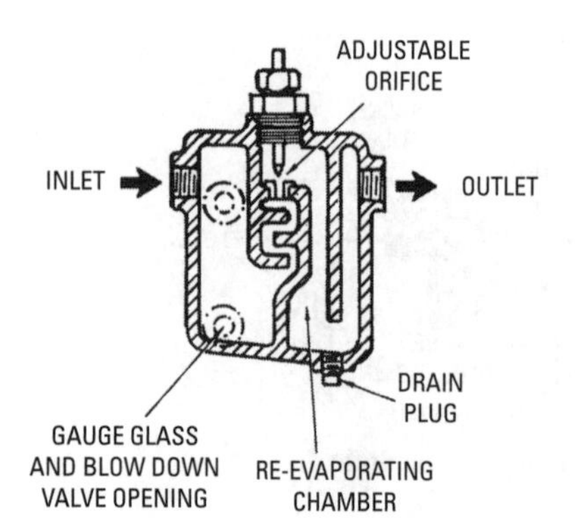

Figure 10-34 Flash trap.

the pressure required to lift the valve is greater than the reduced pressure in the control cylinder, and consequently the valve opens, allowing a free discharge of condensation. As the remaining condensation approaches steam temperature, flashing results, flow through the valve orifice is choked, and the pressure builds up in the control chamber, closing the valve.

Tilting Traps

The operation of a *tilting trap* (see Figure 10-36) is intermittent in nature. With this type of trap, condensation enters a bowl and rises until its weight overbalances that of a counterweight, and the bowl sinks to the bottom. As the bowl sinks, a valve is opened, thus admitting live steam pressure on the surface of the water, and the trap then discharges. After the water is discharged, the counterweight sinks and raises the bowl, which in turn closes the valve, and the cycle begins again.

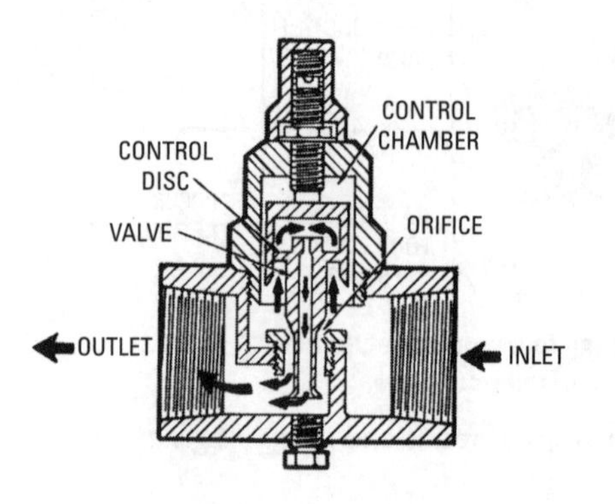

Figure 10-35 Impulse trap.

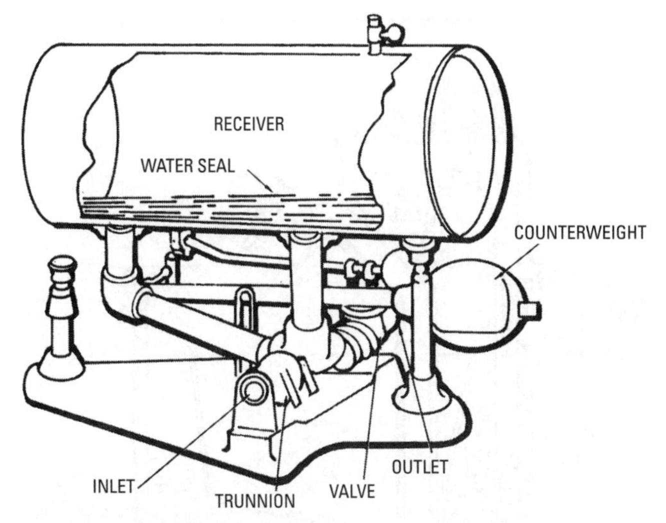

Figure 10-36 Tilting trap. *(Courtesy 1960 ASHRAE Guide)*

Lifting Traps

A *lifting trap* (see Figure 10-37) is an adaptation of the upright bucket trap and is available for pressures ranging from vacuum to 150 psig.

Condensation in the chamber of the trap accumulates until it reaches a level high enough to cause the steam valve in the high-pressure inlet to open. Steam then enters the auxiliary high-pressure inlet on the top of the trap at a pressure higher than the trap inlet pressure. This high-pressure steam forces the condensation to a point above the trap and against a back pressure higher than that which is possible with normal steam pressure. As the condensation is pushed out of the trap, the float or bucket descends to the bottom and causes the valve in the high-pressure inlet to close, shutting off the steam supply. Condensation then begins to refill the float chamber, and the cycle is repeated.

Boiler Return Traps

A *boiler return trap* (or *alternating receiver*) is a device used in some vapor-steam heating systems to return condensation to the boiler under varying pressure conditions of operation up to the working limit of the boiler. A vapor-steam heating system in which a boiler return trap is used is sometimes referred to as a *return-trap system.* A typical installation in which a boiler return trap is used is shown in Figure 10-38.

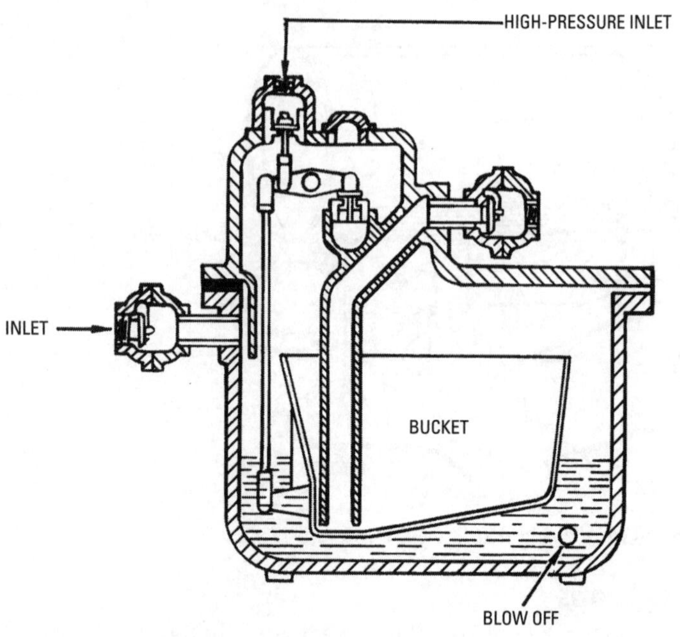

Figure 10-37 Lifting trap.

As shown in Figure 10-39, a boiler return trap consists of a chamber containing a float, which is linked to two valves. These valves control the openings to two connections on the top of the trap. One of these connections (the steam inlet) is connected to the steam header and direct boiler pressure. The other connection is vented to the atmosphere.

Condensation returning from the radiators is unable to enter the boiler by ordinary gravity flow, because the higher pressure of the boiler keeps the check valve closed. As a result, the condensation is forced to back up in the vertical pipe connected to the boiler return trap. As the condensation rises, it fills the bottom of the trap and lifts the float. At a certain level, the float causes the air valve to close and the steam valve to open, allowing steam at boiler pressure to enter the top of the trap. This steam at boiler pressure plus the gravity head (a boiler return trap *must* be located at least 6 inches above the water level in the boiler) is sufficient to force the condensation back down the pipe, through the check valve, and into the boiler.

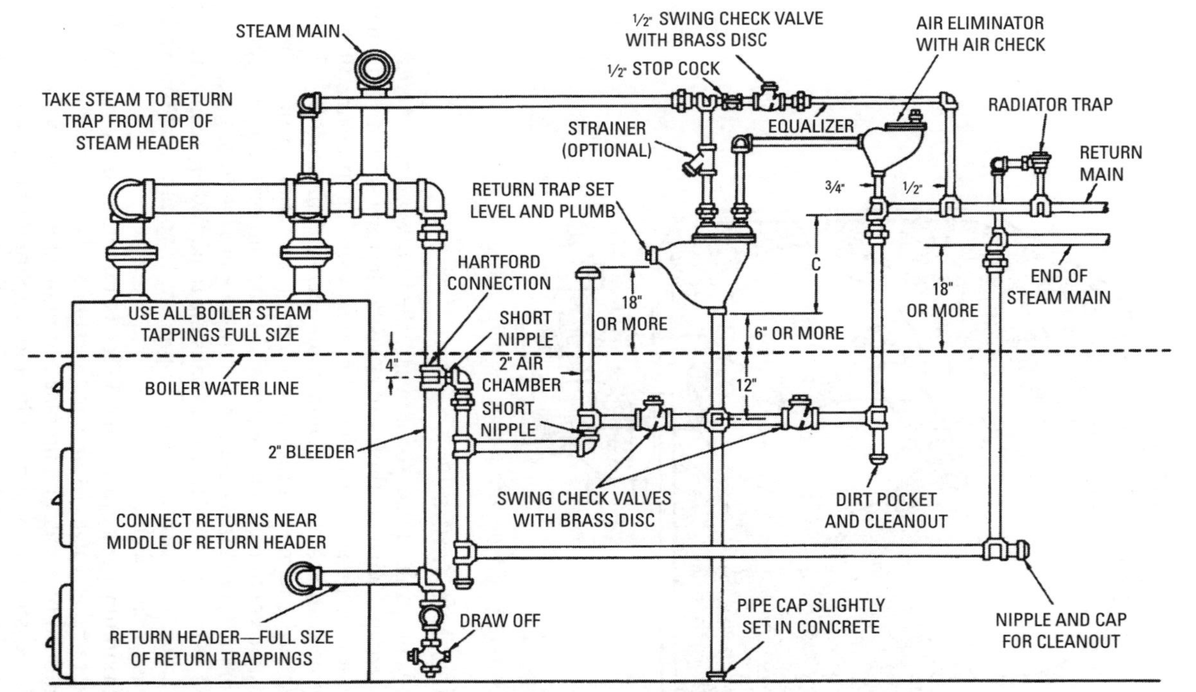

Figure 10-38 Boiler piping of a return-trap steam heating system. *(Courtesy Dunham-Bush, Inc.)*

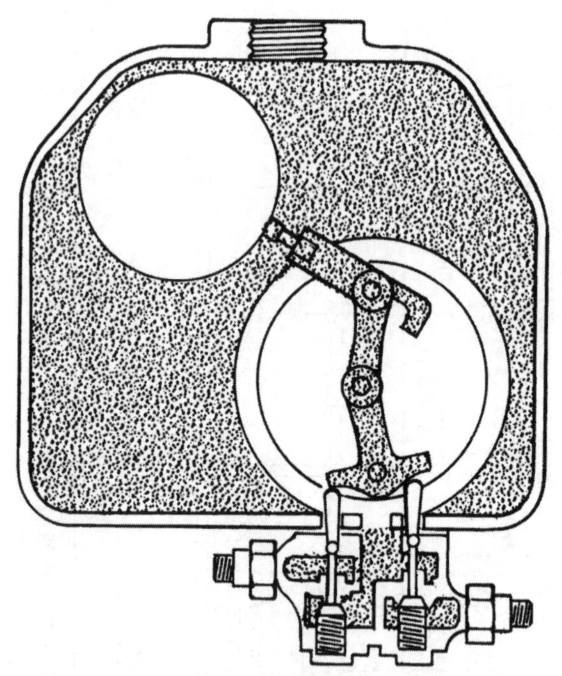

Figure 10-39 Boiler return trap. *(Courtesy Spirax Sarco Co.)*

Expansion Tanks

Expansion tanks (also sometimes called *compression tanks*) are installed in hydronic (hot-water) space heating systems to limit increases in pressure to the allowable working pressure of the equipment and to maintain minimum operating pressures.

When the temperatures rise during the operation of the system, the water volume also increases and builds up pressure. The pressure in the system is relieved to a certain extent by the storage of the excess water volume in the expansion tank. When temperatures drop, there is a corresponding drop in water volume and the water returns to the system.

Maximum pressure at the boiler is maintained by an ASME pressure-relief valve. Minimum pressure in the system is generally maintained by either an automatic or manual water-fill valve.

Closed steel expansion tanks and diaphragm tanks are used to contain the expanding volume of heated water in residential and light-commercial hydronic heating systems. Some typical installations using ITT Bell & Gossett expansion tanks are illustrated in Figures 10-40 through 10-42.

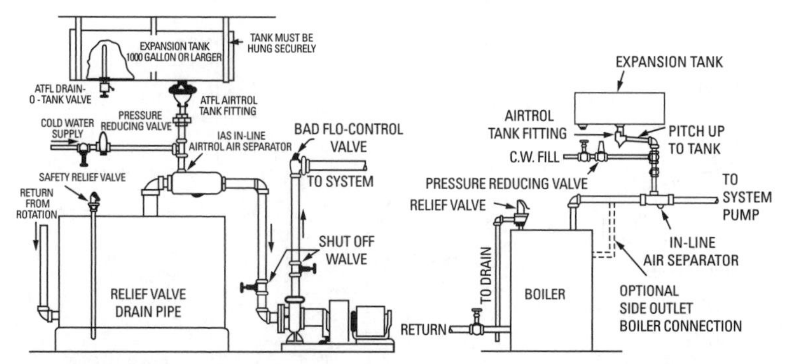

Figure 10-40 Typical installation with an inline Airtrol air separator.
(Courtesy ITT Bell & Gossett)

Closed Steel Expansion Tanks

The closed steel expansion tank has no moving parts (see Figure 10-43). It is normally two-thirds filled with water and one-third with air. As heated water expands and its excess volume enters the tank, it compresses the air at the top of the tank. The compression of the air in the tank results in an increase of system pressure, which is indicated on the boiler pressure gauge.

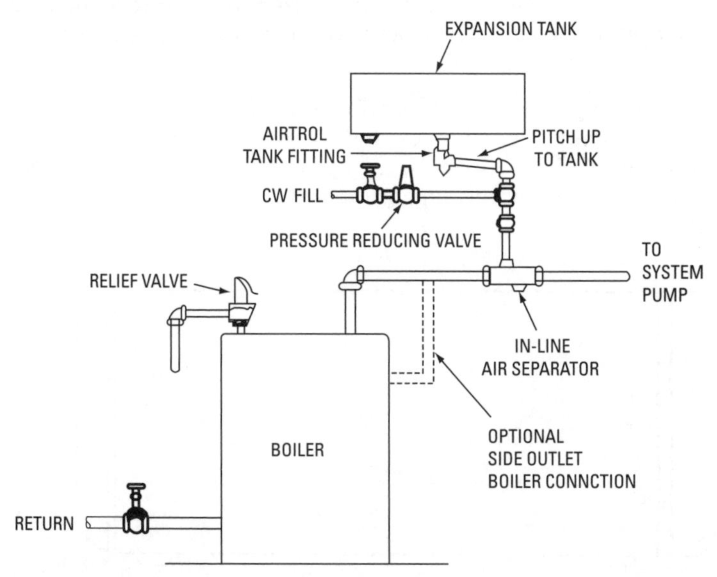

Figure 10-41 Inline air separator with conventional expansion tank.
(Courtesy ITT Bell & Gossett)

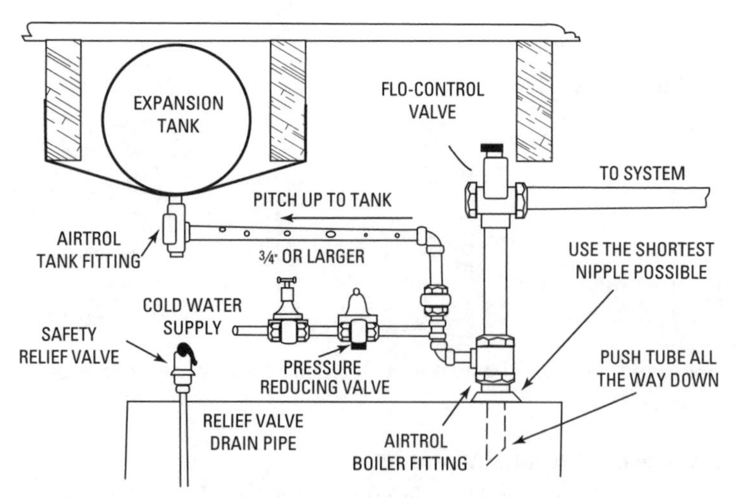

Figure 10-42 Typical installations on top outlet boilers.
(Courtesy ITT Bell & Gossett)

Note

If the expansion tank is properly sized, the pressure increase should be not more than about a pound before the system high-limit temperature is reached.

When the system water cools down, its volume contracts, and the air in the tank expands back to its original volume, causing system pressure to fall. To sum it all up, the rise and fall of system pressure is created by the expansion and contraction of the air in the expansion tank.

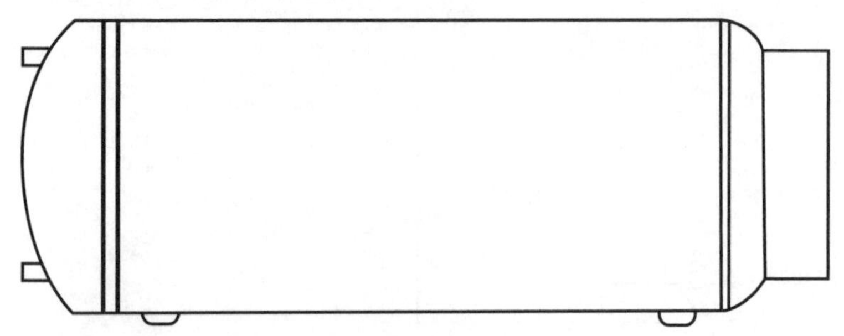

Figure 10-43 Closed steel expansion tank. *(Courtesy ITT Bell & Gossett)*

One problem encountered with a closed steel expansion tank directly connected into the system is that the system water can absorb the air and send it to the radiators and convectors by gravity circulation. Installing a gravity-flow check valve on the expansion tank will prevent gravity circulation.

Diaphragm Expansion Tanks

The air in a diaphragm-type expansion tank is separated from the water by a flexible rubber membrane (see Figure 10-44). These tanks are smaller than the closed steel tanks and come from the manufacturer precharged with compressed air. When the tank arrives at the site, the diaphragm is fully expanded against its inside surfaces. When the tank is installed and connected to the system piping, water enters the other side of the tank chamber and presses down on the diaphragm.

As a rule, diaphragm tank manufacturers will precharge their tanks to 12 psi, which is sufficient to match the water-fill pressure requirements of the typical house or small commercial building.

Sizing Expansion Tanks

Many problems are caused by using an expansion tank of inadequate size. Table 10-3 lists recommended sizes for expansion tanks in both open and closed tank systems.

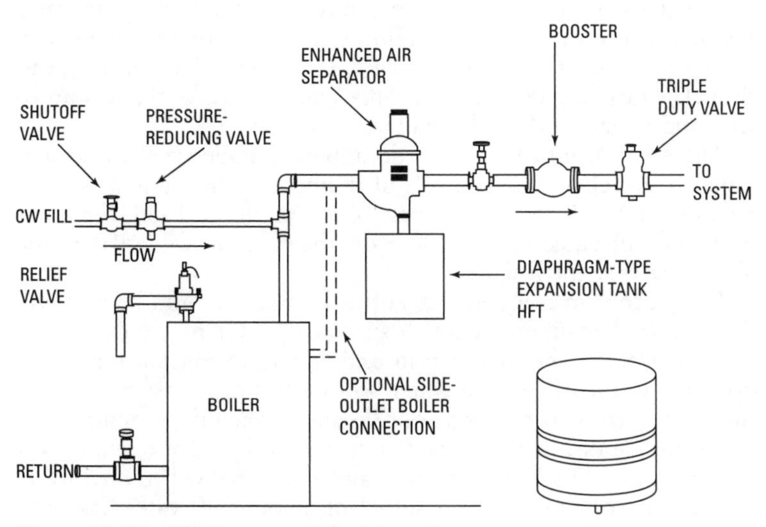

Figure 10-44 Diaphragm expansion tank. *(Courtesy ITT Bell & Gossett)*

Table 10-3 Recommended Sizes for Expansion Tanks

Open System	
Nominal Capacity—Gallons	***Square Feet of Radiation***
10	300
15	500
20	700
26	950
Closed System	
Nominal Capacity—Gallons	***Square Feet of Radiation***
18	350
21	450
24	650
30	900
35	900
35	1100

Troubleshooting Expansion Tanks

An undersized expansion tank or one that is completely filled up with water will cause the boiler pressure to increase when the water heats. Because the expansion tank is too small or too filled with water to absorb the excess pressure, the relief valve will begin to drip. The dripping relief valve is only symptomatic of the real problem, and replacing the valve will in no way solve it.

There is not much you can do about an undersized expansion tank except replace it. As a rule-of-thumb, expansion tanks should be sized at 1 gal. for every 23 ft^2 of radiation, or 1 gal. for every 3500 Btu of radiation installed on the job. In Table 10-3, the allowance is slightly higher.

If the problem is a completely filled tank, it should be partially drained so that there is enough space to permit future expansion under pressure. The first step in draining an expansion tank is to open the drain valve. The water will gush out at first in a heavy flow and then tend to gurgle out because a vacuum is building up inside the tank. Inserting a tube into the drain valve opening will admit air and break the vacuum, and the water will return to its normal rate of flow. After a sufficient amount of water has been removed, the drain valve can be closed.

Air Eliminators

Sometimes air pockets will form in the pipelines of steam or hot-water heating and cooling systems and retard circulation. One method of eliminating these air pockets is to install one or more air eliminators at suitable locations in the pipeline.

An *air eliminator* (or air vent) is a device designed to permit automatic venting of air. These automatic venting devices are available in a number of sizes, shapes, and designs. Not only are air eliminators used for venting convectors, baseboard radiators, and other heat-emitting devices; they are also frequently used for this purpose on overhead mains and circulating lines.

Three types of air eliminators (air vents) used in steam or hot-water heating and cooling systems are:

- Float-type air vents
- Thermostatic air vents
- Combination float and thermostatic air vents

A *float-type air vent* (see Figure 10-45) consists of a chamber (body) containing a float attached to a discharge valve by a lever assembly. The float-controlled discharge valve vents air through the large orifice at the top. The float action prevents the escape of any fluid, because the float closes the valve tightly when it rises. When the float drops, the lever assembly pulls the valve from its seat, and the unit discharges air.

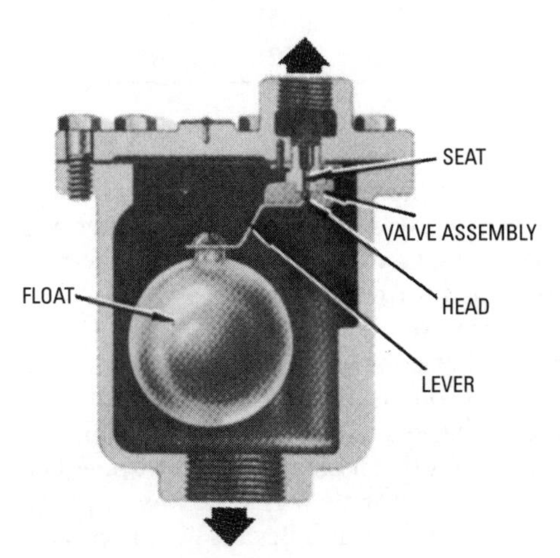

Figure 10-45 Float-type air vent. *(Courtesy Spirax Sarco Co.)*

Float-type air vents are available for hot-water heating and cooling systems to 300 psi and low-pressure steam heating systems to 15 psi. A float-type air vent used in a steam heating system should be equipped with a check valve, which prevents air return under vacuum.

Steam cannot be maintained at its saturated temperature when air is present in the system. As shown in Table 10-4, the temperature of the steam decreases as the percentage of air increases. A *thermostatic air vent* is specifically designed for removing air from a steam system. The one shown in Figure 10-46 consists of a valve head attached to a bellows, operating in conjunction with a thermostatic element. Its operating principle resembles that of a thermostatic steam trap. When air is present, the temperature of the steam drops. The temperature drop is sensed by the thermostatic element, which causes the valve in the vent to open and discharge the air. When the air has been discharged, the temperature of the steam rises, and the valve closes tightly.

Table 10-4 Effects of Air on Temperature of a Steam and Air Mixture

Mixture Pressure (psig)	*Pure Steam*	*5% Air*	*10% Air*	*15% Air*
2	219°	216°	213°	210°
5	227°	225°	222°	219°
10	239°	237°	233°	230°
20	259°	256°	252°	249°

(Courtesy Spirax Sarco Co.)

In some special applications, it is necessary to use an air eliminator that will close when the vent body contains steam or water and opens when it contains air or gases. Combination float and thermostatic air vents have been designed for this purpose.

A *combination float and thermostatic air vent* (see Figure 10-47) consists of a vent body or chamber containing a float attached to a valve assembly. The float rests on a thermostatic element that responds to the temperature of the steam. The operation of this element is similar to the one used in a thermostatic steam trap. When the vent body is filled with air or gas, the float is at its lowest point, causing the thermostatic bellows to contract. Because the float is at a low point in the vent body, the head is moved off the valve seat and the vent discharges the air or gas. The head moves up and

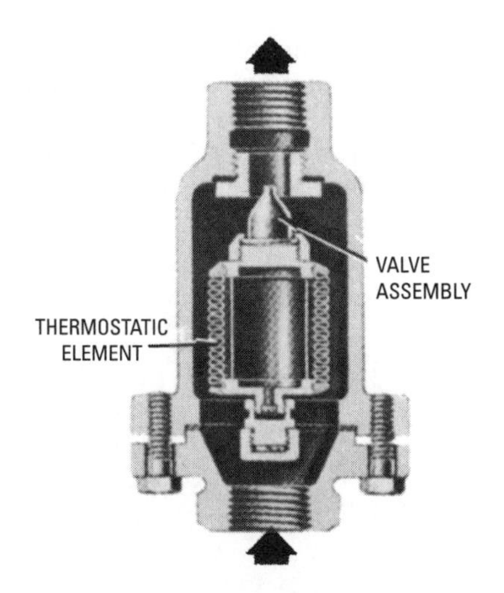

Figure 10-46 Thermostatic air vent. *(Courtesy Spirax Sarco Co.)*

closes the valve when either water or steam enters the vent body. The entry of water into the vent body forces the float upward and eventually closes off the valve. The entry of steam, on the other hand, causes the thermostatic bellows to expand and force the float upward, closing the valve.

Pipeline Valves and Controls

Pipeline valves and controls are used to regulate the temperature, pressure, or flow rate of the steam or water in the lines. Some valves (e.g., check valves) deal with only one of these functions; other valves are designed to handle more than one function.

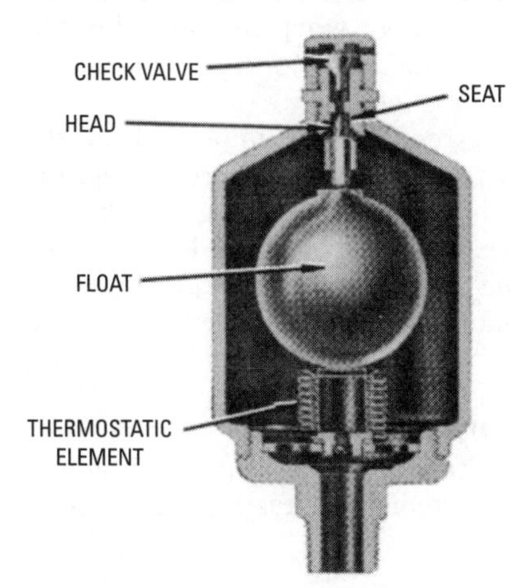

Figure 10-47 Combination float and thermostatic air vent.

Details about the design and construction of valves and the methods used for servicing, repairing, and installing them are found in Chapter 9 ("Valves and Valve Installation"). This chapter is limited to a description of seven of the more common valves and control devices used in steam and hydronic pipelines. They are:

- Temperature regulators
- Electric control valves
- Water-tempering valves
- Hot-water heating control
- Flow control valves
- Electric zone valve
- Balancing valves, valve adaptors, and filters

Temperature Regulators

Temperature regulators are used for many heating and cooling applications, including small-flow instantaneous heaters or coolers (shell-and-tube or shell-and-coil heat exchangers), small storage or tank heaters, and similar installations.

The Spirax Sarco Type 25T temperature regulator shown in Figure 10-48 is a diaphragm-operated valve used for regulating temperature in a variety of different process applications.

Before start-up, the main valve is normally in a closed position and the pilot valve is held open by spring force. The steam enters the orifice inlet, passes through the pilot valve and into the diaphragm chamber, and then out the control orifice. Control pressure builds up in the diaphragm chamber when the flow through the pilot valve exceeds the flow through the control orifice. This build-up in pressure opens the main valve.

The bulb of the temperature regulator is immersed in the medium being heated. At a predetermined temperature setting, the liquid in the bulb expands through the capillary tubing into the bellows and throttles the pilot valve. The main valve will deliver the required steam flow as long as the control pressure is maintained in the diaphragm chamber. The main valve closes when heat is no longer required.

Electric Control Valves (Regulators)

An *electric control valve* (regulator) is designed to provide remote electric on-off control in heating systems and steam process applications (see Figure 10-49).

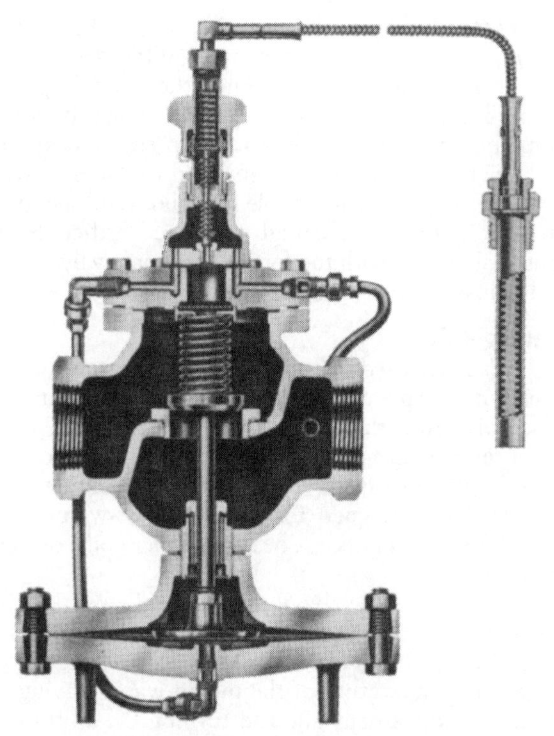

Figure 10-48 Temperature regulator. *(Courtesy Spirax Sarco Co.)*

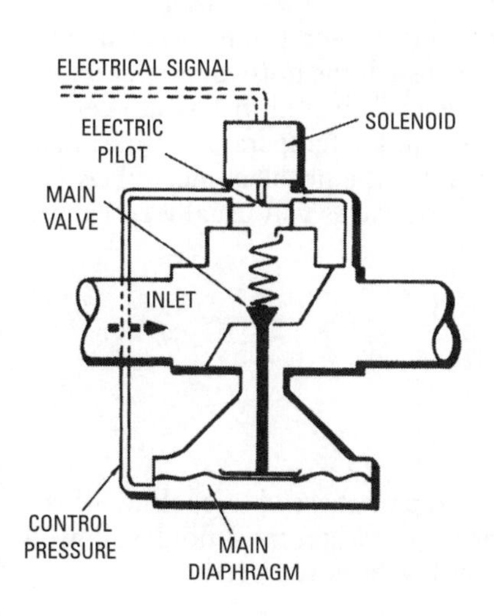

Figure 10-49 Diagram of an electric on-off regulator. *(Courtesy Spirax Sarco Co.)*

The solenoid pilot at the top of the valve is connected to a room thermostat, automatic time clock, or some similar device from which it can receive an electrical signal. When the solenoid pilot is electrically energized, the pilot valve opens, and pressure builds up in the diaphragm chamber. As a result, control pressure is applied to the bottom of the main valve diaphragm, and the main valve is opened. The pilot valve closes when the solenoid pilot is de-energized, and control pressure is relieved through the bleed orifice. Steam pressure acting in conjunction with the force of the main valve-return spring combine to close the main valve.

Water-Tempering Valves

Water-tempering valves are used in hot-water space heating systems where it is necessary to supply a domestic hot-water supply at temperatures considerably lower than those of the water in the supply mains. The water-tempering valve automatically mixes hot and cold water to a desired temperature, thus preventing scalding at the fixtures. These valves are designed for use with hot-water space heating boilers equipped with tankless heaters, boiler coils, or high-temperature water heaters.

The A.W. Cash Type TMA-2 valve illustrated in Figure 10-50 is a thermostatic water-tempering valve that is shipped from the factory preset to operate at 140°F. These valves can also be field-adjusted to change the temperature of the mixed water leaving the valve by loosening the adjustment nut and turning the adjustment screw either clockwise (for colder water) or counterclockwise (for hotter water).

If turning the adjustment screw does not produce the desired mixed-water temperature, carefully touch the hot-water inlet on the valve to make sure hot water is being delivered. If you are certain hot water is getting into the valve and a temperature adjustment still fails to produce the desired results, the problem most likely lies inside the valve. Problems with these valves can usually be caused by one of the following:

- Binding of bonnet to push-rod
- Sticking of push-rod to O-ring
- Binding of piston
- Sticking of body O-ring

Before attempting to service or repair these valves, close off the hot, cold, and mixed water connections. Water must not be allowed to enter the valve when the bonnet has been removed.

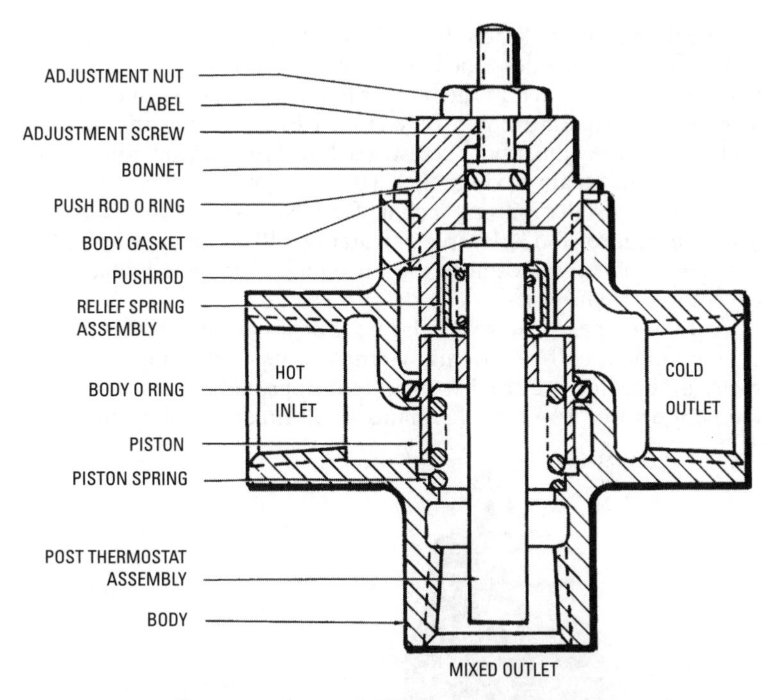

Figure 10-50 Diagram of a model TMA-2 water-tempering valve.
(Courtesy A.W. Cash Valve Mfg. Corp.)

Access to the internal parts of the hot-water tempering valve illustrated in Figure 10-50 is gained by unscrewing the bonnet (i.e., turning it counterclockwise) and removing it. If the bonnet is binding to the push-rod, pull the push-rod out of the bonnet and wipe it off with a crocus cloth. Do the same with the inside of the bonnet. If the push-rod O-ring is sticking, it should be removed and replaced. Reassembly is in reverse order; place the pushrod O-ring, reinsert the push-rod, and then screw the bonnet back on.

A binding piston should be removed, cleaned (with a crocus cloth), and lubricated. Access to the piston is also gained by unscrewing the bonnet. A body O-ring that is sticking should be removed and replaced. Access to the body O-ring is gained by unscrewing and removing the bonnet (leaving the push-rod in the bonnet). Push the piston and piston spring up and out through the top of the tempering valve. Lift out the post thermostat assembly. When reassembling, be sure to lubricate both the body O-ring and piston.

The design of the Watts No. N170 Series water-tempering valve differs from the one described above in that the discharge or mixed-water orifice is located in the bonnet (see Figures 10-51 and 10-52). The water temperature can be changed by loosening the locknut and turning the adjustment screw. Each *full* turn of the adjustment screw is equal to approximately a 10°F change in temperature.

A typical installation in which a Watts No. N170 valve is used is shown in Figure 10-53. Tempered water at 140°F can be delivered to the system. The thermostat in the valve makes trapping unnecessary except in extreme cases.

A two-temperature recirculating hot-water supply system is shown in Figure 10-54. In this system, a water-tempering valve and recirculating line are used to maintain approximate fixture water temperatures of 140°F in the mains at all times. A relatively small

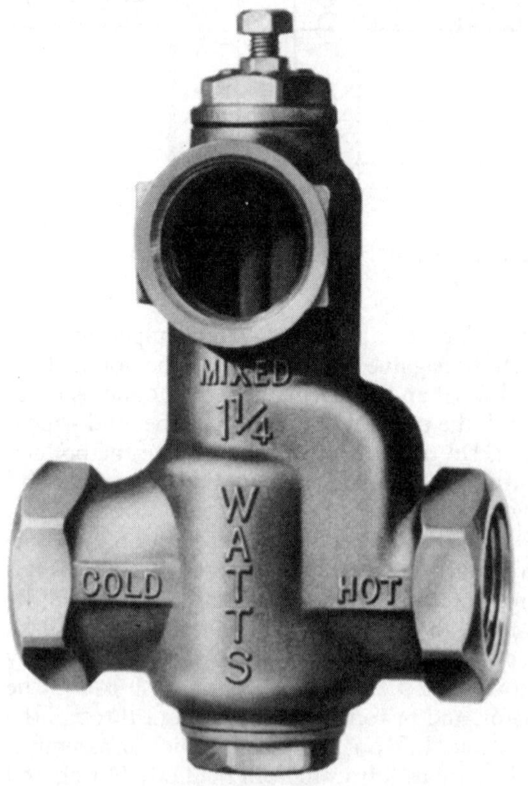

Figure 10-51 Model N170 water-tempering valve. *(Courtesy Watts Regulator Co.)*

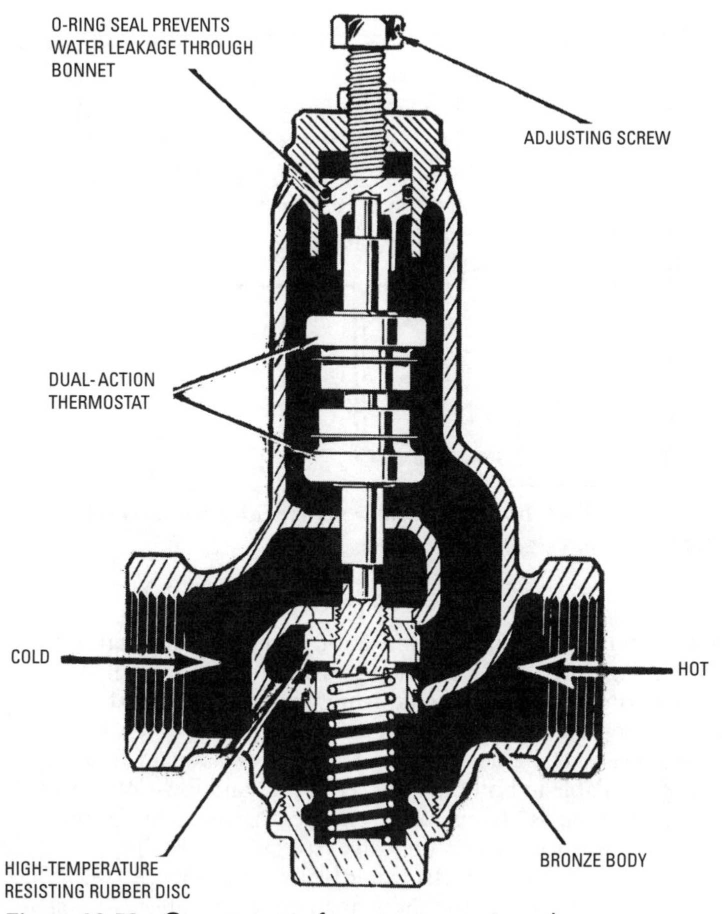

Figure 10-52 Components of a water-tempering valve.
(Courtesy Watts Regulator Co.)

capacity recirculator is used in the recirculating return piping, because very little hot water is required to maintain the low temperature in the mains. Long runs of recirculating piping should be insulated to reduce the heat loss from the piping.

Tempering valves cannot compensate for rapid pressure fluctuations in the system. Where such water pressure fluctuations are expected to occur, a pressure equalizing valve should be installed.

The Watts 70A Series tempering valve, shown in Figure 10-55, is designed for small domestic water-supply systems and tankless

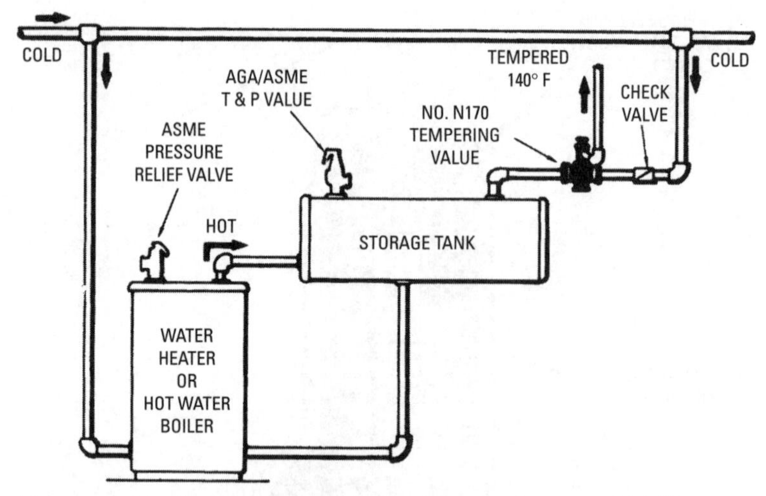

Figure 10-53 Basic hot-water supply system using model N170 tempering valve. *(Courtesy Watts Regulator Co.)*

heater installations. Piping connections for these applications are shown in Figure 10-56. A balancing valve should be installed below the tempering valve in the cold-water line to compensate for the pressure drop through the heater.

This valve is available with both threaded and sweat connections. It is also available in both high- (120–160°F) and low- (100–130°F) temperature models. Temperature changes are made by turning the dial-type adjustment cap on the valve (see Figure 10-57).

The Spirax Sarco Type MB water blender (see Figure 10-58) has a 55°F adjustment range for supplying tempered water to a system. It is a three-way double-ported balancing valve, essentially resembling the Watts and A.W. Cash valves in construction, except for an extended bonnet containing spirals. This type of construction allows a certain degree of pressure fluctuation between the hot- and cold-water inlets without disturbing the control of the tempered water.

Hot-Water Heating Control

A *hot-water heating control* consists of an outdoor liquid expansion-type bulb connected by a capillary system to a double-ported three-way automatic mixing valve (see Figure 10-59). It is designed to blend the hot water from the boiler with the cooler return water

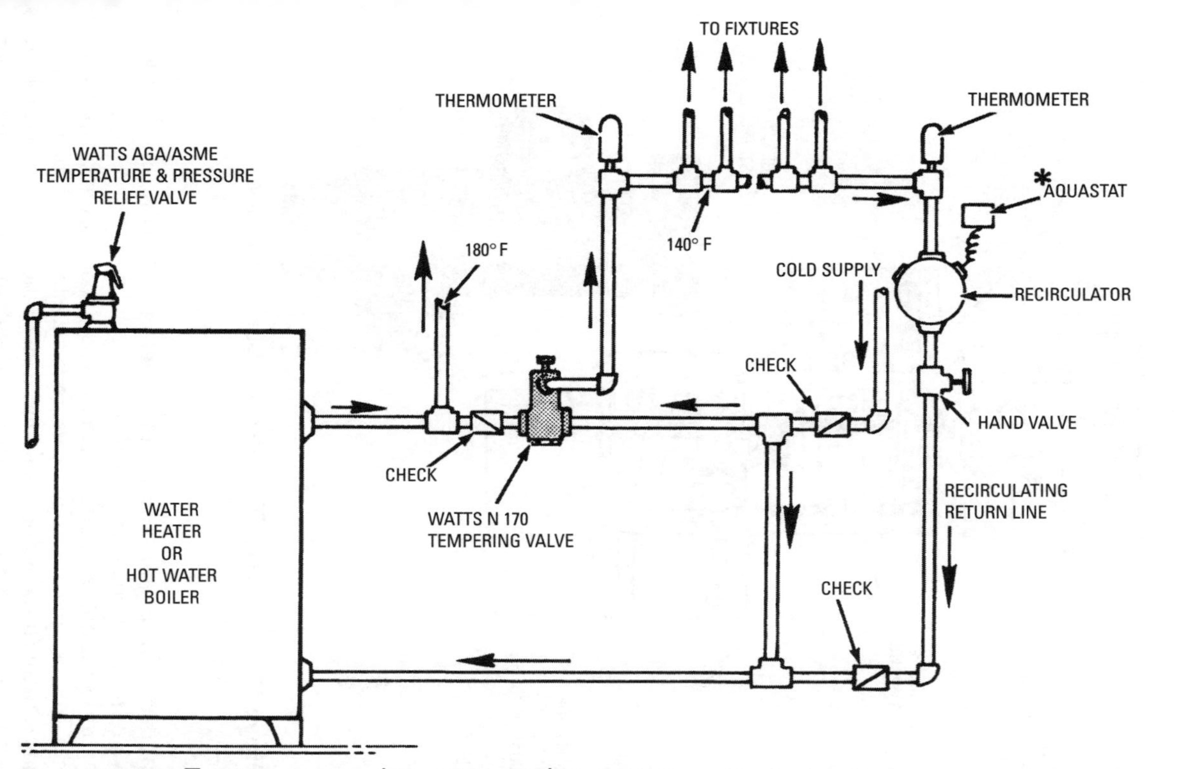

Figure 10-54 Two-temperature hot-water supply system. *(Courtesy Watts Regulator Co.)*

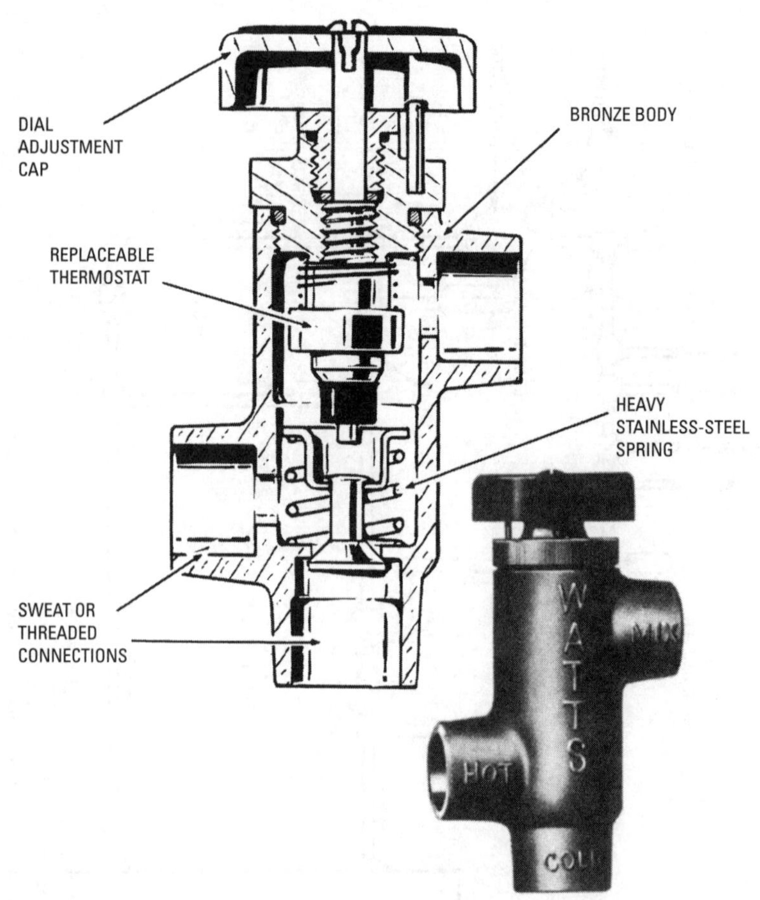

Figure 10-55 Model N70A series water-tempering valve.
(Courtesy Watts Regulator Co.)

in inverse proportion to the outside water and deliver the blended water to the circulating system.

In operation, the outdoor bulb reacts to changes in temperature and creates pressure. This pressure is transferred through a capillary system to the indoor bulb and then to the mixing valve, which is positioned to increase or decrease the amount of hot water from the boiler. Temperature-range adjustments can be made by turning an adjustment on top of the valve. A typical installation in which a hot-water heating control is used is shown in Figure 10-60.

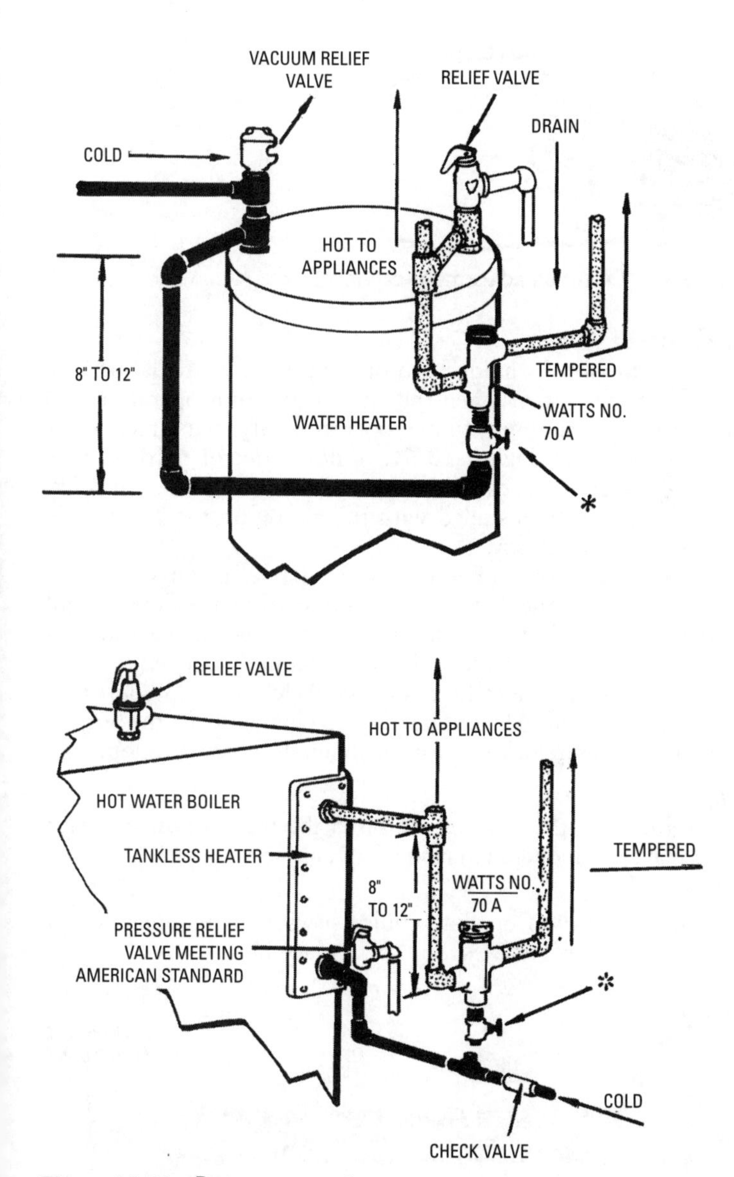

Figure 10-56 Piping connections. *(Courtesy Watts Regulator Co.)*

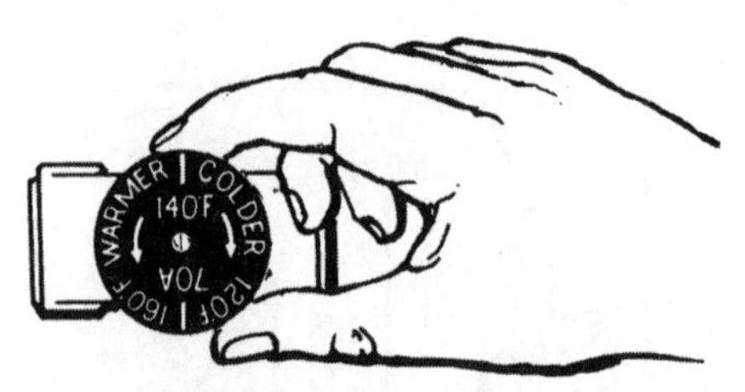

Figure 10-57 Dial-type adjustment cap. *(Courtesy Watts Regulator Co.)*

Flow Control Valve

A flow control valve is used (1) to prevent gravity circulation in a hydronic heating system when the circulator is not operating and (2) to permit the summer–winter operation of an indirect water heater. As shown in Figure 10-61, a flow control valve can be installed in either a vertical or horizontal position. Either location will require that it be installed with the arrow on the valve body facing the direction of flow.

When the circulator is operating, water passes through the flow control valve. When the circulator stops operating, the flow control valve remains closed and blocks the gravity flow of the water. A knob on top of the valve allows it to be manually opened to drain the system or to bypass it if there is a loss of electricity and only partial heating is possible. The knob is turned clockwise for the normal position and counterclockwise for the manual-bypass position.

Warning

> Do not allow the valve to remain in the manual position when normal system operation resumes. Doing so will result in uncontrolled heat.

Flow control valves do not require service or maintenance. A defective flow control valve should be replaced.

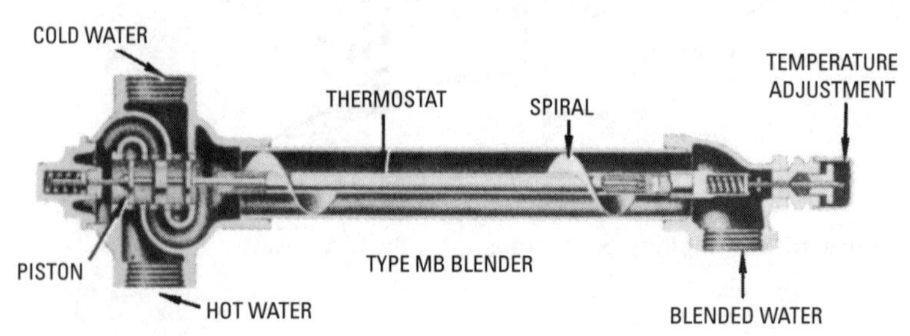

Figure 10-58 Spirax Sarco Type MB water blender. *(Courtesy Spirax Sarco Co.)*

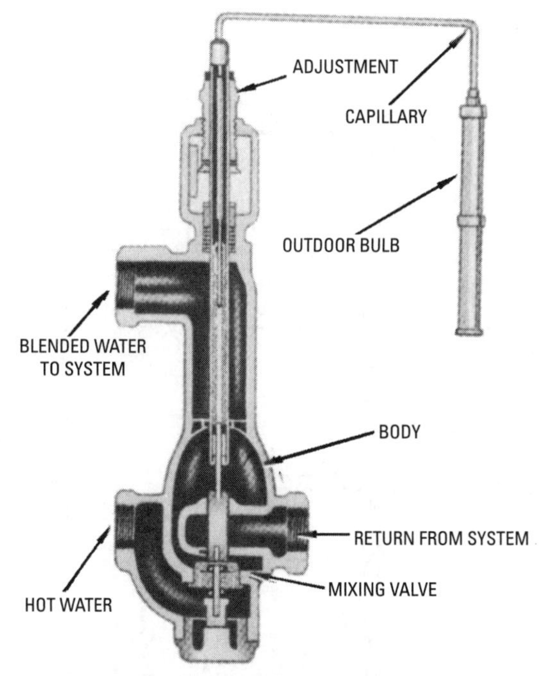

Figure 10-59 Hot-water heating control. *(Courtesy Spirax Sarco Co.)*

Electric Zone Valve

Some hydronic heating systems use electrically operated valves to control the flow of water into each zone. In these zoned hydronic systems, the valves operate in conjunction with a single circulator. The construction details of a typical electric zone valve are illustrated in Figure 10-62.

When the room thermostat calls for heat, an electrical current is sent to the valve operator. The current flows through a normally closed switch and around a coil called a heat motor. The heat created in the heat motor causes a piston to move out and push against a spring-loaded lever that normally holds the valve closed. This action opens the zone valve. At the same time, the piston extends a bit further and trips an end switch that sends a signal through a relay back to the circulator, turning it on and sending water into the zone. The piston extends a bit further and reverses the sequence. This in-and-out movement of the valve piston will continue for as long as the room thermostat calls for heat.

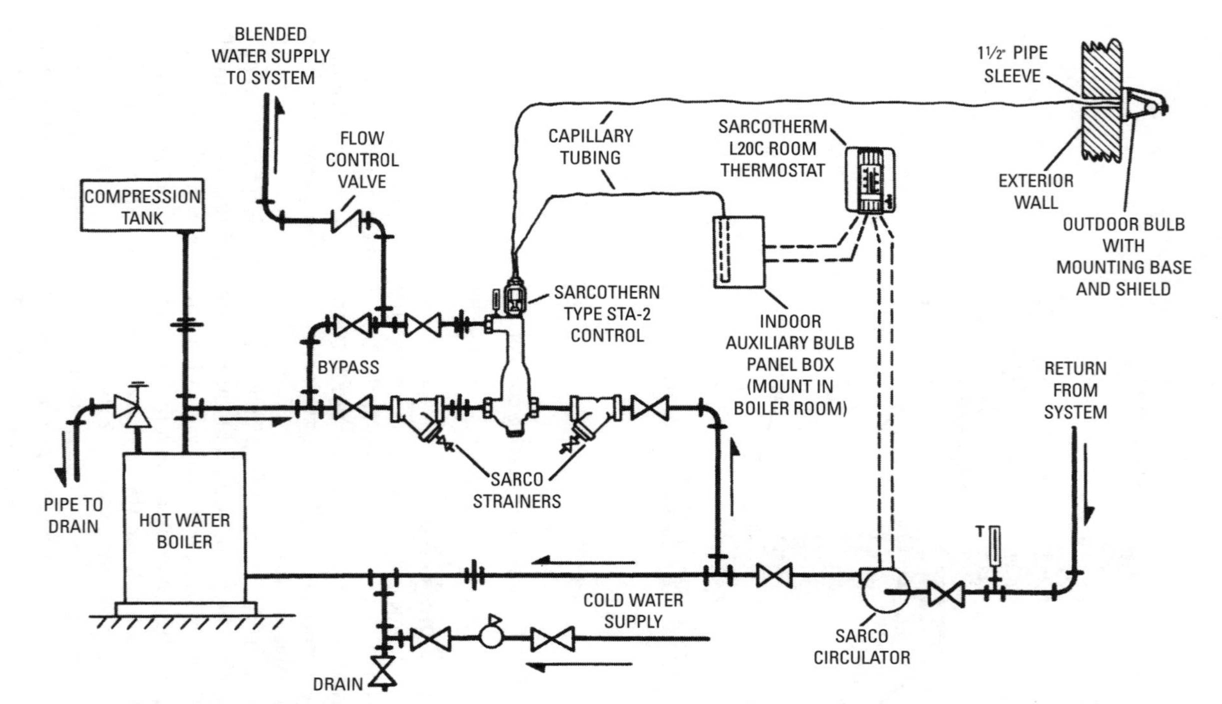

Figure 10-60 Hot-water heating control used to control heating-system water in accordance with outside temperature. *(Courtesy Spirax Sarco Co.)*

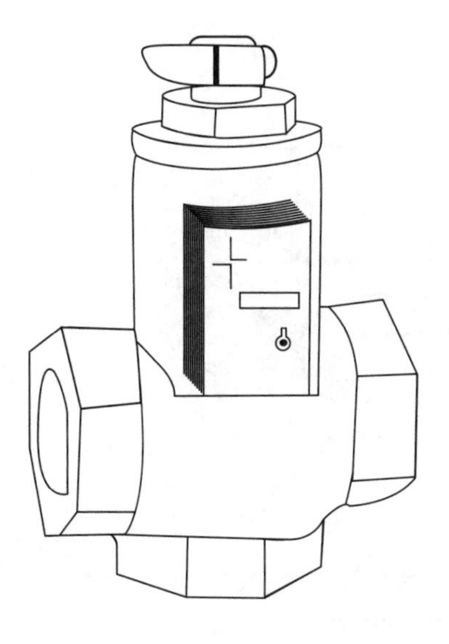

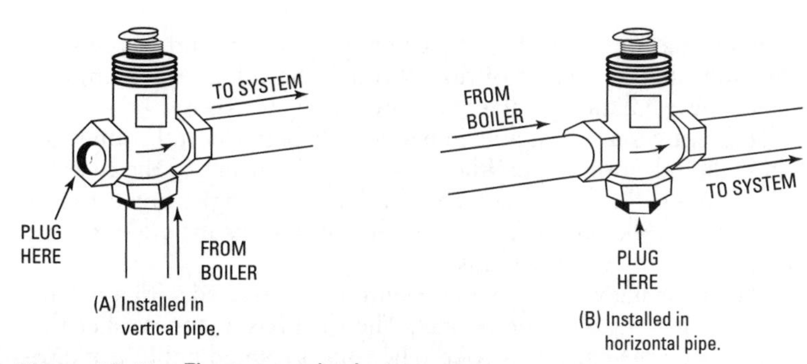

Figure 10-61 Flow control valve. *(Courtesy ITT Bell & Gossett)*

Balancing Valves, Valve Adapters, and Filters

Hot-water heating and cooling systems often require additional balancing not foreseen in the preliminary planning. An effective method of balancing a heating or cooling system is to install balancing valves, valve adapter units, or balancing fittings at suitable locations in the pipelines in order to regulate water flow through the radiators, convectors, baseboard panels, radiant coils, return mains, and branches.

A *balancing valve* is a control device that functions as a combination balancing, indicating, and shutoff valve. This valve is used

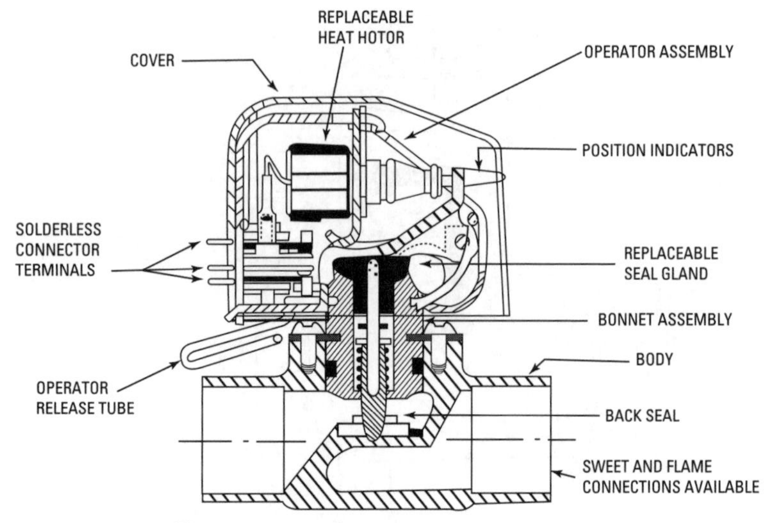

Figure 10-62 Electric zone valve. *(Courtesy ITT Bell & Gossett)*

to balance a hot-water heating or cooling system and at the same time indicate the percent of flow through the valve. An example of one of these balancing valves is shown in Figure 10-63.

These valves are available in many body patterns and connection types, the selection depending on the requirements of the installation. Among the body patterns available are angle, angle union, globe, and globe union. The different connections include screwed, sweat, male, or female unions.

The balancing valve shown in Figure 10-63 is fitted with a balancing yoke that fits over the bonnet. The stem has a stop washer that shoulders on the balancing yoke. The yoke is rotated until the indicator on the calibrated dial points to the percent of flow required at a particular setting. At this point, it is locked in place with an Allen setscrew. The valve can then be opened until the stop washer contacts the top of the yoke, thereby permitting correct percentage of flow. For service work, the valve can be shut and, when opened, will never open beyond its set point. In addition, the valve is equipped with an O-ring that seals off against the bonnet to form a positive back seat, and the valve can be repacked under full line pressure.

Balanced fittings, such as the ones shown in Figure 10-64, are used for balancing branch or circuit resistance of radiators, convectors,

Figure 10-63 Balancing valves. *(Courtesy Spirax Sarco Co.)*

heating or cooling coils, unit heaters, and other heat-transfer equipment employing hot or chilled water. Some are available with an integral manual air vent. They can also be obtained with a thread connection or with a sweat connection for nominal copper tube. When a balancing fitting is used, a stop valve must be installed with it to allow shutdown for necessary service.

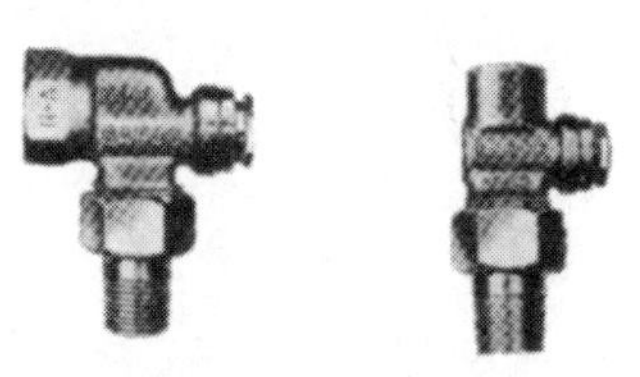

Figure 10-64 Balancing valve fittings. *(Courtesy Spirax Sarco Co.)*

Valve adapters are devices used to convert copper, bronze, cast brass, or cast-iron tees to balancing valves. These adapter devices can be threaded into cast-iron tees, or soldered or sweat-fitted into copper, bronze, or brass tees. They can also be inserted in a side outlet or run of tee of the same size to complete either a straightway or angle balancing valve. Because there is no inside reduction of pipe diameter, there is no water restriction except for the balancing required. Balancing is accomplished by using a screwdriver on the adjustment screw at the top of the adapter device.

Manifolds

A *manifold* (sometimes called a *zone manifold*) is a device used to connect multiple tubing lines to a single supply or return line in a hydronic radiant floor heating system (see Figure 10-65). Each heating system has at least two manifolds: a supply manifold and a return manifold. A supply manifold receives water from the boiler through a single supply pipe and then distributes it through a number of different tubing lines to the rooms and spaces in the structure. A return manifold provides the opposite function. It receives the return water from each of the rooms and spaces through as many tubing lines and sends it back to the boiler by a single return

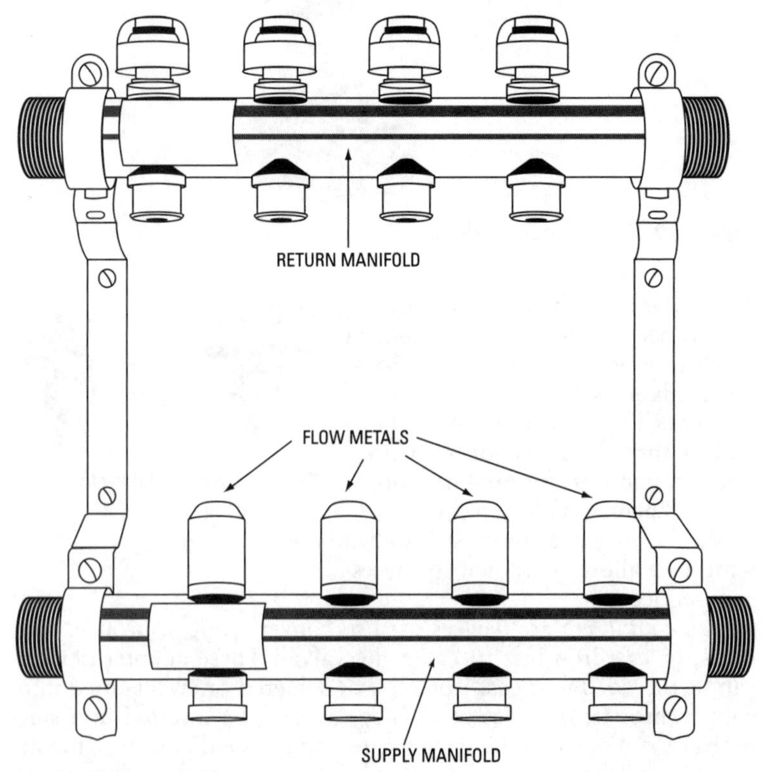

Figure 10-65 Manifold.

pipe. A supply manifold and a return manifold are sometimes referred to jointly as a *manifold station*. Manifolds are described in greater detail in Chapter 1 ("Radiant Heating") in Volume 2.

Pipeline Strainers

Rust, dirt, metal chips, scale, and other impurities are commonly found in both new and old pipelines. Unless these impurities are captured and removed from the pipes, they can damage valves, traps, and other equipment. Protection against these potentially damaging impurities is provided by installing a *strainer* or *scrape strainer* ahead of each mechanical device in the pipeline. Strainers are constructed with a screen socket placed at an angle to the normal direction of flow. The impurities are captured by the screen.

Two typical examples of pipeline strainers are shown in Figures 10-66 and 10-67. These devices are constructed from bronze, semisteel, and steel, and are available in a variety of sizes. The screen socket is tapered and positioned to collect the impurities suspended in the steam or hot water. Both standard and specially designed screens are available from manufacturers.

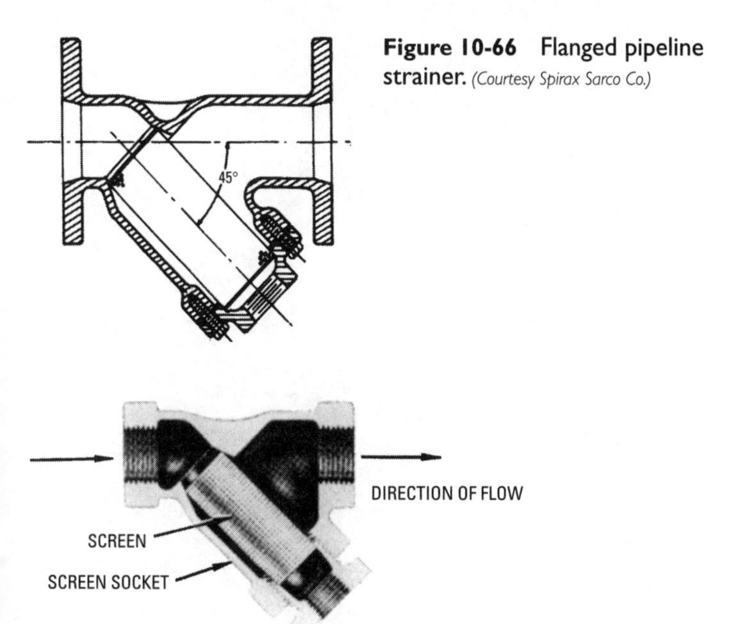

Figure 10-66 Flanged pipeline strainer. *(Courtesy Spirax Sarco Co.)*

Figure 10-67 Pipeline strainer with thread connections. *(Courtesy Spirax Sarco Co.)*

Appendix A

Professional and Trade Associations

Many professional and trade associations have been formed to develop and provide research materials, services, and support for those working in the heating, ventilating, and air conditioning trades. The materials, services, and support include:

1. Formulating and establishing specifications and professional standards.
2. Certifying that equipment and materials meet or exceed minimum standards.
3. Certifying that technicians have met education and training standards.
4. Conducting product research.
5. Promoting interest in the product.
6. Providing education and training.
7. Publishing books, newsletters, articles, and technical papers.
8. Conducting seminars and workshops.

A great deal of useful information can be obtained by contacting these associations. With that in mind the names and addresses of the principal organizations have been included in this Appendix. They are listed in alphabetical order.

Air-Conditioning and Refrigeration Wholesalers International (ARWI)

(*See* Heating, Air Conditioning & Refrigeration Distributors International)

Air-Conditioning and Refrigeration Institute (ARI)

4100 North Fairfax Drive, Suite 200

Arlington, Virginia

Phone: (703) 524-8800

Fax: (703) 528-3816

Email: ari@ari.org

Web site: www.ari.org

A national trade association of manufacturers of central air conditioning, warm-air heating, and commercial and industrial refrigeration equipment, ARI publishes ARI standards and guidelines, which can be downloaded free from its web site. ARI is an approved certifying organization for the EPA Technician Certification Exam. ARI also provides a study manual for those taking the EPA Technician Certification Exam. ARI developed the Curriculum Guide in collaboration with HVACR instructors, manufacturing training experts, and other industry professionals for use in all school programs that educate and train students to become competent, entry-level HVACR technicians.

Air Conditioning Contractors of America (ACCA)

2800 Shirlington Road
Suite 300
Arlington, Virginia 22206
Phone: (703) 575-4477
Fax: (703) 575-4449
Email: info@acca.org
Web site: www.acca.org

A national trade association of heating, air conditioning, and refrigeration systems contractors, ACCA (until 1978, the National Environmental Systems Contractors Association) publishes a variety of different manuals useful for those working in the HVAC trades, including residential and commercial equipment load calculations, residential duct system design, and system installation. ACCA also publishes training and certification manuals. The ACCA publications can be purchased by both members and nonmembers. Check the web site for a list of the ACCA publications, because it is very extensive.

Air Diffusion Council (ADC)

1000 E. Woodfield Road
Suite 102
Schaumburg, Illinois 60173
Phone: (847) 706-6750
Fax: (847) 706-6751
Email: info@flexibleduct.org
Web site: www.flexibleduct.org

The Air Diffusion Council (ADC) was formed to promote the interests of the manufacturers of flexible air ducts and related air distribution equipment. The ADC supports the maintenance and development of uniform industry standards for the installation, use, and performance of flexible duct products. It encourages the use of those standards by various code writing groups, government agencies, architects, engineers, and heating and air conditioning contractors.

Air Filter Institute

(*See* Air-Conditioning and Refrigeration Institute)

Air Movement and Control Association International, Inc. (AMCA)

30 West University Drive
Arlington Heights, Illinois 60004
Phone: (847) 394-0150
Fax: (847) 253-0088
Email: amca@amca.org
Web site: www.amca.org

The Air Movement and Control Association International, Inc. (AMCA) is a trade association of the manufacturers, wholesalers, and retailers of air movement and control equipment (fans, louvers, dampers, and related air systems equipment). The AMCA Certified Ratings Programs are an important function of the association. Their purpose is to give the buyer, specification writer, and user of air movement and control equipment assurance that published ratings are reliable and accurate. The AMCA publishes current test standards for fans, louvers, dampers, and shutters. It also issues a variety of AMCA certified rating seals for different types of air movement and control equipment. It publishes a newsletter various technical specifications for members and those who work with air movement and control systems.

American Boiler Manufacturers Association (ABMA)

4001 North 9th Street
Suite 226
Arlington, Virginia 22203
Phone: (703) 522-7350
Fax: (703) 522-2665

Email: randy@abma.com
Web Site: www.abma.com

The American Boiler Manufacturers Association (ABMA) is a national association representing the manufacturers of commercial, industrial, and utility steam generating and fuel burning equipment, as well as suppliers to the industry. The primary goal of ABMA is topromote the common business interests of the boiler manufacturing industry and to promote the safe, environmentally friendly use of the products and services of its members. Publishes technical guides and manuals.

American Gas Association (AGA)

151400 North Capitol Street, N.W.
Washington, DC 20001
Phone: (202) 824-7000
Fax: (202) 824-7115
Email: Fax: krogers@aga.org
Web site: www.aga.org

The AGA develops residential gas operating and performance standards for distributors and transporters of natural, manufactured, and mixed gas.

American Society of Heating, Refrigeration, and Air-Conditioning Engineers (ASHRAE)

1791 Tullie Circle NE
Atlanta, GA 30329
Phone: (800) 527-4723 (toll free)
Phone: (404) 636-8400
Fax: (404) 321-5478
Email: ashrae@ashrae.org
Web site: www.ashrae.org

The American Society of Heating, Refrigeration, and Air-Conditioning Engineers (ASHRAE) is an international professional association concerned with the advancement of the science and technology of heating, ventilation, air conditioning, and refrigeration through research, standards writing, continuing education, and publications. Membership in ASHRAE is open to any person associated with heating, ventilation, air conditioning, or refrigeration. There are several different types of membership depending on the individual's background and experience in the different HVAC and refrigeration fields. An important benefit of belonging to ASHRAE is access to numerous technical publications.

American Society of Mechanical Engineers (ASME)

Three Park Avenue

New York, New York 10016

Phone: (800) 843-2763 or (973) 882-1167

Fax: (973) 882-1717

Email: infocentral@asme.org

Web site: www.asme.org

A nonprofit technical and education association, ASME develops safety codes and standards and has an extensive list of technical publications covering pressure vessels, piping, and boilers. ASME offers educational and training services and conducts technology seminars and on-site training programs.

Better Heating-Cooling Council

(*See* Hydronics Institute)

Fireplace Institute

(Merged with Wood Energy Institute in 1980 to form the Wood Heating Alliance. *See* Wood Heating Alliance) Gas Appliance Manufacturers Association (GAMA)

2701 Wilson Boulevard, Suite 600

Arlington, Virginia 22201

Phone: (703) 525-7060

Fax: (703) 525-6790

Email: information@gamanet.org.

Web Site: www.gamanet.org

GAMA is a national trade association of manufacturers of gas-fired appliances, and certain types of oil-fired and electrical appliances, used in residential, commercial, and industrial applications. An important service provided by GAMA to its members is a testing and certification program. GAMA will test the rated efficiency and capacity of a manufacturer's product and, if it passes the testing criteria, certify it. The manufacturer can then market the product with the appropriate certification label. Program participants and their products are listed in the ratings directories.

Heating and Piping Contractors National Association

(*See* Mechanical Contractors Association of America)

Heating, Airconditioning & Refrigeration Distributors International (HARDI)

1389 Dublin Road

Columbus, Ohio 43215

Phone: (888) 253-2128 (toll free)

Phone: (614) 488-1835

Fax: (614) 488-0482

Email: HARDImail@HARDInet.org

Web site: www.hardinet.org

Heating, Airconditioning & Refrigeration Distributors International (HARDI) is national trade association of wholesalers and distributors of air conditioning and refrigeration equipment. It was formed by merging the Northamerican Heating, Refrigeration & Airconditioning Wholesalers (NHRAW) and the Air-conditioning & Refrigeration Wholesalers International (ARWI). Among products and services provided to its members are self-study training materials, statistical studies, as well as training and reference materials. See Appendix B (Education, Training, and Certification) for a description of the HARDI Home Study Institute. The HARDI publications are available to both members and nonmembers.

Home Ventilating Institute (HVI)

30 West University Drive

Arlington Heights, Illinois 60004

Phone: (847) 394-0150

Fax: (847) 253-0088

Email: hvi@hvi.org

Web site: www.hvi.org

The Home Ventilating Institute (HVI) is a nonprofit trade association representing national and international manufacturers of residential ventilation products. HVI is primarily concerned with developing performance standards for residential ventilating equipment. It has created a number of certified ratings programs that provide a fair and credible method of comparing ventilation performance of similar products. HVI publishes a number of interesting and informative articles on ventilation that can be downloaded from its Web site.

The Hydronics Foundation, Inc. (THFI)

The Hydronics Foundation, Inc. (THFI) was chartered in 1997 as a nonprofit organization to disseminate knowledge about hydronic equipment and technology. Manufacturers of HVAC equipment also contribute material from installation manuals, specification sheets, and product reviews.

119 East King Street
P.O. Box 1671
Johnson City, TN 37606
Phone: (800) 929-8548
Fax: (800) 929-9506
Email: jdhowell@jdhowell.com
Web site: www.hydronics.com or www.hydronics.org

Hydronic Heating Association (HHA)

P.O. Box 388
Dedham, MA 02026
Phone: (781) 320-9910
Fax: (781) 320-9906
Email: info@comfortableheat.net
Web site: www.comfortableheat.net

The Hydronic Heating Association is an organization of independent contractors, wholesalers, and manufacturers established to promote the latest hydronic technology, set uniform industrial standards, educate HVAC contractors, and inform the public of the benefits of having a quality hot-water system installed. Their Web site contains useful articles and essays on hydronic equipment and systems. It also offers many useful links to manufacturers of hydronic system products.

The Hydronics Institute Division of GAMA

P.O. Box 218

Berkley Heights, New Jersey 07922

Phone: (908) 464-8200

Fax: (908) 464-7818

Email: information@gamanet.org

Web site: www.gamanet.org

The Hydronics Institute was originally formed by a merger of the Better Heating-Cooling Council and the Institute of Boiler and Radiator Manufacturers. It represents manufacturers, suppliers, and installers of hot-water and steam heating and cooling equipment. It is now a division of the Gas Appliance Manufacturers Association. The Hydronics Institute represents and promotes the interests of the manufacturers of hydronic heating equipment. It also provides training materials for hydronic heating courses in schools and technical publications for technicians in the field.

Institute of Boiler and Radiator Manufacturers

(*See* Hydronics Institute)

Mechanical Contractors Association of America, Inc. (MCAA)

1385 Piccard Drive

Rockville, MD 20850

Phone: 301-869-5800

Fax: 301-990-9690

Web site: www.mcaa.org

The Mechanical Contractors Association of America, Inc. (MCAA) is a national trade association for contractors of piping and related equipment used in heating, cooling, refrigeration, ventilating, and air conditioning.

National Association of Plumbing Heating Cooling Contractors (PHCC)

180 S. Washington Street

P.O. Box 6808

Falls Church, VA 22040

Phone: (800) 533-7694 (toll free) or (703) 237-8100
Fax: (703) 237-7442
Email: naphcc@naphcc.org
Web site: www.phccweb.org

The National Association of Plumbing Heating Cooling Contractors (PHCC) is a trade association of local plumbing, heating, and cooling contractors. There are 12 regional chapters.

National Environmental Systems Contractors Association

(*See* Air Conditioning Contractors of America)

National Warm Air Heating and Air Conditioning Association

(*See* Air Conditioning Contractors of America)

North American Heating Refrigerating Air conditioning Wholesalers (NHRAW)

(*See* Heating, Airconditioning & Refrigeration Distributors International)

Radiant Panel Association (RPA)

P.O. Box 717
1399 South Garfield Avenue
Loveland, CO 80537
Phone: (800) 660-7187 (toll free) or (970) 613-0100
Fax: (970) 613-0098
Email: info@rpa-info.com
Web site: www.radiantpanelassociation.org

The Radiant Panel Association provides downloadable technical papers and notes on a variety of different topics concerning radiant panel heating and cooling systems. Links to several manufacturers of radiant heating equipment are also available at their Web site.

Refrigeration and Air Conditioning Contractors Association

(*See* Air Conditioning Contractors of America)

Refrigeration Service Engineers Society (RSES)

1666 Rand Road

Des Plaines, Illinois 60016

Phone: (800) 297-5660 (toll free) or (847) 297-6464

Email: general@rses.org.

Web site: www.rses.org

The Refrigeration Services Engineers Society (RSES) is an international association of refrigeration, air conditioning, and heating equipment installers, service persons, and sales persons. The Society conducts educational meetings, seminars, workshops, technical qualification and examination programs, instructor-led and self-study training courses. It offers a variety of certification program for technicians.

Sheet Metal and Air Conditioning Contractors National Association (SMACNA)

4201 Lafayette Center Dr.

Chantilly, Virginia 20151

Phone: (703) 803-2980

Fax: (703) 803-3732

Email: info@smacna.org

Web site: www.smacna.org

The SMACNA is an international trade association of union contractors who install ventilating, warm-air heating, and air-handling equipment and systems. SMACNA publishes technical papers, answers technical question, and provides distance learning courses for its members. American National Standards Institute has accredited SMACNA as a standards-setting organization matter.

Steam Heating Equipment Manufacturers Association

(*defunct*)

Steel Boiler Institute

(*defunct*)

Underwriters Laboratories, Inc. (UL)

Northbrook Division

Corporate Headquarters

333 Pfingsten Road

Northbrook, Illinois 60062
Phone: (847) 272-8800
Fax: (847) 272-8129
Email: northbrook@us.ul.com
Web site: www.ul.com

The Underwriters Laboratories is an independent, nonprofit product safety testing and certification organization. It promotes safety standards for equipment through independent testing.

Wood Energy Institute

(Merged with Fireplace Institute in 1980 to form Wood Heating Alliance.)

Other National and International Professional and Trade Associations

The following associations also provide support, services, technical publications, and/or training to its members who are involved in the manufacture, sale, or installation and repair of heating, ventilating, and air conditioning systems and equipment. Because space is limited, only their names are listed. Contact information can be obtained by accessing the Internet and entering the association name or by visiting the reference room of your local library and using the *Encyclopedia of Associations*.

Air Distribution Institute (ADI)

American National Standards Institute (ANSI)

American Society for Testing and Materials (ASTM)

Australian Home Heating Association (AHHA)

Australian Institute of Refrigeration, Air Conditioning and Heating (AIRAH)

Heating Alternatives, Inc

Heating, Refrigeration and Air Conditioning Contractors of Canada (HRAC)

Heating, Refrigeration and Air Conditioning Institute of Canada (HRAE)

Institute of Heating & Air Conditioning, Inc (IHACI).

Insulation Contractors of America (ICA)

International Energy Association (IEA)—Solar Heating and Cooling Programme

National LP-Gas Association (NLPGA)
National Oil Fuel Institute
National Oilheat Research Alliance
Plumbing-Heating-Cooling Contractors—National Association (PHCC)
Plumbing-Heating-Cooling Information Bureau (PHCIB)
Wood Heating Alliance (WHA)

Appendix B

Manufacturers

Adams Manufacturing Company

9790 Midwest Avenue

Cleveland, OH 44125

(216) 587-6801

(216) 587-6807 (Fax)

www.gamanet.org

Amana Refrigeration, Inc.

1810 Wilson Parkway

Fayetteville, TN 37334

(800) 843-0304

(931) 433-6101

(931) 433-1312

www.amana.com

American Standard Companies Inc.

One Centennial Ave.

Piscataway, NJ 08855

(732) 980-6000

(732) 980-3340 (Fax)

www.americanstandard.com

A.O. Smith Water Products Company

600 E. John Carpenter Freeway #200

Irving, TX 75062-3990

(972) 719-5900

(972) 719-5960 (Fax)

www.hotwater.com

Bacharach Inc.

621 Hunt Valley

New Kensington, PA 15068

(724) 334-5000

(724) 334-5001 (Fax)
www.bacharach-inc.com

Bard Manufacturing Co
P.O. Box 607
Bryan, OH 43506
(419) 636-1194
(419) 636-2640 (Fax)
www.bardhvac.com

R.W. Beckett Corporation
P.O. Box 1289
Elyria, OH 44036
(800) 645-2876
(440) 327-1060
(440) 327-1064 (Fax)
www. beckett.com

Bell & Gossett
(*See* ITT Bell & Gossett)

Bryan Boilers/Bryan Steam Corporation
P.O. Box 27
783 N. Chili Avenue
Peru, IN 46970
(765) 473-6651
(765) 473-3074 (Fax)
www.bryanboilers.com

Burnham Hydronics
U.S. Boiler Co., Inc.
P.O. Box 3079
Lancaster, PA 17604
(717) 397-4701
(717) 293-5827 (Fax)
www.burnham.com

Carrier Corporation
World Headquarters
One Carrier Place
Farmington, CT 06034
(860) 674-3000
www.carrier.com

Cash Acme
2400 7th Avenue S.W.
Cullman, Alabama 35055
(256) 775-8200
(256) 775-8238 (Fax)
www.cashacme.com

Coleman Corporation
Unitary Products Group
5005 York Dr.
Norman, OK 73069
(405) 364-4040
www.colemanac.com

Columbia Boiler Company
P.O. Box 1070
Pottstown, PA 19464
(610) 323-2700
(610) 323-7292 (Fax)
www.columbiaboiler.com

Danfoss A/S
DK-6430 Nordborg
Denmark
+45 7488 2222
+45 7449 0949 (Fax)
www.danfoss.com

Domestic Pump
(*See* ITT Domestic Pump)

Dornback Furnace
9545 Granger Road
Garfield Heights, OH 44125
(216) 662-1600
(216) 587-6807
www.gamanet.org

Ernst Gage Co.
250 S. Livingston Ave.
Livingston, NJ 07039 4089
973-992-1400
888-229-4243
973-992-0036 (Fax)

General Filters Inc.
43800 Grand River Ave.
Novi, MI
(248) 476-5100
(248) 349-2366 (Fax)
www.generalfilters.com

Goodman Manufacturing Corp
2550 North Loop West #400
Houston, TX 77092
(713) 861-2500
(888) 593-9988
www.goodmanmfg.com

Heat Controller, Inc.
1900 Wellworth Avenue
Jackson, MI 49203
(517) 787-2100
(517) 787-9341
www.heatcontroller.com

Hoffman Specialty
(*See* ITT Hoffman Specialty)

Honeywell, Inc.

101 Columbia Road
Morristown, NJ 07962
(973) 455-2000
(800) 328-5111
(983) 455-4807 (Fax)
www.honeywell.com

Hydro Therm, A Division of Mastek, Inc.

260 North Elm Street
Westfield, MA 01085
(413) 564-5515
www.hydrotherm.com

Invensys Building Systems, Inc.

1354 Clifford Ave.
P.O. Box 2940
Loves Park, IL 61132-2940
(815) 637-3000
(815) 637-5350 (Fax)
www.invensys.com

ITT Bell & Gossett

8200 North Austin Avenue
Morton Grove, IL 60053
(847) 966-3700
(847) 966-9052
www.bellgossett.com

ITT Domestic Pump

8200 N. Austin Ave.
Morton Grove, IL 60053
(847) 966-3700
(847) 966-9052
www.domesticpump.com

ITT Hoffman Specialty
3500 N. Spaulding Avenue
Chicago, IL 60618
(773) 267-1600
(773) 267-0991
www.hoffmanspecialty.com

ITT McDonnell & Miller
3500 N. Spaulding Avenue
Chicago, IL 60618
(723) 267-1600
(773) 267-0991
www.mcdonnellmiller.com

Janitrol Air Conditioning and Heating
www.janitrol.com

(*See* Goodman Manufacturing Company)

S.T. Johnson Company
Innovative Combustion Technologies, Inc.
925 Stanford Avenue
Oakland, CA 94608
(510) 652-6000
(510) 652-4302 (Fax)
www.johnsonburners.com

Johnson Controls, Inc.
5757 North Green Bay Avenue
Milwaukee, WI 53209
(262) 524-3285
www.johnsoncontrols.com
www.jci.com

Lennox Industries Inc.
2100 Lake Park Boulevard
Richardson, TX 75080
(972) 497-5000

(972) 497-5392 (Fax)
www.davelennox.com

Marathon Electric, Inc.
P.O. Box 8003
Wausau, WI 54402
(715) 675-3359
(715) 675-8050 (Fax)

McDonnell & Miller
(*See* ITT McDonnell & Miller)

Midco International Inc.
4140 West Victoria Street
Chicago, IL 60646-6790
(773) 604-8700
(773) 604-4070 (Fax)
www.midco-intl.com

Nordyne
P.O. Box 8809
O'Fallon, MO 63366
(636) 561-7300
(800) 222-4328
(636) 561-7365 (Fax)
www.nordyne.com

Raypak
2151 Eastman Ave.
Oxnard, CA 93030
(805) 278-5300
(805) 278-5468 (Fax)
www.raypak.com

RBI, Mestek Canada, Inc.
1300 Midway Blvd.
Mississauga Ontario L5T 2G8
(905) 670-5888
www.rbimestek.com

Rheem Manufacturing
5600 Old Greenwood Road
Fort Smith, AR 72908
(479) 646-4311
(479) 648-4812 (Fax)
www.rheemac.com

Robertshaw
(See Invensys)

Smith Cast Iron Boilers
260 North Elm Street
Westfield, MA 01085
(413) 562-9631
www.smithboiler.com

SpacePak
125 North Elm Street
Westfield, MA 01085
(413) 564-5530
www.spacepak.com

Spirax Sarco Inc.
Northpoint Business Park
1150 Northpoint Blvd
Blythewood, SC 29016
(803) 714-2000
(803) 714-2222
www.spiraxsarco.com

Sterling Hydronics
260 North Elm Street
Westfield, MA 01085
(413) 564-5535
www.sterlingheat.com

Sterling HVAC
125 North Elm Street
Westfield, MA 01085

(413) 564-5540
www.sterlinghvac.com

Suntec Industries Incorporated
2210 Harrison Ave
P.O. Box 7010
Rockford IL 61125-7010
(815) 226-3700
(815) 226-3848 (Fax)
www.suntecpumps.com

Thermo Pride
P.O. Box 217
North Judson, IN 46366
(574) 896-2133
(574) 896-5301
www.thermopride.com

Trane
P.O. Box 9010
Tyler, TX 75711-9010
(903) 581-3200
www.trane.com

Triangle Tube/Phase III Company, Inc.
Blackwood, NJ
(856) 228-1881
(856) 228-3584 (Fax)
www.triangletube.com

Vulcan Radiator (Mastec)
515 John Fitch Blvd
South Windsor, CT 06074
(413) 568-9571
www.mestec.com

Water Heater Innovations, Inc.
3107 Sibley Memorial Highway
Eagan, MN 55121

(800) 321-6718
www.marathonheaters.com

Watts Regulator Company
815 Chestnut Street
North Andover, MA 01845
(976) 688-1811
(978) 794-848 (Fax)
www.wattsreg.com

Wayne Combustion Systems
801 Glasgow Ave.
Fort Wayne, IN 46803
(800) 443-4625
www.waynecombustion.com

Weil-McLain
500 Blaine St.
Michigan City, IN 46360
(219) 879-6561
(219) 879-4025
www.weil-mclain.com

White-Rodgers, Div. of Emerson Electric Co.
9797 Reavis Rd.
St. Louis, MO 63123
(314) 577-1300
(314) 577-1517
www.white-rodgers.com

Wm. Powell Company
2503 Spring Grove Avenue
Cincinnati, OH 45214
(513) 852-2000
(513) 852-2997 (Fax)
www.powellvalves.com

York International Corporation
P.O. Box 1592-232F
York, PA 17405

(717) 771-7890
(717) 771-7381 (Fax)
www.york.com

Yukon-Eagle Wood & Multifuel Furnaces
10 Industrial Blvd.
P.O. Box 20
Palisade, MN 56469
(800) 358-0060
(800) 440-1994 (Fax)
www.yukon-eagle.com

John Zink Company
Gordon-Piatt Group
11920 East Apache
Tulsa, OK 74116
(800) 638-6940
www.johnzink.com

Appendix C

Data Tables

Table D-1 Equivalent Length of New Straight Pipe for Valves and Fittings for Turbulent Flow

Fittings			Pipe Size																				
			1/4	3/8	1/2	3/4	1	1 1/4	1 1/2	2	2 1/2	3	4	5	6	8	10	12	14	16	18	20	24
Regular 90° Ell	Screwed	Steel	2.3	3.1	3.6	4.4	5.2	6.6	7.4	8.5	9.3	11	13	—	—	—	—	—	—	—	—	—	—
		C. I.	—	—	—	—	—	—	—	—	—	9	11	—	—	—	—	—	—	—	—	—	—
	Flanged	Steel	—	—	0.92	1.2	1.6	2.1	2.4	3.1	3.6	4.4	5.9	7.3	8.9	12	14	17	18	21	23	25	30
		C. I.	—	—	—	—	—	—	—	—	—	3.6	4.8	—	7.2	9.8	12	15	17	19	22	24	28
Long Radius 90° Ell	Screwed	Steel	1.5	2	2.2	2.3	2.7	3.2	3.4	3.6	3.6	4	4.6	—	—	—	—	—	—	—	—	—	—
		C. I.	—	—	—	—	—	—	—	—	—	3.3	3.7	—	—	—	—	—	—	—	—	—	—
	Flanged	Steel	—	—	1.1	1.3	1.6	2	2.3	2.7	2.9	3.4	4.2	5	5.7	7	8	9	9.4	10	11	12	14
		C. I.	—	—	—	—	—	—	—	—	—	2.8	3.4	—	4.7	5.7	6.8	7.8	8.6	9.6	11	11	13
Regular 45° Ell	Screwed	Steel	0.34	0.52	0.71	0.92	1.3	1.7	2.1	2.7	3.2	4	5.5	—	—	—	—	—	—	—	—	—	—
		C. I.	—	—	—	—	—	—	—	—	—	3.3	4.5	—	—	—	—	—	—	—	—	—	—
	Flanged	Steel	—	—	0.45	0.59	0.81	1.1	1.3	1.7	2	2.6	3.5	4.5	5.6	7.7	9	11	13	15	16	18	22
		C. I.	—	—	—	—	—	—	—	—	—	2.1	2.9	—	4.5	6.3	8.1	9.7	12	13	15	17	20
Tee-Line Flow	Screwed	Steel	0.79	1.2	1.7	2.4	3.2	4.6	5.6	7.7	9.3	12	17	—	—	—	—	—	—	—	—	—	—
		C. I.	—	—	—	—	—	—	—	—	—	9.9	14	—	—	—	—	—	—	—	—	—	—
	Flanged	Steel	—	—	0.69	0.82	1	1.3	1.5	1.8	1.9	2.2	2.8	3.3	3.8	4.7	5.2	6	6.4	7.2	7.6	8.2	9.6
		C. I.	—	—	—	—	—	—	—	—	—	1.9	2.2	—	3.1	3.9	4.6	5.2	5.9	6.5	7.2	7.7	8.8
Tee-Branch Flow	Screwed	Steel	2.4	3.5	4.2	5.3	6.6	8.7	9.9	12	13	17	21	—	—	—	—	—	—	—	—	—	—
		C. I.	—	—	—	—	—	—	—	—	—	14	17	—	—	—	—	—	—	—	—	—	—
	Flanged	Steel	—	—	2	2.6	3.3	4.4	5.2	6.6	7.5	9.4	12	15	18	24	30	34	37	43	47	52	62
		C. I.	—	—	—	—	—	—	—	—	—	7.7	10	—	15	20	25	30	35	39	44	49	57

180° Return Bend	Screwed	Steel	2.3	3.1	3.6	4.4	5.2	6.6	7.4	8.5	9.3	11	13	—	—	—	—	—	—	—	—	—	—
		C. I.	—	—	—	—	—	—	—	—	—	9	11	—	—	—	—	—	—	—	—	—	—
	Reg. Flanged	Steel	—	—	0.92	1.2	1.6	2.1	2.4	3.1	3.6	4.4	5.9	7.3	8.9	12	14	17	18	21	23	25	30
		C. I.	—	—	—	—	—	—	—	—	—	3.6	4.8	—	7.2	9.8	12	15	17	19	22	24	28
	Long Rad. Flanged	Steel	—	—	1.1	1.3	1.6	2	2.3	2.7	2.9	3.4	4.2	5	5.7	7	8	9	9.4	10	11	12	14
		C. I.	—	—	—	—	—	—	—	—	—	2.8	3.4	—	4.7	5.7	6.8	7.8	8.6	9.6	11	11	13
Globe Valve	Screwed	Steel	21	22	22	24	29	37	42	54	62	79	110	—	—	—	—	—	—	—	—	—	—
		C. I.	—	—	—	—	—	—	—	—	—	65	86	—	—	—	—	—	—	—	—	—	—
	Flanged	Steel	—	—	38	40	45	54	59	70	77	94	120	150	190	260	310	390	—	—	—	—	—
		C. I.	—	—	—	—	—	—	—	—	—	77	99	—	150	210	270	330	—	—	—	—	—
Gate Valve	Screwed	Steel	0.32	0.45	0.56	0.67	0.84	1.1	1.2	1.5	1.7	1.9	2.5	—	—	—	—	—	—	—	—	—	—
		C. I.	—	—	—	—	—	—	—	—	—	1.6	2	—	—	—	—	—	—	—	—	—	—
	Flanged	Steel	—	—	—	—	—	—	—	2.6	2.7	2.8	2.9	3.1	3.2	3.2	3.2	3.2	3.2	3.2	3.2	3.2	3.2
		C. I.	—	—	—	—	—	—	—	—	—	2.3	2.4	—	2.6	2.7	2.8	2.9	3	3	3	3	3
Angle Valve	Screwed	Steel	12.8	15	15	15	17	18	18	18	18	18	18	—	—	—	—	—	—	—	—	—	—
		C. I.	—	—	—	—	—	—	—	—	—	15	15	—	—	—	—	—	—	—	—	—	—
	Flanged	Steel	—	—	15	15	17	18	18	21	22	28	38	50	63	90	120	140	160	190	210	240	300
		C. I.	—	—	—	—	—	—	—	—	—	23	31	—	52	74	98	120	150	170	200	230	280
Swing Check Valve	Screwed	Steel	7.2	7.3	8	8.8	11	13	15	19	22	27	38	—	—	—	—	—	—	—	—	—	—
		C. I.	—	—	—	—	—	—	—	—	—	22	31	—	—	—	—	—	—	—	—	—	—
			—	—	3.8	5.3	7.2	10	12	17	21	27	38	50	63	90	120	140	—	—	—	—	—
	Flanged	Steel	—	—	—	—	—	—	—	—	—	22	31	—	52	74	98	120	—	—	—	—	—
		C. I.	0.14	0.18	0.21	0.24	0.29	0.36	0.39	0.45	0.47	0.53	0.65	—	—	—	—	—	—	—	—	—	—

(continued)

Table D-1 (continued)

Fittings			Pipe Size ¼	⅜	½	¾	1	1¼	1½	2	2½	3	4	5	6	8	10	12	14	16	18	20	24
Coupling or Union	Screwed	Steel																					
		C. I.																					
	Bell Mouth Inlet	Steel	—	—	—	—	—	—	—	—	—	0.44	0.62	—	—	—	—	—	—	—	—	—	—
		C. I.	0.04	0.07	0.1	0.13	0.18	0.26	0.31	0.43	0.52	0.67	0.95	1.3	1.6	2.3	2.9	3.5	4	4.7	5.3	6.1	7.6
	Square Mouth Inlet	Steel	—	—	—	—	—	—	—	—	—	0.55	0.77	—	1.3	1.9	2.4	3	3.6	4.3	5	5.7	7
		C. I.	0.44	0.68	0.96	1.3	1.8	2.6	3.1	4.3	5.2	6.7	9.5	13	16	23	29	35	40	47	53	61	76
			—	—	—	—	—	—	—	—	—	5.5	7.7	—	13	19	24	30	36	43	50	57	70
	Reentrant Pipe	Steel	0.88	1.4	1.9	2.6	3.6	5.1	6.2	8.5	10	13	19	25	32	45	58	70	80	95	110	120	150
		C. I.	—	—	—	—	—	—	—	—	—	11	15	—	26	37	49	61	73	86	100	110	140
	Y-Strainer		—	4.6	5	6.6	7.7	18	20	27	29	34	42	53	61								

V1 V2 Sudden Enlargement $h = \frac{(V_1 - V_2)^2}{(2g)}$ Feet of Liquid; If $V_2 = 0$ $h = \frac{V^2}{(2g)}$ Feet of Liquid

Courtesy The Hydraulic Institute (reprinted from the Standards of the Hydraulic Institute, Eleventh Edition, Copyright 1965)

Table D-2 Schedule 80 Pipe Dimensions

	Diameters		Nominal	Transverse Areas			Length of Pipe Per Square Foot of				
Size in	External in	Internal in	Thickness in	External in²	Internal in²	Metal in²	External Surface ft	Internal Surface ft	Cubic Feet per ft of Pipe	Weight per ft Pounds	Number Threads per in of Screw
⅛	0.405	0.215	0.095	0.129	0.036	0.093	9.431	17.75	0.00025	0.314	27
¼	0.54	0.302	0.119	0.229	0.072	0.157	7.073	12.65	0.0005	0.535	18
⅜	0.675	0.423	0.126	0.358	0.141	0.217	5.658	9.03	0.00098	0.738	18
½	0.84	0.546	0.147	0.554	0.234	0.32	4.547	7	0.00163	1	14
¾	1.05	0.742	0.154	0.866	0.433	0.433	3.637	5.15	0.003	1.47	14
1	1.315	0.957	0.179	1.358	0.719	0.639	2.904	3.995	0.005	2.17	11½
1¼	1.66	1.278	0.191	2.164	1.283	0.881	2.301	2.99	0.00891	3	11½
1½	1.9	1.5	0.2	2.835	1.767	1.068	2.01	2.542	0.01227	3.65	11½
2	2.375	1.939	0.218	4.43	2.953	1.477	1.608	1.97	0.02051	5.02	11½
2½	2.875	2.323	0.276	6.492	4.238	2.254	1.328	1.645	0.02943	7.66	8
3	3.5	2.9	0.3	9.621	6.605	3.016	1.091	1.317	0.04587	10.3	8
3½	4	3.364	0.318	12.56	8.888	3.678	0.954	1.135	0.06172	12.5	8
4	4.5	3.826	0.337	15.9	11.497	4.407	0.848	0.995	0.0798	14.9	8
5	5.563	4.813	0.375	24.3	18.194	6.112	0.686	0.792	0.1263	20.8	8
6	6.625	5.761	0.432	34.47	26.067	8.3	0.576	0.673	0.181	28.6	8
8	8.625	7.625	0.5	58.42	46.663	12.76	0.442	0.501	0.3171	43.4	8

(continued)

Table D-2 (continued)

	Diameters			*Transverse Areas*			*Length of Pipe Per Square Foot of*				
Size *in*	**External** *in*	**Internal** *in*	**Nominal Thickness** *in*	**External** *in²*	**Internal** *in²*	**Metal** *in²*	**External Surface** *ft*	**Internal Surface** *ft*	**Cubic Feet** *per ft of Pipe*	**Weight** *per ft* **Pounds**	**Number Threads** *per in of Screw*
10	10.75	9.564	0.593	90.76	71.84	18.92	0.355	0.4	0.4989	64.4	8
12	12.75	11.376	0.687	127.64	101.64	26	0.299	0.336	0.7058	88.6	
14	14	12.5	0.75	153.94	122.72	31.22	0.272	0.306	0.8522	107	
16	16	14.314	0.843	201.05	160.92	40.13	0.238	0.263	1.112	137	
18	18	16.126	0.937	254.85	204.24	50.61	0.212	0.237	1.418	171	
20	20	17.938	1.031	314.15	252.72	61.43	0.191	0.208	1.755	209	
24	24	21.564	1.218	452.4	365.22	87.18	0.159	0.177	2.536	297	

Courtesy Sarco Company, Inc.

Table D-3 Schedule 40 Pipe Dimensions

	Diameters			Transverse Areas			Length of Pipe Per ft² of				
Size in	External in	Internal in	Nominal Thickness in	External in²	Internal in²	Metal in²	External Surface ft	Internal Surface ft	Cubic Feet per ft of Pipe	Weight per ft Pounds	Number Threads per in of Screw
⅛	0.405	0.269	0.068	0.129	0.057	0.072	9.431	14.199	0.00039	0.244	27
¼	0.54	0.364	0.088	0.229	0.104	0.125	7.073	10.493	0.00072	0.424	18
⅜	0.675	0.493	0.091	0.358	0.191	0.167	5.658	7.747	0.00133	0.567	18
½	0.84	0.622	0.109	0.554	0.304	0.25	4.547	6.141	0.00211	0.85	14
¾	1.05	0.824	0.113	0.866	0.533	0.333	3.637	4.635	0.0037	1.13	14
1	1.315	1.049	0.133	1.358	0.864	0.494	2.904	3.641	0.006	1.678	11½
1¼	1.66	1.38	0.14	2.164	1.495	0.669	2.301	2.767	0.01039	2.272	11½
1½	1.9	1.61	0.145	2.835	2.036	0.799	2.01	2.372	0.01414	2.717	11½
2	2.375	2.067	0.154	4.43	3.355	1.075	1.608	1.847	0.0233	3.652	11½
2½	2.875	2.469	0.203	6.492	4.788	1.704	1.328	1.547	0.03325	5.793	8
3	3.5	3.068	0.216	9.621	7.393	2.228	1.091	1.245	0.05134	7.575	8
3½	4	3.548	0.226	12.56	9.886	2.68	0.954	1.076	0.06866	9.109	8
4	4.5	4.026	0.237	15.9	12.73	3.174	0.848	0.948	0.0884	10.79	8
5	5.563	5.047	0.258	24.3	20	4.3	0.686	0.756	0.1389	14.61	8
6	6.625	6.065	0.28	34.47	28.9	5.581	0.576	0.629	0.2006	18.97	8
8	8.625	7.981	0.322	58.42	50.02	8.399	0.442	0.478	0.3552	28.55	8

(continued)

Table D-3 (continued)

	Diameters		Nominal	Transverse Areas			Length of Pipe Per ft² of				
Size in	External in	Internal in	Thickness in	External in²	Internal in²	Metal in²	External Surface ft	Internal Surface ft	Cubic Feet per ft of Pipe	Weight per ft Pounds	Number Threads per in of Screw
10	10.75	10.02	0.365	90.76	78.85	11.9	0.355	0.381	0.5476	40.48	8
12	12.75	11.938	0.406	127.64	111.9	15.74	0.299	0.318	0.7763	53.6	
14	14	13.125	0.437	153.94	135.3	18.64	0.272	0.28	0.9354	63	
16	16	15	0.5	201.05	176.7	24.35	0.238	0.254	1.223	78	
18	18	16.874	0.563	254.85	224	30.85	0.212	0.226	1.555	105	
20	20	18.814	0.593	314.15	278	36.15	0.191	0.203	1.926	123	
24	24	22.626	0.687	452.4	402.1	50.3	0.159	0.169	2.793	171	

Courtesy Sarco Company, Inc.

Table D-4 Properties of Saturated Steam

	Gauge Pressure psig	Temper-ature °F	Heat in Btu/lb			Specific Volume ft³/lb	Gauge Pressure psig	Temper-ature °F	Heat in Btu/lb			Specific Volume ft³ per lb
			Sensible	Latent	Total				Sensible	Latent	Total	
	25	134	102	1017	1119	142	150	366	339	857	1196	2.74
	20	162	129	1001	1130	73.9	155	368	341	885	1196	2.68
In Vac.	15	179	147	990	1137	51.3	160	371	344	853	1197	2.6
	10	192	160	982	1142	39.4	165	373	346	851	1197	2.54
	5	203	171	976	1147	31.8	170	375	348	849	1197	2.47
	0	**212**	**180**	**970**	**1150**	**26.8**	**175**	**377**	**351**	**847**	**1198**	**2.41**
	1	215	183	968	1151	25.2	180	380	353	845	1198	2.34
	2	219	187	966	1153	23.5	185	382	355	843	1198	2.29
	3	222	190	964	1154	22.3	190	384	358	841	1199	2.24
	4	224	192	962	1154	21.4	195	386	360	839	1199	2.19
	5	**227**	**195**	**960**	**1155**	**20.1**	**200**	**388**	**362**	**837**	**1199**	**2.14**
	6	230	198	959	1157	19.4	205	390	364	836	1200	2.09
	7	232	200	957	1157	18.7	210	392	366	834	1200	2.05
	8	233	201	956	1157	18.4	215	394	368	832	1200	2
	9	237	205	954	1159	17.1	220	396	370	830	1200	1.96
	10	**239**	**207**	**953**	**1160**	**16.5**	**225**	**397**	**372**	**828**	**1200**	**1.92**
	12	244	212	949	1161	15.3	230	399	374	827	1201	1.89
	14	248	216	947	1163	14.3	235	401	376	825	1201	1.85

(continued)

Table D-4 (continued)

Gauge Pressure psig	Temperature °F	Heat in Btu/lb Sensible	Heat in Btu/lb Latent	Heat in Btu/lb Total	Specific Volume ft³/lb	Gauge Pressure psig	Temperature °F	Heat in Btu/lb Sensible	Heat in Btu/lb Latent	Heat in Btu/lb Total	Specific Volume ft³ per lb
16	252	220	944	1164	13.4	240	403	378	823	1201	1.81
18	256	224	941	1165	12.6	245	404	380	822	1202	1.78
20	**259**	**227**	**939**	**1166**	**11.9**	**250**	**406**	**382**	**820**	**1202**	**1.75**
22	262	230	937	1167	11.3	255	408	383	819	1202	1.72
24	265	233	934	1167	10.8	260	409	385	817	1202	1.69
26	268	236	933	1169	10.3	265	411	387	815	1202	1.66
28	271	239	930	1169	9.85	270	413	389	814	1203	1.63
30	**274**	**243**	**929**	**1172**	**9.46**	**275**	**414**	**391**	**812**	**1203**	**1.6**
32	277	246	927	1173	9.1	280	416	392	811	1203	1.57
34	279	248	925	1173	8.75	285	417	394	809	1203	1.55
36	282	251	923	1174	8.42	290	418	395	808	1203	1.53
38	284	253	922	1175	8.08	295	420	397	806	1203	1.49
40	**286**	**256**	**920**	**1176**	**7.82**	**300**	**421**	**398**	**805**	**1203**	**1.47**
42	289	258	918	1176	7.57	305	423	400	803	1203	1.45
44	291	260	917	1177	7.31	310	425	402	802	1204	1.43
46	293	262	915	1177	7.14	315	426	404	800	1204	1.41
48	295	264	914	1178	6.94	320	427	405	799	1204	1.38
50	**298**	**267**	**912**	**1179**	**6.68**	**325**	**429**	**407**	**797**	**1204**	**1.36**
55	300	271	909	1180	6.27	330	430	408	796	1204	1.34

Table D-4 *(continued)*

Gauge Pressure psig	Temperature °F	Heat in Btu/lb			Specific Volume ft³/lb	Gauge Pressure psig	Temperature °F	Heat in Btu/lb			Specific Volume ft³ per lb
		Sensible	Latent	Total				Sensible	Latent	Total	
60	307	277	906	1183	5.84	335	432	410	794	1204	1.33
65	312	282	901	1183	5.49	340	433	411	793	1204	1.31
70	316	286	898	1184	5.18	345	434	413	791	1204	1.29
75	**320**	**290**	**895**	**1185**	**4.91**	**350**	**435**	**414**	**790**	**1204**	**1.28**
80	324	294	891	1185	4.67	355	437	416	789	1205	1.26
85	328	298	889	1187	4.44	360	438	417	788	1205	1.24
90	331	302	886	1188	4.24	365	440	419	786	1205	1.22
95	335	305	883	1188	4.05	370	441	420	785	1205	1.2
100	**338**	**309**	**880**	**1189**	**3.89**	**375**	**442**	**421**	**784**	**1205**	**1.19**
105	341	312	878	1190	3.74	380	443	422	783	1205	1.18
110	344	316	875	1191	3.59	385	445	424	781	1205	1.16
115	347	319	873	1192	3.46	390	446	425	780	1205	1.14
120	350	322	871	1193	3.34	395	447	427	778	1205	1.13
125	**353**	**325**	**868**	**1193**	**3.23**	**400**	**448**	**428**	**777**	**1205**	**1.12**
130	356	328	866	1194	3.12	450	460	439	766	1205	1
140	361	333	861	1194	2.92	500	470	453	751	1204	0.89
145	363	336	859	1195	2.84	550	479	464	740	1204	0.82
						600	489	475	728	1203	0.74

Courtesy Sarco Company, Inc.

Table D-5 Friction Loss for Water in Feet per 100 Feet of Schedule 40 Steel Pipe

U.S. gal/min	*Velocity ft/sc*	*hf Friction*	*U.S. gal/min*	*Vel. ft/sec*	*hf Friction*
	3/8" Pipe			*1/2" Pipe*	
1.4	2.25	9.03	2	2.11	5.5
1.6	2.68	11.6	2.5	2.64	8.24
1.8	3.02	14.3	3	3.17	11.5
2	3.36	17.3	3.5	3.7	15.3
2.5	4.2	26	4	4.22	19.7
3	5.04	36.6	5	5.28	29.7
3.5	5.88	49	6	6.34	42
4	6.72	63.2	7	7.39	56
5	8.4	96.1	8	8.45	72.1
6	10.08	136	9	9.5	90.1
7	11.8	182	10	10.56	110.6
8	13.4	236	12	12.7	156
9	15.1	297	14	14.8	211
10	16.8	364	16	16.9	270
	3/4" Pipe			*1" Pipe*	
4	2.41	4.85	6	2.23	3.16
5	3.01	7.27	8	2.97	5.2
6	3.61	10.2	10	3.71	7.9
7	4.21	13.6	12	4.45	11.1
8	4.81	17.3	14	5.2	14.7
9	5.42	21.6	16	5.94	19
10	6.02	26.5	18	6.68	23.7
12	7.22	37.5	20	7.42	28.9
14	8.42	50	22	8.17	34.8
16	9.63	64.8	24	8.91	41
18	10.8	80.9	26	9.65	47.8
20	12	99	28	10.39	55.1
22	13.2	120	30	11.1	62.9
24	14.4	141	35	13	84.4
26	15.6	165	40	14.8	109
28	16.8	189	45	16.7	137
			50	18.6	168

(continued)

Table D-5 (continued)

U.S. gal/min	*Velocity ft/sc*	*hf Friction*	*U.S. gal/min*	*Vel. ft/sec*	*hf Friction*
	1¼" Pipe			***1½" Pipe***	
12	2.57	2.85	16	2.52	2.26
14	3	3.77	18	2.84	2.79
16	3.43	4.83	20	3.15	3.38
18	3.86	6	22	3.47	4.05
20	4.29	7.3	24	3.78	4.76
22	4.72	8.72	26	4.1	5.54
24	5.15	10.27	28	4.41	6.34
26	5.58	11.94	30	4.73	7.2
28	6.01	13.7	35	5.51	9.63
30	6.44	15.6	40	6.3	12.41
35	7.51	21.9	45	7.04	15.49
40	8.58	27.1	50	7.88	18.9
45	9.65	33.8	55	8.67	22.7
50	10.7	41.4	60	9.46	26.7
55	11.8	49.7	65	10.24	31.2
60	12.9	58.6	70	11.03	36
65	13.9	68.6	75	11.8	41.2
70	15	79.2	80	12.6	46.6
75	16.1	90.6	85	13.4	52.4
			90	14.2	58.7
			95	15	65
			100	15.8	71.6
	2" Pipe			***2½" Pipe***	
25	2.39	1.48	35	2.35	1.15
30	2.87	2.1	40	2.68	1.47
35	3.35	2.79	45	3.02	1.84
40	3.82	3.57	50	3.35	2.23
45	4.3	4.4	60	4.02	3.13
50	4.78	5.37	70	4.69	4.18
60	5.74	7.58	80	5.36	5.36
70	6.69	10.2	90	6.03	6.69
80	7.65	13.1	100	6.7	8.18
90	8.6	16.3	120	8.04	11.5

(continued)

Table D-5 (continued)

U.S. gal/min	*Velocity ft/sc*	*hf Friction*	*U.S. gal/min*	*Vel. ft/sec*	*hf Friction*
	2" Pipe			***2½" Pipe***	
100	4.34	2.72	200	5.04	12.61
120	11.5	28.5	160	10.7	20
140	13.4	38.2	180	12.1	25.2
160	15.3	49.5	200	13.4	30.7
			220	14.7	37.1
			240	16.1	43.8
	3" Pipe			***4" Pipe***	
50	2.17	0.762	100	2.52	0.718
60	2.6	1.06	120	3.02	1.01
70	3.04	1.4	140	3.53	1.35
80	3.47	1.81	160	4.03	1.71
90	3.91	2.26	180	4.54	2.14
100	3.34	2.75	200	5.04	2.61
120	5.21	3.88	220	5.54	3.13
140	6.08	5.19	240	6.05	3.7
160	6.94	6.68	260	6.55	4.3
180	7.81	8.38	280	7.06	4.95
200	8.68	10.2	300	7.56	5.63
220	9.55	12.3	350	8.82	7.54
240	10.4	14.5	400	10.1	9.75
260	11.3	16.9	450	11.4	12.3
280	12.2	19.5	500	12.6	14.4
300	13	22.1	550	13.9	18.1
350	15.2	30	600	15.1	21.4
	5" Pipe			***6" Pipe***	
160	2.57	0.557	220	2.44	0.411
180	2.89	0.698	240	2.66	0.482
200	3.21	0.847	260	2.89	0.56
220	3.53	1.01	300	3.33	0.733
240	3.85	1.19	350	3.89	0.98
260	4.17	1.38	400	4.44	1.25
300	4.81	1.82	450	5	1.56

(continued)

Table D-5 *(continued)*

U.S. gal/min	***Velocity ft/sc***	***hf Friction***	***U.S. gal/min***	***Vel. ft/sec***	***hf Friction***
350	5.61	2.43	500	5.55	1.91
400	6.41	3.13	600	6.66	2.69
450	7.22	3.92	700	7.77	3.6
500	8.02	4.79	800	8.88	4.64
600	9.62	6.77	900	9.99	5.81
700	11.2	9.13	1000	11.1	7.1
800	12.8	11.8	1100	12.2	8.52
900	14.4	14.8	1200	13.3	10.1
1000	16	18.2	1300	14.4	11.7
			1400	15.5	13.6

Courtesy Sarco Company, Inc.

Table D-6 Flow of Water through Schedule 40 Steel Pipe

Pressure Drop 1000 Feet of Schedule 40 Steel Pipe, in Pounds per Square Inch

Discharge gal/min	Velocity ft/sec	Pressure Drop	Velocity ft/sec	Pressure Drop	Velocity ft/sec	Pressure Drop	Velocity ft/sec	Pressure Drop	Velocity ft/sec	Pressure Drop	Velocity ft/sec	Pressure Drop	Velocity ft/sec	Pressure Drop	Velocity ft/sec	Pressure Drop	Velocity ft/sec	Pressure Drop
	1"																	
1	0.37	0.49	1¼"															
2	0.74	1.7	0.43	0.45	1½"													
3	1.12	3.53	0.64	0.94	0.47	0.44												
4	1.49	5.94	0.86	1.55	0.63	0.74	2"											
5	1.86	9.02	1.07	2.36	0.79	1.12												
6	2.24	12.25	1.28	3.3	0.95	1.53	0.57	0.46										
8	2.98	21.1	1.72	5.52	1.26	2.63	0.76	0.75	2½"									
10	3.72	30.8	2.14	8.34	1.57	3.86	0.96	1.14	0.67	0.48								
15	5.6	64.6	3.21	17.6	2.36	8.13	1.43	2.33	1	0.99	3"							
20	7.44	110.5	4.29	29.1	3.15	13.5	1.91	3.86	1.34	1.64	0.87	0.59	3½"					
25			5.36	43.7	3.94	20.2	2.39	5.81	1.68	2.48	1.08	0.67	0.81	0.42				
30			6.43	62.9	4.72	29.1	2.87	8.04	2.01	3.43	1.3	1.21	0.97	0.6	4"			
35			7.51	82.5	5.51	38.2	3.35	10.95	2.35	4.49	1.52	1.58	1.14	0.79	0.88	0.42		
40					6.3	47.8	3.82	13.7	2.68	5.88	1.74	2.06	1.3	1	1.01	0.53		
45					7.08	60.6	4.3	17.4	3	7.14	1.95	2.51	1.46	1.21	1.13	0.67		
50					7.87	74.7	4.78	20.6	3.35	8.82	2.17	3.1	1.62	1.44	1.26	0.8		
60							5.74	29.6	4.02	12.2	2.6	4.29	1.95	2.07	1.51	1.1	5"	
70							6.69	38.6	4.69	15.3	3.04	5.84	2.27	2.71	1.76	1.5	1.12	0.48

Table D-6 (continued)

Pressure Drop 1000 Feet of Schedule 40 Steel Pipe, in Pounds per Square Inch																		
Discharge gal/min	Velocity ft/sec	Pressure Drop	Velocity ft/sec	Pressure Drop	Velocity ft/sec	Pressure Drop	Velocity ft/sec	Pressure Drop	Velocity ft/sec	Pressure Drop	Velocity ft/sec	Pressure Drop	Velocity ft/sec	Pressure Drop	Velocity ft/sec	Pressure Drop	Velocity ft/sec	Pressure Drop
80		6"					7.65	50.3	5.37	21.7	3.48	7.62	2.59	3.53	2.01	1.87	1.28	0.63
90							8.6	63.6	6.04	26.1	3.91	9.22	2.92	4.46	2.26	2.37	1.44	0.8
100	1.11	0.39					9.56	75.1	6.71	32.3	4.34	11.4	3.24	5.27	2.52	2.81	1.6	0.95
125	1.39	0.56							8.38	48.2	5.45	17.1	4.05	7.86	3.15	4.38	2	1.48
150	1.67	0.78							10.06	60.4	6.51	23.3	4.86	11.3	3.78	6.02	2.41	2.04
175	1.94	1.06		8"					11.73	90	7.59	32	5.67	14.7	4.41	8.2	2.81	2.78
200	2.22	1.32									8.68	39.7	6.48	19.2	5.04	10.2	3.21	3.46
225	2.5	1.66	1.44	0.44							9.77	50.2	7.29	23.1	5.67	12.9	3.61	4.37
250	2.78	2.05	1.6	0.55							10.85	61.9	8.1	28.5	6.3	15.9	4.01	5.14
275	3.06	2.36	1.76	0.63							11.94	75	8.91	34.4	6.93	18.3	4.41	6.22
300	3.33	2.8	1.92	0.75							13.02	84.7	9.72	40.9	7.56	21.8	4.81	7.41
325	3.61	3.29	2.08	0.88									10.53	45.5	8.18	25.5	5.21	8.25
350	3.89	3.62	2.24	0.97									11.35	52.7	8.82	29.7	5.61	9.57
375	4.16	4.16	2.4	1.11									12.17	60.7	9.45	32.3	6.01	11
400	4.44	4.72	2.56	1.27									13.78	77.8	10.7	41.5	6.82	14.1
425	4.72	5.34	2.27	1.43									12.97	68.9	10.08	39.7	6.41	12.9
450	5	5.96	2.88	1.6		10"							14.59	87.3	11.33	46.5	7.22	15
475	5.27	6.66	3.04	1.69	1.93	0.3									11.96	51.7	7.62	16.7

(continued)

Table D-6 (continued)

Pressure Drop 1000 Feet of Schedule 40 Steel Pipe, in Pounds per Square Inch

Discharge gal/min	Velocity ft/sec	Pressure Drop	Velocity ft/sec	Pressure Drop	Velocity ft/sec	Pressure Drop	Velocity ft/sec	Pressure Drop	Velocity ft/sec	Pressure Drop	Velocity ft/sec	Pressure Drop	Velocity ft/sec	Pressure Drop	Velocity ft/sec	Pressure Drop	Velocity ft/sec	Pressure Drop
500	5.55	7.39	3.2	1.87	2.04	0.63									12.59	57.3	8.02	18.5
550	6.11	8.94	3.53	2.26	2.24	0.7									13.84	69.3	8.82	22.4
600	6.66	10.6	3.85	2.7	2.44	0.86	12"								15.1	82.5	9.62	26.7
650	7.21	11.8	4.17	3.16	2.65	1.01											10.42	31.3
700	7.77	13.7	4.49	3.69	2.85	1.18	2.01	0.48									11.22	36.3
750	8.32	15.7	4.81	4.21	3.05	1.35	2.15	0.55									12.02	41.6
800	8.88	17.8	5.13	4.79	3.26	1.54	2.29	0.62	14"								12.82	44.7
850	9.44	20.2	5.45	5.11	3.46	1.74	2.44	0.7	2.02	0.43							13.62	50.5
900	10	22.6	5.77	5.73	3.66	1.94	2.58	0.79	2.14	0.48							14.42	56.6
950	10.55	23.7	6.09	6.38	3.87	2.23	2.72	0.88	2.25	0.53							15.22	63.1
1,000	11.1	26.3	6.41	7.08	4.07	2.4	2.87	0.98	2.38	0.59							16.02	70
1,100	12.22	31.8	7.05	8.56	4.48	2.74	3.16	1.18	2.61	0.68	16"						17.63	84.6
1,200	13.32	37.8	7.69	10.2	4.88	3.27	3.45	1.4	2.85	0.81	2.18	0.4						
1,300	14.43	44.4	8.33	11.3	5.29	3.86	3.73	1.56	3.09	0.95	2.36	0.47						
1,400	15.54	51.5	8.97	13	5.7	4.44	4.02	1.8	3.32	1.1	2.54	0.54						
1,500	16.65	55.5	9.62	15	6.1	5.11	4.3	2.07	3.55	1.19	2.73	0.62						
1,600	17.76	68.1	10.26	17	6.51	5.46	4.59	2.36	3.8	1.35	2.91	0.71	18"					
1,800	19.98	79.8	11.54	21.6	7.32	6.91	5.16	2.98	4.27	1.71	3.27	0.85	2.58	0.48				

Table D-6 (continued)

Pressure Drop 1000 Feet of Schedule 40 Steel Pipe, in Pounds per Square Inch

Discharge gal/min	Velocity ft/sec	Pressure Drop	Velocity ft/sec	Pressure Drop	Velocity ft/sec	Pressure Drop	Velocity ft/sec	Pressure Drop	Velocity ft/sec	Pressure Drop	Velocity ft/sec	Pressure Drop	Velocity ft/sec	Pressure Drop	Velocity ft/sec	Pressure Drop	Velocity ft/sec	Pressure Drop
2,000	22.2	98.5	12.83	25	8.13	8.54	5.73	3.47	4.74	2.11	3.63	1.05	2.88	0.56				
2,500			16.03	39	10.18	12.5	7.17	5.41	5.92	3.09	4.54	1.63	3.59	0.88	20"			
3,000			19.24	52.4	12.21	18	8.6	7.31	7.12	4.45	5.45	2.21	4.31	1.27	3.45	0.73		
3,500			22.43	71.4	14.25	22.9	10.03	9.95	8.32	6.18	6.35	3	5.03	1.52	4.03	0.94	24"	
4,000			25.65	93.3	16.28	29.9	11.48	13	9.49	7.92	7.25	3.92	5.74	2.12	4.61	1.22	3.19	0.51
4,500					18.31	37.8	12.9	15.4	10.67	9.36	8.17	4.97	6.47	2.5	5.19	1.55	3.59	0.6
5,000					20.35	46.7	14.34	18.9	11.84	11.6	9.08	5.72	7.17	3.08	5.76	1.78	3.99	0.74
6,000					24.42	67.2	17.21	27.3	14.32	15.4	10.88	8.24	8.62	4.45	6.92	2.57	4.8	1
7,000					28.5	85.1	20.08	37.2	16.6	21	12.69	12.2	10.04	6.06	8.06	3.5	5.68	1.36
8,000							22.95	45.1	18.98	27.4	14.52	13.6	11.48	7.34	9.23	4.57	6.38	1.78
9,000							25.8	57	21.35	34.7	16.32	17.2	12.92	9.2	10.37	5.36	7.19	2.25
10,000							28.63	70.4	23.75	42.9	18.16	21.2	14.37	11.5	11.53	6.63	7.96	2.78
12,000							34.38	93.6	28.5	61.8	21.8	30.9	17.23	16.5	13.83	9.54	9.57	3.71
14,000									33.2	84	25.42	41.6	20.1	20.7	16.14	12	11.18	5.05
16,000											29.05	54.4	22.96	27.1	18.43	15.7	12.77	6.6

Courtesy Sarco Company, Inc.

Table D-7 Warmup Load in Pounds of Steam per 100 Feet of Steam Main (Ambient Temperature 70°F)[a]*

Steam Pressure (psig)	Main Size														0°F Correction Factor[a]
	2"	2½"	3"	4"	5"	6"	8"	10"	12"	14"	16"	18"	20"	24"	
0	6.2	9.7	12.8	18.2	24.6	31.9	48	68	90	107	140	176	207	208	1.5
5	6.9	11	14.4	20.4	27.7	35.9	48	77	101	120	157	198	233	324	1.44
10	7.5	11.8	15.5	22	29.9	38.8	58	83	109	130	169	213	251	350	1.41
20	8.4	13.4	17.5	24.9	33.8	43.9	66	93	124	146	191	241	284	396	1.37
40	3.9	15.8	20.6	29.3	39.7	51.6	78	110	145	172	225	284	334	465	1.32
60	11	17.5	22.9	32.6	44.2	57.3	86	122	162	192	250	316	372	518	1.29
80	12	19	24.9	35.3	47.9	62.1	93	132	175	208	271	342	403	561	1.27
100	12.8	20.3	26.6	37.8	51.2	66.5	100	142	188	222	290	366	431	600	1.26
125	13.7	21.7	28.4	40.4	54.8	71.1	107	152	200	238	310	391	461	642	1.25
150	14.5	23	30	42.8	58	75.2	113	160	212	251	328	414	487	679	1.24
175	15.3	24.2	31.7	45.1	61.2	79.4	119	169	224	265	347	437	514	716	1.23
200	16	25.3	33.1	47.1	63.8	82.8	125	177	234	277	362	456	537	748	1.22
250	17.2	27.3	35.8	50.8	68.9	89.4	134	191	252	299	390	492	579	807	1.21
300	25	38.3	51.3	74.8	104	142.7	217	322	443	531	682	854	1045	1182	1.2
400	27.8	42.6	57.1	83.2	115.7	158.7	241	358	493	590	759	971	1163	1650	1.18
500	30.2	46.3	62.1	90.5	125.7	172.6	262	389	535	642	825	1033	1263	1793	1.17
600	32.7	50.1	67.1	97.9	136	186.6	284	421	579	694	893	1118	1367	1939	1.16

*Loads based on Schedule 40 pipe for pressures up to and including 250 psig and on Schedule 80 pipe for pressures above 250 psig.

[a]For outdoor temperature of 0°F, multiply load value in table for each main size by correction factor corresponding to steam pressure.

Courtesy Sarco Company, Inc.

Table D-8 Condensation Load in Pounds per Hour per 100 Feet of Insulated Steam Main (Ambient Temperature 70°F; Insulation 80% Efficient)[a]*

Steam Pressure (psig)	Main Size														0°F Correction Factor[a]
	2"	2½"	3"	4"	5"	6"	8"	10"	12"	14"	16"	18"	20"	24"	
10	6	7	9	11	13	16	20	24	29	32	36	39	44	53	1.58
30	8	9	11	14	17	20	26	32	38	42	48	51	57	68	1.5
60	10	12	14	18	24	27	33	41	49	54	62	67	74	89	1.45
100	12	15	18	22	28	33	41	51	61	67	77	83	93	111	1.41
125	13	16	20	24	30	36	45	56	66	73	84	90	101	121	1.39
175	16	19	23	26	33	38	53	66	78	86	98	107	119	142	1.38
250	18	22	27	34	42	50	62	77	92	101	116	126	140	168	1.36
300	20	25	30	37	46	54	68	85	101	111	126	138	154	184	1.35
400	23	28	34	43	53	63	80	99	118	130	148	162	180	216	1.33
500	27	33	39	49	61	73	91	114	135	148	170	185	206	246	1.32
600	30	37	44	55	68	82	103	128	152	167	191	208	232	277	1.31

**Chart loads represent losses due to radiation and convection for saturated steam.*

[a]For outdoor temperature of 0°F, multiply load value in table for each main size by correction factor corresponding to steam pressure.

Courtesy Sarco Company, Inc.

Table D-9 Flange Standards (All dimensions are in inches)

	125-lb Cast Iron		*ASA B16.1*												
Pipe Size	***½***	***¾***	***1***	***1¼***	***1½***	***2***	***2½***	***3***	***3½***	***4***	***5***	***6***	***8***	***10***	***12***
Diameter of Flange			4¼	4⅝	5	6	7	7½	8½	9	10	11	13½	16	19
Thickness of Flange (min)[1]			7/16	½	9/16	⅝	11/16	¾	13/16	15/16	15/16	1	1⅛	1 3/16	1¼
Diameter of Bolt Circle			3⅛	3½	3⅞	4¾	5½	6	7	7½	8½	9½	11¾	14¼	17
Number of Bolts			4	4	4	4	4	4	8	8	8	8	8	12	12
Diameter of Bolts			½	½	½	⅝	⅝	⅝	⅝	⅝	¾	¾	¾	⅞	⅞

	250-lb Cast Iron		ASA B16.2												
Pipe Size	**1/2**	**3/4**	**1**	**1 1/4**	**1 1/2**	**2**	**2 1/2**	**3**	**3 1/2**	**4**	**5**	**6**	**8**	**10**	**12**
Diameter of Flange			4 7/8	5 1/4	6 1/8	6 1/2	7 1/2	8 1/4	9	10	11	12 1/2	15	17 1/2	20 1/2
Thickness of Flange (min)[2]			11/16	3/4	13/16	7/8	1	1 1/8	1 3/16	1 1/4	1 3/8	1 7/16	1 5/8	1 7/8	2
Diameter of Raised Face			2 11/16	3 1/16	3 9/16	4 3/16	4 15/16	6 5/16	6 5/16	6 15/16	8 5/16	9 11/16	11 15/16	14 1/16	16 7/16
Diameter of Bolt Circle			3 1/2	3 7/8	4 1/2	5	5 7/8	6 5/8	7 1/4	7 7/8	9 1/4	10 5/8	13	15 1/4	17 3/4
Number of Bolts			4	4	4	8	8	8	8	8	8	12	12	16	16
Diameter of Bolts			5/8	5/8	3/4	5/8	3/4	3/4	3/4	3/4	3/4	3/4	7/8	1	1 1/8

(continued)

Table D-9 (continued)

	150-lb Bronze		ASA B16.24												
Pipe Size	**1/2**	**3/4**	**1**	**1 1/4**	**1 1/2**	**2**	**2 1/2**	**3**	**3 1/2**	**4**	**5**	**6**	**8**	**10**	**12**
Diameter of Flange	3 1/2	3 7/8	4 1/4	4 5/8	5	6	7	7 1/2	8 1/2	9	10	11	13 1/2	16	19
Thickness of Flange (min)[3]	5/16	11/32	3/8	13/32	7/16	1/2	9/16	5/8	11/16	11/16	3/4	13/16	15/16	1	1 1/16
Diameter of Bolt Circle	2 3/8	2 3/4	3 1/8	3 1/2	3 7/8	4 3/4	5 1/2	6	7	7 1/2	8 1/2	9 1/2	11 3/4	14 1/4	17
Number of Bolts	4	4	4	4	4	4	4	4	8	8	8	8	8	12	12
Diameter of Bolts	1/2	1/2	1/2	1/2	1/2	5/8	5/8	5/8	5/8	5/8	3/4	3/4	3/4	7/8	7/8

	300-lb Bronze		ASA B16.24												
Pipe Size	**½**	**¾**	**1**	**1¼**	**1½**	**2**	**2½**	**3**	**3½**	**4**	**5**	**6**	**8**	**10**	**12**
Diameter of Flange	3¾	4⅝	4⅞	5¼	6⅛	6½	7½	8¼	9	10	11	12½	15		
Thickness of Flange (min)[4]	½	17/32	19/32	⅝	11/16	¾	13/16	29/32	31/32	1 1/16	1⅛	1 3/16	1⅜		
Diameter of Bolt Circle	2⅝	3¼	3½	3⅞	4½	5	5⅞	6⅝	7¼	7⅞	9¼	10⅝	13		
Number of Bolts	4	4	4	4	4	8	8	8	8	8	8	12	12		
Diameter of Bolts	½	⅝	⅝	⅝	¾	⅝	¾	¾	¾	¾	¾	¾	⅞		

(continued)

Table D-9 (continued)

Pipe Size	150-lb Steel		ASA B16.5												
	1/2	**3/4**	**1**	**1 1/4**	**1 1/2**	**2**	**2 1/2**	**3**	**3 1/2**	**4**	**5**	**6**	**8**	**10**	**12**
			4 1/4	4 5/8	5	6	7	7 1/2	8 1/2	9	10	11	13 1/2	16	19
Thickness of Flange (min)[5]			7 1/6	1/2	9/16	5/8	13/16	3/4	13/16	15/16	15/16	1	1 1/8	1 3/16	1 1/4
Diameter of Raised Face			2	2 1/2	2 7/8	3 5/8	4 1/8	5	5 1/2	6 3/16	7 5/16	8 1/2	10 5/8	12 3/4	15
Diameter of Bolt Circle			3 1/8	3 1/2	3 7/8	4 3/4	5 1/2	6	7	7 1/2	8 1/2	9 1/2	11 3/4	14 1/4	17
Number of Bolts			4	4	4	4	4	4	8	8	8	8	8	12	12
Diameter of Bolts			1/2	1/2	1/2	5/8	5/8	5/8	5/8	5/8	3/4	3/4	3/4	7/8	7/8

Pipe Size	300-lb Steel ½	¾	ASA B16.5 1	1¼	1½	2	2½	3	3½	4	5	6	8	10	12
Diameter of Flange	4⅞	5¼	6⅛	6½	7½	8¼	9	10	11	12½	15	17½	20½		
Thickness of Flange (min)[6]	11/16	¾		13/16	⅞	1	1⅛	1 3/16	1¼	1⅜	1 7/16	1⅝	1⅞	2	
Diameter of Raised Face	2	2½	2⅞	3⅝	4⅛	5	5½	6 3/16	7 5/16	8½	10⅝	12¾	15		
Diameter of Bolt Circle	3½	3⅞	4½	5	5⅞	6⅝	7¼	7⅞	9¼	10⅝	13	15¼	17¾		
Number of Bolts	4	4	4	8	8	8	8	8	8	12	12	16	16		
Diameter of Bolts	⅝	⅝	¾	⅝	¾	¾	¾	¾	¾	¾	⅞	1	1⅛		

(continued)

Table D-9 (continued)

	400-lb Steel		**ASA B16.5**												
Pipe Size	**½**	**¾**	**1**	**1¼**	**1½**	**2**	**2½**	**3**	**3½**	**4**	**5**	**6**	**8**	**10**	**12**
Diameter of Flange	3¾	4⅝	4⅞	5¼	6⅛	6½	7½	8¼	9	10	11	12½	15	17½	20½
Thickness of Flange (min)[7]	9/16	⅝	11/16	13/16	⅞	1	1⅛	1¼	1⅜	1⅜	1½	1⅝	⅞	2⅛	2¼
Diameter of Raised	1⅜	1 11/16	2	2½	2⅞	3⅝	4⅛	5	5½	6 3/16	7 5/16	8½	10⅝	12¾	15
Diameter of Bolt Circle	2⅝	3¼	3½	3⅞	4½	5	5⅞	6⅝	7¼	7⅞	9¼	10⅝	13	15¼	17¾
Number of Bolts	4	4	4	4	4	8	8	8	8	8	8	12	12	16	16
Diameter of Bolts	½	⅝	⅝	⅝	¾	¾	⅞	⅞	⅞	⅞	1	1⅛	1¼		

	600-lb Steel		ASA B16.5												
Pipe Size	**1/2**	**3/4**	**1**	**1 1/4**	**1 1/2**	**2**	**2 1/2**	**3**	**3 1/2**	**4**	**5**	**6**	**8**	**10**	**12**
Diameter of Flange	3 3/4	4 5/8	4 7/8	5 1/4	6 1/8	6 1/2	7 1/2	8 1/4	9	10 3/4	13	14	16 1/2	10	12
Thickness of Flange (min)[8]	9/16	5/8	11/16	13/16	7/8	1	1 1/8	1 1/4	1 3/8	1 1/2	1 3/4	1 7/8	2 3/16	2 1/2	2 5/8
Diameter of Raised	1 3/8	1 11/16	2	2 1/2	2 7/8	3 5/8	4 1/8	5	5 1/2	6 3/16	7 5/16	8 1/2	10 5/8	12 3/4	15
Diameter of Bolt Circle	2 5/8	3 1/4	3 1/2	3 7/8	4 1/2	5	5 7/8	6 5/8	7 1/4	8 1/2	10 1/2	11 1/2	13 1/4	17	19 1/2
Number of Bolts	4	4	4	4	4	8	8	8	8	8	8	12	12	16	20
Diameter of Bolts	1/2	5/8	5/8	5/8	3/4	5/8	3/4	3/4	7/8	7/8	1	1	1 1/8	1 1/4	1 1/4

[1] *125-lb. flanges have plain faces.*

[2] *250-lb. flanges have a 1/16" raised face, which is included in the flange thickness dimensions.*

[3] *150-lb. bronze flanges have plain faces with two concentric gasket-retaining grooves between the port and the bolt holes.*

[4] *300-lb. bronze flanges have plain faces with two concentric gasket-retaining grooves between the port and the bolt holes.*

[5] *150-lb. steel flanges have a 1/16" raised face, which is included in the flange thickness dimensions.*

[6] *300-lb. steel flanges have a 1/16" raised face, which is included in the flange thickness dimensions.*

[7] *400-lb. steel flanges have a 1/4" raised face, which is NOT included in the flange dimensions.*

[8] *600-lb. steel flanges have a 1/4" raised face, which is NOT included in the flange dimensions.*

Table D-10 Pressure Drop in Schedule 40 Pipe

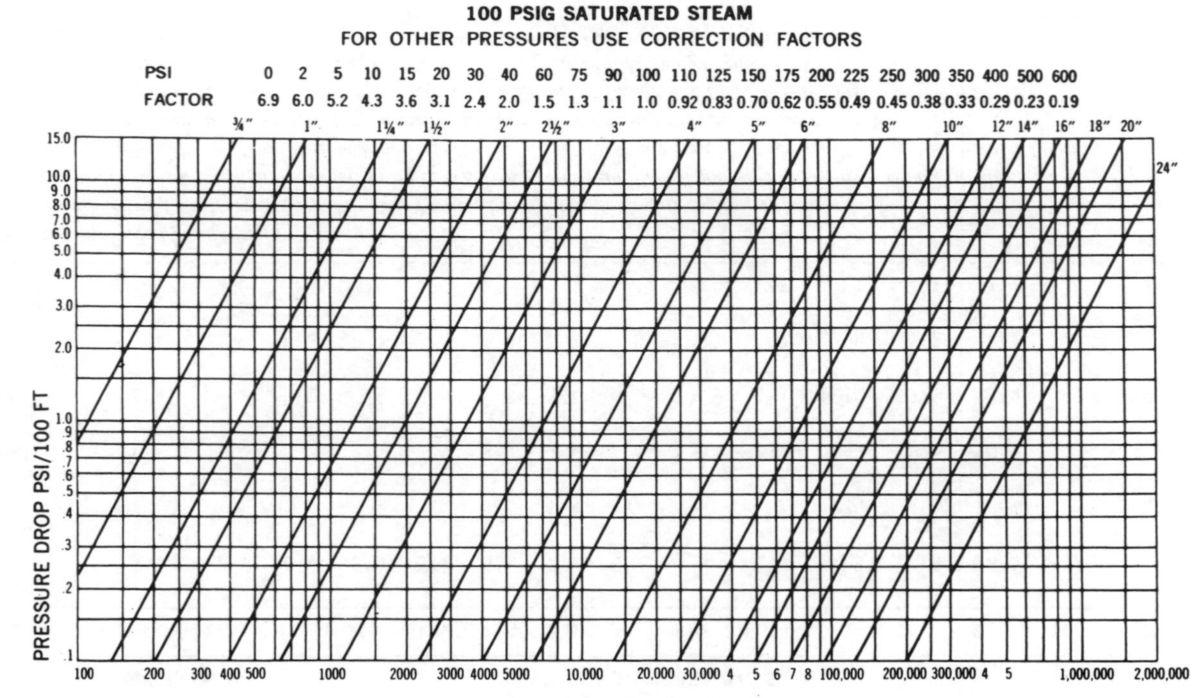

Table D-11 Steam Velocity Chart

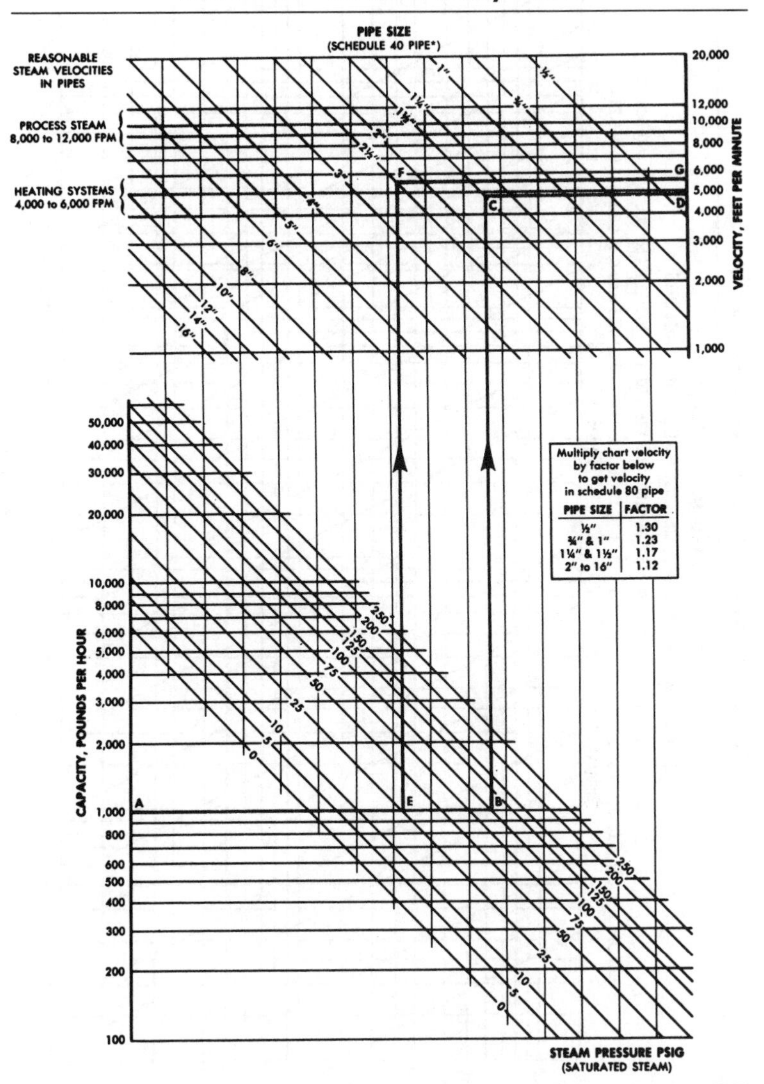

PIPE SIZE	FACTOR
½"	1.30
¾" & 1"	1.23
1¼" & 1½"	1.17
2" to 16"	1.12

Table D-12 Pressure Drop in Schedule 80 Pipe

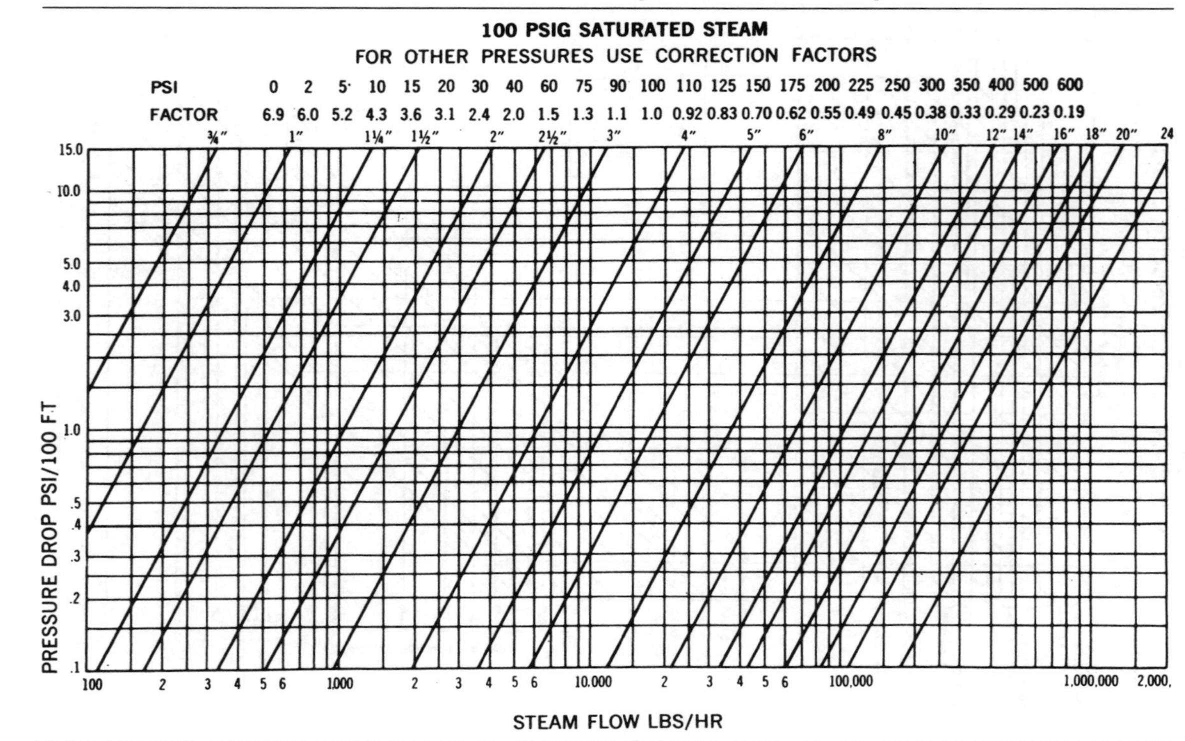

Table D-13 Chimney Connector and Vent Connector Clearance from Combustible Materials

Description of Appliance	***Minimum Clearance, Inches****
Residential Appliances	
Single-Wall, Metal Pipe Connector	
Electric, gas, and oil incinerators	18
Oil and solid-fuel appliances	18
Oil appliances listed as suitable for use with Type L venting system, but only when connected to chimneys	9
Type L Venting-System Piping Connectors	
Electric, gas, and oil incinerators	9
Oil and solid-fuel appliances	9
Oil appliances listed as suitable for use with Type L venting systems	[a]
Commercial and Industrial Appliances	
Low-Heat Appliances	
Single-Wall, Metal Pipe Connectors	
Gas, oil, and solid-fuel boilers, furnaces, and water heaters	18
Ranges, restaurant type	18
Oil unit heaters	18
Other low-heat industrial appliances	18
Medium-Heat Appliances	
Single-Wall, Metal Pipe Connectors	
All gas, oil, and solid-fuel appliances	36

**These clearances apply except if the listing of an appliance specifies different clearances, in which case the listed clearance takes precedence.*

[a]*If listed Type L venting-system piping is used, the clearance may be in accordance with the venting-system listing.*

If listed Type B or Type L venting-system piping is used, the clearance may be in accordance with the venting-system listing.

The clearances from connectors to combustible materials may be reduced if the combustible material is protected in accordance with Table 1C.

Courtesy National Oil Fuel Institute

Table D-14 Standard Clearances for Heat-Producing Appliances in Residential Installations

*Residential-Type Appliances for Installation in Rooms That Are Large**		*Appliance*				
		Above Top of Casing or Appliance	*From Top and Sides of Warm-Air Bonnet or Plenum*	*From Front*[a]	*From Back*	*From Sides*
Boilers and Water Heaters						
Steam boilers—15 psi; water boilers—250°F; water heaters—200°F; all water walled or jacketed	Automatic oil or combination gas-oil	6	—	24	6	6
Furnaces—Central						
gravity, upflow, downflow, horizontal and duct. warm-air—250°F max.	Automatic oil or combination gas-oil	6[b]	6[c]	24	6	6

Table D-14 (continued)

Residential-Type Appliances for Installation in Rooms That Are Large*			Appliance				
			Above Top of Casing or Appliance	From Top and Sides of Warm-Air Bonnet or Plenum	From Front[a]	From Back	From Sides
Furnaces—Floor							
	For mounting in combustible floors	Automatic oil or combination gas-oil	36	—	12	12	12

Rooms that are large in comparison to the size of the appliance are those having a volume equal to at least 12 times the total volume of a furnace and at least 16 times the total volume of a boiler. If the actual ceiling height of a room is greater than 8 ft, the volume of a room shall be figured on the basis of a ceiling height of 8 ft.

[a]*The minimum dimension should be that necessary for servicing the appliance including access for cleaning and normal care, tube removal, etc.*

[b]*For a listed oil, combination gas-oil, gas, or electric furnace, this dimension may be 2 in if the furnace limit control cannot be set higher than 250°F or 1 in if the limit control cannot be set higher than 200°F.*

Courtesy National Oil Fuel Institute

Table D-15 Standard Clearances for Heat-Producing Appliances in Commercial and Industrial Installations

Commercial-Industrial Type Low Heat Appliance (Any and All Physical Sizes Expect As Noted)		***Appliance***				
		Above Top of Casing or Appliance*	***From Top and Sides of Warm-Air Bonnet or Plenum***	***From Front***	***From Back****	***From Sides****
Boiler and Water Heaters						
100 ft^3 or less, any psi, steam	All fuels	18	—	48	18	18
50 psi or less, any size	All fuels	18	—	48	18	18
Unit Heaters						
Floor mounted or suspended—	Steam or hot water	1	—	—	1	1
Suspended—100 ft^3 or less	Oil or combination gas-oil	6	—	24	18	18
Suspended —over 100 ft^3	All fuels	18	—	48	18	18
Floor mounted any	All fuels	18	—	48	18	18

**If the appliance is encased in brick, the 18 in clearance above and at sides and rear may be reduced to not less than 12 in*

Courtesy National Oil Fuel Institute

Table D-16 Clearance (in Inches) with Specified Forms of Protection

Input heat units	Efficiency %	Usable Btu/h	gph 100°F rise	Tank size, gal	Available Hot-Water Storage Plus Recovery: 100°F Rise, 15 min	30 min	45 min	60 min	Continuous Draw, gph
Electricity, kW									
1.5	92.5	4,750	5.7*	20	21.4	22.8	24.3	25.7	5.7
2.5	92.5	7,900	9.5*	20	32.4	34.8	37.1	39.5	9.5
4.5	92.5	14,200	17.1*	30	44.3	48.6	52.9	57.1	17.1
4.5	92.5	14,200	17.1*	50	54.3	58.6	62.9	67.1	17.1
6	92.5	19,000	22.8*	66	71.6	77.2	82.8	88.8	22.8
7	92.5	22,100	26.5	80	86.6	93.2	99.8	106.5	26.5
Gas, Btu/h									
34,000	75	25,500	30.6	30	37.7	45.3	53	55.6[a]	25.6[a]
42,000	75	31,600	38	30	39.5	49	58.8	61.7[a]	31.7[a]
50,000	75	37,400	45	40	51.3	62.6	73.9	77.6[a]	37.6[a]
60,000	75	45,000	54	50	63.5	77	90.5	95.0[a]	45.0[a]
Oil, gph									
0.5	75	52,500	63	30	45.8	61.6	77.4	82.5[a]	52.5[a]
0.75	75	78,700	94.6	30	53.6	77.2	100.8	109.0[a]	79.0[a]
0.85	75	89,100	107	30	57.7	83.4	110.1	119.1[a]	89.0[a]
1	75	105,000	126	50	81.5	13	144.5	155.0[a]	105.0[a]
1.2	75	126,000	151.5	50	87.9	125.8	163.7	176.0[a]	126.0[a]
1.35	75	145,000	174	50	93.5	137	180.5	195.0[a]	145.0[a]
1.5	75	157,000	188.5	85	132.1	179.2	226.3	242.0[a]	157.0[a]
1.65	75	174,000	204.5	85	136.1	187.2	238.4	259.0[a]	174.0[a]

*Assumes simultaneous operation of upper and lower elements.

[a]Based on 50 minute-per-hour operation.

Courtesy National Oil Fuel Institute

Appendix D

Conversion Tables

To Convert	Into	Multiply By
Acres	square feet	43,560.0
Atmospheres	centimeters of mercury	76.0
Atmospheres	feet of water (at 4°C)	33.90
Atmospheres	inches of mercury (at 0°C)	29.92
Atmospheres (atm)	kilograms/ square centimeter	1.0333
Atmospheres	pounds/square inch	14.70
Barrels (U.S., liquid)	gallons	31.5
Barrels (oil) (bbl)	gallons (oil)	42.0
Btu	foot-pounds	778.3
Btu	gram-calories	252.0
Btu	horsepower-hour	3.931×10^{-4}
Btu	kilowatt-hours	2.928×10^{-4}
Btu per hour (Btu/h)	horsepower	3.931×10^{-4}
Btu/h	watts	0.2931
Calories, gram (mean)	Btu (mean)	3.9685×10^{-3}
Degrees centigrade or Celsius (°C)	degrees Fahrenheit	$(°C + 40)\ 9/5 - 40$
Centimeters (cm)	feet	3.281×10^{-2}
Centimeters	inches	0.3937
Centimeters	mils	393.7
Centimeters of mercury (cm Hg)	atmospheres	0.01316
Centimeters of mercury	feet of water	0.4461
Centimeters of mercury	pounds/square inch	0.1934
Circumference	radians	6.283
Cubic centimeters (cm^3)	cubic feet	3.531×10^{-5}
Cubic centimeters	cubic inches	0.06102
Cubic centimeters	gallons (U.S. liquid)	2.642×10^{-4}
Cubic feet (ft^3)	cubic centimeters	28,320.0
Cubic feet	cubic inches	1,728.0

(continued)

To Convert	*Into*	*Multiply By*
Cubic feet	gallons (U.S. liquid)	7.481
Cubic feet	liters	28.32
Cubic feet	quarts (U.S. liquid)	29.92
Cubic feet per minute	gallons/second	0.1247
Cubic feet per minute	pounds of water/ minute	62.43
Cubic inches (in^3)	cubic centimeters	16.39
Cubic inches	gallons	4.329×10^{-3}
Cubic inches	quarts (U.S. liquid)	0.01732
Cubic meters	cubic feet	35.31
Cubic meters	gallons (U.S. liquid)	264.2
Cubic yards (yd^3)	cubic feet	27.0
Cubic yards	cubic meters	0.7646
Cubic yards	gallons (U.S. liquid)	202.0
Degrees (angle)	radians	0.01745
Drams (apothecaries' or troy)	ounces (avoirdupois)	0.13714
Drams (apothecaries' or troy)	ounces (troy)	0.125
Drams (U.S., fluid or apothecary)	cubic centimeter	3.6967
Drams	grams	1.772
Drams	grains	27.3437
Drams	ounces	0.0625
Degrees Fahrenheit (°F)	degrees centigrade	(°F + 40) 5/9 − 40
Feet (ft)	centimeters	30.48
Feet	kilometers	3.048×10^{-4}
Feet	meters	0.3048
Feet	miles (nautical)	1.645×10^{-4}
Feet	miles (statute)	1.894×10^{-4}
Feet of water (ft H_2O)	atmospheres	0.02950
Feet of water	inches of mercury	0.8826
Feet of water	kilograms/ square centimeter	0.03045
Feet of water	kilograms/ square meter	304.8

Feet of water	pounds/square foot	62.43
Feet of water	pounds/square inch	0.4335
Foot-pounds (ft-lb)	Btu	1.286×10^{-3}
Foot-pounds	gram-calories	0.3238
Foot-pounds	horsepower-hours	5.050×10^{-7}
Foot-pounds	kilowatt-hours	3.766×10^{-7}
Foot-pounds/minute (ft-lb/min)	Btu/ minute	1.286×10^{-3}
Foot-pounds/minute	horsepower	3.030×10^{-5}
Foot-pounds/second (ft-lb/sec)	Btu per hour	4.6263
Furlongs	miles (U.S.)	0.125
Furlongs	feet	660.0
Gallons (gal)	cubic centimeter	3,785.0
Gallons	cubic feet	0.1337
Gallons	cubic inches	231.0
Gallons	cubic meters	3.785×10^{-3}
Gallons	cubic yards	4.951×10^{-3}
Gallons	Liters	3.785
Gallons (British Imperial liquid)	gallons (U.S. liquid)	1.20095
Gallons (U.S.)	gallons (Imperial)	0.83267
Gallons of water	pounds of water	8.3453
Gallons per minute (gal/min, gpm)	cubic feet/ second	2.228×10^{-3}
Gallons per minute	liters/second	0.06308
Gallons per minute	cubic feet/hour	8.0208
Grains (troy) (gr)	grains (avdp.)	1.0
Grains (troy)	grams	0.06480
Grains (troy)	ounces (avdp.)	2.286×10^{-3}
Grains (troy)	pennyweight (troy)	0.04167
Grains per U.S. gallon	parts/million	17.118
Grains per U.S. gallon	pounds/ million gallons	142.86
Grains oer Imperial gallon	parts/million	14.286

(continued)

To Convert	*Into*	*Multiply By*
Grams (g)	grains	15.43
Grams	ounces (avdp.)	0.03527
Grams	ounces (troy)	.03215
Grams	poundals	0.07093
Grams	pounds	2.205×10^{-3}
Grams per liter (g/L)	parts/million	1,000.0
Gram-calories (g-cal)	Btu	3.9683×10^{-3}
Gram-calories	foot-pounds	3.0880
Gram-calories	kilowatt-hour	1.1630×10^{-6}
Gram-calories	watt-hour	1.1630×10^{-3}
Horsepower (hp)	Btu/minute	42.40
Horsepower	foot-pounds/minute	33,000.0
Horsepower	foot-pounds/second	550.0
Horsepower (metric) (542.5 ft-lb/sec)	horsepower (550 ft-lb/sec)	0.9863
Horsepower (550 ft-lb/sec)	horsepower (metric) (542.5 ft-lb/sec)	1.014
Horsepower	kilowatts	0.7457
Horsepower	watts	745.7
Horsepower (boiler)	Btu/hour	33.520
Horsepower (boiler)	kilowatts	9.803
Horsepower-hour (hp-hr)	Btu	2,547
Horsepower-hour	foot-pounds	1.98×10^{6}
Horsepower-hour	kilowatt-hours	0.7457
Inches (in)	centimeters	2.540
Inches	meters	25.40
Inches	millimeters	2.540×10^{-2}
Inches	yards	22.778×10^{-2}
Inches of mercury (in Hg)	atmospheres	0.03342
Inches of mercury	feet of water	1.133
Inches of mercury	kilograms/ square centimeters	0.03453

Inches of mercury	kilograms/ square meter	345.3
Inches of mercury	pounds/square foot	70.73
Inches of mercury	pounds/square inch	0.4912
Inches of water (at 4°C) (in H_2O, in wg, in W.C.)	atmospheres	2.458×10^{-3}
Inches of water (at 4°C)	inches of mercury	0.07355
Inches of water (at 4°C)	kilograms/ square centimeter	2.538×10^{-3}
Inches of water (at 4°C)	ounces/ square inch	0.5781
Inches of water (at 4°C)	pounds/ square foot	5.204
Inches of water (at 4°C)	pounds/ square inch	0.03613
Joules (J)	Btu	9.480×10^{-4}
Kilograms (kg)	grams	1,000.0
Kilograms	pounds	2.205
Kilograms per cubic meter (kg/m^3)	pounds/ cubic foot	0.06243
Kilograms per cubic meter	pounds/ cubic inch	3.613×10^{-5}
Kilograms per square centimeter (kg/cm^2)	atmospheres	0.9678
Kilograms per square centimeter	feet of water	32.84
Kilograms per square centimeter	inches of mercury	28.96
Kilograms per square centimeter	pounds/ square foot	2,048.0
Kilograms per square centimeter	pounds/ square inch	14.22
Kilograms per square meter (kg/m^2)	atmospheres	9.678×10^{-5}
Kilograms per square meter	feet of water	3.281×10^{-3}
Kilograms per square meter	inches of mercury	2.896×10^{-3}

(continued)

To Convert	*Into*	*Multiply By*
Kilograms per square meter	pounds/ square foot	0.2048
Kilograms per square meter	pounds/ square inch	1.422×10^{-3}
Kilograms per square millimeter	kilograms/ square meter	10^{6}
Kilogram-calories	Btu	3.968
Kilogram-calories	foot-pounds	3,088.0
Kilogram-calories	horsepower-hour	1.560×10^{-3}
Kilogram-calories	kilowatt-hour	1.163×10^{-3}
Kilogram meters	Btu	9.294×10^{-3}
Kilometers	centimeters	10^{5}
Kilometers	feet	3,281.0
Kilometers	miles	0.6214
Kilowatts	Btu/minute	56.87
Kilowatts	foot-pounds/minute	4.426×10^{4}
Kilowatts	foot-pounds/second	737.6
Kilowatts	horsepower	1,341.0
Kilowatts	watts	1,000.0
Kilowatt-hour	Btu	3,413.0
Kilowatt-hour	foot-pounds	2.655×10^{6}
Kilowatt-hour	horsepower-hour	1,341
Knots	statute miles/hour	1.151
Liters (L)	cubic centimeters	1,000.0
Liters	cubic feet	0.03531
Liters	cubic inches	61.02
Liters	gallons (U.S. liquid)	0.2642
Meters (m)	centimeters	100.0
Meters	feet	3.281
Meters	inches	39.37
Meters	millimeters	1,000.0
Meters	yards	1.094
Microns (μ, μm)	inches	39.37×10^{-6}
Microns	meters	1×10^{-6}
Miles (statute) (mi)	feet	5,280.0
Miles (statute)	kilometers	1.609
Miles per hour (mi/hr, mph)	centimeters/second	44.70

Miles per hour	feet/minute	88.0
Mils	inches	0.001
Mils	yards	2.778×10^{-5}
Nepers	decibels	8.686
Ohms (ω)	megohms	10^{-6}
Ohms	microhms	10^{6}
Ounces (avoirdupois) (oz)	drams	16.0
Ounces (avoirdupois)	grains	437.5
Ounces (avoirdupois)	grams	28.35
Ounces (avoirdupois)	pounds	0.0625
Ounces (avoirdupois)	ounces (troy)	0.9115
Ounces (troy)	grains	480.0
Ounces (troy)	grams	31.10
Ounces (troy)	ounces (avdp.)	1.09714
Ounces (troy)	pounds (troy)	0.08333
Parts per million (ppm)	grains/ U.S. gallon	0.0584
Parts per million	grains/ Imperial gallon	0.07016
Parts per million	pounds/ million gallons	8.33
Pounds (avoirdupois) (lb)	ounces (troy)	14.58
Pounds (avoirdupois)	drams	256.0
Pounds (avoirdupois)	grains	7,000.0
Pounds (avoirdupois)	grams	28.35
Pounds (avoirdupois)	kilograms	0.02835
Pounds (avoirdupois)	ounces	16.0
Pounds (avoirdupois)	tons (short)	0.0005
Pounds (troy)	ounces (avdp.)	13.1657
Pounds of water	cubic feet	0.01602
Pounds of water	cubic inches	27.68
Pounds of water	gallons	0.1198
Pounds of water per minute	cubic feet/ second	2.670×10^{-4}
Pounds per cubic foot (lb/ft^3)	kilogram/ cubic meter	0.01602
Pounds per cubic foot	grams/ cubic centimeter	16.02

(continued)

To Convert	*Into*	*Multiply By*
Pounds per cubic foot	pounds/ cubic inch	5.787×10^{-4}
Pounds per cubic inch	pounds/ cubic foot	1.728.0
Pounds per square foot (lb/ft^2)	atmospheres	4.725×10^{-4}
Pounds per square foot	feet of water	0.01602
Pounds per square foot	inches of mercury	0.01414
Pounds per square inch (lb/in^2, psi)	atmospheres	0.06804
Pounds per square inch	feet of water	2.307
Pounds per square inch	inches of mercury	2.036
Pounds per square inch	kilograms/ square meter	703.1
Pounds per square inch	pounds/ square foot	144.0
Radians	degrees	57.30
Revolutions per minute (rpm)	degrees/ second	6.0
Revolutions per minute	radians/ second	0.1047
Revolutions per minute	revolutions/ second	0.01667
Square centimeters (cm^2)	square feet	1.076×10^{-3}
Square centimeters	square inches	0.1550
Square centimeters	square meters	0.0001
Square centimeters	square millimeters	100.0
Square feet (ft^2)	acres	2.296×10^{-5}
Square feet	square centimeters	929.03
Square feet	square inches	144.0
Square feet	square miles	3.587×10^{-8}
Square inches (in^2)	square centimeters	6.452
Square inches	square feet	6.944×10^{-3}
Square inches	square yards	7.716×10^{-4}
Square meters	square feet	10.76

Square meters	square inches	1,550.0
Square meters	square millimeters	10^6
Square meters	square yards	1.196
Square millimeters (mm^2)	square inches	1.550×10^{-3}
Square yards (yd^2)	square feet	9.0
Square yards	square inches	1,296.0
Square yards	square meters	0.8361
Temperature (°C) + 273.15	absolute temperature (kelvins or °C)	1.0
Temperature (°C) + 17.78	temperature (°F)	1.8
Temperature (°F) + 459.67	absolute temperature (°Rankine)	1.0
Temperature (°F) − 32	temperature (°C)	5/9
Tons (long)	kilograms	1,016.0
Tons (long)	pounds	2,240.0
Tons (long)	tons (short)	1.120
Tons (metric)	kilograms	1,000.0
Tons (metric)	pounds	2,205.0
Tons (short)	kilograms	907.2
Tons (short)	pounds	2,000.0
Tons (short)	tons (long)	0.89287
Tons of water per 24 hours	pounds of water/hour	83.333
Tons of water per 24 hours	gallons/minute	0.16643
Tons of water per 24 hours	cubic feet/hour	1.3349
Watts (W)	Btu/h	3.4129
Watts	Btu/minute	0.05688
Watts	horsepower	1.341×10^{-3}
Watts	horsepower (metric)	1.360×10^{-3}
Watts	kilowatts	0.001
Watts (abs.)	Btu (mean)/minute	0.056884
Watt-hours	Btu	3.413
Watt-hours	horsepower-hours	1.341×10^{-3}
Yards (yd)	centimeters	91.44
Yards	kilometers	9.144×10^{-4}
Yards	meters	0.9144

Index

A

actuators and damper motors (duct system)
  actuator motor characteristics, 316
  installing, 316–320
  modutrol motors, 314–316
  troubleshooting, 320–321
  types of, 313–314
  uses of, 313
air conditioning systems and duct furnaces, 340–341, 342–343
air distribution systems, proprietary, 336–338
air eliminators for steam heating systems, 545–547
air supply in atmospheric injection gas burners, 57, 58, 60–61
air switches for fans, 236–237
air velocities, recommended and maximum for duct systems, 327
aluminum, 2S, table of thicknesses, gauges, and weights, 301
American National Standards Institute (ANSI) Codes and Standards, 451
American Petroleum Institute (API), 451
American Society for Testing Materials (ASTM) Material Specifications, 452
American Society of Heating, Refrigerating, and Air-Condition Engineers (ASHRAE), 81, 295, 384–385
American Society of Mechanical Engineers (ASME) Boiler Construction and Unfired Pressure Vessel Code, 452
angle valves, 454–456
anthracite coal, sizes of, 78–79
ASHRAE Guide (1960), 81
ASHRAE *Handbook of Fundamentals*, 295
atmospheric injection gas burners
  air supply in, 57, 58, 60–61
  flame color and, 60–61
  gas velocity and, 60–61
  primary air ratios, 60
  secondary air in, 60
automatic controls
  adjusting Honeywell step controller, 270–271
  auxiliary switches, 269
  balancing relays, 268
  capacitors and, 284–287
  contactors, cleaning, 280
  contactors, defined and described, 275–277
  contactors, replacing, 280–281
  contactors, troubleshooting, 277–280
  control panels (heating and cooling), 293
  for fans, 233–244
  for gas burners, 61–62
  heating relays/time-delay relays, 261–263
  high-pressure cutout switches, 288–289
  inherent protectors, 282–283
  low-pressure cutout switches, 289–290
  manual switches, 269
  motor starters, 281
  overload relay heaters, 281–282
  pilot duty motor protectors, 283–284
  potential (start) relays, 263–265
  pressure switches, 265–266
  sail switches, 266–267
  sequence controllers, 269–275
  time-delay relays/heating relays, 261–263

automatic controls *(continued)*
time-delay switching, 271, 273, 275
transformers, 290–293
automatic control systems
controlled device for, 99
controller for, 99
humidistats, 134–137, 140–143
temperature control circuits for, 100, 101
thermostats, 100–134
automatic pilot safety valves, 174–178
automatic valves
actuators for, 476
described, 475
lift adjustments and, 486–487
linkages for, 483–487
motorized operators for, 477
pneumatic operators, 476
safety precautions for, 478
self-actuating steam heating systems and, 482
servicing, 478–480
solenoid operators, 476
springless pneumatic operators for, 476–477
uses for, 476
valve operators, 476
zoniing and, 477, 480–481
auxiliary switches, 269

B

Bacharach smoke scale, for oil burners, 44
balancing relays, 268
balancing valves, valve adapters, and filters for steam heating systems, 561–564
bituminous coal, firing methods for, 80–81
brass pipes, 369
brass tubing, 369
Bunsen burners, 57, 59, 60–61
burners, conversion
gas, 63
gas piping for, 68–71
burners, gas
atmospheric injection, 57–61
automatic controls and, 61–62
conversion burners, 63
defined, 57
electrical circuits and, 61
flame sensors,, 214-217
gas conversion, described, 66–68
gas conversion, piping for, 68–71
inshot, 64
integral-type, 65–66
main, 63–64, 66
mercury flame sensors in, 216
operating principles for, 57–61
pilot, 63–64
precedence of local electrical codes over manufacturer instructions, 61
safety precautions and, 72–73
troubleshooting, 73–75
upshot, 64
venting system and, 71–72
burners, oil
adjustments to, 43–48
air delivery and blower adjustment, 42–43
air systems in, 38–39
burner flame detection and, 222
cadmium cell primaries, 220–223
and carbon dioxide in flue gas, 45–48
combination oil and gas, 19, 21
combination primary control and aquastat, 227–231
combustion chamber dimensions, 37
combustion testing of, 43–48
defined, 1
delayed ignition, 50
electrodes, 36
external warning signals for maintenance of, 48–49
flame retention head, 16
fuel pump capacity and, 26
fuel pump nozzles for, 32–36
fuel pump pressure, adjusting, 29–30
fuel pump priming, 29
fuel pump service and maintenance, 26–29
fuel pump troubleshooting, 31–32
fuel temperature and, 22

fuels used in, 22
gun-type (atomizing), 1–2, 3–16
high-static, 16
installing, 40–42
measuring carbon dioxide in flue gas of, 46
net stack temperature and, 45
noise, excessive, 49
odor, excessive, 50
overfire drafts and, 46, 48
primary controls, 219–220
primary safety control services, 39–40
puffbacks, 36–38
rotary, 2–3, 16–18
single-stage fuel pumps and, 21, 23–24
smoke, excessive, 49–50
stack detector primary controls, 223–227
stack thermometer, use of, 45
starting, 42
troubleshooting, 48–56
troubleshooting electrodes in, 38
troubleshooting primary controls in, 231–232
two-stage fuel pumps and, 25
vaporizing (pot-type), 2, 18–20
burners, pilot (gas)
aerated, 196
flame adjustment, 202
flame positioning, 200
installing, 198–200
lighting, 201–202
and main burner ignition, 202–203
nonaerated, 197
operation of, 194–196
pilot bracket in, 197
pilot ports in, 197–198
pilot orifices, 198, 200–201
butterfly valves, 465–467

C

capacitors
capacitor-start, capacitor-run motors, 285–287
capacitor-start motors, 284–285
discharging, 287
replacing, 287
testing, 287
troubleshooting, 287
capacitor-start motors, 284–285
chain wrenches, 406, 407
check valves
described, 458–460
horizontal-lift 460–461, 467
lift-check, 460–461, 467
swing-check, 460, 464, 465, 466
vertical, 461
Watts Blackflow Preventer and, 461–462
circulators for steam heating systems
end suction pumps in, 510–511
inline centrifugal, 510
installation of, 515–516
lubrication requirements for, 509
operation of, 516–517
pressure drop, 514
pump head, 514
safety precautions for, 508, 509
selection of, 511–515
three-piece booster pumps, 506–509
troubleshooting, service, and maintenance, 517–518
volute in, 507
wet-rotor, 509–510
coal firing methods
anthracite coal and, 78–80
automatic controls for stokers, 86–90
bituminous coal and, 80–81
for boiler, 77
central cone, 81
draft requirements for, 77–78
for furnace, 77
hand-firing, 77
semibituminous coal and, 81
stoker firing, 81–86
coal, sizes of anthracite, 78–79
coal stokers
adjustment of coal feed, 94
air adjustment, manual, 92
air control, automatic, 92
air ducts and, 85
ashpits for, 86
automatic controls for, 86–90
bin-fed, 82–83

coal stokers *(continued)*
  changing coal feeds for, 92
  classes of, 81–82
  clinker removal, 94
  and coal burning capacity, 81–82
  coal feeds, changing, 92
  and coal selection, 91
  coal storage methods and, 82–84
  construction of, 84–86
  Drawz, 82, 83, 86
  electric motors and, 85–86
  fans and, 85
  feed worms and, 86
  hold-fire controls and, 87
  limit controls and, 87
  lubrication, 93
  motor overload protection for, 92–93
  natural stack draft, 91–92
  obstruction removal, 93
  operating instructions, 90–91
  retorts and, 85
  stack switches (stack thermostats) and, 89
  starting the fire, 91
  summer service for, 93–94
  thermostats and, 87
  timers and, 88–89
  transmission overload protection for, 93
  transmissions for, 86
  troubleshooting, 94–97
  underfed, 82–83
codes and standards for ducts and duct systems, 295
combination gas valve, 62
combustion chambers for gas conversion burners, 67–68
combustion testing and adjustments, oil burner, 43–48
composite tubing (PAX), 369
compressors, size and motor current ratings of, 277
condensate pumps for steam heating systems
  centrifugal, 499
  duplex, 499
  for gravity flow systems, 497–498
  as mechanical lifts, 501
  operation of, 498–499
  types of, 498
  vertical, 501
connections, valve pipe
  butt-welding, 495–496
  for flanged end valves, 488, 489
  fusion welding for, 492
  for grooved end valves, 488
  installation pointers, 489–492
  silver brasing, 493–495
  socket-welding, 496
  soldering, 493–495
  types of, 487
contactors
  cleaning, 280
  defined, 275
  described, 275–277
  replacing, 280–281
  troubleshooting, 277–280
control panels (heating and cooling), 293
controls, gas
  automatic pilot safety valves, 174–178
  burner primary, 145–152
  combination gas valves, 187–194
  continuous pilot dual automatic valves, 191–194
  diaphragm valves, 164–166
  direct-acting heat motor valves, 163–164
  electrical circuits and, 146
  electronic ignition modules, 203–211
  falling-pressure switches, 170, 173
  igniters, 211–214
  for LP gas, 176
  manual reset pressure switches, 173
  pilot burners, 194–203
  pilot-operated diaphragm valves, 185–186
  pilot-pressure switches, 203
  pressure regulators, 166–174
  pressure switches, 170–174
  rising-pressure switches, 173–174
  solenoid coils, 158–162
  solenoid gas valves, 153–158
  standing pilot combination valves, 187–191

thermocouples, 181–184
thermopiles, 184–185
thermopilot valves, 178–180
universal electronic ignition combination valves, 194
valves in, 153
controls, oil
cadmium cell primary, 220–223
combination primary control and aquastat, 227–231
delayed-discharge, 219
flame detection cells, 221–222
oil burner primary, 219–220
solenoid valves, 217–219
stack detector primary, 223–227
types of, 217
valves, 217–219
conversion burners
gas, 63
gas piping for, 68–71
copper and brass pipes and tubing
for hydronic radiant heating panels, 366
pipes versus tubes, 363, 366
table of sizes and dimensions of water tubes, 367
uses for, 363
copper water tubes, table of sizes and dimensions of, 367
cross-linked polyethylene tubing (PEX), 368

D

damper motors and actuators (duct system)
actuator motor characteristics, 316
installing, 316–320
modutrol motors, 314–316
troubleshooting, 320–321
types of, 313–314
uses of, 313
dampers, duct, 305–313
designing duct systems, 322–323
diaphragm valves, 166
dimensions of pipe, table of
double extra-strong, 362
extra-strong, 360–361
standard, 358–359
DIN (Deutsche [German] Industry Norm) rating system, 366
direct-acting heat motor valves (gas), 163–164
direct-spark ignition modules (gas) 207
dirt pockets in pipe systems, 434
discharging capacities of standard pipe (table of), 388
draft diverters for gas heating equipment, 71–72
draft hoods for gas heating equipment, 71–72
drips in steam heating systems, 432–434
duct furnaces
and air conditioning systems, 340–341, 342–343
duct maintenance and, 346–347
examples of, 338–339
installing, 340–347
maintenance of, 346–347
uses of, 338
duct heaters, electric
components of, 347–353
description of, 347–348
safety controls for, 350–351
vibration controls for, 353
ducts
air leakage, 325
equal friction method of sizing, 326–331
heat gain in, 324–325
heat loss in, 324–325
insulation of, 325–326
maintenance of, 331–332
materials for, 298–299, 300
sizing of, 326–331
duct systems
actuators and damper motors, 313–321
air leakage in, 325
air pressure and, 323–324
air supply and venting of, 305
air velocities, table of recommended and maximum, 327
balancing, 331
blowers for, 322–321
boots and boot fittings, 310–311

duct systems *(continued)*
calculations for, 323–324
components of, 299–305
crawl-space plenum systems, 297–298
damper motors and actuators, 313–321
dampers, 305–313
designing, 322–323
diffusers, 301, 305
duct insulation and, 325–326
electric heaters for, 347–353
extended plenum, 297, 298
fans for, 322–321
furnaces for, 338–347
grilles for, 301, 303
heat loss and gain in, 324–325
ideal, 323
information sources for, 296
maintenance of, 331–332
mobile home, 333–336
perimeter, 295, 296–297
proprietary, 336–338
registers for, 304
return air and exhaust air inlets, 302
roof plenum units, 332
run fittings, 303–305
sizing of ducts in, 326–331
supply air registers, 301, 304
venting and air supply, 305
zoning and, 314

E

elecrical circuits for gas burners, 61
electrical circuits for gas controls
burners, effect of relay positions on, 150
line voltage, 148
low voltage, 146, 147
millivolt, 149
primary control, 146–147
relay positions and effect on burners, 150
electric heaters for duct systems, 347–353
electric zone valves for steam heating systems, 559, 562
electrodes in fuel oil burners
puffbacks and, 36–37
servicing, 38
troubleshooting, 38
electronic ignition modules
described, 203–204
direct spark ignition modules, 207
flame sensors and, 214–217
hot-surface ignition modules, 207–211
intermittent pilot ignition modules, 205–207
end suction pumps for steam heating systems, 510–511
equal friction method of sizing ducts, 326–331
expansion of pipe, table of, 378
expansion tanks in steam and hot water heating systems
closed steel, 541–543
described, 540
diaphragm, 543
sizing, 543–544
troubleshooting, 544

F

fan and limit controls, 62
fan center, 239–241
fan controls
air switches, 236–237
combination fan and limit control devices, 251–256
fan center, 239–241
fan manager, 241
fan timer switches, 241–242, 243–244
impedance relays, 259–561
limit control devices for furnace plenums, 244–247
on/off devices, 233–236
relays, 237–239
safety cutoff switch, 242
secondary high-limit switches, 248–251
switching relays, 256–259
temperature-driven on/off devices, 233–236
upper fan controls, 236
fan relays, 237–239

fittings, pipe
 branching, 380–381
 bushings, 379–380
 caps, 382
 classification of, 369–370
 couplings, 375
 crosses, 381
 defined, 369
 directional, 380
 elbows, 380
 elbows with side outlets, 381
 enlarging, 379–380
 extension, 370–378
 flanges, 382
 flow of liquids in, 389
 hot water pipes and tubing, sizing of, 391
 hydronic pipes and tubing, sizing of, 391
 illustrated table of, 371–375
 joining, 370–378
 joints, 375
 locknuts, 375
 makeup, 382
 nipples, 370, 376–377
 nipples, standard lengths, table of, 376–377
 offsets, 375
 pipes, hot water, sizing of, 391
 plugs, 382
 reducing, 379–380
 return bends, 380
 return bends with back or side outlets, 381
 shutoff or closing, 382
 sizing hot-water (hydronic) pipes and tubing, 391
 and steam pipe runs, 385, 390–391
 tubing, hot water, sizing of, 391
 unions, 379, 382
 Y branches, 381
flame color in atmospheric injection gas burners, 60–61
flame-retention head burners, 16
flame sensors, 62, 214–217
flow control valves for steam heating systems, 558, 561
flue gas (oil burner)
 carbon dioxide content and, 45–48
 measuring carbon dioxide in, 46
 overfire drafts and, 46, 48
fuel pumps, oil burner
 capacity. 26
 nozzles, 32–36
 pressure adjustments, 29–30
 priming, 29
 service and maintenance of, 26–29
 troubleshooting, 31–32
fluids, valves for
 angle, 454–456
 automatic, 475–487
 bolted flanged connets, 447–448
 bonnets, 446–448
 butterfly, 465–467
 check, 458–463
 components of, 445, 446
 connections to pipes, 487–489
 damaged stems in, 475
 described, 445
 discs in, 450–451
 externally controlled, 445
 gate, 450, 456–458
 globe, 454–456
 for high-pressure service, 448
 information sources for, 453–454
 inside screw stems in, 450
 leakage in, 474–475
 for low-pressure service, 448
 matching service use and materials in, 453
 materials for, 451, 453
 nonrising stems in, 450
 operation of, 445
 operators for, 475–487
 outside screw stems in, 449–450
 rising stems in, 450
 screwed bonnets, 447
 seat leakage in, 474–475
 selection of, 469–472
 standards and specifications for, 451
 stems in, 449–450, 475
 stop, 463–465
 stuffing-box leakage in, 474
 terminology for, 445–446
 three-way, 469
 troubleshooting, 472–475

fluids, valves for *(continued)*
  two-way, 467–469
  union bonnets, 447
  valve stems in, 445–446
  wedges in, 450
  Y globe valves, 469
  Y valves, 469
furnace plenums, limit controls for, 244–247
furnaces, duct
  and air conditioning systems, 340–341, 342–343
  duct maintenance and, 346–347
  examples of, 338–339
  installing, 340–347
  maintenance of, 346–347
  uses of, 338

G

galvanized pipe, 363
gas burners
  atmospheric injection, 57–61
  automatic controls and, 61–62
  conversion burners, 63
  defined, 57
  electrical circuits and, 61
  flame sensors, 214–217
  gas conversion, described, 66–68
  gas conversion, piping for, 68–71
  inshot, 64
  integral-type, 65–66
  main, 63–64, 66
  mercury flame sensors in, 216
  operating principles for, 57–61
  pilot, 63–64
  precedence of local electrical coes over manufacturer instructions, 61
  safety precautions and, 72–73
  troubleshooting, 73–75
  upshot, 64
  venting system and, 71–72
gas controls
  automatic pilot safety valves, 174–178
  combination gas valves, 187–194
  continuous pilot dual automatic gas valves, 191–194
  diaphragm valves, 164–166
  direct-acting heat motor valves, 163–164
  electrical circuits and, 146
  electronic ignition modules, 203–211
  falling-pressure switches, 170, 173
  gas burner primary, 145–152
  igniters, 211–214
  for LP gas, 176
  manual reset pressure switches, 173
  pilot burners, 194–203
  pilot-operated diaphragm valves, 185–186
  pilot-pressure switches, 203
  pressure regulators, 166–174
  pressure switches, 170–174
  rising-pressure switches, 173–174
  solenoid coils, 158–162
  solenoid gas valves, 153–158
  standing pilot combination gas valves, 187–191
  thermocouples, 181–184
  thermopiles, 184–185
  thermopilot valves, 178–180
  universal electronic ignition combination valves, 194
  valves, 153
gas conversion burners, 66–71
gas pipe
  capacity table, 69
  installation of, 429–430
  installation requirements for, 70
  threading specifications, 70
gas-pressure regulators
  balanced, 170
  and other controls, 168–169
gas valves
  balanced diaphragm solenoid, 155-156
  combination, 187–194
  continuous pilot dual automatic, 191–194
  diaphragm, 164–166
  direct acting heat motor, 163–164
  manual, 153
  pilot-operated diaphragm, 185–186
  pilot shutoff, 153

pilot valve, 157
shutoff, 153
solenoid, 153–158
solenoid coils, 158–162
standing pilot combination, 187–191
thermopilot, 178–180
three way solenoid, 157–158
types of, 153
universal electronic ignition combination, 194
gas velocity in atmospheric injection gas burners, 60–61
gate valves
double-disc, 456–457
outside screw and yoke, 457–458
wedge, 456
globe valves, 454–456
gun-type oil burners
burner control, 5–6
burner motor and coupling, 7–8
combustion air blowers and, 10
construction details, 3–5
fuel pumps and, 8
gun assembly, 6–7, 9
ignition transformer, 7
operating principles of, 10–16
primary safety control for, 6
turbulators and, 11–12

H

heating relays/time-delay relays, 261–263
high-pressure cutout switches, 288–289
high-static oil burners, 16
Honeywell step controllers, adjusting, 270–271
hot-surface ignition modules (gas)
self-diagnostic capabilities in some, 208–210
operation of, 207–208
principal components of, 207
safety precautions for, 210
wiring for, typical, 209
hot-water heating controls for steam heating systems, 554–557
hot-water (hydronic) pipes/tubing, sizing, 391–392
humidistats
described, 135
electronic, 135–136
location of, 140–142
pneumatic, 136, 140
troubleshooting, 142–143
hydronic (hot water) heating systems
air eliminators, 545–547
balancing valves, 561–564
circulator installation, 515–516
circulators for, 505–518
circulator selection for, 511–515
coils for, making up, 435–438
condensate pumps for, 497–502
drips in, 432–434
electric control valves, 548–550
electric zone valves for, 559, 562
expansion tanks in, 540–544
filters for, 561–564
flow control valves for, 558, 561
Hartford connections for, 434
hot-water heating controls for, 554–557
maifolds for, 564–565
pipeline strainers in, 565
pipeline valves and controls for, 547–548
pipe stress in, relieving, 438–439
pressure tests, 444
pumps for, 497–518
regulators, 548–550
siphons in, 434
steam traps in, 518–540
swivels and offsets, 439
temperature regulators, 548
vacuum pumps, 502–505
valve adapters for, 561–564
water circulating pumps, 505–518
water pockets, eliminating, 439–444
water-tempering valves for, 550–554
hydronic (hot-water) pipes/tubing, sizing, 391
hydronic radiant heating panels, pipes and tubing for, 366

I

igniters (gas)
  ignition control lockout times for, 213
  operation of, 212–213
  replacing hot-surface type, 214
ignition control lockout times chart, 213
impedance relay fan controls, 259–261
inherent protectors, 282–283
inline centrifugal circulators for steam heating systems, 510
inshot gas burners, 64
installation of oil burners, 40–42
Institute of Boiler and Radistor Manufacturers (IBR), 384–385
insulating pipes, 430–431
integral-type gas burners, 65–66
intermittent pilot ignition modules (gas)
  definitions related to, 206–207
  described, 205
  operating sequence for, 205–206

J

joint compound, 405–405

L

leakage in valves, 474–475
limit controls for furnace plenums, 244–247
liquid flow, resistance of valves and fittings to, 389
LP gas and automatic pilot valves, 176
low-pressure cutout switches, 289–290

M

main gas burner, 63–64
main gas valve, 62
maintenance of duct systems, 331–332
manifolds for steam heating systems, 564–565
manual switches, 268
Manufacturers' Standardization Society of the Valve and Fitting Industry (MSS) Standard Practices, 452
mercury flame sensors, 216–217
mobile home duct systems, 333–336
monkey wrenches, 405

N

National Electrical Code, 265, 290
natural gas manifold pressures, table of, 346

O

offsets in pipe systems, calculating, 397–404
oil burner nozzles
  described, 32–33
  hollow cone pattern, 34
  solid cone pattern, 34
  spray angles of, 35–36
oil burners
  air delivery and blower adjustment, 42–43
  air systems in, 38–39
  burner flame detection and, 222
  cadmium cell primaries, 220–223
  and carbon dioxide in flue gas, 45–48
  combination oil and gas, 19, 21
  combination primary control and aquastat, 227–231
  combustion chamber dimensions, 37
  combustion testing and adjustments, 43–48
  defined, 1
  delayed ignition, 50
  electrodes, 36
  external warning signals and, 48–49
  flame retention head, 16
  fuel pump capacity and, 26
  fuel pump nozzles, 32–36
  fuel pump pressure, adjusting, 29–30
  fuel pump priming, 29

fuel pump service and maintenance, 26–29
fuel pump troubleshooting, 31–32
fuels used in, 22
fuel temperature and, 22
gun-type (atomizing), 1–2, 3–16
high-static, 16
installing, 40–42
measuring carbon dioxide in flue gas of, 46
net stack temperature and, 45
noise, excessive, 49
odor, excessive, 50
overfire drafts and, 46, 48
primary controls, 219–220
primary safety control services, 39–40
puffbacks and, 36–38
rotary, 2–3, 16–18
single-stage fuel pumps and, 21, 23–24
smoke, excessive, 49–50
stack detector primary controls, 223–227
stack thermometer, use of, 45
starting, 42
troubleshooting, 48–56
troubleshooting electrodes in, 38
troubleshooting primary controls in, 231–232
two-stage fuel pumps and, 25
vaporizing (pot-type), 2, 18–20
oil controls
cadmium cell primary, 220–223
combination primary control and aquastat, 227–231
delayed-discharge, 219
flame detection cells, 221–222
oil burner primary, 219–220
solenoid valves, 217–219
stack detector primary, 223–227
types of, 217
valves, 217–219
oil valves, 217–219
open-end wrenches, 408
operating principles for gas burners, 57–61
overload relay heaters, 281–282

P

PEX (cross-linked polyethylene) tubing, 368
pilot burners (gas)
aerated, 196
flame adjustment, 202
flame positioning, 200
installing, 198–200
lighting, 201–202
and main burner ignition, 202–203
nonaerated, 197
operation of, 194–196
pilot bracket, 197
pilot orifices, 198, 200–201
pilot ports in, 197–198
pilot duty motor protectors, 283–284
pilot duty thermostats, 283–284
pilot generators, 184–185
pilot-pressure switches, 203
pilot safety valves, automatic, 174–178
pipe dimensions, table of
double-extra-strong, 362
extra-strong, 360–361
standard, 358–359
pipe dope, 404–405
pipe expansion, 382–384
pipe fitting measurements
hot-water mains, 393
mains and branches, 393
steam mains, 392
using, 392, 394, 397
pipe fitting methods and procedures, ferrous metal pipes
assembly and makeup, 415–419
bending, 414–415
cleaning, 414
cutting, 409–412
reaming, 414
tapping, 414
threading, 412–413
pipe fitting methods and procedures, nonferrous pipes
for brass and copper, 419–420
braze welding, 425–426
brazing, 424–425

pipe fitting methods and procedures, nonferrous pipes *(continued)*
pressure ratings for soldered joints, table of, 423
sequence of operations for, 420
soldering, 420–424
welding, 426–426–429
pipe fittings
branching, 380–381
bushings, 379–380
calculating offsets and, 397–404
caps, 382
classification of, 369–370
closing, 382
couplings, 375
crosses, 381
defined, 369
directional, 380
elbows, 380
elbows with side outlets, 381
enlarging, 379–380
extension, 370–378
flanges, 382
flow of liquids in, 389
illustrated table of, 371–375
installation methods for, 409–419
joining, 370–378
joints, 375
lift, 432
locknuts, 375
makeup of, 382
measurements, 392–397
nipples, 370, 376–377
nipples, standard lengths, table of, 376–377
offsets, 375, 397–404
offsets, calculating, 397–404
plugs, 382
reducing, 379–380
return bends, 380
return bends with back or side outlets, 381
shutoff, 382
sizing hot-water (hydronic) pipes and tubing, 391
and steam pipe runs, 385, 390–391
unions, 379, 382
Y branches, 381
pipe fitting wrenches, 405–408
pipeline strainers for steam heating systems, 565
pipeline valves and controls for steam heating systems
described, 547–548
electric control valves, 548–550
temperature regulators, 548
pipe materials
brass, 363, 366
bronze, 363, 366
composites, 369
copper, 363, 366
corrosion and ferrous metal, 356
expansion of pipe and, table, 378
galvanized steel, 363
plastic, 366–368
synthetic rubber, 368–369
wrought iron, 356–357
wrought steel, 357, 363
pipes
brass, versus brass tubing, 369
connections to valves, 487–489
convectors, connecting to, 431–432
copper water tubes, table of sizes and dimensions of, 367
dimensions of double extra-strong, table of, 362
dimensions of extra-strong, table of, 360–361
dimensions of standard pipe, table of, 358–359
drips in steam, 432–434
expansion of, table, 378
gas, for conversion burners, 68–71
installation procedures for gas, 429–430
installation requirements for gas, 70
insulating, 430–431
lift fittings and, 432
mains, connecting risers to, 431
radiators, connecting to, 431–432
relative discharging capacities of, 388

risers, connecting to mains, 431
sizing, 384–385
specific gravities, table of multipliers for, 69
steam mains, table of sizes of, 392
steam, sizing, 385, 390–391
supports, 404
threading specifications for, 70, 384
threads, standard, table of, 386–387
pipe sizing, 384–385, 390–391
pipe supports, 404
pipe systems
dirt pockets in, 434
drips in steam, 432–434
siphons in, 434
pipe tongs, 406, 407
pipe vices, 408–409
pipe wrench, 405
plastic tubing
cross-linked polyethylene (PEX), 368
failures and, 367
polybutylene, 368
temperatures and, 368
uses for, 366
potential (start) relays, 263–265
pressure regulators, gas, 62
defined, 166
external bleed systems and, 168
internal bleed systems and, 167–168
use of, 166
vent hole orifice and, 167, 168
venting and, 166–168
pressure switches, 265–266
falling-pressure (gas), 170, 173
manual reset (gas), 173
pilot-pressure, 203
rising-pressure (gas), 173–174
proprietary air distribution systems, 336–338
pumps for steam heating systems
circulators, 505–518
condensate, 497–502
vacuum, 502–505

R

relative discharging capacities of standard pipe, 388
rotary boil burners, 16–18

S

safety precautions
for capacitors, 287
for cast-iron head (oil) burners, 16
for circulating pumps in steam heating systems, 515, 516
for cross-linked polyethylene tubing (PEX), 368
for duct furnaces, 343–345
for electrically operated valves, 478
for flow control valves in steam heating systems, 558
for gas burners, 72–73
for gas-fired and oil fired equipment, 145
for gas ignition modules, 204
for gas piping, 429
for hot surface ignition systems, 210
for oil burners, 40
for servicing gas burner primary controls, 151–152
for stop valves, 465
for three-piece booster pumps, 508, 509
for transformers, 290, 291
for water-tempering valves for steam heating systems, 550
sail switches, 266–267
semibituminous coal, 81
sequence controllers, 269–275
servicing gas burner primary controls, 151–152
sheet metal, table of thicknesses, gauges, and weights, 300
single-stage fuel pumps, oil burner
described, 21, 23–24
fuel and, 22
siphons in steam pipe systems, 434
smoke scale, Bacharach (for oil burners), 44

soldered joints, pressure ratings table, 423
solenoid coils (for gas valves)
construction of, 161–162
servicing, 160–162
temperature and, 158
sprayers, fuel oil, 1
steam heating systems
air eliminators, 545–547
balancing valves, 561–564
circulator installation, 515–516
circulators, 505–518
circulator selection for, 511–515
coils for, making up, 435–438
condensate pumps, 497–502
drips in, 432–434
electric control valves, 548–550
electric zone valves for, 559, 562
expansion tanks in, 540–544
filters for, 561–564
flow control valves for, 558, 561
Hartford connections for, 434
hot-water heating controls for, 554–557
manifolds for, 564–565
pipeline strainers, 565
pipeline valves and controls for, 547–548
pipe stress, relieving, 438–439
pressure tests, 444
pumps for, 497–518
regulators, 548–550
siphons in, 434
steam traps in, 518–540
swivels and offsets, 439
temperature regulators, 548
water circulating pumps, 505–518
water pockets, eliminating, 439–444
vacuum pumps, 502–505
valve adapters for, 561–564
water-tempering valves for, 550–554
steam mains, table of size of, 392
steam traps in steam heating systems
automatic heat-up and, 520
balanced-pressure thermostatic, 525–526
boiler return, 537–539
bucket, 530–534
description of, 518–519
flash, 534
float and thermostatic, 526–528
float traps, 523–524
impulse, 534, 536
information on, 519
installing, 522–523
lifting, 537. 538
maintenance of, 520
sizing, 519–520, 521, 522
thermodynamic, 529–530
thermostatic traps, 524
tilting, 536–537
Stillson wrenches, 406, 407
stokers, coal
adjustment of coal feed, 94
air adjustment, manual, 92
air control, automatic, 92
air ducts and, 85
ashpits for, 86
automatic controls for, 86–90
bin-fed, 82–83
changing coal feeds for, 92
classes of, 81–82
clinker removal, 94
and coal burning capacity, 81–82
and coal selection, 91
coal storage methods and, 82–84
construction of, 84–86
Drawz, 82, 83, 86
electric motors and, 85–86
fans and, 85
feed worms and, 86
hold-fire controls and, 87
limit controls and, 87
lubrication of, 93
motor overload protection for, 92–93
natural stack draft, 91–92
obstruction removal, 93
operating instructions, 90–91
retorts and, 85
stack switches (stack thermostats) and, 89
starting the fire, 91
summer service for, 93–94
thermostats and, 87

timers and, 88–89
transmission overload protection, 93
transmissions for, 86
troubleshooting, 94–97
underfed, 82–83
stoker firing, 81–86
stop valves, 463–465
strap wrenches, 408
switching relays for fans, 256–259
synthetic rubber hose, 368–369

T

tables, gas pipe capcity, 69
thermocouples, 62
functions of, 181
testing of, 181–183
troubleshooting, 183–184
thermopiles, 184–185
thermopilot valves, 178–180
thermostats
anticipators, 111–112, 114–119
base wiring for, 111
bimetalic strip, 102–103
boiler, 129
calibration, 123–125
components of, 105–109
cylinder, 127, 129
defined, 100, 102
electrical resistance, 104–105
and gas burners, 61–62
immersion, 126–127
insertion, 125–126
installing room type, 121–123
mercury switch. 102–104
outdoor, 132–134
programmable, 125
proportional, 132
remote bulb, 129–132
room, 119–121
switching combinations for, 112
terminal identification standards for, 109, 114
types of, 102–105
threading specifications for gas pipe, 70
threads, pipe, 384, 386–387
three-way valves, 469
time-delay relays/heating relays. 261–263
transformers
defined, 290
installing, 291–293
local electrical codes and, 292
sizing, 291
voltage circuits and, 290
troubleshooting
capacitors, 287
coal stockers, 94–97
contactors, 277–280
damper motors, 320–321
expansion tanks, 544
fuel oil pumps, 31–32
gas burners, 73–75
humidistats, 142–143
oil burner electrodes, 38
oil burner primary control, 231–232
oil burners, 48–56
thermocouples, 183–184
thermostats, 138–139
valves for fluids, 472–475
tubes, copper water, table of sizes and dimensions of, 367
tubing, brass, versus brass pipes, 369
tubing, plastic
cross-linked polyethylene (PEX), 368
failures and, 367
polybutylene, 368
temperatures and, 368
uses for, 366
two-stage fuel pumps, oil burner
described, 25
fuel and, 22
two-way valves, 467–469

U

upshot gas burners, 64

V

vacuum pumps for steam heating systems
dry, 505
operation of, 502–505, 506, 507
use of, 502
wet, 505

valve pipe connections
  butt-welding, 495–496
  for flanged end valves, 488, 489
  fusion welding for, 492
  for grooved end valves, 488
  installation pointers, 489–492
  socket-welding, 496
  soldering and silver brazing, 493–495
  types of, 487
valves, 384
valves and fittings, resistance to flow of liquids, 389
valves, check
  described, 458–460
  horizontal-lift check, 460–461, 467
  lift-check, 460–461, 467
  swing-check, 460, 464, 465, 466
  vertical, 461
  Watts Blackflow Preventer and, 461–462
valves for fluids
  angle, 454–456
  automatic, 475–487
  bolted flanged bonnets, 447–448
  bonnets, 446–448
  butterfly, 465–467
  check, 458–463
  components of, 445, 446
  connections to pipes, 487–489
  damaged stems in, 475
  described, 445
  discs in, 450–451
  externally controlled, 445
  gate, 450, 456–458
  globe, 454–456
  for high-pressure service, 448
  information sources for, 453–454
  inside screw stems in, 450
  leakage in, 474–475
  for low-pressure service, 448
  matching of service use and materials in, 453
  materials for, 451, 453
  nonrising stems in, 450
  operation of, 445
  operators for, 475–487
  outside screw stems in, 449–450
  rising stems in, 450
  screwed bonnets, 447
  seat leakage in, 474–475
  selection of, 469–472
  standards and specifications for, 451
  stems in, 449
  stop, 463–465
  stuffing-box leakage in, 474
  terminology for, 445–446
  three-way, 469
  troubleshooting, 472–475
  two-way, 467–469
  union bonnets, 447
  valve stems in, 445–446
  wedges in, 450
  Y globe valves, 469
  Y valves, 469
valves, gas
  balanced diaphragm solenoid, 155-156
  combination, 187–194
  continuous pilot dual automatic, 191–194
  diaphragm, 164–166
  direct acting heat motor, 163–164
  manual, 153
  oil, 217–219
  pilot-operated diaphragm, 185–186
  pilot shutoff, 153
  pilot valve, 157
  shutoff, 153
  solenoid, 153–158
  solenoid coils, 158–162
  standing pilot combination, 187–191
  thermopilot, 178–180
  three way solenoid, 157–158
  types of, 153
  universal electronic ignition combination, 194
valves, oil, 217–219
vaporizing (pot-type) oil burners, 18–20
venting systems for gas burners, 71–73

vibration controls for electric duct heaters, 353
vices, pipe, 408–409

W

water-circulating pumps for steam heating systems
- end suction pumps, 510–511
- inline centrifugal, 510
- installation of, 515–516
- lubrication requirements and, 509
- operation of, 516–517
- pressure drop, 514
- pump head, 514
- safety precautions for, 508, 509
- selection of, 511–515
- three-piece booster pumps, 506–509
- troubleshooting, service, and maintenance, 517–518
- volute in, 507
- wet-rotor, 509-510

water pockets in steam heating systems, 439–444
water-tempering valves for steam heating systems
- described, 550
- limitations of, 553
- safety precautions for, 550
- servicing, 550–551
- uses of, 552–553, 554

Watts Backflow Preventer, 461–462
wet-rotor circulators for steam hearing systems, 509–510
wrenches, pipe fitting, 405–408
wrought-iron pipe
- corrosion and, 356
- grades of, 356–357
- marking of, 356
- table of dimensions of, 358–362

wrought steel pipe
- bursting and working pressure for, table of, 364–365
- weights of, 357, 363

Y

Y globe valves, 469
Y valves, 469